RF/Microwave Engineering and Applications in Energy Systems

Abdullah Eroglu
North Carolina A&T State University, NC, USA
Purdue University, IN, USA

This edition first published 2022
© 2022 John Wiley & Sons Ltd

The right of Abdullah Eroglu to be identified as the author of this work has been asserted in accordance with law.

Registered Offices
John Wiley & Sons, Inc., 111 River Street, Hoboken, NJ 07030, USA
John Wiley & Sons Ltd, The Atrium, Southern Gate, Chichester, West Sussex, PO19 8SQ, UK

Editorial Office
The Atrium, Southern Gate, Chichester, West Sussex, PO19 8SQ, UK

For details of our global editorial offices, customer services, and more information about Wiley products visit us at www.wiley.com.

Wiley also publishes its books in a variety of electronic formats and by print-on-demand. Some content that appears in standard print versions of this book may not be available in other formats.

Library of Congress Cataloging–in–Publication Data Applied for:

HB ISBN: 9781119268796

Cover Design: Wiley
Cover Image: © wacomka/Getty Images

Set in 9.5/12.5pt STIXTwoText by Straive, Pondicherry, India
Printed and bound by CPI Group (UK) Ltd, Croydon, CR0 4YY

C9781119268796 _180322

Dedicated to my inspiration, G. Dilek

I hear and I forget. I see and I remember. I do and I understand.

Confucius

Contents

Preface

RF (radiofrequency)/microwave engineering became a part of our everyday life in this era. Its theory turns ideas into devices that make our life practical. We can now use smart devices that can be used as portable computers, phones, and visual communication tools. It is now possible to drive your car with keyless entry and control the appliances in your kitchen using RF communication. Energy transfer is now possible and gives us the ability to charge our devices wirelessly. It is possible to remotely control vehicles using satellite signals. And there are many more examples of how RF technology has become inseparable from our daily activities.

Although RF/microwave engineering theory and its principles have been widely available in textbooks since the end of World War II, its applications were not commonly used for commercial applications as they are now until students, engineers, and scientists were able to share the information and knowledge around the world using online tools. Furthermore, the advance of semiconductor technology and the implementation of wide-bandgap semiconductors accelerated the implementation of RF/microwave theory in areas and ways that were not possible before.

The speed of the advance of RF technology means that students require a fundamental knowledge of RF/microwave engineering so that they are well prepared to take part as contributors in this technology. The readiness of students also requires them to be familiar with the applications of RF/microwave theory. Hence, impactful RF/microwave engineering education blends theory and practice. In the RF/microwave world, computer-aided design (CAD) tools are critically important as they reduce design-to-implementation time and cost. Students need to be able to use these tools before they become part of the engineering workforce.

This book is specifically designed to be a unique textbook and resource for senior level undergraduate and introductory graduate level courses in RF/microwave engineering and to meet the needs of instructors and students by providing the fundamental theory for each subject and blending it with real-world engineering application examples that include design, simulation, and prototyping stages. Each chapter is supplemented with engineering application examples and end-of-chapter problems. There are also design challenges that can be assigned by the instructor for students as a project for each topic to implement their analytical skills using CAD tools.

The scope of each chapter in this book can be summarized as follows. Chapter 1 is on the fundamentals of electromagnetics and presents vectors, theorems, and Maxwell's equations. Chapter 2 gives details on passive and active components in microwave engineering. In Chapter 3, transmission lines are discussed and Smith charts are introduced. Network and scattering parameters are covered in Chapter 4. Impedance matching is detailed in Chapter 5. Chapter 6 presents resonator circuits. In Chapter 7, couplers, combiners, and dividers are discussed. RF filter design methods are given in Chapter 8. Waveguides and their theories are given in Chapter 9. In Chapter 10, power amplifier design methods are introduced. Chapter 11 discusses wire antenna and microstrip type antenna design techniques. Chapter 12 presents emerging technologies using RF/microwave engineering principles. Chapter 13 details applications of RF technology in energy systems focused on energy harvesting and heating, ventilation and air conditioning systems.

Overall, the content of this book will enable students to have a better understanding of the theory through real-world engineering application examples.

Abdullah Eroglu
Greensboro, NC, USA

May 2021

Biography

Abdullah Eroglu received his PhD degree in electrical engineering from the Electrical Engineering & Computer Science Department at Syracuse University, in 2004. He was a Senior RF Design Engineer with MKS Instruments, from 2000 to 2008, where he was involved in the design of RF amplifier systems. He joined Purdue University Fort Wayne in 2008 and worked as Professor and Chair of the Electrical and Computer Engineering Department until 2018. Since then, he has been Emeritus Professor at Purdue University and Professor and Chair at the Electrical and Computer Engineering Department, North Carolina A&T State University. He was a Faculty Fellow with the Fusion Energy Division, Oak Ridge National Laboratory, in 2009. His current research interests include RF/microwave/THz circuit design and applications, RF amplifiers and topologies, antennas, RF metrology, wave propagation, and radiation characteristics of anisotropic, gyrotropic, and metamaterials. He has published over 140 journal and conference publications, and has several patents in his area of expertise. He is the author of five books and co-editor of one book. He is a reviewer and on the editorial board of several journals.

Acknowledgments

Spending a lifetime with a researcher sometimes can be challenging. My wife and my children do just that, and they are always with me at every step along the way. They support me for every project – and gave me the encouragement I needed to complete this book, too.

I would also like to thank my editor, Juliet Booker, for her patience and support of this book. She was very understanding and gave me the time I needed to complete the book.

About the Companion Website

RF/Microwave Engineering and Applications in Energy Systems is accompanied by a companion website:

www.wiley.com/go/eroglu/rfmicrowave

The website includes:

- Solutions to End of Chapter Problems

1

Fundamentals of Electromagnetics

1.1 Introduction

The fundamentals of electromagnetics constitute steps toward the advancement of technology in communication, radar, and energy and power applications. In this chapter, some important theorems and mathematical concepts are discussed with examples.

1.2 Line, Surface, and Volume Integrals

1.2.1 Vector Analysis

A scalar quantity gives us a single value of some variable for every point in space such as voltage, current, energy, and temperature. A vector is a quantity which has both a magnitude and a direction in space. Velocity, momentum, acceleration, and force are examples of vector quantities. A mathematical representation of a vector is given in (1.1).

$$\overline{A} = \hat{a}A \tag{1.1}$$

A vector can be represented as a directed line segment, as shown in Figure 1.1.

In Figure 1.1, the magnitude of the vector is given by $|A|$. The unit vector which defines the direction of the vector is given by

$$\hat{a} = \frac{\overline{A}}{|\overline{A}|} \tag{1.2}$$

You may consider falling snowflakes as an example for a vector which has direction and magnitude.

1.2.1.1 Unit Vector Relationship

It is frequently useful to resolve vectors into components along the axial direction using general unit vectors, as given by (1.3)

$$\text{Unit vectors} \rightarrow \hat{a}_u = (\hat{a}_{u1}, \hat{a}_{u2}, \hat{a}_{u3}) \tag{1.3}$$

RF/Microwave Engineering and Applications in Energy Systems, First Edition. Abdullah Eroglu.
© 2022 John Wiley & Sons Ltd. Published 2022 by John Wiley & Sons Ltd.
Companion website: www.wiley.com/go/eroglu/rfmicrowave

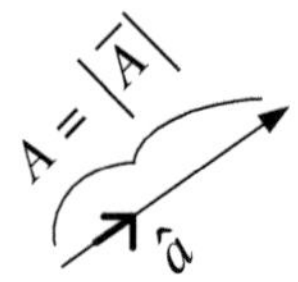

Figure 1.1 Vector representation.

Then

$$\hat{a}_{u1} \times \hat{a}_{u2} = \hat{a}_{u3} \tag{1.4a}$$

$$\hat{a}_{u2} \times \hat{a}_{u3} = \hat{a}_{u1} \tag{1.4b}$$

$$\hat{a}_{u3} \times \hat{a}_{u1} = \hat{a}_{u2} \tag{1.4c}$$

In Cartesian coordinate system, unit vectors are defined as

$$(\hat{a}_{u1}, \hat{a}_{u2}, \hat{a}_{u3}) = (\hat{x}, \hat{y}, \hat{z}) \tag{1.5}$$

So

$$\hat{x} \times \hat{y} = \hat{z} \tag{1.6a}$$

$$\hat{y} \times \hat{z} = \hat{x} \tag{1.6b}$$

$$\hat{z} \times \hat{x} = \hat{y} \tag{1.6c}$$

In a cylindrical coordinate system, unit vectors are defined as

$$(\hat{a}_{u1}, \hat{a}_{u2}, \hat{a}_{u3}) = (\hat{r}, \hat{\phi}, \hat{z}) \tag{1.7}$$

Hence

$$\hat{r} \times \hat{\phi} = \hat{z} \tag{1.8a}$$

$$\hat{\phi} \times \hat{z} = \hat{r} \tag{1.8b}$$

$$\hat{z} \times \hat{r} = \hat{\phi} \tag{1.8c}$$

In a spherical coordinate system, unit vectors are defined as

$$(\hat{a}_{u1}, \hat{a}_{u2}, \hat{a}_{u3}) = (\hat{R}, \hat{\theta}, \hat{\phi}) \tag{1.9}$$

which leads to

$$\hat{R} \times \hat{\theta} = \hat{\phi} \tag{1.10b}$$

$$\hat{\theta} \times \hat{\phi} = \hat{R} \tag{1.10b}$$

$$\hat{\phi} \times \hat{R} = \hat{\theta} \tag{1.10c}$$

1.2.1.2 Vector Operations and Properties
Dot Product

The dot product between two vectors results in a scalar quantity. One simple application example for a dot product is the work that is done by a force for displacement from one point to another. Consider vectors $\vec{A}$ and $\vec{B}$ shown in Figure 1.2a. The dot product of these two vectors are then expressed as

$$A \cdot B = |A||B| \cos\theta = B \cdot A \tag{1.11}$$

If θ between two vectors is 90°, then the dot product of these two vectors is equal to zero.

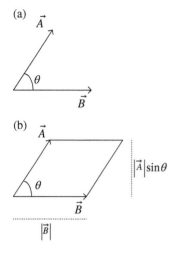

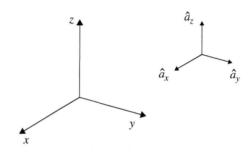

Figure 1.2 (a) Representation of vectors $\vec{A}$ and $\vec{B}$ for dot product. (b) Representation of vectors $\vec{A}$ and $\vec{B}$ for cross product.

Figure 1.3 Unit vector representation in a Cartesian coordinate system.

Cross Product

The cross product between two vectors is also a vector. The magnitude of the cross product of $\vec{A}$ and $\vec{B}$ is equal to the area of the parallelogram which is formed by these vectors, as shown in Figure 1.2b. Consider the two vectors given in Figure 1.2b. The cross product of these two vectors can be written as

$$A \times B = |A||B| \sin\theta \ \hat{a}_n \tag{1.12}$$

$\hat{a}_n$ is the unit vector along the normal to the plane containing A and B. It is important to note that

$$\vec{A} \times \vec{B} = -\vec{B} \times \vec{A} \tag{1.13}$$

Vector Operation Properties for Dot and Cross Products

Some of the properties for dot and cross products are given below

$$\text{Commutative} \rightarrow \vec{A} \cdot \vec{B} = \vec{B} \cdot \vec{A} \tag{1.14a}$$

$$\text{Commutative} \rightarrow \vec{A} \times \vec{B} = -\vec{B} \times \vec{A} \tag{1.14b}$$

$$\text{Distributive} \rightarrow \vec{A} \cdot \left(\vec{B} + \vec{C}\right) = \vec{A} \cdot \vec{B} + \vec{A} \cdot \vec{C} \tag{1.15a}$$

$$\text{Distributive} \rightarrow \vec{A} \times \left(\vec{B} + \vec{C}\right) = \vec{A} \times \vec{B} + \vec{A} \times \vec{C} \tag{1.15b}$$

$$\text{Associative} \rightarrow \vec{A} \cdot \vec{B}\vec{C} \cdot \vec{D} = \left(\vec{A} \cdot \vec{B}\right)\left(\vec{C} \cdot \vec{D}\right) \tag{1.16a}$$

$$\text{Associative} \rightarrow \vec{A} \cdot \vec{B}\vec{C} = \left(\vec{A} \cdot \vec{B}\right)\vec{C} \tag{1.16b}$$

$$\text{Associative} \rightarrow \vec{A} \times \vec{B} \cdot \vec{C} = \left(\vec{A} \times \vec{B}\right) \cdot \vec{C} \tag{1.16c}$$

$$\text{Associative} \rightarrow \vec{A} \times \left(\vec{B} \times \vec{C}\right) \neq \left(\vec{A} \times \vec{B}\right) \times \vec{C} \tag{1.16d}$$

1.2.2 Coordinate Systems

Three coordinate systems are defined in this section: the rectangular or Cartesian coordinate system, the cylindrical coordinate system, and the spherical coordinate system.

1.2.2.1 Cartesian Coordinate System

Unit vector representation for a Cartesian coordinate system is shown in Figure 1.3.

A vector representation in a Cartesian coordinate system is given by (1.17).

$$\vec{A} = \hat{x}A_x + \hat{y}A_y + \hat{z}A_z \tag{1.17}$$

The magnitude of vector $\vec{A}$ is found from

$$\left|\vec{A}\right| = \sqrt{\vec{A} \cdot \vec{A}} = \sqrt{A_x^2 + A_y^2 + A_z^2} \tag{1.18}$$

The vector $\vec{A}$ can be illustrated in the coordinate system, as shown in Figure 1.4. The base vector properties can be given by

$$\hat{x} \cdot \hat{x} = \hat{y} \cdot \hat{y} = \hat{z} \cdot \hat{z} = 1$$
$$\hat{x} \cdot \hat{y} = \hat{y} \cdot \hat{z} = \hat{z} \cdot \hat{x} = 0 \tag{1.19}$$

$$\hat{x} \times \hat{y} = \hat{z}$$
$$\hat{y} \times \hat{z} = \hat{x} \tag{1.20}$$
$$\hat{z} \times \hat{x} = \hat{y}$$

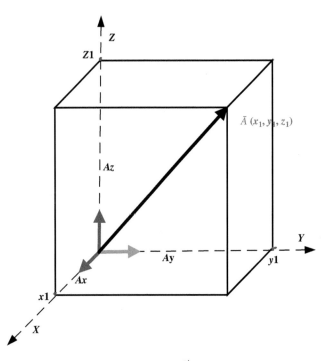

Figure 1.4 Representation of vector $\vec{A}$ in a Cartesian coordinate system.

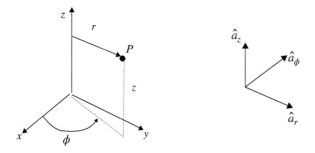

Figure 1.5 Representation of vector $\vec{A}$ in a cylindrical coordinate system.

The vector operations for dot and cross products for vectors $\vec{A}$ and $\vec{B}$ are given by

$$\vec{A} \cdot \vec{B} = A_x B_x + A_y B_y + A_z B_z \tag{1.21}$$

$$\vec{A} \times \vec{B} = \begin{vmatrix} \hat{x} & \hat{y} & \hat{z} \\ A_x & A_y & A_z \\ B_x & B_y & B_z \end{vmatrix} \tag{1.22}$$

1.2.2.2 Cylindrical Coordinate System

Unit vector representation for a cylindrical coordinate system is shown in Figure 1.5.

A vector representation for a Cartesian coordinate system is given by (1.23).

$$\vec{A} = \hat{r} A_r + \hat{\phi} A_\phi + \hat{z} A_z \tag{1.23}$$

The magnitude of vector $\vec{A}$ is found from

$$\left| \vec{A} \right| = \sqrt{\vec{A} \cdot \vec{A}} = \sqrt{A_r^2 + A_\phi^2 + A_z^2} \tag{1.24}$$

The vector $\vec{A}$ can be illustrated in the cylindrical coordinate system, as shown in Figure 1.6. In Figure 1.6, the range of r, ϕ, and z are given as

$$r - \text{radial distance in } xy \text{ plane} \rightarrow 0 \le r \le \infty$$
$$\phi - \text{azimuthal angle measured from } x \text{ axis} \rightarrow 0 \le \phi < 2\pi$$
$$z - - \infty < 2\pi < \infty$$

The vector operations for dot and cross products for vectors $\vec{A}$ and $\vec{B}$ are given by

$$\vec{A} \cdot \vec{B} = A_r B_r + A_\phi B_\phi + A_z B_z \tag{1.25}$$

$$\vec{A} \times \vec{B} = \begin{vmatrix} \hat{r} & \hat{\phi} & \hat{z} \\ A_r & A_\phi & A_z \\ B_r & B_\phi & B_z \end{vmatrix} \tag{1.26}$$

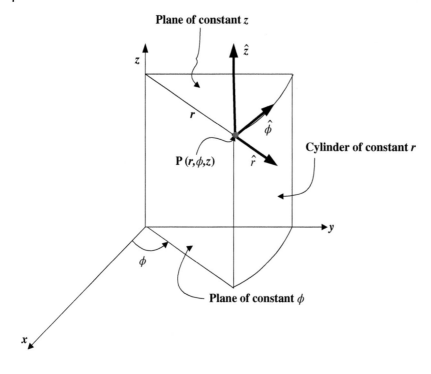

Plane of constant z

Cylinder of constant r

$P(r,\phi,z)$

Plane of constant ϕ

Figure 1.6 Representation of vector $\vec{A}$ in a cylindrical coordinate system.

The base vector properties can be obtained as

$$\hat{r} \times \hat{\phi} = \hat{z}$$
$$\hat{\phi} \times \hat{z} = \hat{r}$$
$$\hat{z} \times \hat{r} = \hat{\phi}$$

(1.27)

1.2.2.3 Spherical Coordinate System

Unit vector representation for a spherical coordinate system is shown in Figure 1.7.

A vector representation for a Cartesian coordinate system is given by (1.28).

$$\vec{A} = \hat{R}A_r + \hat{\theta}A_\theta + \hat{\phi}A_\phi$$

(1.28)

The magnitude of vector $\vec{A}$ is found from

$$\left|\vec{A}\right| = \sqrt{\vec{A} \cdot \vec{A}} = \sqrt{A_R^2 + A_\theta^2 + A_\phi^2}$$

(1.29)

The vector $\vec{A}$ can be illustrated in the spherical coordinate system, as shown in Figure 1.8.
In Figure 1.8, the range of R, θ, and ϕ are given as

$$R - \text{radial distance in } xy \text{ plane} \quad \rightarrow 0 \leq R \leq \infty$$
$$\theta - \text{elevation angle measured from } z \text{ axis} \quad \rightarrow 0 \leq \theta \leq \pi$$
$$\phi - \text{azimuthal angle measured from } x \text{ axis} \quad \rightarrow 0 \leq \phi < 2\pi$$

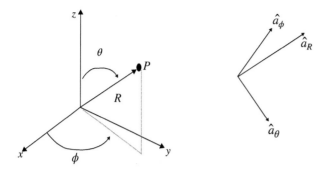

Figure 1.7 Representation of vector $\vec{A}$ in a spherical coordinate system.

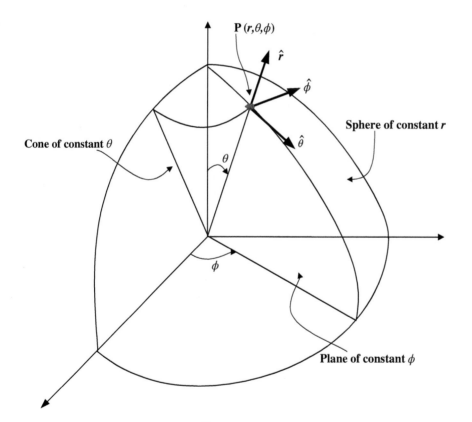

Figure 1.8 Representation of vector $\vec{A}$ in a spherical coordinate system.

The vector operations for dot and cross products for vectors $\vec{A}$ and $\vec{B}$ are given by

$$\vec{A} \cdot \vec{B} = A_R B_R + A_\theta B_\theta + A_\phi B_\phi \tag{1.30}$$

$$\vec{A} \times \vec{B} = \begin{vmatrix} \hat{R} & \hat{\theta} & \hat{\phi} \\ A_R & A_\theta & A_\phi \\ B_R & B_\theta & B_\phi \end{vmatrix} \tag{1.31}$$

The base vector properties are then equal to

$$\hat{R} \times \hat{\theta} = \hat{\phi}$$
$$\hat{\theta} \times \hat{\phi} = \hat{R}$$
$$\hat{\phi} \times \hat{R} = \hat{\theta}$$

(1.32)

1.2.3 Differential Length (<u>dl</u>), Differential Area (<u>ds</u>), and Differential Volume (<u>dv</u>)

In vector analysis, line, surface, and volume integrals are expressed using differential lengths, areas, and volumes.

1.2.3.1 dl, ds, and dv in a Cartesian Coordinate System

Differential length represents infinitesimal change in any direction of the axis in the coordinate system and is represented by $\vec{dl}$. In a Cartesian coordinate system, the differential length is given by

$$\vec{dl} = dx \cdot \hat{a}x + dy \cdot \hat{a}y + dz \cdot \hat{a}z$$

(1.33)

Its magnitude is defined by

$$dl = \sqrt{dx^2 + dy^2 + dz^2}$$

(1.34)

The infinitesimal change for the surface area is defined by the differential area. The unit vector of a differential area points normal to the surface. The differential areas in a Cartesian coordinate system for the surface areas shown in Figure 1.9 are defined as

$$\vec{ds}_x = \hat{x}dydz$$
$$\vec{ds}_y = \hat{y}dxdz$$
$$\vec{ds}_z = \hat{z}dxdy$$

(1.35)

The infinitesimal change for the volume is defined by

$$dv = dxdydz$$

(1.36)

The illustration of the changes in length, area, and volume for a Cartesian coordinate system are given in Figure 1.9.

1.2.3.2 dl, ds, and dv in a Cylindrical Coordinate System

In a cylindrical coordinate system, the differential length is given by

$$\vec{dl} = dr \cdot \hat{a}r + r \cdot d\phi \cdot \hat{a}_\phi + dz \cdot \hat{a}z$$

(1.37)

Its magnitude is defined by

$$dl = \sqrt{dr^2 + r^2d\phi^2 + dz^2}$$

(1.38)

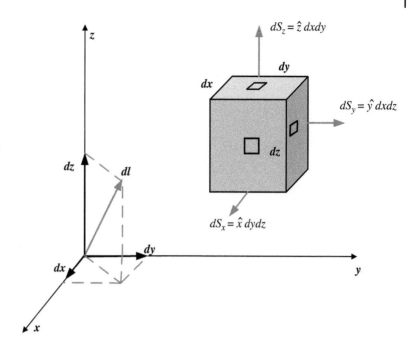

Figure 1.9 Illustration of differential length, area, and volume in a Cartesian coordinate system.

The differential areas in a cylindrical coordinate system for the surface areas shown in Figure 1.10 are defined as

$$d\vec{s}_r = \hat{r}rd\phi dz$$
$$d\vec{s}_\phi = \hat{\phi}drdz \quad (1.39)$$
$$d\vec{s}_z = \hat{z}rdrd\phi$$

The infinitesimal change for the volume is defined by

$$dv = rdrd\phi dz \quad (1.40)$$

The illustration of the changes in length, area, and volume for a cylindrical coordinate system is given in Figure 1.10.

1.2.3.3   *dl*, *ds*, and *dv* in a Spherical Coordinate System

In a spherical coordinate system, the differential length is given by

$$d\vec{l} = dr \cdot \hat{r} + rd\theta \cdot \hat{\theta} + r\sin\theta d\phi \cdot \hat{\phi} \quad (1.41)$$

Its magnitude is defined by

$$dl = \sqrt{dr^2 + r^2 d\theta^2 + r^2 \sin^2\theta d\phi^2} \quad (1.42)$$

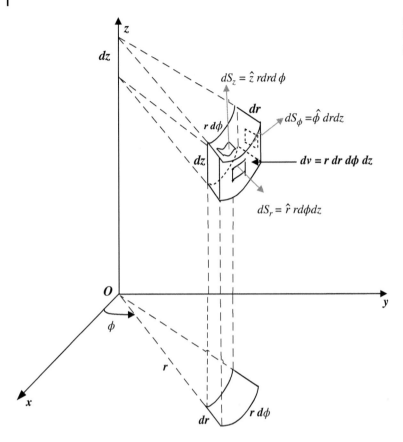

The differential areas in a spherical coordinate system for the surface areas shown in Figure 1.11 are defined as

$$d\vec{s}_R = \hat{R}R^2 \sin\theta d\theta d\phi$$
$$d\vec{s}_\theta = \hat{\theta}R \sin\theta dRd\phi \tag{1.43}$$
$$d\vec{s}_\phi = \hat{\phi}RdRd\theta$$

The infinitesimal change for the volume is defined by

$$dv = R^2 \sin\theta dRd\theta d\phi \tag{1.44}$$

The illustration of the changes in length, area, and volume for a spherical coordinate system is given in Figure 1.11.

1.2.4 Line Integral

A line integral is defined as the integral of the tangential components of a vector along the curve, such as the one shown in Figure 1.12. A line integral is then expressed as

Figure 1.11 Illustration of differential length, area, and volume in a spherical coordinate system.

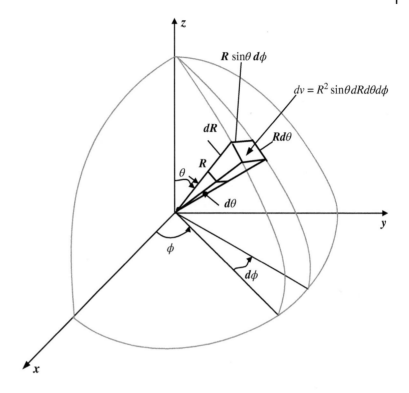

$$\int_L \vec{F} \cdot d\vec{l} = \int_a^b (F_x dx + F_y dy + F_z dz) = \int_a^b \left|\vec{F}\right| \cos(\theta)\, dl \qquad (1.45)$$

One of the practical examples for the line integral application is finding work that is required to move a charge from point a to point b, similar to the one shown in Figure 1.12. This can be expressed in terms of the line integral of the electric field, $\vec{E}$.

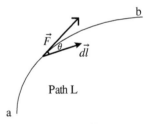

Figure 1.12 Vector $\vec{F}$ along the curve.

Example 1.1 Line Integral

Find the line integral of $\vec{E} = 2\hat{x}x + \hat{y}y$ along the path given in Figure 1.13.

Solution
We need to find

$$\int_P \vec{E} \cdot d\vec{l} = \int_{P_1}^{P_2} \vec{E} \cdot d\vec{l}$$

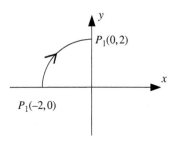

Figure 1.13 Path for line integral Example 1.1.

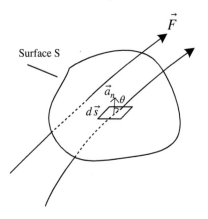

Figure 1.14 Illustration of surface integral.

Since the geometry is circular, it is easier to work with a cylindrical coordinate system, hence we transform vector $\vec{E}$ to cylindrical coordinates as

$$\vec{E} = \hat{r}2(\cos^2\phi - \sin^2\phi) + \hat{\phi}2r\cos\phi\sin\phi$$

In a cylindrical coordinate system, the differential length along the path is defined as

$$d\vec{l} = \hat{\phi}rd\phi$$

where $r = 2$ along the path. So

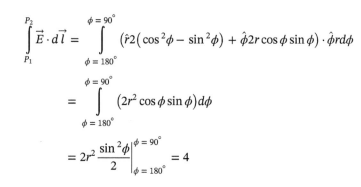

$$\int_{P_1}^{P_2} \vec{E}\cdot d\vec{l} = \int_{\phi=180°}^{\phi=90°} (\hat{r}2(\cos^2\phi - \sin^2\phi) + \hat{\phi}2r\cos\phi\sin\phi)\cdot\hat{\phi}rd\phi$$

$$= \int_{\phi=180°}^{\phi=90°} (2r^2\cos\phi\sin\phi)d\phi$$

$$= 2r^2\frac{\sin^2\phi}{2}\Big|_{\phi=180°}^{\phi=90°} = 4$$

1.2.5 Surface Integral

Consider the region given in Figure 1.14. The surface integral is then defined as the flux of the vector $\vec{F}$ through surface S and expressed as

$$\int_S \vec{F}\cdot\hat{a}_n ds = \int\int |\vec{F}|\cos(\theta)\,ds \tag{1.46}$$

The normal vector $\hat{a}_n$ is normal on S at any point. The outward flux of the vector $\vec{F}$ shown for a closed surface is then defined by

$$\psi = \oint_S \vec{F}\cdot d\vec{s} \tag{1.47}$$

1.2.6 Volume Integral

The volume integral for scalar quantities such as charge densities over the given volume is defined as

$$\int_V \sigma_v dv \tag{1.48}$$

The volume integral given in (1.48) is a triple integral and can be solved in the corresponding coordinate system when geometry of the problem is given.

1.3 Vector Operators and Theorems

In this section, vector differential operators del, gradient, and curl are discussed. In addition, divergence and Stokes' theorem are presented.

1.3.1 Del Operator

The del operator, ∇, is used as a differential operator and can be used to find the gradient of a scalar, divergence of a vector, curl of a vector, or Laplacian of a scalar in electromagnetics. These operations can be expressed as

∇F for finding the gradient of a scalar

$\nabla.\vec{F}$ for finding the divergence of a vector

$\nabla \times \vec{F}$ for finding the curl of a vector

$\nabla^2 F$ for finding the Laplacian of a scalar

The del operator, in three different coordinate systems, can be written in the following forms

$$\text{Cartesian coordinate} \rightarrow \nabla = \frac{\partial}{\partial x}\hat{x} + \frac{\partial}{\partial y}\hat{y} + \frac{\partial}{\partial z}\hat{z} \tag{1.49}$$

$$\text{Cylindrical coordinate} \rightarrow \nabla = \frac{\partial}{\partial r}\hat{r} + \frac{1}{r}\frac{\partial}{\partial \phi}\hat{\phi} + \frac{\partial}{\partial z}\hat{z} \tag{1.50}$$

$$\text{Spherical coordinate} \rightarrow \nabla = \frac{\partial}{\partial R}\hat{R} + \frac{1}{R}\frac{\partial}{\partial \theta}\hat{\theta} + \frac{1}{R\sin\theta}\frac{\partial}{\partial \phi}\hat{\phi} \tag{1.51}$$

1.3.2 Gradient

The gradient is used to identify the maximum change in direction and magnitude for a scalar field. Consider a scalar field $\phi(x, y, z)$. The gradient of $\phi(x, y, z)$ is defined as

$$\text{grad } \phi = \overline{\nabla}\phi = \hat{x}\frac{\partial \phi}{\partial x} + \hat{y}\frac{\partial \phi}{\partial y} + \hat{z}\frac{\partial \phi}{\partial z} \tag{1.52}$$

The change in ϕ from $\vec{r}$ to $\vec{r} + d\vec{r}$ can be found from

$$\overline{\nabla}\phi \cdot d\vec{r} = \left(\hat{x}\frac{\partial \phi}{\partial x} + \hat{y}\frac{\partial \phi}{\partial y} + \hat{z}\frac{\partial \phi}{\partial z}\right) \cdot (\hat{x}dx + \hat{y}dy + \hat{z}dz) \tag{1.53}$$

$$= \frac{\partial \phi}{\partial x}dx + \frac{\partial \phi}{\partial y}dy + \frac{\partial \phi}{\partial z}dz = d\phi \tag{1.54}$$

If we assume $\bar{r} = \bar{r}(s)$ and s is the arc along a curve, then

$$\frac{d\phi}{ds} = \overline{\nabla}\phi \cdot \frac{d\bar{r}}{ds} = \overline{\nabla}\phi \cdot \hat{u} \tag{1.55}$$

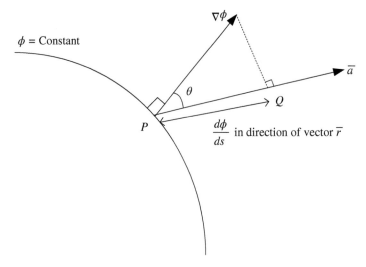

Figure 1.15 Illustration of gradient.

$\hat{u}$ is the unit vector tangent to the curve. Then, we can find the rate of change in ϕ for the distances with respect to vector $\bar{r}$ shown in Figure 1.15 as

$$\frac{d\phi}{ds} = \overline{\nabla}\phi \cdot \hat{a} = |\overline{\nabla}\phi| \cos\theta \qquad (1.56)$$

in a cylindrical coordinate system as

$$\nabla T = \hat{r}\frac{\partial T}{\partial r} + \hat{\phi}\frac{1}{r}\frac{\partial T}{\partial \phi} + \hat{z}\frac{\partial T}{\partial z} \qquad (1.57)$$

and in a spherical coordinate system as

$$\nabla T = \hat{R}\frac{\partial T}{\partial R} + \hat{\theta}\frac{1}{R}\frac{\partial T}{\partial \theta} + \hat{\phi}\frac{1}{R\sin\theta}\frac{\partial T}{\partial \phi} \qquad (1.58)$$

Example 1.2 Gradient

If function $\phi = x^2 y + yz$ is given at point $(1,2-1)$. (a) Find its rate of change for a distance in the direction $\bar{r} = \hat{x} + 2\hat{y} + 3\hat{z}$. (b) What is the greatest possible rate of change with distance and in which direction does it occur at the same point?

Solution

a)

$$\overline{\nabla}\phi = \hat{x}\frac{\partial \phi}{\partial x} + \hat{y}\frac{\partial \phi}{\partial y} + \hat{z}\frac{\partial \phi}{\partial z} = 2xy\hat{x} + (x^2 + z)\hat{y} + y\hat{z}$$

Then, at point $(1,2,-1)$

$$\overline{\nabla}\phi = 4\hat{x} + 2\hat{z}$$

So, its rate of change is found from

$$\frac{d\phi}{ds} = \overline{\nabla}\phi \cdot \hat{r} = \frac{1}{\sqrt{14}}(4 + 6) = \frac{10}{\sqrt{14}}$$

where

$$\hat{r} = \frac{\bar{r}}{|\bar{r}|} = \frac{1}{\sqrt{14}}(\hat{x} + 2\hat{y} + 3\hat{z})$$

b) The greatest possible rate of change at $(1,2,-1)$ is found from

$$\left(\frac{d\phi}{ds}\right)_{\text{max}} = |\nabla\phi| = \sqrt{20}$$

where $\bar{\nabla}\phi = 4\hat{x} + 2\hat{z}$ at $(1,2,-1)$.

1.3.3 Divergence

The divergence of a vector field at a point is a measure of the net outward flux of the same vector per unit volume. The divergence of vector $\bar{T}$ is defined as

$$\nabla \cdot \bar{T} = \left(\hat{x}\frac{\partial}{\partial x} + \hat{y}\frac{\partial}{\partial y} + \hat{z}\frac{\partial}{\partial z}\right) \cdot (\hat{x}T_x + \hat{y}T_y + \hat{z}T_z)$$

$$= \frac{\partial T_x}{\partial x} + \hat{y}\frac{\partial T_y}{\partial y} + \hat{z}\frac{\partial T_z}{\partial z} \tag{1.59}$$

The divergence can be represented in a cylindrical coordinate system as

$$\nabla \cdot \bar{T} = \frac{1}{r}\frac{\partial}{\partial r}(rT_r) + \frac{1}{r}\frac{\partial T_\phi}{\partial \phi} + \frac{\partial T_z}{\partial z} \tag{1.60}$$

In a spherical coordinate system, the divergence is given as

$$\nabla \cdot \bar{T} = \frac{1}{r^2}\frac{\partial}{\partial r}(r^2 T_r) + \frac{1}{r\sin\theta}\frac{\partial}{\partial\theta}(T_\theta\sin\theta) + \frac{1}{r\sin\theta}\frac{\partial T_\phi}{\partial\phi} \tag{1.61}$$

The divergence of a vector field gives a scalar result. In addition, the divergence of a scalar is not a valid operation.

Example 1.3   Divergence

In cylindrical coordinate system, it is given that

$$\bar{T} = r\cos\phi\hat{r} + \left(\frac{z}{r}\right)\sin\phi\hat{z}$$

Calculate $\nabla \cdot \bar{T}$.

Solution

$$\nabla \cdot \bar{T} = \frac{1}{r}\frac{\partial}{\partial r}(rT_r) + \frac{1}{r}\frac{\partial T_\phi}{\partial\phi} + \frac{\partial T_z}{\partial z} = \frac{1}{r}\frac{\partial}{\partial r}(r^2\cos\phi) + \frac{\partial\left(\left(\frac{z}{r}\right)\sin\phi\right)}{\partial z}$$

$$= 2\cos\phi + \frac{1}{r}\sin\phi$$

1.3.4 Curl

The curl of a vector is used to identify how much the vector v curls around a reference point. The curl of a vector is expressed as

$$\nabla x\overline{T} = \begin{vmatrix} \hat{x} & \hat{y} & \hat{z} \\ \dfrac{\partial}{\partial x} & \dfrac{\partial}{\partial y} & \dfrac{\partial}{\partial z} \\ T_x & T_y & T_z \end{vmatrix} = \hat{x}\left(\frac{\partial T_z}{\partial y} - \frac{\partial T_y}{\partial z}\right) + \hat{y}\left(\frac{\partial T_x}{\partial z} - \frac{\partial T_z}{\partial x}\right) + \hat{z}\left(\frac{\partial T_y}{\partial x} - \frac{\partial T_x}{\partial y}\right) \tag{1.62}$$

Curl can be given in a cylindrical or spherical coordinate system as

$$\nabla x\overline{T} = \frac{1}{r}\begin{vmatrix} \hat{r} & r\hat{\phi} & \hat{z} \\ \dfrac{\partial}{\partial r} & \dfrac{\partial}{\partial \phi} & \dfrac{\partial}{\partial z} \\ T_r & rT_\phi & T_z \end{vmatrix} = \hat{r}\left(\frac{1}{r}\frac{\partial T_z}{\partial \phi} - \frac{\partial T_\phi}{\partial z}\right) + \hat{\phi}\left(\frac{\partial T_r}{\partial z} - \frac{\partial T_z}{\partial r}\right) + \hat{z}\frac{1}{r}\left(\frac{\partial}{\partial r}(rT_\phi) - \frac{\partial T_r}{\partial \phi}\right) \tag{1.63}$$

and

$$\nabla x\overline{T} = \frac{1}{r^2 \sin\theta}\begin{vmatrix} \hat{R} & R\hat{\theta} & R\sin\theta\hat{\phi} \\ \dfrac{\partial}{\partial R} & \dfrac{\partial}{\partial \theta} & \dfrac{\partial}{\partial \phi} \\ T_r & RT_\theta & R\sin\theta T_\phi \end{vmatrix} = \hat{R}\frac{1}{R\sin\theta}\left(\frac{\partial}{\partial\theta}(T_\phi \sin\theta) - \frac{\partial T_\theta}{\partial\phi}\right) + \hat{\theta}\frac{1}{R}\left(\frac{1}{\sin\theta}\frac{\partial T_R}{\partial\phi} - \frac{\partial}{\partial R}(RT_\phi)\right)$$
$$+ \hat{\phi}\frac{1}{R}\left(\frac{\partial}{\partial R}(RT_\theta) - \frac{\partial T_R}{\partial\theta}\right)$$

$$\tag{1.64}$$

It is important to note that curl operation on any vector results in another vector. Curl cannot operate on a scalar quantity. In addition, the following two properties follow for curl operation.

$$\nabla \cdot \left(\nabla x\overline{T}\right) = 0 \tag{1.65}$$

$$\nabla x\nabla T = 0 \tag{1.66}$$

Equation (1.65) implies that the curl of a vector does not diverge and the curl of a gradient of a scalar does not exist as expected.

1.3.5 Divergence Theorem

The divergence theorems states that the volume integral of the divergence of a vector is equal to the surface integral of the same vector enclosing that volume. It is mathematically given by

$$\int_V \nabla \cdot \overline{T}\, dv = \oint_S \overline{T} \cdot d\overline{s} \tag{1.67}$$

While this theorem can be applied for an electric flux density, it is valid for any vector. When there are adjoining incremental small volumes, an arbitrary shape is considered to form a larger volume which is enclosed by surface S. Then, flux leaving incremental volume enters the adjacent incremental volume, as shown in Figure 1.16 [1]. Hence, the net flux contribution for a surface integral due to interior surfaces is zero. The contributions that are nonzero occur only for the surfaces enclosing the outer surface of the volume.

Figure 1.16 Illustration of divergence theorem.

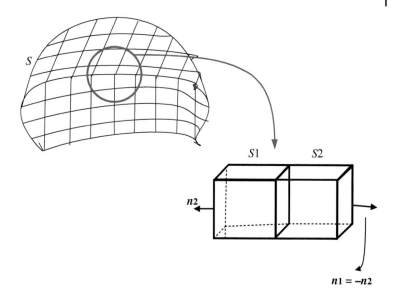

Example 1.4   Divergence Theorem

Verify the divergence theorem for vector

$$\overline{T} = x\hat{x} + y\hat{y} + z\hat{z} \tag{1.68}$$

and the geometry of the pyramid given in Figure 1.17. Surfaces are defined by 1 – aob, 2 – aoc, 3 – boc, 4 – acb.

Solution

Let's calculate the right-hand side of Eq. (1.67).

$$\oint_S \overline{T}.d\overline{s} = \oint_{S_1} \overline{T}.d\overline{s}_1 + \oint_{S_2} \overline{T}.d\overline{s}_2 + \oint_{S_3} \overline{T}.d\overline{s}_3 + \oint_{S_4} \overline{T}.d\overline{s}_4 \tag{1.69}$$

In (1.69), surface areas are defined by

$$d\overline{s}_1 = -\hat{z}dxdy \tag{1.70a}$$

$$d\overline{s}_2 = -\hat{y}dxdz \tag{1.70b}$$

$$d\overline{s}_3 = -\hat{x}dydz \tag{1.70c}$$

To be able to find the differential surface area vector $d\overline{s}_4$ for surface 4, we need to determine the normal vector for that surface. The normal vector can be found using the equation

$$\hat{n} = \frac{\nabla f}{|\nabla f|} \tag{1.71}$$

In (1.71), f is the equation that defines the surface. The equation that defines the surface is given as

$$C = x + Ay + Bz \tag{1.72}$$

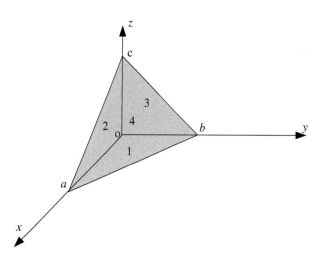

Figure 1.17 Geometry of Example 1.4.

Constants A, B, and C are obtained from the geometry at the intercept points which are (a,0,0), (0,b,0), and (0,0,c). When intercept point a is on the axis with $y = 0$, and $z = 0$ are substituted into (1.72), we obtain

$$C = x + A(0) + B(0) \Rightarrow C = a \tag{1.73}$$

Similarly, from intercept point (0,b,0)

$$a = (0) + A(b) + B(0) \Rightarrow A = \frac{a}{b} \tag{1.74}$$

and from intercept point (0,0,c)

$$a = x(0) + A(0) + B(c) \Rightarrow B = \frac{a}{c} \tag{1.75}$$

Then, function f is defined by

$$f = x + \frac{a}{b}y + \frac{a}{c}z - a = 0 \tag{1.76}$$

Hence, from (1.71) and (1.76)

$$\hat{n} = \frac{\nabla f}{|\nabla f|} = \frac{\hat{x} + \frac{a}{b}\hat{y} + \frac{a}{c}\hat{z}}{\sqrt{1 + \left(\frac{a}{b}\right)^2 + \left(\frac{a}{c}\right)^2}} \tag{1.77}$$

We can then calculate the surface area as

$$d\bar{s}_4 = \hat{n}\frac{dxdy}{\cos\alpha} \tag{1.78}$$

where

$$\cos\alpha = \hat{n}\cdot\hat{z} = \frac{\left(\frac{a}{c}\right)}{\sqrt{1 + \left(\frac{a}{b}\right)^2 + \left(\frac{a}{c}\right)^2}} \tag{1.79}$$

α in (1.79) is the angle between the xy plane and surface 4. Substituting (1.79) into (1.78) gives

$$d\bar{s}_4 = \hat{n}\frac{dxdy}{\cos\alpha} = dxdy\frac{c}{a}\left(\hat{x} + \frac{a}{b}\hat{y} + \frac{a}{c}\hat{z}\right) \tag{1.80}$$

Then, from (1.69)

$$\oint_S \overline{T}\cdot d\bar{s} = \int_{x=0}^{a}\int_{y=0}^{b}(x\hat{x} + y\hat{y} + z\hat{z})\cdot(-\hat{z}dxdy) + \int_{x=0}^{a}\int_{z=0}^{c}(x\hat{x} + y\hat{y} + z\hat{z})\cdot(-\hat{y}dxdz) + \int_{y=0}^{b}\int_{z=0}^{c}(x\hat{x} + y\hat{y} + z\hat{z})\cdot(-\hat{x}dydz)$$

$$+ \int_{x=0}^{a}\int_{y=0}^{b\left(1-\frac{x}{a}\right)}(x\hat{x} + y\hat{y} + z\hat{z})\cdot\left(dxdy\frac{c}{a}\left(\hat{x} + \frac{a}{b}\hat{y} + \frac{a}{c}\hat{z}\right)\right)$$

$$\tag{1.81}$$

In (1.81), the first three terms are zero, hence (1.81) reduces to

$$\oint_S \overline{T}\cdot d\bar{s} = \int_{x=0}^{a}\int_{y=0}^{b\left(1-\frac{x}{a}\right)}(x\hat{x} + y\hat{y} + z\hat{z})\cdot\left(dxdy\frac{c}{a}\left(\hat{x} + \frac{a}{b}\hat{y} + \frac{a}{c}\hat{z}\right)\right)$$

or

$$\oint_S \overline{T} \cdot d\overline{s} = \int\limits_{x=0}^{a} \int\limits_{y=0}^{b\left(1-\frac{x}{a}\right)} \left(x + \frac{a}{b}y + \frac{a}{c}z\right)\frac{c}{a}\,dxdy \tag{1.82}$$

From (1.76)

$$a = x + \frac{a}{b}y + \frac{a}{c}z \tag{1.83}$$

Hence, (1.82) can be written as

$$\oint_S \overline{T} \cdot d\overline{s} = \int\limits_{x=0}^{a} \int\limits_{y=0}^{b\left(1-\frac{x}{a}\right)} (a)\frac{c}{a}\,dxdy = \int\limits_{x=0}^{a} \int\limits_{y=0}^{b\left(1-\frac{x}{a}\right)} c\,dxdy \tag{1.84}$$

(1.84) leads to

$$\oint_S \overline{T} \cdot d\overline{s} = \int\limits_{x=0}^{a} cb\left(1 - \frac{x}{a}\right)dx = \frac{abc}{2} \tag{1.85}$$

The left-hand side of the equation now can be found as

$$\int_V \nabla \cdot \overline{T}\,dv = 3\underbrace{\left(\frac{1}{3}\frac{abc}{2}\right)}_{\text{volume of pyramid}} = \frac{abc}{2} \tag{1.86}$$

1.3.6 Stokes' Theorem

Stokes' theorem states that the line integral of a vector around a closed contour is equal to the surface integral of the curl of a vector over an open surface. It is expressed by

$$\int_S (\nabla \times \overline{T}) \cdot d\overline{s} = \oint_C \overline{T} \cdot d\overline{l} \tag{1.87}$$

This can be illustrated in Figure 1.18. The small interior contours have adjacent sides that are in opposite directions yielding no net line integral contribution, as shown in Figure 1.18 [1]. The net nonzero contribution occurs due to contours with having a side on the open boundary L. Then, the total result of adding the contributions for all the contours can be represented as given in (1.87) by Stokes' theorem.

Example 1.5 Stokes' Theorem
Please verify Stokes' theorem for the geometry given in Figure 1.19 for the vector $T = 4y^2\hat{y} + 2yz\hat{z}$

Solution
We first calculate the left-hand side of Eq. (1.87) as

$$\nabla \times \overline{T} = \left(\frac{\partial}{\partial x}\hat{x} + \frac{\partial}{\partial y}\hat{y} + \frac{\partial}{\partial z}\hat{z}\right) \times (4y^2\hat{y} + 2yz\hat{z})$$

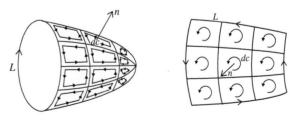

Figure 1.18 Illustration of Stokes' theorem.

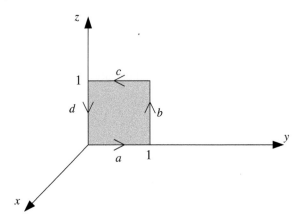

Figure 1.19 Geometry of Example 1.5.

which leads to

$$\nabla \times \overline{T} = \hat{x} \frac{\partial}{\partial y}\left(4yz^2\right) = \hat{x}2z$$

The differential surface area is found from

$$d\bar{s} = \hat{x}dydz$$

Then

$$\int_S \left(\nabla \times \overline{T}\right) \cdot d\bar{s} = \int_{z=0}^{1} \int_{y=0}^{1} 2zdydz = 1$$

Now, we calculate the right-hand side of the equation for each segment as

a) $x = 0, z = 0, d\bar{l} = \hat{y}dy, \int(4y^2\hat{y} + 2yz\hat{z}) \cdot \hat{y}dy = \int_{y=0}^{1} 4y^2dy = \dfrac{4}{3}$

b) $x = 0, y = 1, d\bar{l} = \hat{z}dz, \int(4y^2\hat{y} + 2yz\hat{z}) \cdot \hat{z}dz = \int_{z=0}^{1} 2zdz = 1$

c) $x = 0, z = 1, d\bar{l} = \hat{y}dy, \int(4y^2\hat{y} + 2yz\hat{z}) \cdot \hat{y}dy = \int_{y=1}^{0} 4y^2dy = -\dfrac{4}{3}$

d) $x = 0, y = 0, d\bar{l} = \hat{z}dz, \int(4y^2\hat{y} + 2yz\hat{z}) \cdot \hat{z}dz = 0$

Hence

$$\oint_C \overline{T}.d\bar{l} = \frac{4}{3} + 1 - \frac{4}{3} + 0 = 1$$

As a result, it is confirmed that

$$\int_S \left(\nabla \times \overline{T}\right) \cdot d\bar{s} = \oint_C \overline{T} \cdot d\bar{l} = 1$$

for Example 1.5.

1.4 Maxwell's Equations

1.4.1 Differential Forms of Maxwell's Equations

Maxwell's equations in differential forms in the presence of an impressed magnetic current density $\overline{M}$ and an electric current density $\overline{J}$ can be written as [2]

$$\nabla \times \overline{E} = -\frac{\partial \overline{B}}{\partial t} - \overline{M} \text{ (Faraday's law)} \tag{1.88}$$

$$\nabla \times \overline{H} = \frac{\partial \overline{D}}{\partial t} + \overline{J} \text{ (Ampere's law)} \tag{1.89}$$

$$\nabla \cdot \overline{B} = 0 \text{ (Gauss's law for magnetic field)} \tag{1.90}$$

$$\nabla \cdot \overline{D} = \rho \text{ (Gauss's law for electric field)} \tag{1.91}$$

where $\overline{E}$ is electrical field intensity vector in volts/meter (V/m), $\overline{H}$ is magnetic field intensity vector in amperes/meter (A/m), $\overline{J}$ is electric current density in amperes/meter^2 (A/m^2), $\overline{M}$ is magnetic current density in amperes/meter^2 (A/m^2), $\overline{B}$ is magnetic flux density in webers/meter^2 (W/m^2), $\overline{D}$ is electric flux density in coulombs/meter^2 (C/m^2), and ρ is electric charge density in coulombs/meter^3 (C/m^3).

The continuity equation is derived by taking the divergence of Eq. (1.89) and using Eq. (1.91) as

$$\nabla \cdot \left(\nabla \times \overline{H} \right) = \nabla \cdot \left(\frac{\partial \overline{D}}{\partial t} + \overline{J} \right) \tag{1.92}$$

Since $\nabla \cdot \left(\nabla \times \overline{H} \right) = 0$, then

$$\nabla \cdot \overline{J} = -\frac{\partial \rho}{\partial t} \tag{1.93}$$

Equation (1.93) also represents the fundamental law of physics which is known as conservation of an electric charge.

In an isotropic medium, the material properties do not depend on the direction of the field vectors. In other words, the electric field vector is in parallel with the electric flux density and the magnetic field vector is in parallel with the magnetic flux density.

$$\overline{D} = \varepsilon \overline{E} \tag{1.94}$$

$$\overline{B} = \mu \overline{H} \tag{1.95}$$

ε is the permittivity of the medium and represents its electrical properties and μ is the permeability of the medium and represents its magnetic properties. They are both scalar in the existence of an isotropic medium. However, when the medium is anisotropic this is no longer the case. The electrical and magnetic properties of the medium depend on the direction of the field vectors. Electric and magnetic field vectors are not in parallel with electric and magnetic flux. So, the constitutive relations get the following forms for an anisotropic medium.

$$\overline{D} = \varepsilon_0 \overline{\overline{\varepsilon}} \cdot \overline{E} \tag{1.96}$$

$$\overline{B} = \mu_0 \overline{\overline{\mu}} \cdot \overline{H} \tag{1.97}$$

The permittivity and permeability of anisotropic medium are now tensors. They are expressed as

$$\overline{\overline{\varepsilon}} = \begin{bmatrix} \varepsilon_{11} & \varepsilon_{12} & \varepsilon_{13} \\ \varepsilon_{21} & \varepsilon_{22} & \varepsilon_{23} \\ \varepsilon_{31} & \varepsilon_{32} & \varepsilon_{33} \end{bmatrix} \tag{1.98}$$

and

$$\overline{\overline{\mu}} = \begin{bmatrix} \mu_{11} & \mu_{12} & \mu_{13} \\ \mu_{21} & \mu_{22} & \mu_{23} \\ \mu_{31} & \mu_{32} & \mu_{33} \end{bmatrix} \tag{1.99}$$

Then, (1.96) and (1.97) take the following form in a rectangular coordinate system.

$$D_x(B_x) = \varepsilon_0(\mu_o)\left[\varepsilon_{11}(\mu_{11})E_x(H_x) + \varepsilon_{12}(\mu_{12})E_y(H_y) + \varepsilon_{13}(\mu_{13})E_z(H_z)\right] \tag{1.100}$$

$$D_y(B_y) = \varepsilon_0(\mu_o)\left[\varepsilon_{21}(\mu_{21})E_x(H_x) + \varepsilon_{22}(\mu_{22})E_y(H_y) + \varepsilon_{23}(\mu_{23})E_z(H_z)\right] \tag{1.101}$$

$$D_z(B_z) = \varepsilon_0(\mu_o)\left[\varepsilon_{31}(\mu_{31})E_x(H_x) + \varepsilon_{32}(\mu_{32})E_y(H_y) + \varepsilon_{33}(\mu_{33})E_z(H_z)\right] \tag{1.102}$$

1.4.2 Integral Forms of Maxwell's Equations

Let's begin our analysis with Eq. (1.89). Taking the surface integral of both sides of Eq. (1.89) over surface S with contour C gives

$$\int_S (\nabla \times \overline{H}) \cdot \hat{n} dS = \int_S \overline{J} \cdot \hat{n} dS + \int_S \frac{\partial \overline{D}}{\partial t} \cdot \hat{n}\, dS \tag{1.103}$$

We now apply Stokes' theorem as described in Section 1.3.6 for the left-hand side of the equation in (1.103) as

$$\int_S \underbrace{(\nabla \times \overline{H}) \cdot}_{\substack{\text{circulation per} \\ \text{unit area of } S}} d\overline{s} = \underbrace{\oint_C \overline{H} \cdot d\overline{l}}_{\substack{\text{circulation on} \\ \text{boundary of } S}} \tag{1.104}$$

From (1.103) and (1.105), we can now write as

$$\oint_C \overline{H} \cdot d\overline{l} = \int_S \overline{J} \cdot \hat{n} dS + \int_S \frac{\partial \overline{D}}{\partial t} \cdot \hat{n} dS \tag{1.105}$$

or

$$\oint_C \overline{H} \cdot d\overline{l} = i_s + \int_S \frac{\partial \overline{D}}{\partial t} \cdot \hat{n} dS \tag{1.106}$$

In (1.106) i_s is the current through the surface S. Equation (1.106) is the integral form of the equation given in (1.89).

Similarly, we can express the integral form of Eq. (1.88) as

$$\oint_C \overline{E} \cdot d\overline{l} = -\int_S \frac{\partial \overline{B}}{\partial t} \cdot \hat{n}\, dS \tag{1.107}$$

As a note, it is assumed that the magnetic current source does not exist. Taking the volume integral of both sides over volume V and surface S gives

$$\int_V (\nabla \cdot \overline{D})\, dV = \int_V \rho_v\, dV \tag{1.108}$$

We now apply divergence theorem as described in Section 1.3.5 for the left-hand side of the equation in (1.108) as

$$\int_V \underbrace{(\nabla \cdot \overline{D})}_{\substack{\text{flux per} \\ \text{unit volume}}} dV = \underbrace{\oint_S \overline{D} \cdot \hat{n} \, dS}_{\substack{\text{net flux} \\ \text{out of } S}} \tag{1.109}$$

From (1.108) and (1.109), we can write

$$\oint_S \overline{D} \cdot \hat{n} \, dS = \int_V \rho_v \, dV \tag{1.110}$$

Equation (1.110) is the integral form of the equation given in (1.91). Similarly, the integral form of Eq. (1.90) can be found as

$$\oint_S \overline{B} \cdot \hat{n} dS = 0 \tag{1.111}$$

In summary, the integral forms of Maxwell's equations are

$$\oint_C \overline{E} \cdot d\overline{l} = -\int_S \frac{\partial \overline{B}}{\partial t} \cdot \hat{n} \, dS \tag{1.112}$$

$$\oint_C \overline{H} \cdot d\overline{l} = i_s + \int_S \frac{\partial \overline{D}}{\partial t} \cdot \hat{n} dS \tag{1.113}$$

$$\oint_S \overline{D} \cdot \hat{n} \, dS = \int_V \rho_v \, dV \tag{1.114}$$

$$\oint_S \overline{B} \cdot \hat{n} dS = 0 \tag{1.115}$$

1.5 Time Harmonic Fields

Let's assume we have a sinusoidal function that changes in position and time. This function can be expressed as

$$\begin{aligned} g(\overline{r}, t) &= P(\overline{r}) \cos\left(\omega t + \phi(\overline{r})\right) \\ &= \mathrm{Re}\left\{ P(\overline{r}) e^{j\phi(\overline{r})} e^{j\omega t} \right\} \end{aligned} \tag{1.116}$$

Equation (1.116) can rewritten as

$$g(\overline{r}, t) = \mathrm{Re}\left\{ F(\overline{r}) e^{j\omega t} \right\} \tag{1.117}$$

In (1.117), $F(\bar{r})$ is defined as the phasor form of the function $g(r,t)$ and expressed as

$$F(\bar{r}) = P(\bar{r})e^{j\phi(\bar{r})} \tag{1.118}$$

This can be applied for vectorial function as

$$\overline{F}(\bar{r}) = \overline{P}(\bar{r})e^{j\phi(\bar{r})} \tag{1.119}$$

The representation of the time harmonic functions in phasor form provides several advantages. They convert the time domain differential equations to frequency domain algebraic equations. This can be better understood by studying the derivative property as follows. Let's take derivative function $g(r,t)$ with respect to time as

$$
\begin{aligned}
\frac{\partial g}{\partial t} &= \frac{\partial}{\partial t}\operatorname{Re}\left\{F(\bar{r})e^{j\omega t}\right\} \\
&= \operatorname{Re}\left\{\frac{\partial}{\partial t}\left(F(\bar{r})e^{j\omega t}\right)\right\} \\
&= \operatorname{Re}\left\{j\,\omega\,F(\bar{r})e^{j\omega t}\right\}
\end{aligned}
\tag{1.120}
$$

As a result of (1.120), it can be seen that the time derivative of a harmonic function means multiplying the same function by jω in the frequency domain. This can be shown as

$$\frac{\partial F(\bar{r},t)}{\partial t} \leftrightarrow j\,\omega\,F(\bar{r}) \text{ for scalars} \tag{1.121a}$$

$$\frac{\partial \overline{F}(\bar{r},t)}{\partial t} \leftrightarrow j\,\omega\,\overline{F}(\bar{r}) \text{ for vectors} \tag{1.121b}$$

Example 1.6   Maxwell's Equations
Derive the phasor representation of Maxwell's equations in free space with no source.

Solution
We begin with the equation given in (1.88) as

$$\nabla \times \overline{E}(\bar{r},t) = -\frac{\partial \overline{B}(\bar{r},t)}{\partial t} \tag{1.122}$$

Using the relation given in (1.122), this can be represented as

$$\nabla \times \operatorname{Re}\left\{\overline{E}(\bar{r})e^{j\omega t}\right\} = -\operatorname{Re}\left\{j\,\omega\,\overline{B}(\bar{r})e^{j\omega t}\right\}$$

or

$$\operatorname{Re}\left\{\left(\nabla \times \overline{E} + j\,\omega\,\overline{B}\right)e^{j\omega t}\right\} = \underline{0} \tag{1.123}$$

From Eq. (1.123), we can then express the phasor representation of the Maxwell's equation as

$$\nabla \times \overline{E} + j\,\omega\,\overline{B} = 0$$

or

$$\nabla \times \overline{E} = -j\,\omega\,\overline{B} \tag{1.124}$$

This can be applied to obtain the phasor form of all of the Maxwell's equations. The phasor form of the Maxwell's equations for (1.88)–(1.91) are

$$
\begin{aligned}
\nabla \times \overline{E} &= -j\omega \overline{B} \\
\nabla \times \overline{H} &= \overline{J} + j\omega \overline{D} \\
\nabla \cdot \overline{D} &= \rho_v \\
\nabla \cdot \overline{B} &= 0
\end{aligned}
\tag{1.125}
$$

References

1 Zahn, M. (1987). *Electromagnetic Field Theory: A Problem Solving Approach*. Krieger Pub Co.
2 Eroglu, A. (2010). *Wave Propagation and Radiation in Gyrotropic and Anisotropic Media*. Springer.

Problems

Problem 1.1
If K(1,2,0), L(2,5,0), and M(0,4,7) are given, calculate
(a) $\mathbf{KL} \times \mathbf{KM}$
(b) the angle between $\mathbf{KL}$ and $\mathbf{KM}$

Problem 1.2
Find vector $\mathbf{AB}$ in the Cartesian coordinate system if points $A(2m,\pi,0)$ and $B(2m,3\pi/2,0)$ are given in a cylindrical coordinate system.

Problem 1.3
If $\mathbf{A} = -2\hat{a}_x + 2\hat{a}_y + 2\hat{a}_z$ and $\mathbf{B} = -\hat{a}_x + \hat{a}_y$, then find unit vector nor mal to these two vectors.

Problem 1.4
If a plane is defined by $3x + 4y + 5z = 10$, find the unit vector normal to this plane.

Problem 1.5
If two arbitrary vectors are given as $\mathbf{K} = \hat{a}_x + 4\hat{a}_y + 3\hat{a}_z$ and $\mathbf{L} = 4\hat{a}_x + 2\hat{a}_y - 4\hat{a}_z$, how can you prove they are perpendicular?

Problem 1.6
Calculate the gradients of the following functions.
(a) $f = x^3 y^2 z$
(b) $g = \dfrac{1}{r}$ (cylindrical coordinate)
(c) $h = \dfrac{1}{e^R}$

Problem 1.7
Calculate the divergence of the following vectors
(a) $\mathbf{K} = \hat{a}_x 2 + \hat{a}_y 7x^2 y^2 z^2 - \hat{a}_z z^{-1}$
(b) $\mathbf{L} = \hat{r} \sin\phi + \hat{\phi} \cos\phi + \hat{a}_z z^2$

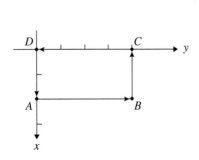

Figure 1.20 Geometry for Problem 1.9.

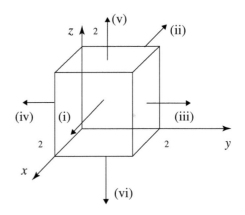

Figure 1.21 Geometry for Problem 1.10.

(c) $\mathbf{M} = \hat{R}r^{-1} + \hat{\theta}\sin\theta\sin\phi - \hat{\phi}\sin\theta\sin\phi$

Problem 1.8
Find the curl of the following vectors (a) $\mathbf{K} = 3xy^2z^{-1}\,\hat{a}_x$ (b) $\mathbf{L} = \hat{a}_r\,(\sin^2\phi)r - \hat{a}_\phi\,r^2\,z\,\cos\phi$ (c) $\mathbf{M} = \hat{a}_r\,R^2\sin\theta + \hat{a}_\theta\,r$ $(\cos\phi)^{-1}$

Problem 1.9
$\mathbf{A} = y^2\hat{a}_x + x^2\hat{a}_y$. Verify Stokes' theorem for the geometry given in Figure 1.20.

Problem 1.10
Prove divergence theorem using the function

$$\mathbf{K} = 3\mathbf{a}_x + \mathbf{a}_y 2xy + \mathbf{a}_z\,8x^2y^3$$

and a cube situated at the origin as in Figure 1.21.

2

Passive and Active Components

2.1 Introduction

In this chapter, the analysis and use of passive and active components in microwave engineering is discussed. High frequency (HF) circuits include combinations of passive and active components that must meet various specifications and so have different requirements. Passive components show linear transfer characteristics, as illustrated in Figure 2.1a. The core passive components used in the high frequency circuit design are resistors, capacitors, and inductors. Transformers can also be considered part of an inductor group and so as critical passive components. Active components exhibit nonlinear transfer characteristics, such as the one shown in Figure 2.1b. Nonlinear devices can still have close to linear behavior in the presence of small signals. The characteristics of these components vary with respect to the operational parameters that these components are designed for, including frequency of operation, power dissipation capacity, size, etc.

2.2 Resistors

Let's consider a piece of conductive wire with a uniform cross section, A, length, l, conductivity, σ, as shown in Figure 2.2. The current density and electric field in the conductor will be in the same direction and can be related through

$$\bar{J} = \sigma \bar{E} \tag{2.1}$$

The voltage between points a and b is found from

$$V_{ab} = - \int \bar{E}.\overline{dl} = El \tag{2.2a}$$

or

$$E = \frac{V_{ab}}{l} \tag{2.2b}$$

The current in the conductor is calculated using

$$I = \int_A \bar{J} \cdot \overline{dA} = JA \tag{2.3}$$

RF/Microwave Engineering and Applications in Energy Systems, First Edition. Abdullah Eroglu.
© 2022 John Wiley & Sons Ltd. Published 2022 by John Wiley & Sons Ltd.
Companion website: www.wiley.com/go/eroglu/rfmicrowave

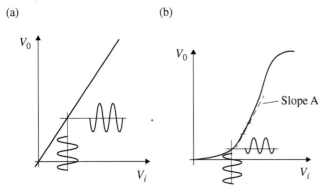

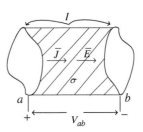

Figure 2.2 Conductor with uniform cross section.

Figure 2.1 Transfer characteristics: (a) passive component; (b) active component.

or

$$J = \frac{I}{A} \tag{2.4}$$

Substituting (2.3) and (2.4) into (2.1) gives

$$\frac{I}{A} = \sigma \frac{V_{ab}}{l} \tag{2.5}$$

Rearranging (2.5) leads to

$$V_{ab} = \left(\frac{l}{\sigma A}\right) I = RI \tag{2.6}$$

So, the DC resistance can be expressed using (2.6) as

$$R = \left(\frac{l}{\sigma A}\right) \tag{2.7}$$

In high frequency applications, the frequency independent linear model for a resistor shown in Figure 2.3a does not accurately describe resistor behavior. Resistors at high frequencies present resonance due to parasitic inductance effects from the leads and parasitic capacitance and interlead capacitance effects. These nonideal effects can be included with the resistor model given in Figure 2.3b. There are several types of resistors, depending on how they are formed. These include carbon composition resistors, wire wound resistors, metal film resistors, and thin film chip resistors.

When a nonideal resistor acts as a capacitor, the frequency can be related as

$$f = \frac{1}{2\pi R C_{Parasitic}} \tag{2.8}$$

When a nonideal resistor acts as an inductor, the frequency can be written as

$$f = \frac{1}{2\pi \sqrt{L_{Lead} C_{Parasitic}}} \tag{2.9}$$

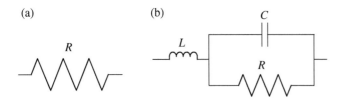

Figure 2.3 Representation of the resistor equivalent model: (a) ideal resistor model; (b) nonideal resistor model.

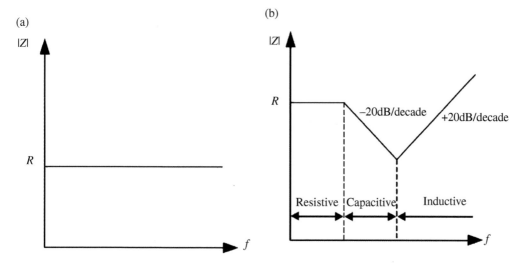

Figure 2.4 Frequency characteristics of (a) ideal resistor and (b) nonideal resistor.

This ideal and nonideal resistor behavior versus frequency is illustrated in Figure 2.4.

The resonance frequency of the nonideal model for the resistor given in Figure 2.3b can be found from the equivalent impedance of the network as

$$Z = j\omega L + \frac{R}{1 + j\omega RC} = \frac{R}{1 + \omega^2 R^2 C^2} + j\frac{\omega^3 R^2 LC^2 + \omega(L - R^2 C)}{1 + \omega^2 R^2 C^2} \qquad (2.10)$$

The resonant frequency is then found when the imaginary part of (2.10) is zero.

$$f_r = \frac{1}{2\pi}\sqrt{\frac{1}{LC} - \frac{1}{R^2 C^2}} \qquad (2.11)$$

Example 2.1   Nonideal Resistor
Assume that the nonideal resistor shown in Figure 2.5 with $L = 10\,\text{nH}$ and $C = 1\,\text{pF}$ is given. Obtain the resonance frequency of the resistor versus R.

Solution
The relation in (2.11) is used to plot the resonance frequency versus R when $L = 10\,\text{nH}$ and $C = 1\,\text{pF}$. This is illustrated in Figure 2.6.

2.3 Capacitors

Consider the two-conductor system shown in Figure 2.6. Application of DC voltage, V_{ab}, initiates a charge transfer from the $+Q$ plate to the $-Q$ plate. This will result in electric field flow, as illustrated with dashed lines in the

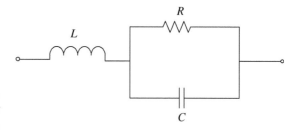

Figure 2.5   Figure for Example 2.1.

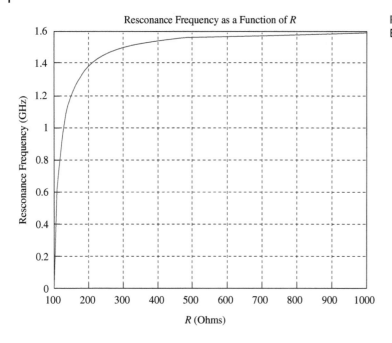

Figure 2.6 Resonance frequency plot for Example 2.1.

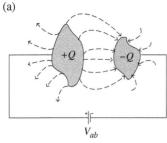

figure. This leads to the formation of capacitance, which can then be related to applied voltage and Q as

$$C = \frac{Q}{V_{ab}} \tag{2.12}$$

If parallel plates are used in the configuration shown in Figure 2.7a, then the capacitor can be expressed as

$$C = \varepsilon_0 \varepsilon_r \frac{A}{d} \tag{2.13}$$

In (2.13), A is the surface area of the plate and d is given as the plate separation. In practice, the capacitors have lead inductances, conductor, and dielectric losses for HF applications. These effects can be taken into account by using the model given in Figure 2.7b.

The HF model of the capacitor given in Figure 2.7b has lead inductance, L, conductor loss, R_s, dielectric loss, R_s, which only become relevant at high frequencies. The characteristics of the capacitor can be obtained by finding the equivalent impedance as

Figure 2.7 (a) Two conductor capacitor. (b) HF model of capacitor.

$$Z = (j\omega L_s + R_s) + \left(\frac{1}{G_d + j\omega C}\right) = \frac{R_s G_d^2 + (\omega C)^2 + G_d}{G_d^2 + (\omega C)^2} + j\frac{\omega L G_d^2 + L\omega(\omega C)^2 - \omega C}{G_d^2 + (\omega C)^2} \tag{2.14}$$

The equation in (2.14) can be expressed as

$$Z = R_s + jX_s \tag{2.15}$$

where

$$R_s = \frac{R_s G_d^2 + (\omega C)^2 + G_d}{G_d^2 + (\omega C)^2} \qquad (2.16)$$

$$X_s = \frac{\omega L G_d^2 + L\omega(\omega C)^2 - \omega C}{G_d^2 + (\omega C)^2} \qquad (2.17)$$

Impedance given in (2.15) can be converted to admittance as

$$Y = Z^{-1} = \frac{R_s}{R_s^2 + X_s^2} + j\frac{-X_s}{R_s^2 + X_s^2} = G + jB \qquad (2.18)$$

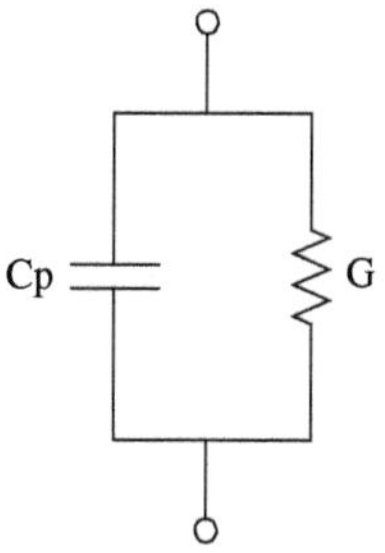

or

$$Y = \frac{\left(R_s G_d^2 + (\omega C)^2 + G_d\right)\left(G_d^2 + (\omega C)^2\right)}{\left(R_s G_d^2 + (\omega C)^2 + G_d\right)^2 + \left(\omega L G_d^2 + L\omega(\omega C)^2 - \omega C\right)^2} + j\frac{\left(\omega C - \omega L G_d^2 - L\omega(\omega C)^2\right)\left(G_d^2 + (\omega C)^2\right)}{\left(R_s G_d^2 + (\omega C)^2 + G_d\right)^2 + \left(\omega L G_d^2 + L\omega(\omega C)^2 - \omega C\right)^2} \qquad (2.19)$$

Then, the capacitor can be represented by a parallel equivalent circuit, as shown in Figure 2.8 where

$$G = \frac{\left(R_s G_d^2 + (\omega C)^2 + G_d\right)\left(G_d^2 + (\omega C)^2\right)}{\left(R_s G_d^2 + (\omega C)^2 + G_d\right)^2 + \left(\omega L G_d^2 + L\omega(\omega C)^2 - \omega C\right)^2} \qquad (2.20)$$

$$B = \frac{\left(\omega C - \omega L G_d^2 - L\omega(\omega C)^2\right)\left(G_d^2 + (\omega C)^2\right)}{\left(R_s G_d^2 + (\omega C)^2 + G_d\right)^2 + \left(\omega L G_d^2 + L\omega(\omega C)^2 - \omega C\right)^2} \qquad (2.21)$$

The resonance frequency for the circuit shown in Figure 2.8 is found when $B = 0$ as

$$f_r = \frac{1}{2\pi}\sqrt{\frac{R_d^2 C - L}{R_d^2 C^2 L}} \qquad (2.22)$$

The quality factor for the parallel network is then obtained from

$$Q = \frac{|B|}{G} = \frac{R_p}{|X_p|} \qquad (2.23)$$

The tangent loss for capacitors is related to the dielectric loss in the material that capacitors are manufactured. It can be found by using Eq. (2.14) while ignoring the lead effects. Then, (2.14) can be expressed as

$$Z = \frac{1}{G_d + j\omega C} \qquad (2.24)$$

In (2.24)

$$G_d = \frac{\sigma_{diel} A}{d} \qquad (2.25)$$

Since loss tangent, $\tan\delta$, is

$$\tan \delta = \frac{\sigma_{diel}}{\omega \varepsilon} \qquad (2.26)$$

Substituting (2.26) into (2.25) gives

$$G_d = \omega C \tan \delta \qquad (2.27)$$

Figure 2.8 Equivalent parallel circuit.

Please note that

$$R_d = \frac{1}{G_d} = \frac{1}{\omega C \tan \delta} \tag{2.28}$$

In practice, datasheets list ESR, which is known as equivalent series resistance, for capacitors to define their loss. ESR can be related to tangent loss as

$$ESR = \frac{\tan \delta}{\omega C} \tag{2.29}$$

Example 2.2 Capacitor

Obtain the high frequency characteristics of a 36 pF capacitor whose loss tangent is 0.001. Assume the total length of the leads 2 cm and the radius is 1 mm. The leads are made of copper with conductivity 59.6×10^6 (S/m). Assume the total lead inductance is 25 nH.

Solution

In the question, C, L are given for the HF model shown in Figure 2.6. We need to calculate the series resistance, R_s, and obtain the loss resistance, R_d. The series resistance is an AC resistance and is calculated from

$$R_s = R_{dc} \frac{r}{2\delta} = \frac{l}{r} \sqrt{\frac{\mu_o f}{\pi \sigma_{copper}}} \tag{2.30}$$

where r is the radius of the lead wire, δ is skin depth, and σ is the conductivity. From (2.30)

$$R_s = 1.65 \sqrt{f} \, \mu\Omega$$

From (2.28)

$$R_d = \frac{1}{\omega C \tan \delta} = \frac{4.42 \times 10^{12}}{f} \, \Omega \tag{2.31}$$

We can then plot $Z(f)$ from (2.14), as shown in Figure 2.9.

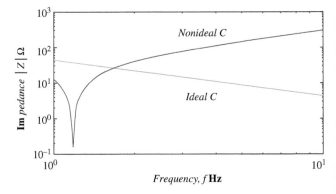

Figure 2.9 Plot of $Z(f)$ versus frequency.

2.4 Inductors

An inductor can be implemented as a discrete component or distributed element depending on the frequency of application. If a current flows through a wire wound, a flux is produced through each turn as a result of magnetic flux density, as shown in Figure 2.10. The relation between the flux density and flux through each turn can be represented as

$$\Psi = \int \overline{B} \cdot d\overline{s} \tag{2.32}$$

If there are N turns, then we define the flux linkage as

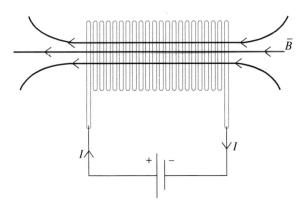

Figure 2.10 Flux through each turn.

$$\lambda = N\Psi = N\int \overline{B} \cdot d\overline{s} \tag{2.33}$$

Inductance is defined as the ratio of flux linkage to current flowing through the windings as defined by

$$L = \frac{\lambda}{I} = \frac{N\Psi}{I} \tag{2.34}$$

The inductance defined by (2.34) is also known as self-inductance of the core and is formed by the windings. The core can be an air core or a magnetic core.

Inductors can be formed as air core inductors or magnetic core inductors, depending on the application. When air core inductors are formed through windings and operated at HF, they present HF characteristics. These include winding resistance and distributed capacitance effects between each turn, as shown in Figure 2.11.

The HF model of the inductor with its equivalent circuit is shown in Figure 2.12.

As a result, the inductor presents inductive characteristics up to a certain frequency and then gets into resonance and exhibits capacitive effects after the resonance frequency. It can be shown that the series equivalent circuit for the HF circuit can be obtained, as shown in Figure 2.13 [1].

We can write equivalent impedance as

$$Z = R_s + jX_s \tag{2.35}$$

where

$$R_s = \frac{R}{(1 - \omega^2 L C_s)^2 + (\omega R C_s)^2} \tag{2.36}$$

and

$$X_s = \frac{\omega(L - R^2 C_s) - \omega^3 L^2 C_s}{(1 - \omega^2 L C_s)^2 + (\omega R C_s)^2} \tag{2.37}$$

The resonance frequency is found when $X_s = 0$ as

$$f_r = \frac{1}{2\pi}\sqrt{\frac{L - R^2 C_s}{L^2 C_s}} \tag{2.38}$$

The quality factor is obtained from

$$Q = \frac{|X_s|}{R_s} \tag{2.39}$$

C_s in Figure 2.12 is the capacitance, including the effects of the distributed capacitance of the inductor and is given by

$$C_s = \frac{2\pi\varepsilon_0 da N^2}{l_W} \tag{2.40}$$

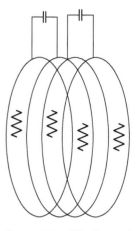

Figure 2.11 HF effects of an RF inductor.

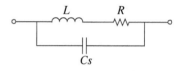

Figure 2.12 HF model of an RF air core inductor.

Figure 2.13 Equivalent series circuit.

Example 2.3 Inductor

Find the HF characteristics of an inductor using the model given in Figure 2.12 when $L = 10\,\text{nH}$, $C_s = 5\,\text{pF}$, and $R = 10\,\Omega$.

High frequency inductor response

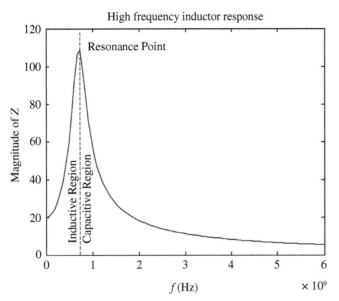

Figure 2.14 The HF characteristics of the inductor in Example 2.3.

Solution
The frequency characteristics are obtained with MATLAB and illustrated in Figure 2.14.

2.4.1 Air Core Inductor Design

In practice, inductors can be implemented as air core or toroidal inductors using magnetic cores, as illustrated in Figure 2.15.

For the air core solenoidal inductor given in Figure 2.15, the inductance can be calculated using the relation

$$L = \frac{d^2 N^2}{18d + 40l} \, \mu H \tag{2.41}$$

In this equation, L is given as inductance in μH, d is the coil inner diameter in inches (in), l is the coil length in inches (in), and N is the number of turns of the coil. The formula given in (2.41) can be extended to include the spacing between each turn of the air coil inductor. Then, Eq. (2.41) can be modified as

$$L = \frac{d^2 N^2}{18d + 40(Na + (N-1)s)} \, \mu H \tag{2.42}$$

In (2.42), a represents the wire diameter in inches and s represents the spacing in inches between each turn.

Example 2.4 Air Core Inductor

Plot the HF characteristics of an air core inductor of 330 nH and calculate its resonant frequency and find the quality factor when 12 AWG is used with a 0.5 in. rod. Neglect the spacing between each turn.

Solution
From Eq. (2.41)

$$L = \frac{d^2 N^2}{18d + 40(Na)} \, \mu H$$

(a)

(b)

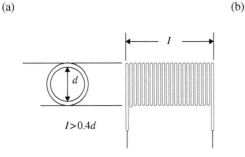

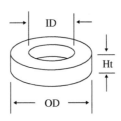

Figure 2.15 (a) Air core inductor. (b) Toroidal inductor.

where $l_{ind} = Na$ is used to find the length of the inductor. This leads to $0.25N^2 - 1.066N - 2.97 = 0$
N is found to be $N = 6.18$. N has to be positive integer. We choose N to be $N = 7$. So

$$N = 7, \qquad L = 0.387\,\mu H, \quad l_{ind} = 0.5656 \text{ in.}$$

Then, the length of the air core inductor is

$$l_{ind} = Na = 7(0.0808) = 0.5656 \text{ in.} = 0.01144 \text{ m}$$

Since $\sigma = 5.96 \times 10^7$ S/m, $l_W = 2\pi r N = 0.045$ m, then the cross section area of the wire can be found as $A = \pi(2.053 \times 10^{-3}/2)^2 = 3.31 \times 10^{-6} \text{m}^2$. So

$$R = \frac{l_W}{\sigma A} = \frac{0.045}{(5.96 \times 10^7)(3.31 \times 10^{-6})} = 0.228 \text{ m}\Omega$$

and

$$C_s = \frac{2\pi\varepsilon_0 daN^2}{l_W} = \frac{2\pi(8.85 \times 10^{-12})(0.0127)(2.053 \times 10^{-3})49}{0.045} = 1.578 \text{ pF}$$

The HF response of the inductor is illustrated in Figure 2.7.
The quality factor versus frequency for this inductor is given in Figure 2.17.
The resonant frequency for this inductor is

$$f_r = \frac{1}{2\pi}\sqrt{\frac{L - R^2 C_s}{L^2 C_s}} = \frac{1}{2\pi}\sqrt{\frac{330 \times 10^{-9} - (0.228 \times 10^{-3})^2(1.578 \times 10^{-12})}{(330 \times 10^{-9})^2(1.578 \times 10^{-12})}} = 2.2 \times 10^8 \text{ Hz}$$

The resonant frequency calculated above matches with the results illustrated in Figures 2.16 and 2.17. Air core inductors inherently have very high Q factors. At the resonance frequency, the quality factor is zero, as expected.

Figure 2.16 HF characteristics' response of air core inductor.

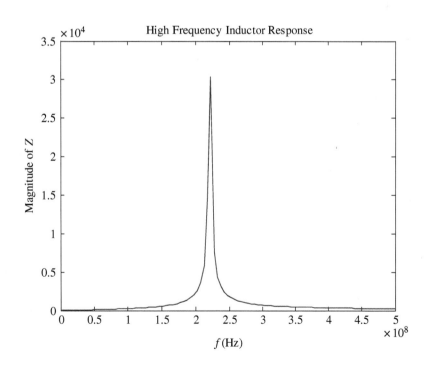

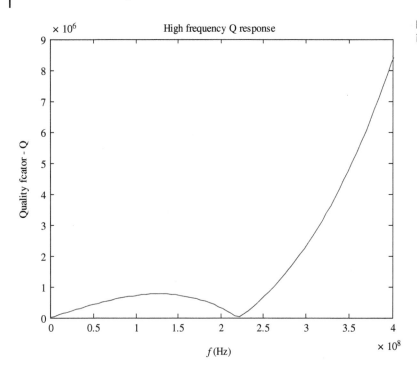

$\times 10^6$

High frequency Q response

f (Hz)

$\times 10^8$

Figure 2.17 Quality factor of air core inductor versus frequency.

2.4.2 Magnetic Core Inductor Design

In several radio frequency (RF) applications, larger inductance values may be required for those areas with restricted space. One solution to increase the inductance value for an air core inductor is to increase the number of turns. However, this increases the size of the air core inductor.

This challenge can be overcome by using magnetic cores. One of the other advantages of using toroidal cores is being able to retain the flux within the core, as shown in Figure 2.18. This provides self-shielding. In air core inductor design, air is used as a nonmagnetic material to wind the wire around it. When air is replaced with a magnetic material such as a toroidal core, the inductance of the formed inductor can be calculated using

$$L = \frac{4\pi N^2 \mu_i A_{Tc}}{l_e} \text{ nH} \tag{2.43}$$

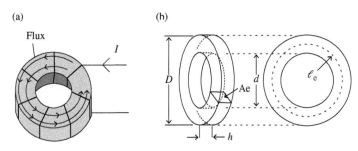

(a)

Flux

(b)

Figure 2.18 (a) Magnetic flux in toroidal magnetic core; (b) geometry of toroid.

In (2.43), L is the inductance in nanohenries (nH), N is the number of turns, μ_i is the initial permeability, A_{Tc} is the total cross-sectional area of the core in cm^2, and l_e is the effective length of the core in cm. The effective length of the core l_e is defined as

$$l_e = \frac{\pi(od - id)}{\ln(od/id)} \text{ cm} \tag{2.44}$$

where od is the outside and id is the inside diameters of the core in cm. The total cross-sectional area of the core, A_{Tc}, is defined as

$$A_{Tc} = \frac{1}{2}(od - id) \times h \times n \ \text{cm}^2 \tag{2.45}$$

h refers the thickness of the core in cm and n is used to define the number of stacked cores.

It is not uncommon to have the information about the inductance index of the core on its datasheet. If the inductance index is given, then Eq. (2.43) can be modified as

$$L = N^2 A_L \ \text{nH} \tag{2.46}$$

where A_L is the inductance index in nanohenries/turn^2. The maximum operational flux density for toroidal core is calculated from

$$B_{op} = \frac{V_{rms} \times 10^8}{4.44 f N A_{Tc}} \ \text{Gauss} \tag{2.47}$$

In Eq. (2.47), B_{op} is the magnetic-flux density in Gauss, V_{rms} is the maximum root mean square (RMS) voltage across the inductor in volts, f is the frequency in Hertz, N is the number of turns, and A_{Tc} is the total cross-sectional area of the core in cm^2. The proper design of a toroidal core inductor requires operational voltage of the inductor and the required inductance value. This helps to identify the right material for the inductor design to prevent saturation.

2.4.3 Planar Inductor Design

Spiral inductors can be good choice when larger inductance values may be required for those areas with restricted space. Spiral type planar inductors are widely used in the design of power amplifiers, oscillators, microwave switches, combiners, and splitters, etc. The inductance value of the spiral inductors at the HF range can be determined using the quasistatic method proposed by Greenhouse [2] and Eroglu [3] with a good level of accuracy (Figure 2.19).

The method proposed by Greenhouse takes into account self-coupling and mutual coupling between each trace. The layout of the two conductors that is used in the inductance calculation is illustrated in Figure 2.19a. GMD is the geometric mean distance between two conductors and AMD represents the arithmetic mean distance between two conductors. The total inductance of the configuration of the spiral inductor is

$$L_T = L_0 + \Sigma M \tag{2.48}$$

where L_T is the total inductance, L_0 is the sum of the self-inductances, and ΣM is the sum of the total mutual inductances. The application of the formulation given by (2.28) can be demonstrated for the spiral inductor illustrated in Figures 2.19b and c as

$$L_T = L_1 + L_2 + L_3 + L_4 + L_5 - 2(M_{13} + M_{24} + M_{35}) + 2M_{15} \tag{2.49}$$

The general relations that can be used in the algorithm for the spiral inductance calculation then become

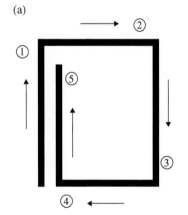

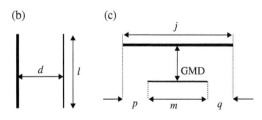

Figure 2.19 (a) Rectangular spiral inductor. (b) Layout of current filaments. (c) Two-parallel filament geometry.

$$L_i = 0.0002l_i \left[\ln \left(2\frac{l_i}{GMD} \right) - 1.25 + \frac{AMD}{l_i} + \frac{\mu}{4}T \right] \tag{2.50}$$

$$M_{ij} = 0.0002l_i Q_i \tag{2.51}$$

$$\ln(GMD_i) = \ln(d) - \frac{1}{12\left(\dfrac{d}{w}\right)^2} - \frac{1}{60\left(\dfrac{d}{w}\right)^4} - \frac{1}{168\left(\dfrac{d}{w}\right)^6} - \frac{1}{360\left(\dfrac{d}{w}\right)^8} - \cdots \tag{2.52}$$

$$Q_i = \ln\left[\frac{l_i}{GMD_i} + \left(1 + \left(\frac{l_i}{GMD_i}\right)^2\right)^{0.5} \right] - \left(1 + \left(\frac{GMD_i}{l_i}\right)^2\right)^{0.5} + \frac{GMD_i}{l_i} \tag{2.53}$$

$$AMD = w + t \tag{2.54}$$

where L_i is the self-inductance of the segment i, M_{ij} is the mutual inductance between segments i and j, l_i is the length of segment l_i, μ is the permeability of the conductor, T is the frequency correction factor, d is the distance between conductor filaments, w is the width of the conductor, t is the thickness of the conductor, Q_i is the mutual inductance parameter of segment i, GMD_i is the geometric distance of segment i, and AMD is the arithmetic mean distance.

2.4.4 Transformers

A transformer is a device that converts one level of voltage to another level of voltage at the same RF. It consists of one or more coils of wire wrapped around a common magnetic core. These coils are not usually connected electrically. However, they are connected through the common magnetic flux confined to the core. Assuming that the transformer has at least two windings, one of them (primary) is connected to a source of RF power and the other (secondary) is connected to the RF loads. The typical toroidal transformer is shown in Figure 2.20.

The relationship between the voltage applied to the primary winding $v_p(t)$ and the voltage produced on the secondary winding $v_s(t)$ can be established as follows. Assume $v_p(t)$ is applied to the primary winding of the toroid in Figure 2.20. The average flux in the primary winding can be written as

$$\overline{\phi} = \frac{1}{N_p} \int v_p(t)dt \tag{2.55}$$

A portion of the flux produced in the primary coil passes through the secondary coil as mutual flux and the rest is lost as leakage flux, which can be shown using the following relation.

$$\overline{\phi}_p = \phi_m + \phi_{Lp} \tag{2.56}$$

In the secondary coil a similar relation holds as

$$\overline{\phi}_s = \phi_m + \phi_{Ls} \tag{2.57}$$

Using Faraday's law, we can write the voltage on the primary as

$$v_p(t) = N_p \frac{d\overline{\phi}_p}{dt} = N_p \frac{d\phi_m}{dt} + N_p \frac{d\phi_{Lp}}{dt} = v_1(t) + v_{1L}(t) \tag{2.58}$$

Then, the secondary voltage can be expressed as

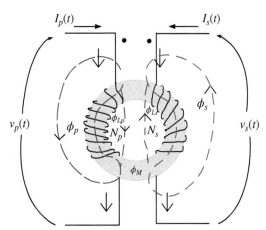

Figure 2.20 Toroidal transformer.

$$v_s(t) = N_s \frac{d\overline{\phi}_s}{dt} = N_s \frac{d\phi_m}{dt} + N_s \frac{d\phi_{Ls}}{dt} = v_2(t) + v_{2L}(t) \tag{2.59}$$

The voltage on the primary and secondary of the transformer due to mutual flux, ϕ_M, can be written as

$$v_1(t) = N_p \frac{d\phi_m}{dt} \tag{2.60}$$

and

$$v_2(t) = N_s \frac{d\phi_m}{dt} \tag{2.61}$$

Taking the ratio of the $v_1(t)$ to $v_2(t)$ using Eqs. (2.60) and (2.61) gives

$$\frac{v_1(t)}{v_2(t)} = \left(N_p \frac{d\phi_m}{dt}\right)\left(\frac{1}{N_s}\frac{dt}{d\phi_m}\right) = \frac{N_p}{N_s} \tag{2.62}$$

As a result, the ratio of the primary voltage to the secondary voltage both caused by the mutual flux is equal to the turns ratio of the transformer. It is desired that transformers have very low leakage flux such that

$$\phi_m >> \phi_{Lp}; \quad \phi_m >> \phi_{Ls} \tag{2.63}$$

Then, we can write

$$\frac{v_p(t)}{v_s(t)} \approx \frac{N_p}{N_s} \tag{2.64}$$

These are the characteristics of ideal transformers. The same derivation techniques can also be used to obtain the relation between the current on the primary and secondary sides as

$$\frac{i_{p(t)}}{i_{s(t)}} \approx \frac{N_s}{N_p} \tag{2.65}$$

2.5 Semiconductor Materials and Active Devices

The selection of a transistor for RF amplification is critical since it affects efficiency, dissipation, power delivery, stability, linearity, etc. Once the transistor is selected for the corresponding amplifier topology, the size of the transistor, die placement, bond pads, bonding of the wires, and lead connections will determine the layout of the amplifier and the thermal management of the system. The most commonly used RF and microwave power devices for commercial purposes are based on silicon (Si) and gallium arsenide (GaAs). There is an intense research in development of high power density devices using wide-bandgap (WBG) materials, such as silicon carbide (SiC) and gallium nitride (GaN). In essence, a device's performance is determined by several parameters, including material energy bandgap, breakdown field, electrons, and holes transport properties, thermal conductivity, saturated electron velocity, and conductivity. The typical values of these parameters for various types of semiconductor materials are given in Table 2.1. Different types of figure of merits can be used to determine the performance of the material for RF application. One of the most commonly used figures of merit is the Johnson figure of merit (JFOM) as it can also provide the frequency limit as given by

$$JFOM = \frac{E_{br} \times V_{sat}}{2\pi} \tag{2.66}$$

where E_{br} is the breakdown voltage and V_{sat} is the saturated electron density. JFOM is given in Table 2.1 for each commonly used material in RF technology. As can be seen from the values, WBG materials such as GaN or SiC have a significantly higher JFOM, which makes them more attractive.

Table 2.1 Typical parameter values for semiconductor materials.

RF High power material	μ (cm²/Vs)	ε_r	E_g (ev)	Thermal conductivity (W/cmK)	E_{br} (MV/cm)	T_{max} (°C)	JFOM = $E_{br}v_{sat}/2\pi$
Si	1350	11.8	1.1	1.3	0.3	300	1.0
GaAs	8500	13.1	1.42	0.46	0.4	300	2.7
SiC	700	9.7	3.26	4.9	3.0	600	20
GaN	2000	9.0	3.39	1.7	3.3	700	27.5

Another commonly used figure of merit is Baliga's figure of merit (BFOM), which is defined by

$$BFOM = \varepsilon_r\varepsilon_0\mu_n E_G^3 \qquad (2.67)$$

In (2.67), μ_n is electron mobility and E_g is known as energy for the bandgap.

2.5.1 Si

Silicon (Si) has been used since the invention of the first transistor. That is part of the reason most of the power semiconductor devices today are still manufactured using Si. Si can be used to implement unipolar and bipolar devices in a cost-effective way. However, it has limitations for high power and HF operations. The introduction of laterally diffused metal oxide semiconductor (LDMOS) in the 1970s improved performance and enabled the use of this material for high power applications up to 4 GHz until 2005. The comparison of the semiconductor material Si versus other materials is illustrated in Table 2.1. As illustrated, although it has been used for decades and is a low-cost choice for devices, due to its critical parameters such as band energy, it lacks the performance of WBG devices.

2.5.2 Wide-Bandgap Devices [4]

Electrons exist at energy levels that combine to form energy bands in solids. This is illustrated in Figure 2.21. Figure 2.21 shows the conduction band, forbidden band, and valence band. The forbidden band is the region between the valence band and the conduction band. No electrons exist in this region. For electrons to move from the valence band to the conduction band or the conduction band to the valence band, they need to be excited with a certain energy level. WBG semiconductors are classified as wide band because of the wider bandgap existing between conduction and valance bands. For instance, Si has a bandgap energy level of 1.12 eV, whereas WBG semiconductors such as SiC, GaN, or diamond have much higher energy levels. The bandgap energy levels of these WBG semiconductors are illustrated in Table 2.1. WBG semiconductors have several advantages, such as high

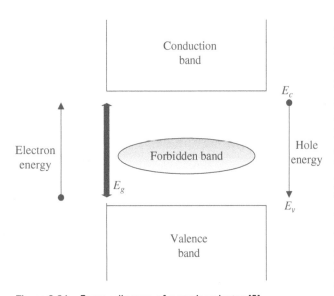

Figure 2.21 Energy diagram of a semiconductor [5].

temperature and radiation hardening. In semiconductors, the thermal energy of the electrons in the valence band increases when the temperature increases. When the temperature reaches a threshold value, electrons have sufficient energy to move to the conduction band. This should be avoided for an uncontrolled conduction. The uncontrolled conduction temperature level for Si is 150 °C. This value for WBG is much higher. For instance, it is 900 °C for SiC and even higher for diamond. Hence, WBG devices manufactured with WBG semiconductors can operate with more heat and radiation without losing their electrical characteristics, which make them more reliable at higher powers during RF applications. Furthermore, WBG devices have larger electric breakdown fields and as result have higher breakdown voltages than Si devices. This enables higher doping levels, which allows device layers to be thinner than Si devices for the breakdown voltage levels with smaller drift resistances.

2.5.2.1 GaAs [5–6]

GaAs technology has been widely used since the 1980s. GaAs transistors can operate from 30 MHz to 250 GHz and over due to higher saturated electron velocity and higher electron mobility, with typical operating voltages of 5 to 7 V for narrowband and wideband applications. Their advantage is small noise and high sensitivity, whereas their power handling capability is limited to 5–10 W. It also has a low breakdown voltage and cannot withstand high temperatures. Higher output power can be obtained by using parallel or push–pull configurations, which sacrifice efficiency. It is also widely used in monolithic microwave integrated circuits (MMICs).

2.5.2.2 GaN [7–8]

GaN devices were first commercialized around 2005 after 15 years of development. The GaN technology progressed rapidly with the advancement in a wireless system. GaN has high power density, high breakdown voltage, and high current handling, in comparison to other existing semiconductor materials. Furthermore, it has high saturated electron drift velocity, and it has improved thermal conductivity when it is epitaxially grown on semi-insulating SiC substrates. These features make it possible to develop RF power devices such as high electron mobility transistors (HEMTs) with larger output power. Some of the comparison from Table 2.1 leads to the following superior characteristics.

- JFOM for GaN is more than 10 times better than GaAs.
- Breakdown voltage for GaN is more than 10 times better than Si and around 10 times the value of GaAs.
- GaN has a higher channel operational channel temperature than Si or GaAs.

One of the disadvantages of GaN is its poor thermal characteristic. That is part of the reason GaN devices are almost always fabricated on low-loss, high-thermal conductivity substrates, such as silicon carbide (GaN-on-SiC), to help with GaN's poor thermal management characteristics. Another alternative implementation is GaN on Si material. However, this configuration has a frequency limitation of 6 GHz. Diamond is a promising option to improve the thermal profile of GaN.

The most commonly used semiconductor materials GaN, GaAs, and Si are compared and given in Figure 2.22 [9].

2.5.3 Active Devices

The power device family tree can be simplified, as shown in Figure 2.23.

The details about some of the most commonly used transistors are discussed in this section.

2.5.3.1 BJT and HBTs

A typical npn bipolar junction transistor (BJT) being formed out of Si and the emitter and collector regions of which being implanted with donors. FET (field effect transistor) and HBT (heterojunction bipolar transistor)

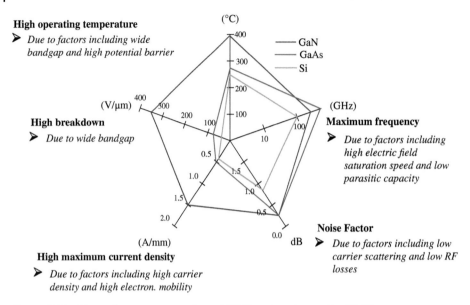

Figure 2.22 The performance comparison of WBG semiconductors for RF [9].

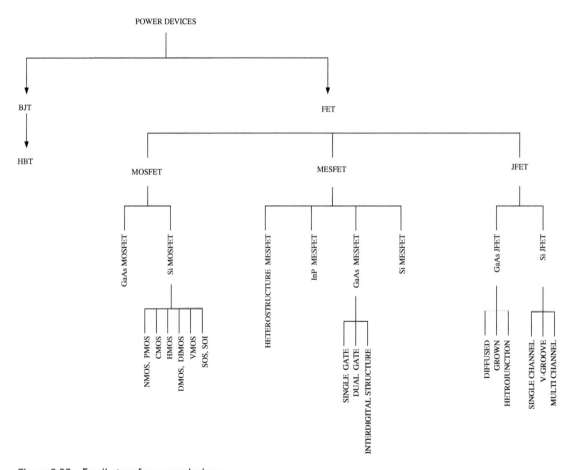

Figure 2.23 Family tree for power devices.

(a)

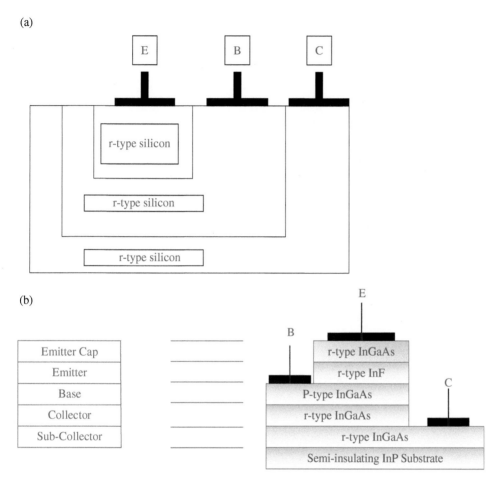

(b)

Figure 2.24 Typical layer structure of (a) Si BJT and (b) InP/InGaAs HBT transistor.

devices have different characteristics. The main difference between a BJT and HBT is the introduction of a hetero-junction at the emitter–base interface in the HBT device [1]. This is illustrated in Figure 2.24.

FET is a planar device, whereas the HBT is a vertical device. The HBT device is an enhanced version of conventional BJTs, as a result of the exploitation of heterostructure junctions. Unlike conventional BJTs, in HBTs the bandgap difference between the emitter and the base materials results in a higher common emitter gain. The base sheet resistance is lower than in ordinary BJTs, and the resulting operating frequency is accordingly higher [3].

2.5.3.2 FETs

The FET family includes a variety of structures, among which are MESFETs, MOSFETs, and HEMTS. They typically consist of a conductive channel accessed by two ohmic contacts, acting as source (S) and drain (D) terminals, respectively. The third terminal, the gate (G), forms a rectifying junction with the channel or a metal oxide semiconductor (MOS) structure. A simplified structure of a metal-semiconductor n-type FET is depicted in Figure 2.25 [3]. FET devices ideally do not draw current through the gate terminal, unlike the BJTs, which conversely require a significant base current, thus simplifying the biasing arrangement. FET devices exhibit a negative temperature coefficient, resulting in a decreasing drain current as the temperature increases. This prevents thermal runaway

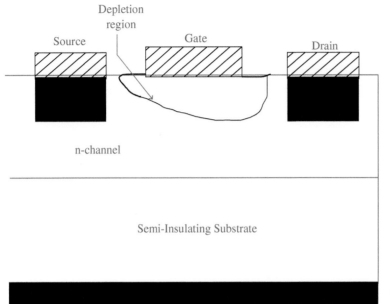

Figure 2.25 Simplified structure of FET.

Table 2.2 RF Power transistors and their applications and frequency of operations.

RF transistor	Drain BV (V)	Frequency (GHz)	Major applications
RF power FET	65	0.001–0.4	VHF power amplifier
GaAs MesFET	16–22, 60	1–30	Radar, satellite, defense
SiC MesFET	100	0.5–2.3	Base station
GaN MesFET	160	1–30	Replacement for GaAs
Si LDMOS (FET)	65	0.5–2	Base station
Si VDMOS (FET)	65–1200	0.001–0.5	HF power Amp and FM broadcasting and MRI

MesFET: metal semiconductor field-effect transistor.

and allows multiple FETs to be connected in parallel without ballasting, a useful property if a corporate or combined device concept has to be adopted for high power amplifier design.

RF metal oxide semiconductor field effect transistor (MOSFET) power transistors and their major applications and frequency of operation are given in Table 2.2.

2.5.3.3 MOSFETs

MOSFETs are widely used in RF power amplifier applications and their parameters are identified by manufacturers at different static and dynamic conditions. Hence, each MOSFET device has been manufactured with different characteristics. The designer selects the appropriate device for the specific circuit under consideration. One of the standard ways commonly used by designers for selection of the right MOSFET device is called a figure of merit (FOM). There are different types of FOMs that are used. FOM in its simplest form compares the gate charge, Q_g, against R_{dsON}. The multiplication of gate charge and drain to source on resistance relates to a certain device technology as it can be related to the required Q_g and R_{dsON} to achieve the right scale for MOSFET. The challenge is

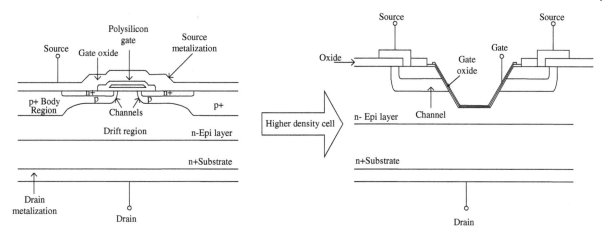

Figure 2.26 FOM comparison of planar and trench MOSFET structures.

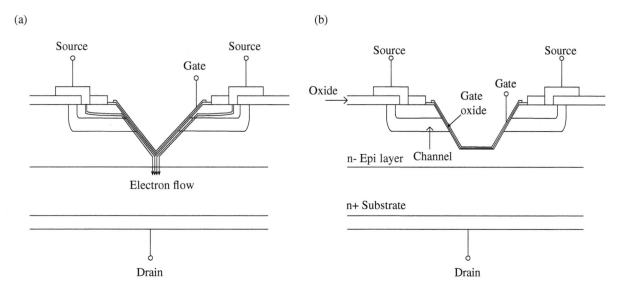

Figure 2.27 Trench MOSFET: (a) Current Crowding in V-Groove Trench MOSFET, (b) Truncated V-Groove.

the relation between Q_g, and R_{dsON} because MOSFET has inherent tradeoffs between ON resistance and gate charge, i.e. the lower the rDS(ON) the higher the gate charge will be. In device design, this translates as a conduction loss versus switching loss tradeoff. The new generation MOSFETs are manufactured to have an improved FOM. The comparison of FOM on MOSFETs manufactured with different processes can be illustrated on planar MOSFET structures and trench MOSFET structures. MOSFETs with a trench structure have a seven times better FOM versus planar structure, as shown in Figure 2.26.

Two variations of the trench power MOSFET are shown Figure 2.27. The trench technology has the advantage of higher cell density but is more difficult to manufacture than the planar device.

There is a best die size for MOSFET devices for a given output power. Optimum die size for minimal power loss (P_{Loss}) depends on load impedance, rated power, and switching frequency. The relation between P_{Loss} and optimized die size is illustrated in Figure 2.28.

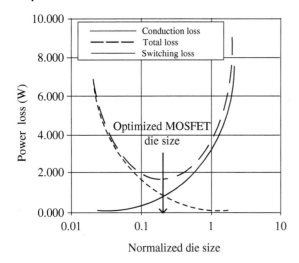

Figure 2.28 Die size versus power loss.

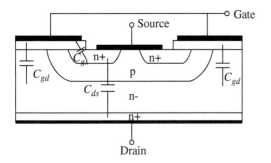

Figure 2.29 MOSFET structure capacitance illustration.

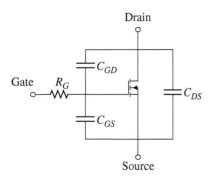

Figure 2.30 Simplest view of MOSFET with intrinsic capacitances.

The typical MOSFET structure with internal capacitances is illustrated in Figure 2.29.

In MOSFET structures, considerable capacitance is observed at the input due to the oxide layer. The simplest view of an n-channel MOSFET is shown in Figure 2.30, where the three capacitors, C_{gd} (gate to drain), C_{ds} (drain to source), and C_{gs} (gate to source), represent the parasitic capacitances. These values can be manipulated to form the input capacitance, output capacitance, and transfer capacitance. C_{gs} is the capacitance due to the overlap of the source and channel regions by the polysilicon gate. It is independent of the applied voltage. C_{gd} consists of a part associated with the overlap of the polysilicon gate and the Si underneath in the junction field effect transistor (JFET) region, which is also independent of the applied voltage and the capacitance associated with the depletion region immediately under the gate, which is a nonlinear function of the applied voltage. This capacitance provides a feedback loop between the output and input circuits. C_{gd} is also called the Miller capacitance because it causes the total dynamic input capacitance to become greater than the sum of the static capacitors. C_{ds} is the capacitance associated with the body drift diode. It varies inversely with the square root of the drain source bias voltage. C_{iss}, input capacitance, and C_{oss}, output capacitance, information is usually given in the manufacturer's datasheet. C_{iss} is made up of the gate to drain capacitance C_{gd} in parallel with the gate to source capacitance C_{gs}, or

$$C_{iss} = C_{gs} + C_{gd} \tag{2.68}$$

The input capacitance must be charged to the threshold voltage before the device begins to turn on, and discharged to the plateau voltage before the device turns off. Therefore, the impedance of the drive circuitry and C_{iss} have a direct effect on the turn on and turn off delays. C_{oss} is made up of the drain to source capacitance C_{ds} in parallel with the gate to drain capacitance C_{gd}, which can be expressed as

$$C_{oss} = C_{ds} + C_{gd} \tag{2.69}$$

Common MOSFET packages are identified as TO (transistor outline), SOT (small outline transistor), and SOP (small outline package). TO packages are early package specifications, such as TO-92, TO-92L, TO-220, TO-247, TO-252, etc., which are plug-in package designs. In recent years, market demand has increased devices that have surface mount and TO packages. SOT packages are lower power SMD transistor packages than TO packages, and are generally used for small power MOSFET. Common SOT packages are SOT-23 SOT-89, and SOT-236. SOP is a surface-mount package. The pin from the package has gull wing leads on both sides (L-shaped). MOSFET manufacturers are trying to improve chip production technology and processes to have better packaging technology. MOSFET packages have important characteristics that will have limitations on device performance. The characteristics of the package include package resistance, package inductance, and thermal impedance. Package resistance depends on bonding wire resistance based on bonding wire type, length, and lead frame resistance. Bonding wire can be copper (Cu), aluminum (Al), Al ribbon, and Cu clip based. A lead frame based package with internal wire bonds introduces parasitic inductance to the gate, source, and drain terminals. During current switching, this inductance produces a large Ldi/dt effect to slow down the turn-on and turn-off of the device. This effect will significantly hinder the performance at high switching frequencies. Parasitic inductances directly affect body diode reverse recovery characteristics and peak voltage spikes. Thermal impedance of the package consists of thermal resistances due to junction-to-printed circuit board (PCB) and junction-to-case.

Equivalent circuits are used to model and represent the characteristics of transistors. Equivalent circuits for transistors are physically measured for electrical parameters using specific measurement setups and fixtures. The measured results are then transformed into a circuit that mimics the electrical behavior of the transistor. This common procedure is applied to any active device used in power amplifiers. The commonly used HF small signal model for MOSFET is given in Figure 2.31. Figure 2.31 illustrates the intrinsic parameters based on the measured parameters for MOSFET. Transistors consist of die, exterior package, bonding wires, bonding pads, etc., designed for specific ratings, and applications as discussed in Section 2.5.3. The die has to be placed inside a package, and internal and external connections are established with binding wires and vias. The effects of the transistor packages, such as lead inductances, and package capacitances are called extrinsic parasitics. The complete MOSFET small signal, HF model including extrinsic and intrinsic parameters is shown in Figure 2.32.

When there is no feedback capacitance or resistance, the HF model in Figure 2.31 simplifies to the one in Figure 2.33 with load resistor as shown.

The circuit parameters, such as current, i_0, and voltage, v_0, can be expressed as

$$i_0 = \frac{g_m v_{GS}}{\sqrt{1 + \omega^2 C_{ds}^2 R_L^2}} \tag{2.70}$$

$$v_0 = \frac{g_m v_{GS} R_L}{\sqrt{1 + \omega^2 C_{ds}^2 R_L^2}} \tag{2.71}$$

The output power is then equal to

$$P_0 = \frac{g_m^2 v_{GS}^2 R_L}{1 + \omega^2 C_{ds}^2 R_L^2} \tag{2.72}$$

Figure 2.31   HF small signal model for MOSFET transistor.

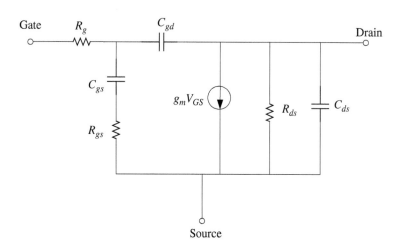

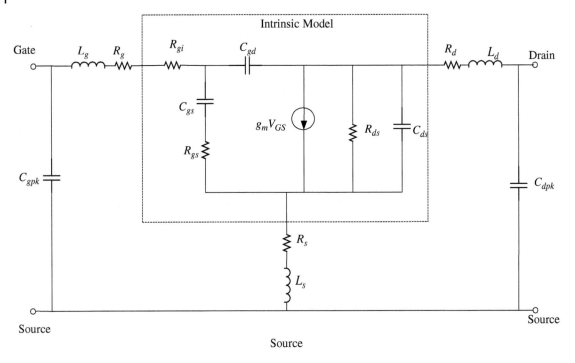

Figure 2.32 HF small signal model for MOSFET transistor with extrinsic parameters.

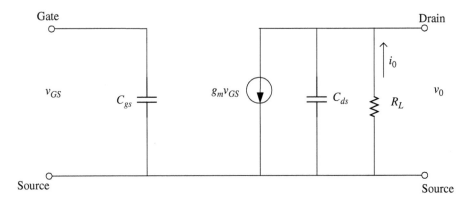

Figure 2.33 Simplified HF model for MOSFET.

The corresponding current and voltage gains are

$$A_I = \frac{g_m}{\omega C_{gs}\sqrt{1 + \omega^2 C_{ds}^2 R_L^2}} \tag{2.73}$$

$$G = \frac{g_m^2}{\omega C_{gs}\left(1 + \omega^2 C_{ds}^2 R_L^2\right)} \tag{2.74}$$

Example 2.5   MOSFET IV Curves

Obtain the IV (current and voltage) curves of Power MOSFET, IXFH12N100Q by using Spice Lib file in Advanced Design System (ADS).

Solution

The Spice Lib file for this example is imported to ADS following the steps below to obtain the IV curve of the transistor.

Open a new schematic as shown above and name it I_V_Spice_File. Set up the simulation as laid out above to get to the screen shown below.

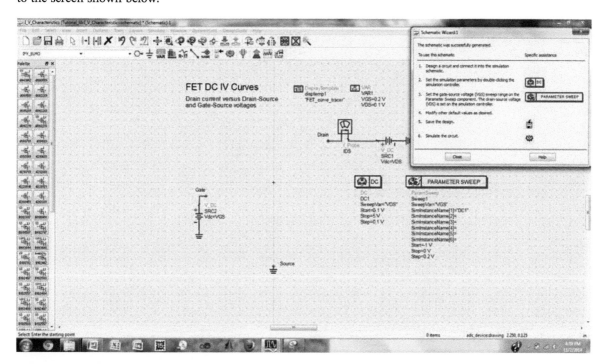

Next, we will be import the .lib file.

Now, Choose File>Import to open the import window.

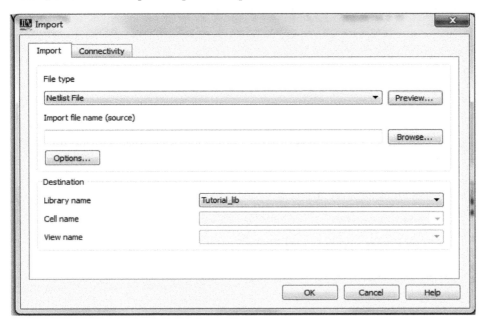

All defaults are correct, just select Browse... and select the desired .lib file.

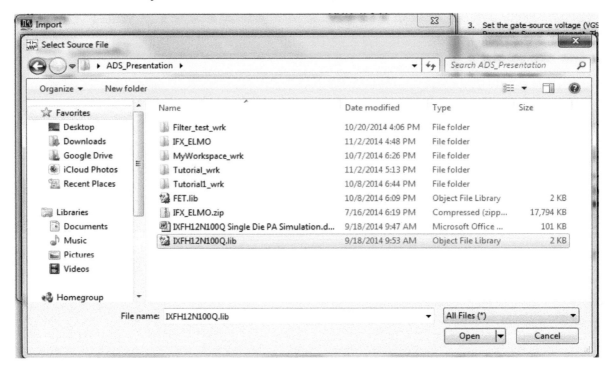

Here we will choose the IXFH12N100Q.lib file, then choose open.

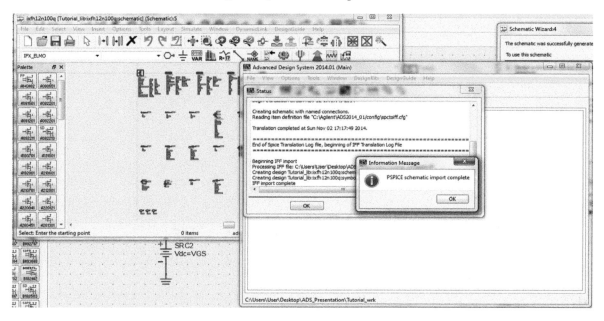

Choose OK on the import menu and the import process will begin and the following will open. Select OK and zoom in on the lower left of the schematic that opened when the PSPICE import completed. The screen should look like this:

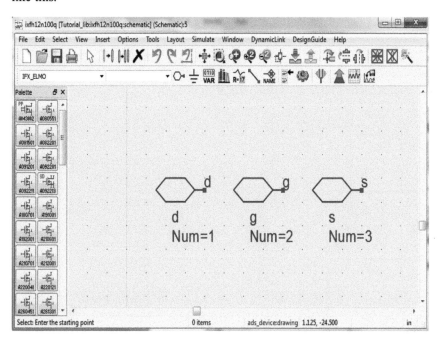

This will show what pin numbers on the model correspond to which pin on the transistor.

Open the screen as above and click and drag the highlighted file across to the schematic. This will place the model into the simulation. Then, using the pin numbers we highlighted before, connect the pins to the correct wires and run the simulation as before to obtain the curves shown below.

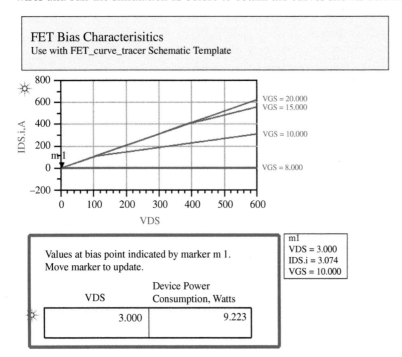

The above settings are with gate source voltage (VGS) from 0 to 20 steps of 5 V. VDS is from 0 to 600 V steps of 1 V.

Figure 2.34 Incremental MOSFET HF small signal model.

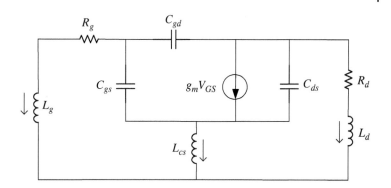

Example 2.6 MOSFET HF Model

Consider the simplified MOSFET HF model given in Figure 2.34. Determine the generalized stability characteristics of the transistor based on the simplified HF model.

Solution

The stability of a MOSFET transistor is based on its small signal response stability. The Routh–Hurwitz stability criterion can be implemented for the model given in Figure 2.34 to find the stability of the circuit [9]. For this, we write a set of state equations for the circuit and find the characteristic polynomial. The circuit stability is then determined by location of the roots of the characteristic polynomial in the *s* plane. The Routh–Hurwitz criterion states that a necessary and sufficient condition for all of the roots of an *n*th order equation to lie in the left half of the *s* plane is that all of the Hurwitz determinants, D_k ($k = 1, 2, ..., n$), must be positive. Application of Kirchhoff's voltage law (KVL) and Kirchhoff's current law (KCL) for the circuit in Figure 2.34 gives

$$\dot{I}_g L_g - \dot{I}_{cs} L_{cs} = -I_g R_g + V_{gs} \tag{2.75}$$

$$\dot{I}_d - \dot{I}_{cs} L_{cs} = -I_d R_d + V_{ds} \tag{2.76}$$

$$V_{gd} + V_{ds} = V_{gs} \tag{2.77}$$

$$C_{gs} V_{gs} + C_{gd} V_{gd} = -I_g \tag{2.78}$$

$$C_{ds} V_{ds} - C_{gd} V_{gd} = -I_g - V_{gs} g_m \tag{2.79}$$

$$I_g + I_d + I_{cs} = 0 \tag{2.80}$$

When Eqs. (2.) through (2.) are put into a matrix form for a solution, the characteristic equation is found from the determinant of the matrix and the roots are found from the characteristic polynomial obtained from

$$|A - sI| = 0 \tag{2.81}$$

The characteristic polynomial found from (2.82) is

$$P(s) = a_0 s^4 + a_1 s^3 + a_2 s^2 + a_3 s + a_4 \tag{2.83}$$

Once the values of the components in Figure 2.34 are known, Eqs. (2.75)–(2.83) can be used to identify the stability based on the locations of the roots.

2.5.3.4 LDMOS [10]

An n-type LDMOS transistor structure is illustrated in Figure 2.35. The first LDMOS structure was proposed in the early 1970s. The structure is similar to a MOSFET structure but can sustain higher voltage and as a result

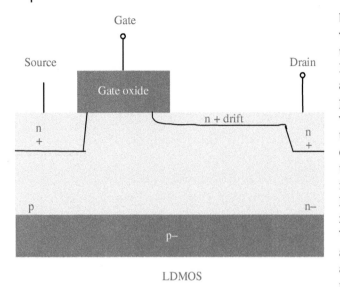

Figure 2.35 Conventional LDMOS structure.

breakdown voltage due to an extended drain region, which reduces the risk of carrier injection. Besides the advantage in relocating the high electric field, the LDMOS offers the possibility of higher breakdown voltage. If the voltage can spread over a larger distance, the probability of impact ionization is lower and the supply voltage can be increased. When no voltage is applied to the gate and the transistor is in the OFF state, the drain voltage is supported by the reversed biased junction between the p and n layers. LDMOS technology is widely used for base station applications, in particular GSM-EDGE (Enhanced Data rates for global system for mobile evolution) applications at 1 and 2 GHz, WCDMA (Wideband Code Division Multiple Access) applications at 2.2 GHz, and more recently for WiMax applications around 2.7 and 3.8 GHz. The main advantage of LDMOS is that the devices are linear when used in back-off and easily linearizable through predistortion methods such as DPD, whereas this is challenging with GaN transistors due to their slow compression gain.

2.5.3.5 High Electron Mobility Transistor (HEMT) [11]

The polarization doped HEMT is a field effect transistor in which two layers of different bandgap and polarization field are grown on each other. Surface charges at the heterointerface are created due to discontinuity in the polarization field. When the induced charge is positive, electrons will tend to compensate the induced charge resulting in the formation of the channel. Since in the HEMT the channel electrons are confined in a quantum well in a very narrow spatial region at the heterointerface, the channel electrons are referred to as a two-dimensional electron gas (2DEG). This confinement grants the electrons high mobilities surpassing the bulk mobility of the material in which the electrons are flowing. Hence, GaN HEMT and aluminum gallium nitride (AlGaN)/GaN HEMTs are promising devices for HF applications with high power and low noise, such as microwave and millimeter wave communications, imaging, and radars. Typical HEMT device structures for open channel and pinch are given in Figure 2.36.

The comparison of Si LDMOS and GaN HEMT devices are given in Figure 2.37 [12]. These two devices are the most commonly used devices for high power and HF RF applications.

The market share for the predicted RF device technologies to 2025 is given in Figure 2.38 [13].

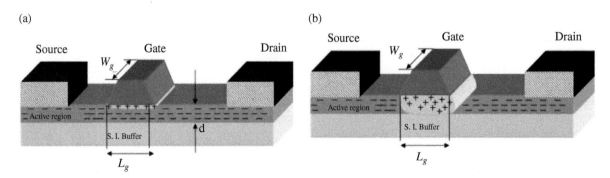

Figure 2.36 HEMT device structure: (a) open channel; (b) pinch off.

Figure 2.37 Comparison of Si LDMOS and GaN HEMT.

	Silicon LDMOS	GaN HEMT
epi	Homogeneous silicon	heterogeneous, GaN
bandgap	1.1 eV	3.4 eV
electron velocity:	–	–
saturated	1×10^5 m/s	1.5×10^5 m/s
peak	1×10^5 m/s	2.7×10^5 m/s
breakdown field	25V/μm	300V/μm
typ BVds	75V	175V
processing	standard CMOS	bespoke fab
mask count	22	13
max frequency	3.8 GHz	>12 GHz
max temperature	225°C	250°C
Johnson's FoM	1	324

Figure 2.38 The market share for device technologies.

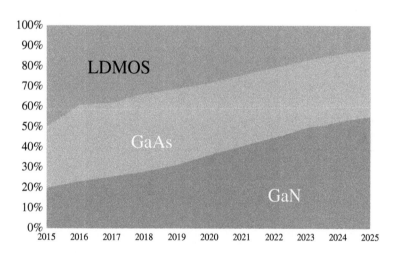

2.6 Engineering Application Examples

In this section, design examples from real world engineering applications are given. One of the design examples is about the design of isolation gate transformer which is widely used in RF amplifiers at the input section to provide the required signal level to transistors. Other application example is about RF autotransformers. RF autotransformers is also used in RF amplifiers to set the required signal level at any section of the amplifiers and matching purposes.

Design Example 2.1   Isolation Gate Transformer
It is required to design and characterize an isolation gate transformer to operate at 13.56 MHz. Current at the primary is given to be 1.73 A_{rms}. The power input is 150 W, and the voltage at the secondary should be 30 V_{rms}. The input impedance that will be presented is 50 Ω.

Solution

Based on the given information, the voltage at the secondary is 30 V, and at the primary is

$$V_p = \sqrt{150(50)} = 86.6\,V_{rms}$$

The turn ratio is then calculated to be

$$\frac{N_1}{N_2} = \frac{V_p}{V_s} = \frac{86.6}{30} = 2.89$$

This can be rounded to 3. Hence, the transformer with ratio 3 : 1 should satisfy the design specifications. Consider the single bead geometry given in Figure 2.39.

The cross-sectional area and magnetic path length are given as

$$A_e = 0.0996\,cm^2, l_e = 1.5\,cm$$

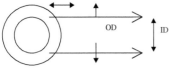

Figure 2.39   Single bead geometry.

The permeability of the material is 125. In our design, we did not initially know the number of turns. So, as a rule of thumb, we targeted impedance on the primary 10 times higher than the impedance that is interfaced.

$$Z = 2\pi fL = 2\pi(13.56 * 10^6) * L = 500\,\Omega$$

From this relation, the target inductance is 5.86 μH. We use the following formulation to obtain the inductance. When six beads are used, the total cross-sectional area becomes $A_{Te} = 6 \times A_e$ and the inductance value becomes 5.63 μH.

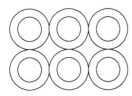

Figure 2.40   Combined bead geometry.

$$L = \frac{0.4\pi\mu_i A_{Te}(cm^2)N^2}{100 le(cm)}\mu H$$

When this is placed back into the impedance formulation, we obtain

$$Z = 2\pi fL = 2\pi(13.56 * 10^6) * L = 479.8\,\Omega$$

3 : 1 transformer with six beads is realized using the bead configuration shown in Figure 2.40.

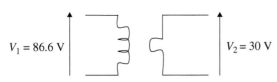

$V_1 = 86.6\,V$ $V_2 = 30\,V$

Figure 2.41   Equivalent circuit for the transformer.

The size of the wires can be determined as follows. Since the current at the primary is 1.73 A, the minimum wire size should be 26AWG. 26AWG wire will be able to handle 2.2 A if it is solid wire. Since the current at the secondary is 5.19 A, the minimum wire size should be 22AWG. 22AWG will be able to handle 5.5 A when a solid wire is used. One of the design practices is to choose the wire insulation properly to prevent arcing between primary and secondary windings. This can be visualized using Figure 2.41.

The voltage difference between the windings is $\Delta V = 86.6 - 30 = 56.6\,V$. Hence the breakdown voltage of the wires is greater than 56.6 V. The Teflon insulation used on the primary winding and enameled wire on the secondary winding has a much higher voltage rating than 56.6 V. The transformer is implemented in Figure 2.42. 14AWG wire is used on the secondary side which has

Figure 2.42   Isolation gate transformer.

Table 2.3 Test results for transformer.

Measured	Primary	Secondary	L
Primary	Open	Open	5.15 µH
Secondary	Open	Open	579.6 nH
Secondary	Short	Open	22 nH

much more higher rating than desired value. This transformer is used to drive the MOSFETs and are connected between gate and source of the transistor.

Characterization of the transformer is done using open and short circuit tests as explained before. The open and short circuit test results for this transformer is shown below (Table 2.3).

From Table 2.3, the measured values are

$$L_p = 5.15\,\mu H, L_s = 579.6\,nH, L_s' = 22\,nH, n = 3$$

Using

$$\left(\frac{L_{1l}^2}{n^2 L_p}\right) - \frac{2L_{1l}}{n^2} + \left(L_s' - L_s + \frac{L_p}{n^2}\right) = 0$$

or

$$\left(\frac{L_{1l}^2}{9(5.15 \times 10^{-6})}\right) - \frac{2L_{1l}}{9} + \left(22 \times 10^{-9} - 579.6 \times 10^{-9} + \frac{5.15 \times 10^{-6}}{9}\right) = 0$$

The solution of the above equation gives the leakage inductance on the primary side as $L_{1l} = 66.22$ nH. The leakage inductance on the secondary side can be found from [1]

$$L_s = L_{2l} + \frac{L_p - L_{1l}}{n^2}$$

The leakage inductance on the secondary is then equal to

$$L_{2l} = L_s - \frac{L_p - L_{1l}}{n^2} = 579.6\,nH - \frac{5150\,nH - 66.22\,nH}{9} = 14.74\,nH$$

Hence, the measured leakage inductances for the isolation gate transformer are

$$L_{1l} = 66.22\,nH \text{ and } L_{2l} = 14.74\,nH$$

The self-inductances on the primary and secondary sides are found from

$$L_{1p} = L_p - L_{1l} = 5150\,nH - 66.22\,nH = 5083.78\,nH$$

$$L_{2s} = L_s - L_{2l} = 579.6\,nH - 14.74\,nH = 564.86\,nH$$

The coupling coefficient is obtained from

$$k = \sqrt{\frac{L_{1p}L_{2s}}{L_p L_s}} = \sqrt{\frac{(5083.78)(564.86)}{(5150)(579.6)}} = 0.98$$

The mutual inductance is then equal to

$$M = k\sqrt{L_p L_s} = 0.98\sqrt{(5150)(579.6)} = 1693.14\,nH$$

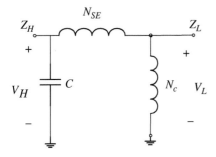

Figure 2.43 Step down transformer for load pull applications.

Design Example 2.2   RF Autotransformer

It is desired to design autotransformers for an RF amplifier load pull application at 2 MHz when load presents (i) $VSWR = 1.5:1$, (ii) $VSWR = 2:1$, (iii) $VSWR = 3:1$, and (iv) $VSWR = 5:1$. It is given that the RF amplifier output power is 4000 W when terminated to a 50 Ω load. Use K-type binocular ferrite core with $\mu = 125$, $l_e = 6.38$cm, $A_e = 2.22$ cm^2 in the construction of the transformers.

Solution

The autotransformer configuration that will be constructed is illustrated in Figure 2.43. In all the configurations, $Z_H = 75\Omega$ and that is the termination that will be interfaced with the amplifier output. The low impedance port, $Z_L = 50\Omega$, is terminated to a high power load which is also 50Ω. As a result, the low impedance side of the transformer is matched with a high power load. For load pulls, autotransformers can be used to make impedance transformation needed to provide the desired voltage standing wave ratio ($VSWR$) ratios. For instance, when $VSWR = 1.5:1$, it is required that $Z_H = 75\Omega$ when $Z_L = 50\Omega$. Our task is to find N_C and N_{SE} to give that impedance ratio. The power and impedance information give in the design problem can be used to determine the winding that will be used in the transformer construction.

The winding should be able to handle $V_{rms} = 447.2$ V and $I_{rms} = 8.94$ A. Copper strip with 350 mil width and 10 mil thickness is a conservative choice that will be able to carry the amount of current calculated for VSWRs that are specified. The insulation for the copper strip is provided with a 1 mil Teflon jacket that will be able to withstand the voltage.

i) $VSWR = 1.5:1$

$VSWR = 1.5:1$ can be obtained when $N_{SE} = 1$ and $N_C = 4$. The low impedance side of the transformer is fixed and can be used as a reference. We aim to have 5–10 times of the terminating impedance for the inductance calculation. So, the impedance at the fixed impedance side is $Z = 10(50) = 500\ \Omega$. From the target impedance, we obtain the inductance value as

$$L \geq \frac{500}{2\pi(2 \times 10^6)} = 39.79\,\mu H$$

The number of turns required to obtain the inductance calculated is found from

$$N^2 = \frac{39.79 \times 100 \times (6.38)}{0.4\pi(125)(2.22)} = 72.8 \rightarrow N > 8$$

Although the number of turns, $N_C = 4$, is less than the minimum desired number of turns for the required inductance value, it is acceptable to use it due to a lower operational frequency and very conservative rule of thumb rule.

When $N_C = 4$, the inductance value is obtained to be 8.75 μH. This circuit that is showing the configuration of the autotransformer that gives $VSWR = 1.5:1$ is shown in Figure 2.44. The compensation capacitor, C, which is connected in shunt at the high impedance side, can be used when needed to reduce the leakage inductance.

Based on the turns ratio, the impedance at the high side is found from

$$Z_H = \left(\frac{N_C + N_{SE}}{N_C}\right)^2 Z_L = \left(\frac{5}{4}\right)^2 50 = 78.125\ \Omega$$

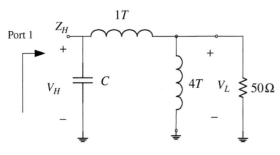

Figure 2.44   *VSWR* = 1.5 : 1 autotransformer.

The *VSWR* is then obtained as

$$VSWR = \frac{78.125}{50} = 1.56$$

This is very close to the required *VSWR* ratio and as a result we can construct the autotransformer based on the turns ratios we determined. The constructed auto-transformer is shown in Figure 2.45. Each autotransformer is constructed similarly with only a different number of turns. The performance of the autotransformer is measured by a network analyzer from 1.8 to 8 MHz. The measurement results show that the transformer performance at 2 MHz is in agreement with the calculated results as desired, as shown in Figure 2.46. The impedance measured at port 1 at 2 MHz is 77.815 Ω. The leakage reactance is significantly reduced with the shunt capacitor, 250 pF. The windings for N_C and N_{SE} are done separately with the copper strips, and connections are made, as illustrated in Figure 2.45.

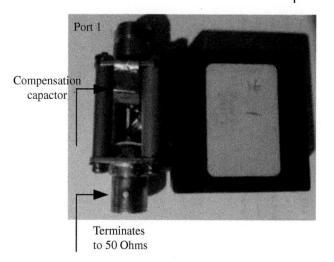

Figure 2.45 Constructed 1.5 : 1 autotransformer.

ii) *VSWR* = 2 : 1

The autotransformer configuration that gives *VSWR* = 2 : 1 is obtained with $N_{SE} = 2$ and $N_C = 5$, as shown in Figure 2.47. Then, the impedance at the high impedance side is

$$Z_H = \left(\frac{N_C + N_{SE}}{N_C}\right)^2 Z_L = \left(\frac{7}{5}\right)^2 50 = 98\,\Omega$$

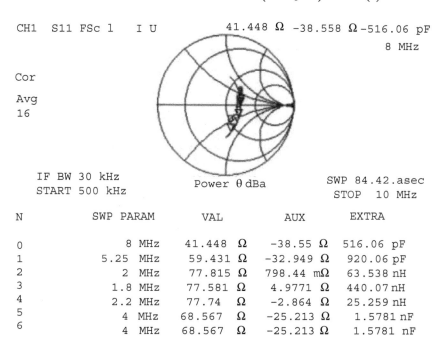

```
CH1   S11 FSc 1    I U          41.448 Ω  -38.558 Ω -516.06 pF
                                                     8 MHz

Cor

Avg
16




IF BW 30 kHz                                SWP 84.42.asec
START 500 kHz      Power θ dBa              STOP   10 MHz
```

N	SWP PARAM	VAL	AUX	EXTRA
0	8 MHz	41.448 Ω	−38.55 Ω	516.06 pF
1	5.25 MHz	59.431 Ω	−32.949 Ω	920.06 pF
2	2 MHz	77.815 Ω	798.44 mΩ	63.538 nH
3	1.8 MHz	77.581 Ω	4.9771 Ω	440.07 nH
4	2.2 MHz	77.74 Ω	−2.864 Ω	25.259 nH
5	4 MHz	68.567 Ω	−25.213 Ω	1.5781 nF
6	4 MHz	68.567 Ω	−25.213 Ω	1.5781 nF

Figure 2.46 Measurement results of a 1.5 : 1 autotransformer.

Figure 2.47 *VSWR* = 2 : 1 autotransformer.

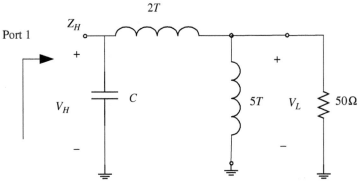

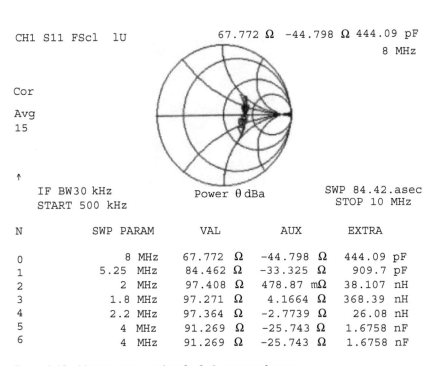

```
CH1 S11 FScl   1U            67.772 Ω  -44.798 Ω 444.09 pF
                                                   8 MHz

Cor

Avg
15


↑
     IF BW30 kHz          Power θ dBa       SWP 84.42.asec
     START 500 kHz                          STOP 10 MHz

N            SWP PARAM        VAL         AUX         EXTRA

0              8 MHz      67.772 Ω    -44.798 Ω    444.09 pF
1           5.25 MHz      84.462 Ω    -33.325 Ω     909.7 pF
2              2 MHz      97.408 Ω    478.87 mΩ    38.107 nH
3            1.8 MHz      97.271 Ω     4.1664 Ω    368.39 nH
4            2.2 MHz      97.364 Ω    -2.7739 Ω     26.08 nH
5              4 MHz      91.269 Ω    -25.743 Ω    1.6758 nF
6              4 MHz      91.269 Ω    -25.743 Ω    1.6758 nF
```

Figure 2.48 Measurement results of a 2 : 1 autotransformer.

This gives the *VSWR* ratio as

$$VSWR = \frac{98}{50} = 1.96$$

The autotransformer is constructed and measured. The measurement results are illustrated in Figure 2.48. The measured impedance at the operational frequency, 2 MHz, is 97.408 Ω. The compensation capacitor in this configuration is 150 pF.

iii) *VSWR* = 3 : 1

The autotransformer configuration that gives *VSWR* = 3 : 1 is obtained with $N_{SE} = 3$ and $N_C = 4$, as shown in Figure 2.49. Then, the impedance at the high impedance side is

Figure 2.49 *VSWR* = 3 : 1 autotransformer.

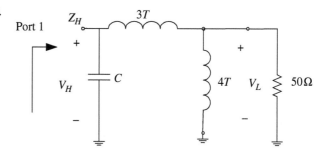

$$Z_H = \left(\frac{N_C + N_{SE}}{N_C}\right)^2 Z_L = \left(\frac{7}{4}\right)^2 50 = 153.125\,\Omega$$

This gives the *VSWR* ratio as

$$VSWR = \frac{153.125}{50} = 3.06$$

The autotransformer is constructed and measured. The measurement results are illustrated in Figure 2.50. The measured impedance at the operational frequency, 2 MHz, is 152.39 Ω. The compensation capacitor in this configuration is 150 pF.

iv) *VSWR* = 5 : 1

The autotransformer configuration that gives *VSWR* = 5 : 1 is obtained with N_{SE} = 5 and N_C = 4, as shown in Figure 2.51. Then, the impedance at the high impedance side is

$$Z_H = \left(\frac{N_C + N_{SE}}{N_C}\right)^2 Z_L = \left(\frac{9}{4}\right)^2 50 = 253.125\,\Omega$$

Figure 2.50 Measurement results of a 3 : 1 autotransformer.

CH1S11 FScl 1U

71.281 Ω −76.255 Ω 260.89 pF

8 MHz

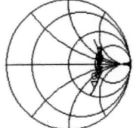

Cor

Avg

16

↑

IF BW30 kHz

START 500 kHz

Power θ dBa

SWP 84.42.msec

STOP 10 MHz

N	SWP PARAM	VAL	AUX	EXTRA
0	8 MHz	71.261 Ω	−76.255 Ω	260.89 pF
1	5.25 MHz	108.61 Ω	−69.345 Ω	437.16 pF
2	2 MHz	152.39 Ω	205.08 mΩ	16.32 nH
3	1.8 MHz	151.8 Ω	9.4473 Ω	835.33 nH
4	2.2 MHz	152.06 Ω	−7.8097 Ω	9.2632 nF
5	4 MHz	129.42 Ω	−54.878 Ω	725.04 nF
6	4 MHz	129.42 Ω	−54.878 Ω	725.04 nF

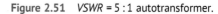

Figure 2.51 *VSWR* = 5 : 1 autotransformer.

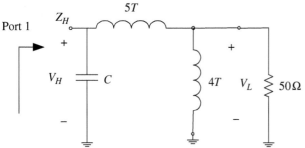

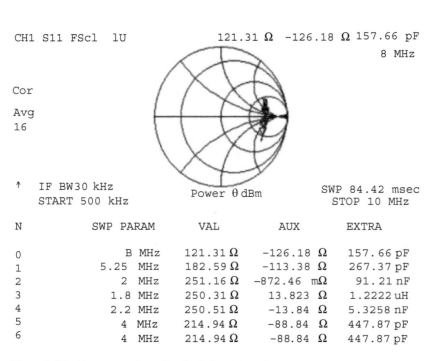

N	SWP PARAM		VAL	AUX	EXTRA
0	B	MHz	121.31 Ω	−126.18 Ω	157.66 pF
1	5.25	MHz	182.59 Ω	−113.38 Ω	267.37 pF
2	2	MHz	251.16 Ω	−872.46 mΩ	91.21 nF
3	1.8	MHz	250.31 Ω	13.823 Ω	1.2222 uH
4	2.2	MHz	250.51 Ω	−13.84 Ω	5.3258 nF
5	4	MHz	214.94 Ω	−88.84 Ω	447.87 pF
6	4	MHz	214.94 Ω	−88.84 Ω	447.87 pF

Figure 2.52 Measurement results of a 5 : 1 autotransformer.

This gives the *VSWR* ratio as

$$VSWR = \frac{253.125}{50} = 5.06$$

The autotransformer is constructed and measured. The measurement results are illustrated in Figure 2.52. The measured impedance at the operational frequency, 2 MHz, is 251.16 Ω. The compensation capacitor in this configuration is 86 pF.

References

1 Eroglu, A. (2013). *RF Circuit Design Techniques for MF-UHF Applications*, 1e. CRC Press. ISBN: 978-1-4398-6165-3.

2 Greenhouse, H.M. (1974). Design of planar rectangular microelectronic inductors. *IEEE Trans. Parts, Hybrids, Packaging* **PHP-10**: 101–109.

3 Eroglu, A. (2017). *Linear and Switch-Mode RF Power Amplifiers: Design and Implementation Methods*. CRC Press. ISBN: 978-1-4987-4576-5.

4 Ozpineci, B. and Tolbert, L.M. (2003). Comparison of wide-bandgap semiconductors for power electronics applications. *Report ORNL/TM-2003/257*. Oak Ridge National Laboratory, Oak Ridge, TN.

5 DeLorenzo, J. and Khandelwal, D. (1982). *GaAs FET Principles and Technology*. Dedham, MA: Artech House.

6 Combes, A. (2020). *Application Note AN-007: A Comparative Review of GaN, LDMOS, and GaAs for RF and Microwave Applications*. NuWaves, https://nuwaves.com/wp-content/uploads/2020/08/AN-007-A-Comparative-Review-of-GaN-LDMOS-and-GaAs-for-RF-and-Microwave-Applications.pdf.

7 Aethercomm Application Note, Gallium Nitride (GaN) Microwave Transistor Technology For Radar Applications https://www.aethercomm.com/wp-content/uploads/2019/07/Aethercomm-White-Paper.pdf (accessed 1 November 2021).

8 Trew, R.J. (2004). Wide bandgap transistor amplifiers for improved performance microwave power and radar applications. In: *15th International Conference on Microwaves, Radar and Wireless Communications, vol. 1*, 18–23.

9 Saucedo, R. and Schiring, E. (1968). *Introduction to Continuous and Digital Control Systems*, 96. Macmillan.

10 Runton, D.W., Trabert, B., Shealy, J.B., and Vetury, R. (2013). History of GaN: high-power RF gallium nitride (GaN) from infancy to manufacturable process and beyond. *IEEE Microw. Mag.* **14** (3): 82–93.

11 Theeuwen, S.J.C.H. and Mollee, H.. (2008). LDMOS transistors in power microwave applications. http://k5tra.net/tech%20library/LDMOS/LDMOS%20transistors%20in%20power%20microwave%20applications.pdf (accessed 1 November 2021).

12 NXP Application Note, "NXP goes with GaN." (2011).

13 Yole Development, RF Market Device Technologies. (2017) http://www.yole.fr/PowerRF_Devices_Applications.aspx#.YT61zJ1Kjb4 (accessed 1 November 2021).

Problems

Problem 2.1

Consider the circuit given in Figure 2.53. In Figure 2.53, C is voltage dependent capacitor just like varactors and is given by

$$C = C_0 \left(1 - \frac{V}{V_o}\right)^{-1/2}$$

Calculate the voltage, V, such that the circuit exhibits a resonance at the frequency of 1 GHz. It is given that $C_0 = 10$ pF, $R = 3\,\Omega$, and $V_o = 0.75$ V.

Problem 2.2

The HF equivalent circuit for a toroidal inductor is given in Figure 2.54. Derive resonant frequency and plot resonance frequency versus R when $L = 10$ nH and $C = 1$ pF.

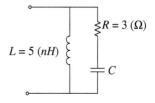

Figure 2.53 Problem 2.1.

Problem 2.3

It is required to design an air core inductor with an inductance of 180 nH using a rod with 0.25 in. diameter. The inductor should be able to handle at 5 A_{rms} of current during the operation where the voltage is 50 V_{rms} at 27.12 MHz.

a) Calculate number of turns.
b) Determine the minimum gauge wire that needs to be used.
c) Obtain the HF characteristic of inductor.

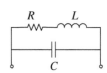

Figure 2.54 Problem 2.2.

d) Identify its resonant frequency.
e) Find its quality factor.

Problem 2.4
Assume the measured transfer characteristics for MOSFET are given by the manufacturer in Figure 2.55 when $T_j = 25\,°C$ and $T_j = 125\,°C$. It is communicated that the full load current is equal to Io $= 5\,A$. Find the threshold voltage and gate plateau voltage using the transfer curves.

Problem 2.5
The following data $C_{GS} = 800\,pF$; $C_{GD} = 150\,pF$; $gf = 4$; $v_{GS}(th) = 3\,V$ is given for a Power MOSFET. It is used to switch a clamped inductive load, as shown in Figure 2.56, of 20 A with a supply voltage $V_D = 200\,V$. The gate

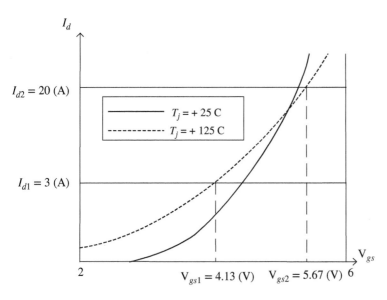

Figure 2.55 Transfer curve for MOSFET in Problem 2.4.

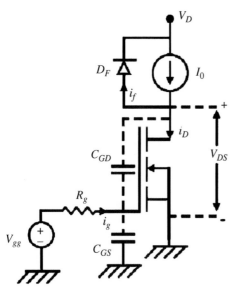

Figure 2.56 Inductive switching circuit.

drive voltage is $v_{gg} = 15\,V$, and gate resistance $R_g = 50\,\Omega$. Find out maximum value of $|di_d/dt|$ and $|dv_{DS}/dt|$ during turn ON.

Problem 2.6

Obtain IV characteristics of *n* channel MOSFET for a given spice.lib file using Orcad/Pspice. The spice.lib file of MOSFET is

```
.SUBCKT MOSFETex1 1 2 3
* External Node Designations
* Node 1 -> Drain
* Node 2 -> Gate
* Node 3 -> Source
M1 9 7 8 8 MM L=100u W=100u
.MODEL MM NMOS LEVEL=1 IS=1e-32
+VTO=4.54 LAMBDA=0.00633779 KP=7.09
+CGSO=2.17164e-05 CGDO=3.39758e-07
RS 8 3 0.0001
D1 3 1 MD
.MODEL MD D IS=5.2e-09 RS=0.00580776 N=1.275 BV=1000
+IBV=10 EG=1.061 XTI=2.999 TT=3.28994e-05
+CJO=2.95707e-09 VJ=1.57759 M=0.9 FC=0.1
RDS 3 1 1e+06
RD 9 1 0.95
RG 2 7 0.4
D2 4 5 MD1
* Default values used in MD1:
*    RS=0 EG=1.11 XTI=3.0 TT=0
*    BV=infinite IBV=1mA
.MODEL MD1 D IS=1e-32 N=50
+CJO=2.05889e-09 VJ=0.5 M=0.9 FC=1e-08
D3 0 5 MD2
* Default values used in MD2:
*    EG=1.11 XTI=3.0 TT=0 CJO=0
*    BV=infinite IBV=1mA
.MODEL MD2 D IS=1e-10 N=0.4 RS=3.00001e-06
RL 5 10 1
FI2 7 9 VFI2 -1
VFI2 4 0 0
EV16 10 0 9 7 1
CAP 11 10 2.05889e-09
FI1 7 9 VFI1 -1
VFI1 11 6 0
RCAP 6 10 1
D4 0 6 MD3
* Default values used in MD3:
*    EG=1.11 XTI=3.0 TT=0 CJO=0
*    RS=0 BV=infinite IBV=1mA
.MODEL MD3 D IS=1e-10 N=0.4
.ENDS MOSFETex1
```

Problem 2.7
Obtain IV curves of ATF-54143, Low Noise Enhancement Mode Pseudomorphic HEMT, device using ADS. The ADS model is available from the vendor's website or can be downloaded from https://jp.broadcom.com/docs/MPUB-1511. Compare the IV curves obtained from simulation with the one in the manufacturer datasheet given in Figure 2.57.

Problem 2.8
Find the total charge needed to turn MOSFET on for the circuit given in Figure 2.58. Figure 2.58a is the total gate charge measurement circuit, whereas Figure 2.58b is the waveforms for $i_G(t)$, $V_{GS}(t)$, and $V_{DS}(t)$.

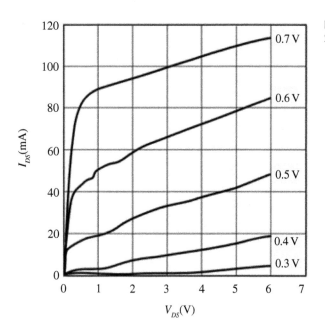

Figure 2.57 IV curves with V_{GS} = 0.1 steps. Source: Data Sheet from Avago.

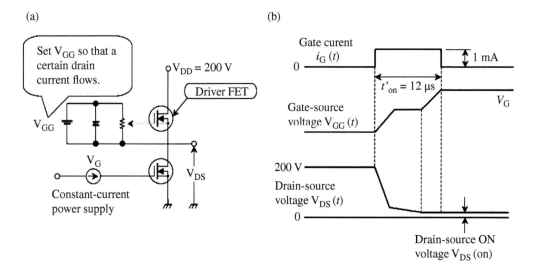

Figure 2.58 (a) Total gate measurement circuit. (b) Waveforms.

Problem 2.9

Consider a representative diagram of the MOSFET device, IRFP450, in a ground referenced gate drive application, as given in Figure 2.59.

The following information is given for the circuit in Figure 2.57.

- VDS, OFF = 380 V – the nominal drain-to-source off state voltage of the device.
- ID = 5A – the maximum drain current at full load.
- TJ = 100 °C – the operating junction temperature.
- VDRV = 13 V – the amplitude of the gate drive waveform.
- RGATE = 5 Ω – the external gate resistance.
- RLO = RHI = 5 Ω – the output resistances of the gate driver circuit.

   In addition, the capacitance information for IRFP450 is given by the manufacturer as

   $C_{iss} = 2600$ pF, $C_{oss} = 720$ pF, $C_{rss} = 340$ pF and $V_{ds} = 25$ V. Calculate actual physical capacitances, C_{GD}, C_{GS}, and C_{DS} for IRFP450.

Problem 2.10

The manufacturer gives the following measured S parameters for the high power TO-247 MOSFET.

   At 3 MHz, low frequency measurement

$$S_{11} = 0.09 - j0.31; S_{12} = 0.89 + j0.01$$

$$S_{21} = 0.89 + j0.01; S_{22} = 0.02 - j0.34$$

AT 300 MHz, HF measurement

$$S_{11} = -0.35 + j0.76; S_{12} = 0.2 + j0.1$$

$$S_{21} = 0.2 + j0.1; S_{22} = -0.4 + j0.85$$

   Calculate the extrinsic and intrinsic parameters of this device by ignoring test fixture effects using the zero bias MOSFET model given in Figure 2.60. Compare your results with exact given high power TO-247 MOSFET extrinsic and intrinsic values which are

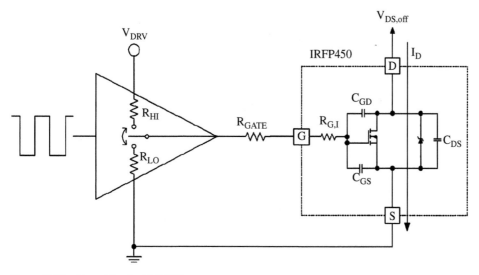

Figure 2.59 Gate drive for MOSFET.

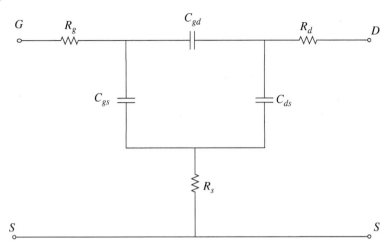

Figure 2.60 Zero bias MOSFET model.

- $C_{iss} = 2700\,\mathrm{pF}$, $Cgs = 2625\,\mathrm{pF}$
- $C_{rss} = Cgd = 75\,\mathrm{pF}$
- $C_{oss} = 350\,\mathrm{pF}$, $Cds = 275\,\mathrm{pF}$
- $L_g = 13\mathrm{nH}$, $Ls = 13\mathrm{nH}$, $Ld = 5\,\mathrm{nH}$
- $Rs = 0.95\,\Omega$, $Rs. = 0.5\,\Omega$, $Rg = 5\,\Omega$

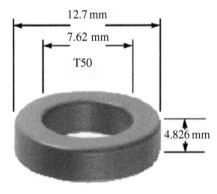

Figure 2.61 Dimension of a toroidal core, T50 material.

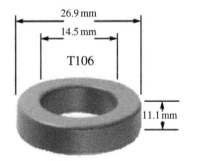

Figure 2.62 The geometry of a T106 core.

Problem 2.11

Design a toroidal inductor using powder iron core material, T50, with a frequency independent inductance value of 330nH. The coating material used has a thickness of 0.0225 mm with permittivity 3.5. (a) Determine the number of turns and (b) the largest solid AWG wire that can be used to carry $10\,\mathrm{A_{rms}}$ current. (c) The maximum rms voltage drop across the inductor is required to be $30\,\mathrm{V_{rms}}$. What is the maximum operational magnetic flux density and (d) total core P_{Loss}. (e) Obtain the HF characteristics of this inductor, identify its resonant frequency, and find its quality factor versus at 13.56 MHz and quality factor versus frequency characteristics. Use the toroidal core given in Figure 2.61.

Problem 2.12

Design a toroidal inductor that is formed by stacking two T106-6 powdered iron cores, as shown in Figure 2.62, with 21 turns using 16 AWG wire at 2 MHz. (a) What is the theoretical inductance value? (b) Calculate the quality factor. Plot calculated quality factor versus frequency. (c) Calculate the resonant frequency and obtain HF characteristics. (d) Calculate the wire length used to wind the core.

Design Challenge 2.1 Toroidal Inductor Design

Build the toroidal inductor discussed in Problem 2.12 by stacking two T106-6 powdered iron cores, as shown in Figure 2.62, with 21 turns using 16 AWG wire at 2 MHz. (a) Measure the inductance value and compare it with the theoretical inductance value. (b) Measure the quality factor of the inductor and compare it with the calculated value. (c) Measure the resonant frequency and obtain HF characteristics and compare them with the theoretical value.

Design Challenge 2.2 Inductor Design for RF Amplifiers

Consider the amplifier network given in Figure 2.63. In this network, it is required to interface 10 Ω differential output of the amplifier 1 (Amp 1) to 100 Ω input impedance of the second amplifier (Amp 2) using an unbalanced L-C network at 100 MHz. The current and voltage at the output of Amp 1 are 12 A_{rms} and 72 V_{rms}, respectively. Design the magnetic core inductor with powdered iron material using stacked core configuration and

a) Identify the core material to be used.
b) Calculate number of turns.
c) Determine the minimum gauge wire that should be used.
d) Obtain the HF characteristic of the inductor.
e) Identify its resonant frequency.
f) Find its quality factor.
g) Determine the length of the wire that will be used.
h) Calculate total core P_{Loss}.
i) What is the maximum operation flux density?

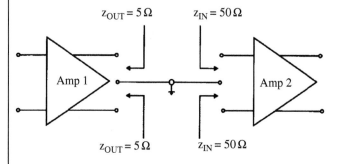

$z_{OUT} = 5\,\Omega$ $z_{IN} = 50\,\Omega$

Amp 1

Amp 2

$z_{OUT} = 5\,\Omega$ $z_{IN} = 50\,\Omega$

Figure 2.63 Design Challenge 2.2.

3

Transmission Lines

3.1 Introduction

In wireless systems, electromagnetic energy and signals can be transferred in free space, such as air. It is also possible to transfer energy and signals from one point to another using guided structures called transmission lines. This is why the transmission line concept is so heavily used and implemented in RF and microwave communication and engineering applications. In this chapter, the analysis of transmission lines is given and transmission line applications discussed. Some of the commonly used transmission lines are illustrated in Figure 3.1

3.2 Transmission Line Analysis

A transmission line is a distributed-parameter network, where voltages and currents can vary in magnitude and phase over the length of the line. Transmission lines usually consist of two parallel conductors that can be represented with a short segment of Δz. This short segment of transmission line can be modeled as a lumped-element circuit, as shown in Figure 3.2 [1].

In Figure 3.2, R is the series resistance per unit length for both conductors, $R(\Omega/m)$, L is the series inductance per unit length for both conductors, $L(H/m)$, G is the shunt conductance per unit length, $G(S/m)$, and C represents the shunt capacitance per unit length, $C(F/m)$, in the transmission line. Application of Kirchhoff's voltage law (KVL) and Kirchhoff's current law (KCL) give

$$v(z,t) - R\Delta z i(z,t) - L\Delta z \frac{\partial i(z,t)}{\partial t} - v(z+\Delta z, t) = 0 \tag{3.1}$$

$$i(z,t) - G\Delta z v(z+\Delta z, t) - C\Delta z \frac{\partial v(z+\Delta z, t)}{\partial t} - i(z+\Delta z, t) = 0 \tag{3.2}$$

Dividing (3.1) and (3.2) by Δz and assuming $\Delta z \rightarrow 0$, we obtain

$$\frac{\partial v(z,t)}{\partial z} = -Ri(z,t) - L\frac{\partial i(z,t)}{\partial t} \tag{3.3}$$

$$\frac{\partial i(z,t)}{\partial z} = -Gv(z,t) - C\frac{\partial v(z,t)}{\partial t} \tag{3.4}$$

Equations (3.3) and (3.4) are known as the time-domain form of the transmission line, or telegrapher, equations. Assuming the sinusoidal steady-state condition with application cosine-based phasors, Eqs. (3.3) and (3.4) take the following forms.

RF/Microwave Engineering and Applications in Energy Systems, First Edition. Abdullah Eroglu.
© 2022 John Wiley & Sons Ltd. Published 2022 by John Wiley & Sons Ltd.
Companion website: www.wiley.com/go/eroglu/rfmicrowave

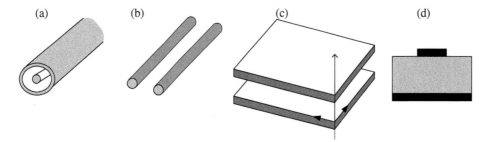

Figure 3.1 Common transmission line structures: (a) coaxial line, (b) two-wire, (c) parallel-plate, and (d) microstrip.

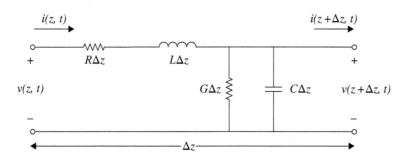

Figure 3.2 Short segment of transmission line.

$$\frac{dV(z)}{dz} = -(R + j\omega L)I(z) \tag{3.5}$$

$$\frac{dI(z)}{dz} = -(G + j\omega C)V(z) \tag{3.6}$$

By eliminating either $I(z)$ or $V(z)$ from (3.5) to (3.6), we obtain the wave equations as

$$\frac{d^2V(z)}{dz^2} = -\gamma^2 V(z) \tag{3.7}$$

$$\frac{d^2I(z)}{dz^2} = -\gamma^2 I(z) \tag{3.8}$$

where

$$\gamma = \alpha + j\beta = \sqrt{(R + j\omega L)(G + j\omega C)} \tag{3.9}$$

In (3.9), γ is the complex propagation constant, α is attenuation constant, and β is known as the phase constant. In transmission lines, phase velocity is defined as

$$v_p = \frac{\omega}{\beta} \tag{3.10}$$

Wavelength can be defined using

$$\lambda = \frac{2\pi}{\beta} \tag{3.11}$$

Traveling wave solutions to the equations obtained in (3.7) and (3.8) are

$$V(z) = V_0^+ e^{-\gamma z} + V_0^- e^{+\gamma z} \tag{3.12}$$

$$I(z) = I_0^+ e^{-\gamma z} + I_0^- e^{+\gamma z} \tag{3.13}$$

Substitution of (3.12) into (3.5) gives

$$I(z) = \frac{\gamma}{R + j\omega L} \left[V_0^+ e^{-\gamma z} + V_0^- e^{+\gamma z} \right] \tag{3.14}$$

From Eq. (3.14), we define the characteristic impedance, Z_0, which is defined as

$$Z_0 = \frac{R + \int j\omega L}{\gamma} = \sqrt{\frac{R + j\omega L}{G + j\omega C}} \tag{3.15}$$

Hence

$$\frac{V_0^+}{I_0^+} = Z_0 = -\frac{V_0^-}{I_0^-} \tag{3.16}$$

and

$$I(z) = \frac{V_0^+}{Z_0} e^{-\gamma z} - \frac{V_0^-}{Z_0} e^{+\gamma z} \tag{3.17}$$

Using the derived formulation, we can find the voltage and current at any point on the transmission line, as shown in Figure 3.3. At the load, $z = 0$

$$V(0) = Z_L I(0) \tag{3.18}$$

$$V_0^+ + V_0^- = \frac{Z_L}{Z_0} \left(V_0^+ - V_0^- \right) \tag{3.19}$$

or

$$V_0^- \left(1 + \frac{Z_L}{Z_0} \right) = V_0^+ \left(\frac{Z_L}{Z_0} - 1 \right) \tag{3.20}$$

which leads to

$$\frac{V_0^-}{V_0^+} = \left(\frac{Z_L - Z_0}{Z_L + Z_0} \right) \tag{3.21}$$

Figure 3.3 Finite terminated transmission line.

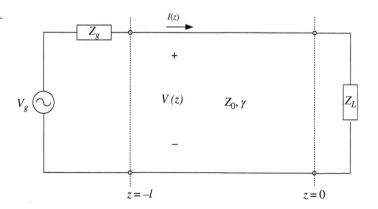

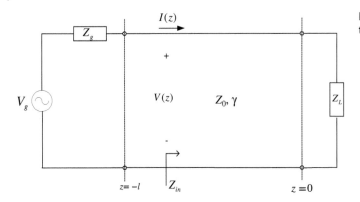

Figure 3.4 Input impedance calculation on the transmission line.

We then define (3.21) as the reflection coefficient at the load and express it as

$$\Gamma_L = \left(\frac{Z_L - Z_0}{Z_L + Z_0}\right) \tag{3.22}$$

We can the express voltage and current in terms of reflection coefficient as

$$V(z) = V_0^+ \left(e^{-\gamma z} + \Gamma_L e^{+\gamma z}\right) \tag{3.23}$$

$$I(z) = \frac{1}{Z_0} V_0^+ \left(e^{-\gamma z} - \Gamma_L e^{+\gamma z}\right) \tag{3.24}$$

We can find the input impedance at any point on the transmission line, as shown in Figure 3.4, from

$$Z_{in}(z) = \frac{V(z)}{I(z)} \tag{3.25}$$

We then have

$$Z_{in}(z) = Z_0 \frac{\left(e^{-\gamma z} + \Gamma_L e^{+\gamma z}\right)}{\left(e^{-\gamma z} - \Gamma_L e^{+\gamma z}\right)} \tag{3.26}$$

which can be expressed as

$$Z_{in}(z) = Z_0 \left(\frac{1 + \left(\dfrac{Z_L - Z_0}{Z_L + Z_0}\right) e^{+2\gamma z}}{1 - \left(\dfrac{Z_L - Z_0}{Z_L + Z_0}\right) e^{+2\gamma z}}\right) = Z_0 \left(\frac{(Z_L + Z_0) + (Z_L - Z_0)\, e^{+2\gamma z}}{(Z_L + Z_0) - (Z_L - Z_0)\, e^{+2\gamma z}}\right)$$

$$= Z_0 \left(\frac{(Z_L + Z_0)e^{-\gamma z} + (Z_L - Z_0)\, e^{+\gamma z}}{(Z_L + Z_0)e^{-\gamma z} - (Z_L - Z_0)\, e^{+\gamma z}}\right) \tag{3.27}$$

We can rewrite (3.27) as

$$Z_{in}(z) = Z_0 \left(\frac{(Z_L + Z_0)e^{-\gamma z} + (Z_L - Z_0)\, e^{+\gamma z}}{(Z_L + Z_0)e^{-\gamma z} - (Z_L - Z_0)\, e^{+\gamma z}}\right) = Z_0 \left(\frac{Z_L(e^{+\gamma z} + e^{-\gamma z}) - Z_0(e^{+\gamma z} - e^{-\gamma z})}{-Z_L(e^{+\gamma z} - e^{-\gamma z}) + Z_0(e^{+\gamma z} + e^{-\gamma z})}\right) \tag{3.28}$$

which can also be expressed as

$$Z_{in}(z) = Z_0 \left(\frac{Z_L - Z_0 \tanh(\gamma z)}{Z_0 - Z_L \tanh(\gamma z)}\right) \tag{3.29}$$

At the input when $z = -l$, the impedance can be found from (3.29) as

$$Z_{in}(z) = Z_0 \left(\frac{Z_L + Z_0 \tanh{(\gamma l)}}{Z_0 + Z_L \tanh{(\gamma l)}} \right)$$ (3.30)

Example 3.1 Coaxial Transmission Line

A 2 m lossless air-spaced transmission line having a characteristic impedance of 50 Ω is terminated with an impedance $40 + j30\,\Omega$ at an operating frequency of 200 MHz. Find the input impedance.

Solution

The phase constant is found from

$$\beta = \frac{\omega}{v_p} = \frac{4}{3}\pi$$

Since it is given that $R_0 = 50\,\Omega$, $Z_L = 40 + j30$, $\ell = 2$m, we obtain the input impedance from (3.30) as

$$Z_i = 50 \frac{(40 + j30) + j50 \cdot \tan\left(\frac{4\pi}{3} \cdot 2\right)}{50 + j(40 + j30) \cdot \tan\left(\frac{4\pi}{3} \cdot 2\right)} = 26.3 - j9.87$$

3.2.1 Limiting Cases for Transmission Lines

There are three cases that can be considered as the limiting case for transmission lines. These are lossless lines, low loss lines, and distortionless lines.

a) Lossless line $(R = G = 0)$

Transmission lines can be considered lossless when $R = G = 0$. When $R = G = 0$, the defining equations for the transmission lines can be simplified as

$$\gamma = \alpha + j\beta = j\omega\sqrt{LC} \Rightarrow \alpha = 0$$ (3.31a)

$$\beta = \omega\sqrt{LC}$$ (3.31b)

$$v_p = \frac{\omega}{\beta} = \frac{1}{\sqrt{LC}}$$ (3.31c)

$$Z_0 = \sqrt{\frac{L}{C}} = R_0 + jX_0 \Rightarrow R_0 = \sqrt{\frac{L}{C}}, X_0 = 0$$ (3.31d)

b) Low-loss line $(R \ll \omega L, G \ll \omega C)$

For low-loss transmission lines, $R \ll \omega L$, $G \ll \omega C$, and defining equations simplify to

$$\gamma = \alpha + j\beta = j\omega\sqrt{LC}\left(1 + \frac{R}{j\omega L}\right)^{1/2}\left(1 + \frac{G}{j\omega C}\right)^{1/2}$$ (3.32a)

$$\alpha \cong \frac{1}{2}\left(R\sqrt{\frac{C}{L}} + G\sqrt{\frac{L}{C}}\right)$$ (3.32b)

$$\beta \cong \omega\sqrt{LC}$$ (3.32c)

$$v_p = \frac{\omega}{\beta} \cong \frac{1}{\sqrt{LC}}$$ (3.32d)

$$Z = R_0 + jX_0 = \sqrt{\frac{L}{C}}\left(1 + \frac{R}{j\omega L}\right)^{1/2}\left(1 + \frac{G}{j\omega C}\right)^{-1/2}$$ (3.32e)

c) Distortionless line ($R/L = G/C$)

In distortionless transmission lines, $R/L = G/C$, and the defining equations can be simplified as

$$\gamma = \alpha + j\beta = \sqrt{\frac{C}{L}}(R + j\omega L) \tag{3.33a}$$

$$\alpha = R\sqrt{\frac{C}{L}} \tag{3.33b}$$

$$\beta = \omega\sqrt{LC} \tag{3.33c}$$

$$v_p = \frac{1}{\sqrt{LC}} \tag{3.33d}$$

$$Z_0 = \sqrt{\frac{L}{C}} \tag{3.33e}$$

3.2.2 Transmission Line Parameters

Transmission lines are characterized by four distributed parameters: R – resistor per unit length, L – inductance per unit length, C – capacitance per unit length, and G – conductance per unit length, as discussed in Section 3.2.1. The transmission line parameters for coaxial and two-wire transmission line structures are given here.

3.2.2.1 Coaxial Line

The configuration of a coaxial cable is given in Figure 3.5. It has an inner conductor with radius a and an outer conductor with radius b. There is a dielectric with relative permittivity constant, ε_r, between the inner and outer conductor. The current is going from the inner conductor and returning to the outer, as shown in Figure 3.5. From Ampere's law

$$\overline{H} = \hat{\phi}\left(\frac{I}{2\pi\rho}\right) \tag{3.34}$$

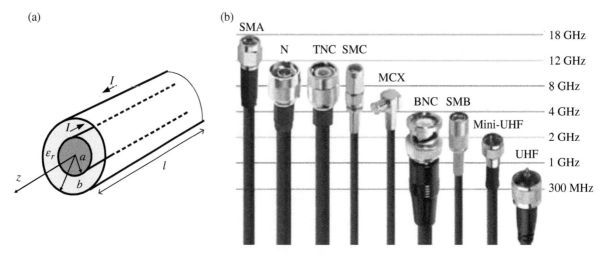

Figure 3.5 (a) The geometry of a coaxial cable and (b) commonly used commercial coaxial cables and frequency ranges.

The flux, ψ, is found from

$$\psi = \mu_0 \mu_r \int_a^b H_\phi \, d\rho \tag{3.35}$$

Then the inductance per unit length for coaxial cable is found from

$$L = \frac{\lambda}{I} = \frac{(Nl)\psi}{I} \quad \text{or} \quad \frac{L}{l} = \frac{\mu_0 \mu_r}{2\pi} \ln\left(\frac{b}{a}\right) \text{ H/m} \tag{3.36}$$

where λ is the flux linkage and N is equal to 1. The capacitance of the coaxial line in Figure 3.5a is found from

$$C = \frac{Q}{V} = \frac{\rho_\ell l}{V} \tag{3.37}$$

The voltage, V, can be obtained from the electric field between the inner and outer conductors as

$$V = \int_a^b E_\rho \, d\rho = \int_a^b \left(\frac{\rho_\ell}{2\pi\varepsilon_0\varepsilon_r\rho}\right) d\rho = \frac{\rho_\ell}{2\pi\varepsilon_0\varepsilon_r} \ln\left(\frac{b}{a}\right) \tag{3.38}$$

where

$$\overline{E} = \hat{\rho}\,\frac{\rho_\ell}{2\pi\varepsilon_0\varepsilon_r\rho} \tag{3.39}$$

Substituting (3.38) into (3.37) gives

$$C = \frac{Q}{V} = \frac{l}{\frac{1}{2\pi\varepsilon_0\varepsilon_r} \ln\left(\frac{b}{a}\right)} \quad \text{or} \quad \frac{C}{l} = \frac{2\pi\varepsilon_0\varepsilon_r}{\ln\left(\frac{b}{a}\right)} \text{ F/m} \tag{3.40}$$

The conductance is calculated from

$$G = \frac{I}{V} = \frac{J[(2\pi a)l]}{V} \tag{3.41}$$

In (3.40)

$$\overline{J} = \sigma\overline{E} \tag{3.42}$$

Substituting (3.42) into (3.41) gives

$$G = \frac{J[(2\pi a)l]}{V} = \frac{(2\pi a)l\sigma\left(\dfrac{\rho_{\ell 0}}{2\pi\varepsilon_0\varepsilon_r a}\right)}{\dfrac{\rho_{\ell 0}}{2\pi\varepsilon_0\varepsilon_r} \ln\left(\dfrac{b}{a}\right)} = \frac{(2\pi)\sigma l}{\ln\left(\dfrac{b}{a}\right)}$$

The conductance per unit length for a coaxial line is

$$\frac{G}{l} = \frac{2\pi\sigma}{\ln\left(\dfrac{b}{a}\right)} \text{ S/m} \tag{3.43}$$

The total resistance per unit length of the coaxial line is

$$\frac{R}{l} = R_a + R_b \tag{3.44}$$

where R_a is the resistance of the inner conductor and R_b is the resistance of the outer conductor and equal to

$$R_a = R_{sa}\left(\frac{1}{2\pi a}\right) \tag{3.45}$$

$$R_b = R_{sb}\left(\frac{1}{2\pi b}\right) \tag{3.46}$$

In (3.45) and (3.46), R_{sa} and R_{sb} are the surface resistances for the inner and outer conductors and are given by

$$R_{sa} = \frac{1}{\sigma_a \delta_a} \tag{3.47}$$

$$R_{sb} = \frac{1}{\sigma_b \delta_b} \tag{3.48}$$

where

$$\delta_a = \sqrt{\frac{2}{\omega \mu_0 \mu_{ra} \sigma_a}} \tag{3.49}$$

$$\delta_b = \sqrt{\frac{2}{\omega \mu_0 \mu_{rb} \sigma_b}} \tag{3.50}$$

Hence, substituting (3.45)–(3.50) into (3.44) and assuming the conductor is the same for both inner and outer conductor, we can find total resistance per unit length of the coaxial line as

$$\frac{R}{l} = \frac{R_s}{2\pi}\left(\frac{1}{a} + \frac{1}{b}\right) \Omega/\text{m} \tag{3.51}$$

In summary, the per unit length of distributed transmission line parameters for a coaxial line is

$$C = \frac{2\pi \varepsilon_0 \varepsilon_r}{\ln\left(\frac{b}{a}\right)} \ \text{F/m} \tag{3.52}$$

$$L = \frac{\mu_0}{2\pi} \ln\left(\frac{b}{a}\right) \ \text{H/m} \tag{3.53}$$

$$G = \frac{2\pi \sigma_d}{\ln\left(\frac{b}{a}\right)} \ \text{S/m} \tag{3.54}$$

$$R = \frac{R_s}{2\pi}\left(\frac{1}{a} + \frac{1}{b}\right) \Omega/\text{m} \tag{3.55}$$

The characteristic impedance of a coaxial line is found from (3.15). For a lossless coaxial line, $R = G = 0$, then

$$Z_0 = \sqrt{\frac{L}{C}} = \eta_0 \sqrt{\frac{\mu_r}{\varepsilon_r}} \frac{1}{2\pi} \ln\left(\frac{b}{a}\right) \Omega \tag{3.56}$$

Commonly used coaxial lines and their frequency ranges are illustrated in Figure 3.5b.

Example 3.2   Coaxial Transmission Line
A semirigid coaxial cable has an inner conductor diameter of 0.036 in., and an inner diameter of the outer conductor equal to 0.12 in. Compute the distributed transmission line parameters $R, L, G,$ and C, and total attenuation and Z_0 for the coaxial line at 1 GHz. Assume the dielectric is Teflon with $\varepsilon_r = 2.08$, and $\tan \delta = 0.0004$. It is also given that the conductor is copper.

Solution

The surface resistivity is found from (3.47) to (3.50) as

$$R_s = \sqrt{\frac{\omega\mu}{2\sigma_{cu}}}$$

Since $\sigma_{cu} = 5.813 \cdot 10^7 \frac{S}{m}$, thus

$$R_s = \sqrt{\frac{2\pi \cdot 1\,\text{GHz} \cdot 4\pi \cdot 10^{-7}\frac{H}{m}}{2.813 \cdot 10^7 \frac{S}{m}}} = 0.0082\,41\ \Omega$$

It is given that $\varepsilon_r = 2.08$ and $\tan\delta = 0.0004$ for Teflon. The lumped element parameters for the coaxial transmission line can be calculated using (3.52)–(3.55) as

$$L = \frac{\mu_0}{2\pi} \ln\left(\frac{b}{a}\right) = \frac{4\pi \cdot 10^{-7}\frac{H}{m}}{2\pi} \ln\left(\frac{3.02\,\text{mm}}{0.91\,\text{mm}}\right) = 0.24\frac{\mu H}{m}$$

$$C = \frac{2\pi\varepsilon_0\varepsilon_r}{\ln(b/a)} = \frac{2\pi \cdot 8.85\frac{pF}{m} \cdot 2.08}{\ln\left(3.02\,\text{mm}/0.91\,\text{mm}\right)} = 96.5\frac{pF}{m}$$

$$R = \frac{R_s}{2\pi}\left(\frac{1}{a} + \frac{1}{b}\right) = \frac{0.0082\,41\Omega}{2\pi}\left(\frac{1}{0.91\,\text{mm}} + \frac{1}{3.02\,\text{mm}}\right) = 1.876\frac{\Omega}{m}$$

$$G = \frac{2\pi\omega\varepsilon_0\varepsilon_r}{\ln(b/a)}\tan\delta = \frac{4\pi^2 \cdot 1\,\text{GHz} \cdot 8.85\frac{pF}{m} \cdot 2.08}{\ln\left(3.02\,\text{mm}/0.91\,\text{mm}\right)}(0.0004) = 2.424 \cdot 10^{-4}\frac{S}{m}$$

We can assume a low-loss case due to very low tangent loss. For a low loss case such as this

$$Z_0 \approx \sqrt{\frac{L}{C}} = \sqrt{\frac{0.24\frac{\mu H}{m}}{96.5\frac{pF}{m}}} = 49.871\ \Omega$$

The total loss, $\alpha_T = (\alpha_c + \alpha_d)$, per unit length can be determined from (3.32b)

$$\alpha \cong \frac{1}{2}\left(R\sqrt{\frac{C}{L}} + G\sqrt{\frac{L}{C}}\right) = \frac{1}{2}\left(1.876\sqrt{\frac{96.5 \times 10^{-12}}{0.24 \times 10^{-6}}} + 2.424 \times 10^{-4}\sqrt{\frac{0.24 \times 10^{-6}}{96.5 \times 10^{-12}}}\right) = 0.025\,\text{Np/m}$$

We can verify this also by individually calculating the conductor loss, α_c, and dielectric loss, α_d

$$\alpha_c = \frac{R_s}{2\eta\ln(b/a)}\left(\frac{1}{a} + \frac{1}{b}\right) = \frac{0.0082\,41}{2 \cdot 261.2\Omega \cdot \ln(3.02\,\text{mm}/0.91\,\text{mm})} = 0.019\,\text{Np/m}$$

where $\eta = \sqrt{\frac{\mu_0}{\varepsilon_0\varepsilon_r}} = 261.2\ \Omega$

$$\alpha_d = \frac{\omega\varepsilon_0\varepsilon_r}{2}\eta \cdot \tan\delta = \frac{(2\pi \cdot 1\text{GHz})(8.85 \times 10^{-12} \times 2.08)}{2}(261.2\ \Omega) \cdot (0.0004) = 0.006\,\text{Np/m}$$

Then, the total loss is

$$\alpha_T = (\alpha_c + \alpha_d) = 0.019 + 0.006 = 0.025\,\text{Np/m}$$

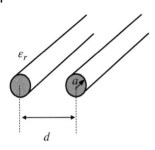

Figure 3.6 The geometry of a two-wire transmission line.

In dB

$$\alpha_T = (\alpha_c + \alpha_d) \cdot 8.686 \ \text{dB}/\text{Np} = (0.019 \ \text{Np}/\text{m} + 0.006 \ \text{Np}/\text{m}) \cdot 8.686 \ \text{dB}/\text{Np}$$

$$\alpha_T = 0.217 \ \text{dB}/\text{m}$$

3.2.2.2 Two-wire Transmission Line

The geometry of a two-wire transmission line is illustrated in Figure 3.6. Two-wire transmission line per unit length distributed parameters can be obtained as

$$C = \frac{\pi \varepsilon_0 \varepsilon_r}{\cosh^{-1}\left(\dfrac{d}{2a}\right)} \quad \text{F/m} \tag{3.57}$$

$$L = \frac{\mu_0}{\pi} \cosh^{-1}\left(\frac{d}{2a}\right) \ \text{H/m} \tag{3.58}$$

and

$$G = \frac{\pi \sigma}{\cosh^{-1}\left(\dfrac{d}{2a}\right)} \quad \text{S/m} \tag{3.59}$$

$$R = 2\left(\frac{1}{\sigma \delta}\right)\left(\frac{1}{2\pi a}\right) \ \Omega/\text{m} \tag{3.60}$$

Then, for a lossless two-wire transmission line, the characteristic impedance is obtained as

$$Z_0 = \sqrt{\frac{L}{C}} = \frac{1}{\pi}\eta_0 \frac{1}{\sqrt{\varepsilon_r}} \cosh^{-1}\left(\frac{d}{2a}\right) \ \Omega \tag{3.61}$$

3.2.2.3 Parallel Plate Transmission Line

Consider the parallel plate transmission line configuration given in Figure 3.7. The width of the plates is w and the separation distance between the plates is assumed to be d. There is a dielectric placed between the plates with ε and μ. The wave is propagating in z direction and polarized in y direction and it is a transverse electromagnetic (TEM) wave. Then, we can write field vectors as

$$\vec{E} = \hat{y} E_0 e^{-j\beta z} \tag{3.62}$$

$$\vec{H} = \hat{z} \times \frac{\overline{E}}{\eta} = -\hat{x} \frac{E_0}{\eta_0} e^{-j\beta z} \tag{3.63}$$

where β is the phase constant and η is the intrinsic impedance. The application of boundary conditions for the lower ($y = 0$) and upper ($y = d$) plate gives a charge density and current density at the lower plate as

$$\rho_{sl} = \varepsilon E_0 e^{-j\beta z} \tag{3.64}$$

$$\vec{J}_{sl} = \hat{z} \frac{E_0}{\eta} e^{-j\beta z} \tag{3.65}$$

Similarly for the upper plate

$$\rho_{su} = -\varepsilon E_0 e^{-j\beta z} \tag{3.66}$$

$$\vec{J}_{su} = -\hat{z} \frac{E_0}{\eta} e^{-j\beta z} \tag{3.67}$$

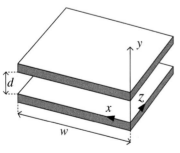

Figure 3.7 Parallel plate transmission line.

From

$$\nabla \times \vec{E} = -j\omega\mu\vec{H} \tag{3.68}$$

we obtain

$$-\frac{dV(z)}{dz} = j\omega\mu J_{su}(z)d \tag{3.69}$$

which leads to

$$j\omega LI(z) = j\omega \left(\mu\frac{d}{\omega}\right)[J_{su}(z)w] \tag{3.70}$$

Comparison of left and right side of the equalities gives

$$L = \mu\frac{d}{w} \text{ H/m} \tag{3.71}$$

Similarly, the capacitance is found from

$$-\frac{dI(z)}{dz} = -j\omega\varepsilon E_y(z)w \tag{3.72}$$

Equation (3.72) can be written as

$$j\omega CV(z) = j\omega\left(\varepsilon\frac{w}{d}\right)\left[-E_y(z)d\right] \tag{3.73}$$

Then

$$C = \varepsilon\frac{w}{d} \text{ F/m} \tag{3.74}$$

The conductance and resistance per unit length of the parallel plate configuration is found from

$$G = \sigma\frac{w}{d} \text{ S/m} \tag{3.75}$$

$$R = \frac{2}{w}\sqrt{\frac{\pi f \mu_c}{\sigma_c}} = 2\left(\frac{R_s}{w}\right) \Omega/\text{m} \tag{3.76}$$

The characteristic impedance of the transmission line is

$$Z_0 = \sqrt{\frac{L}{C}} = \frac{d}{w}\sqrt{\frac{\mu}{\varepsilon}}\Omega \tag{3.77}$$

3.2.3 Terminated Lossless Transmission Lines

Consider the lossless transmission line shown in Figure 3.8. The voltage and current at any point on the line can be written as

$$V(z) = V_0^+ e^{-j\beta z} + V_0^- e^{j\beta z} \tag{3.78}$$

$$I(z) = \frac{V_0^+}{Z_0}e^{-j\beta z} - \frac{V_0^-}{Z_0}e^{+j\beta z} \tag{3.79}$$

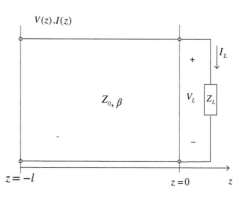

Figure 3.8 Lossless transmission line.

The voltage and current at the load, $z = 0$, in terms of load reflection coefficient is

$$V(z) = V_0^+ \left[e^{-j\beta z} + \Gamma e^{j\beta z} \right] \tag{3.80a}$$

$$I(z) = \frac{V_0^+}{Z_0} \left[e^{-j\beta z} - \Gamma e^{j\beta z} \right] \tag{3.80b}$$

It can be seen that the voltage and current on the line consist of a superposition of an incident and reflected wave, which represents standing waves. When $\Gamma = 0$, then it is a matched condition. The time-average power flow along the line at point z can be written as

$$P_{avg} = \frac{1}{2} \operatorname{Re} \{ V(z) I^*(z) \} = \frac{1}{2} \frac{|V_0^+|^2}{Z_0} \operatorname{Re} \left\{ 1 - \Gamma^* e^{-2j\beta z} + \Gamma e^{2j\beta z} - |\Gamma|^2 \right\} \tag{3.81}$$

or

$$P_{avg} = \frac{1}{2} \frac{|V_0^+|^2}{Z_0} \left(1 - |\Gamma|^2 \right) \tag{3.82}$$

When the load is mismatched, not all of the available power from the generator is delivered to the load. The power that is lost is known as the return loss (RL), and this can be found from

$$\text{RL} = -20 \log |\Gamma| \ \text{dB} \tag{3.83}$$

Under mismatched conditions, the voltage on the line can be written as

$$|V(z)| = |V_0^+| \left| 1 + \Gamma e^{2j\beta z} \right| = |V_0^+| \left| 1 + \Gamma e^{-2j\beta l} \right| = |V_0^+| \left| 1 + |\Gamma| e^{j(\theta - 2\beta l)} \right| \tag{3.84}$$

The minimum and maximum values of the voltage from (3.84) can be found as

$$V_{\max} = |V_0^+| (1 + |\Gamma|) \quad \text{and} \quad V_{\min} = |V_0^+| (1 - |\Gamma|) \tag{3.85}$$

A measure of the mismatch of a line is called the voltage standing wave ratio (VSWR) and can be expressed as the ratio of maximum voltage to minimum voltage as

$$\text{VSWR} = \text{SWR} = \frac{V_{\max}}{V_{\min}} = \frac{1 + |\Gamma|}{1 - |\Gamma|} \tag{3.86}$$

From (3.84), the distance between two successive voltage maxima (or minima) is $l = 2\pi/2\beta = \lambda/2 \, (2\beta l = 2\pi)$, while the distance between a maximum and a minimum is $l = \pi/2\beta = \lambda/4$. From (3.64) with $z = -l$

$$\Gamma(l) = \frac{V_0^- e^{-j\beta l}}{V_0^+ e^{j\beta l}} = \Gamma(0) e^{2j\beta l} \tag{3.87}$$

For the current

$$I(z) = V_0^+ e^{-j\beta z} \left(\frac{1}{Z_0} \right) \left(1 - |\Gamma| e^{+j(\phi - 2\beta l)} \right) \tag{3.88}$$

or

$$|I(z)| = |V_0^+| \left(\frac{1}{Z_0} \right) \left| 1 - |\Gamma| e^{+j(\phi - 2\beta l)} \right| \tag{3.89}$$

Hence, the maximum and minimum values of the current on the line can be written as

$$I_{\max} = |I(z)|_{\max} = |V_0^+|\left(\frac{1}{Z_0}\right)(1 + |\Gamma|) \qquad (3.90)$$

$$I_{\min} = |I(z)|_{\min} = |V_0^+|\left(\frac{1}{Z_0}\right)(1 - |\Gamma|) \qquad (3.91)$$

The current standing wave ratio, ISWR, is

$$\text{ISWR} = \frac{I_{\max}}{I_{\min}} = \frac{1 + |\Gamma|}{1 - |\Gamma|} \qquad (3.92)$$

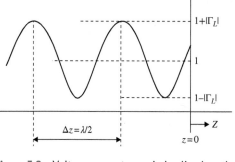

Figure 3.9 Voltage versus transmission line length.

Hence, VSWR = ISWR from (3.86) to (3.92). The voltage waveform versus length of the transmission line along the axis is plotted in Figure 3.9.

At a distance of $l = -z$, the input impedance is equal to

$$Z_{in} = \frac{V(-l)}{I(-l)} = Z_0\frac{Z_L + jZ_0 \tan\beta l}{Z_0 + jZ_L \tan\beta l} \qquad (3.93)$$

Example 3.3 Parameters and Coefficients

For a transmission, it is given that $Z_L = 17.4 - j30\,\Omega$ and $Z_0 = 50\,\Omega$. Calculate

$$\Gamma_L, \text{SWR}, z_{\min}, \ V_{\max}, V_{\min}$$

on the transmission line.

Solution

From the information received, we find the load reflection coefficient as

$$\Gamma_L = \frac{Z_L - Z_0}{Z_L + Z_0} = -0.24 - j0.55 = 0.6\,e^{-j(1.99)}$$

VSWR is found from

$$\text{SWR} = \frac{V_{\max}}{V_{\min}} = \frac{1 + |\Gamma_L|}{1 - |\Gamma_L|} = \frac{1 + 0.6}{1 - 0.6} = 4.0$$

This leads to

$$V_{\max}/|V^+| = 1 + |\Gamma_L| = 1.6$$
$$V_{\min}/|V^+| = 1 - |\Gamma_L| = 0.4$$

Hence, the maximum and minimum values of the voltage are obtained when

$$V_{\max} \text{ when } \phi + 2\beta z = 0, \ -2\pi, \dots$$
$$V_{\min} \text{ when } \phi + 2\beta z = -\pi, \ -3\pi, \dots$$

So, the distance that will give the minimum value of the voltage is found from

$$z_{\min} = \frac{-\pi - \phi}{2\beta} = \frac{(-\pi + 1.99)}{2(2\pi/\lambda)} = -0.092\lambda$$

When the voltage waveform is plotted versus the transmission line length, the results agree with the calculated results, as shown in Figure 3.10.

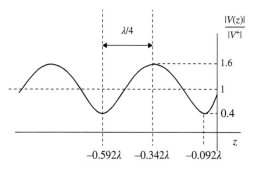

Figure 3.10 Voltage versus transmission length for the example.

Example 3.4   Lossless Transmission Lines

The standing wave ratio (SWR) on a lossless 50 Ω terminated line terminated in an unknown load impedance is 4. The distance between the successive minimum is 30 cm. And the first minimum is located at 6 cm from the load. Determine Γ, Z_L, and l_m.

Solution

From the information received, the wavelength can be found as

$$\frac{\lambda}{2} = 0.3 \Rightarrow \lambda = 0.6 \, \text{m}, \beta = \frac{2\pi}{\lambda} = 3.33\pi$$

The reflection coefficient is equal to

$$|\Gamma| = \frac{4-1}{4+1} = 0.6, z_m' = 0.06 \, \text{m} \Rightarrow l_m = \frac{\lambda}{2} - z_m' = 0.24 \, \text{m}$$

$$\theta_\Gamma = 2\beta z_m' - \pi = -0.6\pi, \Gamma = |\Gamma|e^{j\theta_\Gamma} = 0.6e^{-j0.6\pi} = -0.185 - j0.95$$

The load impedance (Z_L) is then equal to

$$Z_L = Z_0 \frac{1+\Gamma_L}{1-\Gamma_L} = 50 \cdot \frac{1 + (-0.185 - j0.95)}{1 - (-0.185 - j0.95)} = 1.43 - j41.17$$

Example 3.5   Transmission Line Circuits

Calculate the following parameters given below for the transmission line circuit shown in Figure 3.11.

1) The SWR on the line.
2) The reflection coefficient at the load.
3) The load admittance.
4) The input impedance of the line.
5) The distance from the load to the first voltage minimum.
6) The distance from the load to the first voltage maximum.

Solution

See the attached Smith chart. First, we normalize the load impedance.

$$z_L = \frac{Z_L}{Z_0} = \frac{(60 + j50)\Omega}{50\Omega} = 1.2 + j$$

We then calculate the other parameters using the formulation given as

1) SWR = 2.5
2) $\Gamma_L = 0.42\angle 54.5°$
3) $Y_L = \frac{y_L}{Z_0} = \frac{0.5 - j0.42}{50\Omega} = (10 - j8.4) \, \text{mS}$

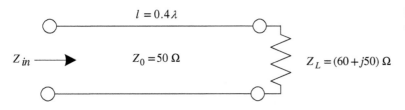

$l = 0.4\lambda$

$Z_{in} \longrightarrow$

$Z_0 = 50 \, \Omega$

$Z_L = (60 + j50) \, \Omega$

Figure 3.11   Transmission line configuration for Example 3.5.

4) $Z_{in} = z_{in} \cdot Z_0 = (0.5 + j0.4) \cdot Z_0 = (25 + j20)\ \Omega$
5) $\ell_{min} = 0.5\lambda - 0.174\lambda = 0.326\lambda$
6) $\ell_{max} = 0.25\lambda - 0.174\lambda = 0.076\lambda$

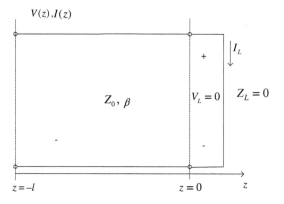

Figure 3.12   Short circuited transmission line.

3.2.4   Special Cases of Terminated Transmission Lines

3.2.4.1   Short-circuited Line

Consider the short-circuited transmission line shown in Figure 3.12. When a transmission line is short-circuited, $Z_L = 0$ → $\Gamma = -1$, then voltage and current can be written as

$$V(z) = V_0^+ \left[e^{-j\beta z} - e^{j\beta z}\right] = -2jV_0^+ \sin\beta z \qquad (3.94)$$

$$I(z) = \frac{V_0^+}{Z_0}\left[e^{-j\beta z} + e^{j\beta z}\right] = 2\frac{V_0^+}{Z_0}\cos\beta z \qquad (3.95)$$

The input impedance when $z = -l$ is equal to

$$Z_{in} = jZ_0\tan\beta l \qquad (3.96)$$

The impedance variation of the line along the z is given in Figure 3.13.

At lower frequencies, Eq. (3.96) can be written as

$$X_{in} \approx Z_0(\beta l) = \sqrt{\frac{L}{C}}\left(\omega\sqrt{LC}\,l\right) = \omega(Ll) \qquad (3.97)$$

Then, the lumped element equivalent model of the transmission line can be represented as shown in Figure 3.14.

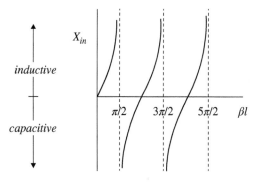

Figure 3.13   Impedance variation for a short-circuited transmission line.

3.2.4.2   Open-circuited Line

Consider the open circuited transmission line shown in Figure 3.15. When a transmission line is short-circuited, $Z_L = \infty$ → $\Gamma = 1$, then voltage and current can be written as

$$V(z) = V_0^+ \left[e^{-j\beta z} + e^{j\beta z}\right] = 2V_0^+ \cos\beta z \qquad (3.98)$$

$$I(z) = \frac{V_0^+}{Z_0}\left[e^{-j\beta z} - e^{j\beta z}\right] = \frac{-2jV_0^+}{Z_0}\sin\beta z \qquad (3.99)$$

The input impedance when $z = -l$ is equal to

$$Z_{in} = -jZ_0\cot\beta l \qquad (3.100)$$

The impedance variation of the line along the z is given in Figure 3.16.

At lower frequencies, Eq. (3.100) can be written as

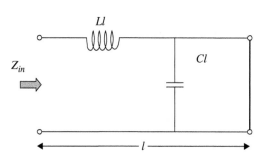

Figure 3.14   Low frequency equivalent circuit of short-circuited transmission line.

$$X_{in} \approx -Z_0/(\beta l) = -\sqrt{\frac{L}{C}}\left(\frac{1}{\omega\sqrt{LC}\,l}\right) = \frac{-1}{\omega(Cl)} \qquad (3.101)$$

Then, the lumped element equivalent model of the transmission line can be represented as shown in Figure 3.17.

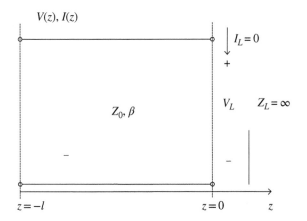

Figure 3.15   Open-circuited transmission line.

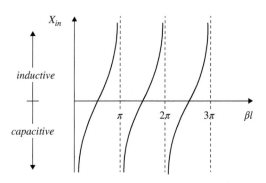

Figure 3.16   Impedance variation for an open-circuited transmission line.

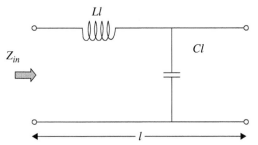

Figure 3.17   Low frequency equivalent circuit of an open-circuited transmission line

3.3   Smith Chart

A Smith chart is a conformal mapping of the normalized complex impedance plane and the complex reflection coefficient plane. It is a graphical method of displaying impedances and all related parameters using a reflection coefficient. It was invented by Phillip Hagar Smith while he was working at Radio Corporation of America (RCA). The process of establishing a Smith chart begins with normalizing the impedance, as shown by Eq. (3.102).

$$z_L = \frac{Z_L}{Z_0} = \frac{R_L + jX_L}{Z_0} \tag{3.102}$$

Consider now the right-hand portion of the normalized complex impedance plane, as illustrated in Figure 3.18. All values of impedance such that $R \geq 0$ are represented by points in the plane. The impedance of all passive devices will be represented by points in the right-half plane.

The complex reflection coefficient may be written as a magnitude and a phase or as real and imaginary parts.

$$\Gamma_L = |\Gamma_L| e^{\angle \Gamma_L} = \Gamma_{Lr} + j\Gamma_{Li} \tag{3.103}$$

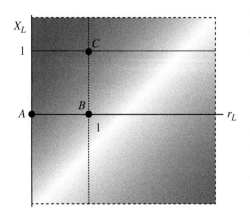

Figure 3.18   Right-hand portion of the normalized complex impedance plane.

The reflection coefficient in terms of the load Z_L terminating line Z_0 is defined as

$$\Gamma_L = \frac{Z_L - Z_0}{Z_L + Z_0} \qquad (3.104)$$

We can rearrange (3.104) to get

$$Z_L = Z_0 \frac{1 + \Gamma_L}{1 - \Gamma_L} \qquad (3.105)$$

In terms of normalized quantities, Eq. (3.105) can be written as

$$z_L = r_L + jx_L = \frac{Z_L}{Z_0} = \frac{1 + \Gamma_L}{1 - \Gamma_L} \qquad (3.106)$$

Substituting in the complex expression for Γ_L and equating real and imaginary parts, we find the two equations which represent circles in the complex reflection coefficient plane as

$$\left(\Gamma_{Lr} - \frac{r_L}{1 + r_L}\right)^2 + (\Gamma_{Li} - 0)^2 = \left(\frac{1}{1 + r_L}\right)^2 \qquad (3.107)$$

$$(\Gamma_{Lr} - 1)^2 + \left(\Gamma_{Li} - \frac{1}{x_L}\right)^2 = \left(\frac{1}{x_L}\right)^2 \qquad (3.108)$$

The first circle is centered at

$$\left(\frac{r_L}{1 + r_L}, 0\right) \qquad (3.109)$$

and the second circle is centered at

$$\left(1, \frac{1}{x_L}\right) \qquad (3.110)$$

The location of the first circle is always inside the unit circle in the complex reflection coefficient plane with corresponding radius is

$$\frac{1}{1 + r_L} \qquad (3.111)$$

Hence, this circle will always be fully contained within the unit circle because the radius can never by greater than unity. This conformal mapping represents the mapping of the real resistance circle and is shown in Figure 3.19 using the mapping equation

$$\left(\Gamma_r - \frac{r}{1 + r}\right)^2 + (\Gamma_L)^2 = \left(\frac{1}{1 + r}\right)^2 \qquad (3.112)$$

The location of the second circle is always outside the unit circle in the complex reflection coefficient plane with the corresponding radius is

$$\left(\frac{1}{x_L}\right) \qquad (3.113)$$

The value radius can vary between 0 and infinity. This conformal mapping represents the mapping of the imaginary reactance circle and is shown in Figure 3.20 using the mapping equation

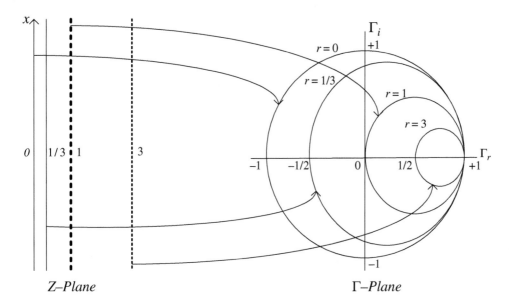

Figure 3.19 Conformal mapping of constant resistances.

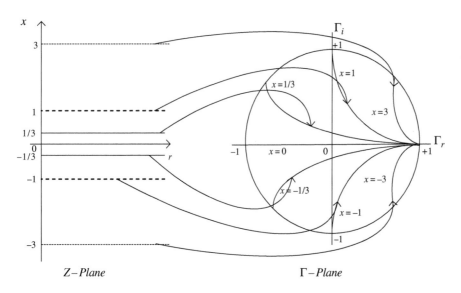

Figure 3.20 Conformal mapping of constant reactances.

$$(\Gamma_r - 1)^2 + \left(\Gamma_i - \frac{1}{x}\right)^2 = \left(\frac{1}{x}\right)^2 \tag{3.114}$$

The circles centered on the real axis represent lines of constant real part of the load impedance (r_L is constant; x_L varies). The circles whose centers reside outside the unit circle represent lines of constant imaginary part of the load impedance (x_L is constant, r_L varies). Mapping them into a single chart gives the display of a complete Smith chart, as shown in Figure 3.21.

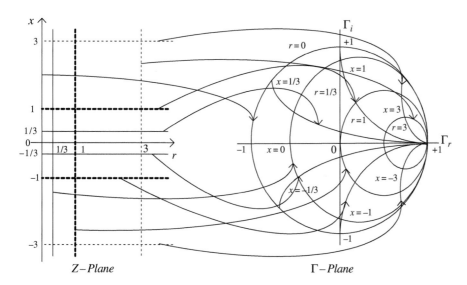

Figure 3.21 Combined conformal mapping lead to the display of a Smith chart.

In summary the properties of the r circles are:

- The centers of all r circles lie on the Γ_r axis.
- The $r = 0$ circle, having a unity radius and centered at the origin, is the largest.
- The r circles become progressively smaller as r increases from 0 to ∞, ending at the ($\Gamma_r = 1, \Gamma_i = 0$) point for open-circuit.
- All r circles pass through the ($\Gamma_r = 1, \Gamma_i = 0$) point.

Similarly, the properties of the x circles are:

- The centers of all x circles lie on the $\Gamma_r = 1$ line, those for $x > 0$ (inductive reactance) lie above the Γ_r axis, and those for $x < 0$ (capacitive reactance) lie below the Γ_r axis.
- The $x = 0$ circle becomes the Γ_r axis.
- The x circle becomes progressively smaller as $|x|$ increases from 0 to ∞, ending at the ($\Gamma_r = 1, \Gamma_i = 0$) point for open-circuit.
- All x circles pass through the ($\Gamma_r = 1, \Gamma_i = 0$) point.

Hence, in the combined display of the Smith chart

- All $|\Gamma|$ circles are centered at the origin, and their radii vary uniformly from 0 to 1.
- The angle, measured from the positive real axis, of the line drawn from the origin through the point representing z_L equals θ_Γ.
- The value of the r circle passing through the intersection of the $|\Gamma|$ circle and the positive real axis equals the SWR.

Example 3.6 Smith Chart
Locate the following normalized impedances on the Smith chart and calculate SWRs and reflection coefficients. (a) $z = 0.2 + j0.5$ (b) $z = 0.4 + j0.7$ (c) $z = 0.6 + j0.1$.

Solution

The generic MATLAB code given here is developed to calculate mark impedance points, draw VSWR circles, and calculate reflection coefficients on the Smith Chart at single frequency.

```
%This program marks impedance points, draws VSWR circle, calculates
%reflection coefficients, and marks them on the Smith Chart at single
%frequency

clear all;
close all;
global Z0;
Set_Z0(1); %Set Z0 to 1
%convert the strings received from the GUI to numbers
valuearray=str2double(answer);

%Give variable names to the received numbers
ZL1=valuearray(1);
ZL2=valuearray(2);
ZL3=valuearray(3);

%part a
gamma1=(ZL1-Z0)/(ZL1+Z0);
VSWR1=(1+abs(gamma1))/(1-abs(gamma1));
[th1,rl1]=cart2pol(real(gamma1),imag(gamma1));
smith; %Call Smith Chart Program
s_point(ZL1);
const_SWR_circle(ZL1,'r--');
hold on;
text(real(gamma1)+0.04,imag(gamma1)-0.03,'\bf\Gamma_1');

%part b
gamma2=(ZL2-Z0)/(ZL2+Z0);
VSWR2=(1+abs(gamma2))/(1-abs(gamma2));
[th2,rl2]=cart2pol(real(gamma2),imag(gamma2));
s_point(ZL2);

%part c
gamma3=(ZL3-Z0)/(ZL3+Z0);
VSWR3=(1+abs(gamma3))/(1-abs(gamma3));
[th3,rl3]=cart2pol(real(gamma3),imag(gamma3));
s_point(ZL3);
const_SWR_circle(ZL3,'r--');
hold on;
text(real(gamma3)+0.04,imag(gamma3)-0.03,'\bf\Gamma_3');
msgbox( sprintf([...
    'Calculated Parameters for Z1 \n'...
    ' Reflection coefficient for Z1:  gamma1 =%f +j(%f)\n'...
```

```
' Reflection Coefficient for Z1 In Polar form :|gamma1|=%f,angle1=%f\n'...
'    Standing Wave Ratio for Z1 : VSWR1=%f \n'...
'\n'...
'Calculated Parameters for Z2 \n'...
'    Reflection coefficient for Z2:  gamma2 =%f +j(%f)\n'...
'    Reflection Coefficient for Z2 In Polar form :|gamma2|=%f,angle1=%f\n'...
'    Standing Wave Ratio for Z2 : VSWR2=%f \n'...
'\n'...
'Calculated Parameters for Z3 \n'...
' Reflection coefficient for Z3:  gamma3 =%f +j(%f)\n'...
' Reflection Coefficient for Z3 In Polar form :|gamma3|=%f,angle3=%f\n'...
' Standing Wave Ratio for Z3 : VSWR3=%f \n'...
    '\n']...,
real(gamma1),imag(gamma1),rl1,th1*180/pi,VSWR1,real(gamma2),imag(gamma2),rl2,
th2*180/pi,VSWR2,real(gamma3),imag(gamma3),rl3,th3*180/pi,VSWR3));
```

When the program is executed, the following graphical user interface (GUI) is displayed for entering impedances and result.

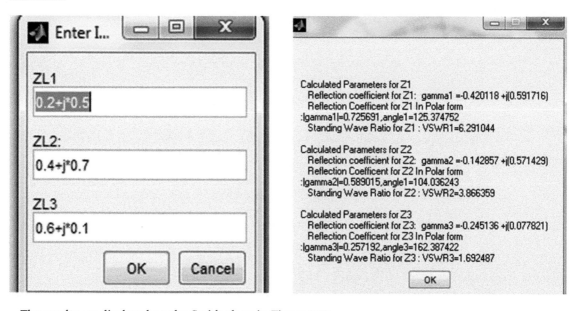

The results are displayed on the Smith chart in Figure 3.22.

3.3.1 Input Impedance Determination with a Smith Chart

It is shown in Section 3.2.3 that the voltage and current at any point on the transmission line can be expressed as

$$V(z') = \frac{I_L}{2}(Z_L + Z_0)e^{\gamma z'}\left[1 + \Gamma e^{-2\gamma z'}\right] \tag{3.115}$$

$$I(z') = \frac{I_L}{2Z_0}(Z_L + Z_0)e^{\gamma z'}\left[1 - \Gamma e^{-2\gamma z'}\right] \tag{3.116}$$

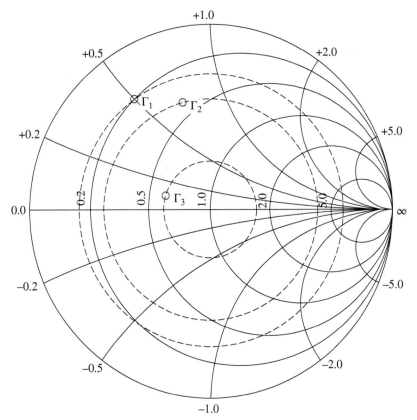

Figure 3.22 Smith chart displaying the calculated values and impedances.

where $z' = 1 - z$. Then, input impedance at a distance d away from the load on the line in terms of reflection coefficient can be obtained as

$$Z(d) = \frac{V(d)}{I(d)} = Z_0 \frac{1 + \Gamma(d)}{1 - \Gamma(d)} \tag{3.117}$$

where

$$\Gamma(d) = \Gamma_L e^{-j2\beta d} \tag{3.118}$$

Example 3.7 Transmission Lines with Smith Chart

A transmission line of characteristic impedance $Z_0 = 50\ \Omega$ and length $d = 0.2\ \lambda$ is terminated into a load impedance of $Z_L = (25 - j50)\ \Omega$. Find Γ_L, $Z_{in}\ (d)$, and SWR using a Smith chart.

Solution

The generic MATLAB code given below is developed to find input impedance by moving toward a generator at any length.

```
%This program find input impedance by moving towards generator
%on the transmission line at single frequency at any length
clear all;
close all;
global Z0;
```

```
%Gui Prompt
prompt = {'Ente Load Impedance ZL :', 'Enter the Length (in lambda) d : ','Enter
Characteristic Impedance Z0:'};
dlg_title = 'Enter Impedance ';
num_lines = 1;
def = {'25-j*50','.2','50'};
answer = inputdlg(prompt,dlg_title,num_lines,def, 'on');

%convert the strings received from the GUI to numbers
valuearray=str2double(answer);

%Give variable names to the received numbers
ZL=valuearray(1);
d=valuearray(2);
Z0=valuearray(3);
Set_Z0(Z0);
gamma_0=(ZL-Z0)/(ZL+Z0);
[th0,mag_gamma_0]=cart2pol(real(gamma_0),imag(gamma_0));
if th0<0
    th0=th0+2*pi;
end
th_in=th0-2*2*pi*d;
if th_in<0
    th_in=th_in+2*pi;
end

[x_gamma_in,y_gamma_in]=pol2cart(th_in,mag_gamma_0);
Zin=Z0*(1+x_gamma_in+j*y_gamma_in)/(1-x_gamma_in-j*y_gamma_in);
SWR=(1+abs(gamma_0))/(1-abs(gamma_0));
smith_chart(0);
hold on;
th=th0:(th_in-th0)/29:th_in;
gamma=mag_gamma_0*ones(1,30);
polar(th,gamma,'k');
hold on
s_point(Zin);
text(x_gamma_in+0.04,y_gamma_in-0.03,'\bfZ_{in}');
s_point(ZL);
text(real(gamma_0)+0.04,imag(gamma_0)-0.03,'\bfZ_{L}');
msgbox( sprintf([...
        'Calculated Parameters for Transmission Line \n'...
        '   Load Reflection coefficient :   gamma_0 =%f +j(%f)\n'...
        '   Magnitude of Load Reflection Coefficient :|gamma_0|=%f,angle=%f\n'...
        '   Input Impedance Zin : Zin=%f +j(%f)\n'...
        '   Standing Wave Ratio : SWR=%f \n'...
        '\n']...
,real(gamma_0),imag(gamma_0),mag_gamma_0,th0*180/pi,real(Zin),
imag(Zin),SWR));
```

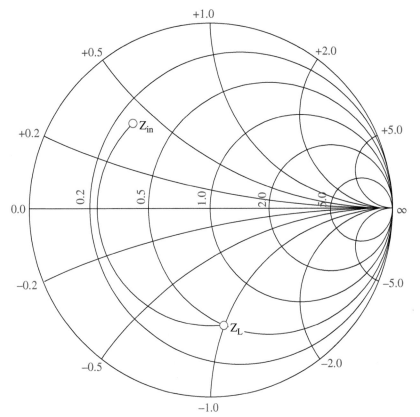

Figure 3.23 Input impedance display using a Smith chart.

When the program is executed, the following GUI and calculated results are displayed. The program also displays input impedance on the Smith chart, as shown in Figure 3.23, with the move toward the generator.

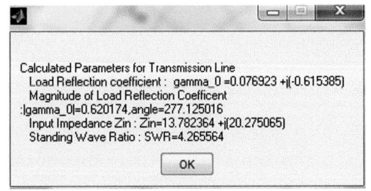

3.3.2 Smith Chart as an Admittance Chart

A Smith chart can also be used as an admittance chart by transforming impedances to admittances. Consider the expression for a normalized impedance at any point on the transmission line in terms of reflection coefficient as

$$Z_{in}(z) = \left(\frac{1 + \Gamma(z)}{1 - \Gamma(z)}\right) \tag{3.119}$$

The normalized admittance is reciprocal of impedance and can be written as

$$Y_{in}(z) = \frac{Y_{in}(z)}{Y_0} = \frac{1/Z_{in}(z)}{1/Z_0} = \frac{1}{Z_{in}(z)/Z_0} = \frac{1}{Z_{in}(z)} \tag{3.120}$$

Then, normalized admittance in terms of reflection coefficient can be expressed as

$$Y_{in}(z) = \left(\frac{1 - \Gamma(z)}{1 + \Gamma(z)}\right) \tag{3.121}$$

which can be written as

$$Y_{in}(z) = \left(\frac{1 + \Gamma'(z)}{1 - \Gamma'(z)}\right) \tag{3.122}$$

where

$$\Gamma'(z) = -\Gamma(z) = \Gamma(z)e^{-j\pi} \tag{3.123}$$

That means a 180° phase shift for the reflection coefficient gives the value of admittance for the corresponding impedance value. When the impedance point is marked on the Smith chart, moving in a 180° clockwise direction gives the value admittance. Instead of repeating this for each impedance point on the Smith chart, we can keep the location of the impedance fixed and rotate the Smith chart by 180°. This gives the admittance chart shown in Figure 3.24. When both the Z and Y chart are plotted together, we obtain the ZY chart shown in Figure 3.25.

3.3.3 *ZY* Smith Chart and Its Applications

A ZY Smith chart gives us the ability to implement both impedances and admittances on a single chart. It is a power chart and it allows designers to make impedance transformations and matchings using a unique graphical display when components are connected in series or in shunt. The effect of adding a single reactive component in a series with a complex impedance results in motion along a constant resistance circle in a ZY chart. If a single reactive component is added with a complex impedance in shunt, then a motion along a constant conductance circle in a ZY chart is required. Whenever an inductor is connected to the network, the direction of movement on a ZY chart is toward the upper half, whereas the addition of a capacitor results in movement toward the lower part of the chart. These component motions on a ZY chart are illustrated in Figure 3.26.

Example 3.8   Input Impedance with Smith Chart
Find the input impedance for the circuit shown in Figure 3.27 at 4 GHz when the load connected is $Z_L = 62.5\ \Omega$ using a Smith Chart.

Solution
The process begins with normalizing the load impedance, $Z_L = R = 62.5\ \Omega$.

$$z_L = \frac{Z_L}{Z_0} = \frac{62.5}{50} = 1.25$$

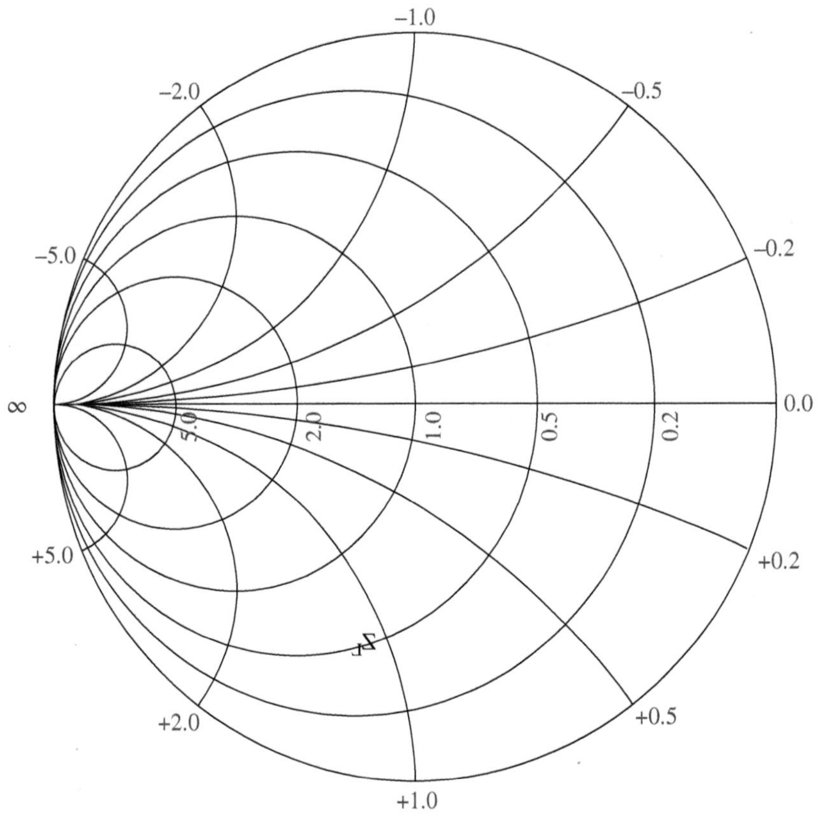

Figure 3.24 Admittance, *Y*, Smith chart.

Since the next component is a shunt connected component, we need to convert this value to a conductance value. That is

$$g_L = \frac{1}{z_L} = \frac{1}{1.25} = 0.8$$

On a *ZY* Smith chart, we mark this point on the conductance circle. The next component is shunt *C* with value of 1.59 pF. The normalized susceptance value of the capacitor at 4 GHz is found from

$$b_C = B_C Z_0 = \omega C Z_0 = (2\pi 4 \times 10^9)(1.59 \times 10^{-12})50 = 2$$

This corresponds to point *B* on the Smith chart shown in Figure 3.28. This is the amount of rotation that needs to be done on the conductance circle, as shown by point *B*. The admittance at point *B* is equal to

$$y_B = 0.8 + j2$$

The next component connected is a series *L* with a value of 8 nH. So, we move from a conductance circle to a resistance circle and read the corresponding impedance value as

$$z_B = 0.17 - j0.43$$

The normalized reactance value of the inductor is equal to

$$x_L = \frac{X_L}{Z_0} = \frac{(2\pi 4 \times 10^9)(8 \times 10^{-9})}{50} = 4$$

Figure 3.25 *ZY* Smith chart.

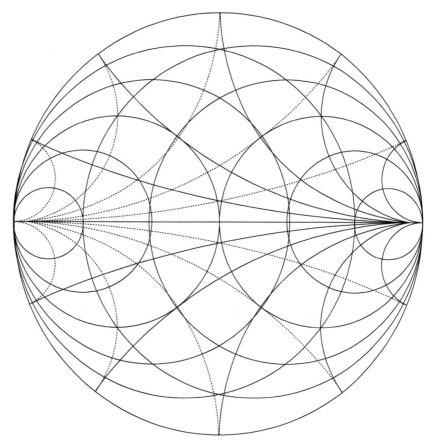

This value needs to be added to the impedance at point *B* to find the impedance value shown as point *C* on the Smith chart.

$$z_C = z_B + x_L = 0.17 - j0.43 + 4 = 0.17 + 3.57$$

De-normalizing impedance z_C gives the input impedance as

$$Z_{in} = z_C Z_0 = (0.17 + j3.57)50 = (8.5 + j178.5)\,\Omega$$

The results are shown on the Smith chart in Figure 3.28.

3.4 Microstrip Lines

Microstrip lines are widely used transmission lines for high frequency applications because of the ease of their fabrication process. The microstrip transmission line structure is illustrated in Figure 3.29. The dielectric with permittivity ε_r and thickness *d* is placed between a metal strip of width *w* and the ground plane. The field distribution illustrated shows that there are axial *E* field components which render pure TEM wave propagation impossible. Hence, the analysis can be done based on quasistatic approximations which allows quasi-TEM mode propagation. This is possible because the dielectric substrate is electrically very thin ($d \ll \lambda$), and we express this as

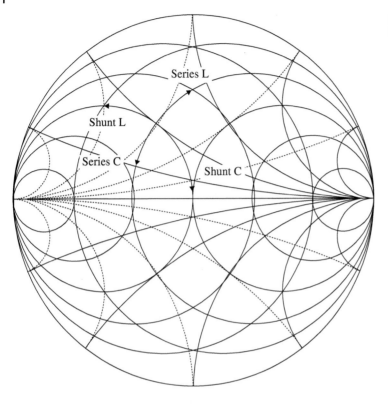

Figure 3.26 Adding component using a *ZY* Smith chart.

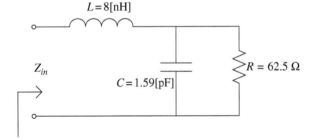

Figure 3.27 Impedance transformation.

$$v_p = \frac{c}{\sqrt{\varepsilon_e}}, \beta = k_0\sqrt{\varepsilon_e}, \quad 1 < \varepsilon_e < \varepsilon_r \tag{3.124}$$

where ε_e is the effective permittivity constant. Electric fields are going through air and dielectric and then terminate at the ground plane. The air–dielectric interface creates an equivalent homogeneous medium with an effective permittivity constant ε_e, as shown in Figure 3.30 [2].

The effective permittivity constant, ε_e, when thickness of the strip is much smaller than the thickness of the dielectric, $t/d < 0.005$, is found from

$$\varepsilon_e = \frac{\varepsilon_r + 1}{2} + \frac{\varepsilon_r - 1}{2}\left(\frac{1}{\sqrt{1 + 12\left(\frac{d}{W}\right)}} + 0.04\left(1 - \frac{W}{d}\right)^2\right) \text{ for } W/d \leq 1 \tag{3.125a}$$

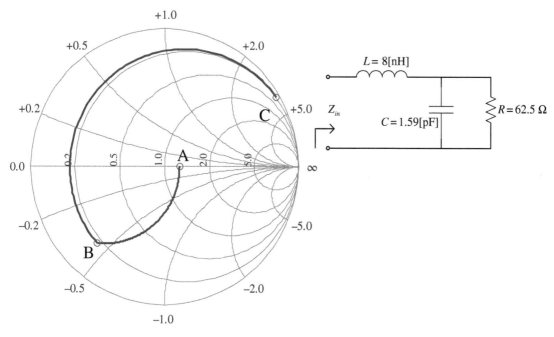

Figure 3.28 Impedance transformation using a Smith chart.

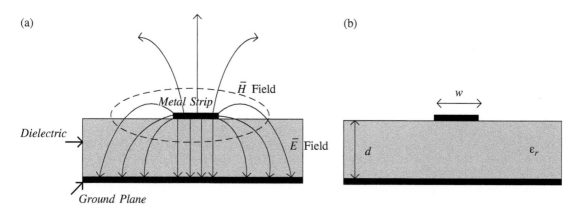

Figure 3.29 Microstrip line: (a) field lines of microstrip line; (b) geometry of microstrip line.

Figure 3.30 Representation of equivalent microstrip circuit.

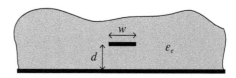

$$\varepsilon_e = \frac{\varepsilon_r + 1}{2} + \frac{\varepsilon_r - 1}{2}\left(\frac{1}{\sqrt{1 + 12\left(\frac{d}{W}\right)}}\right) \quad \text{for } W/d > 1 \tag{3.125b}$$

The design of a microstrip line uses (i) information about physical parameters of the structure such as width and thickness and then calculates the characteristic impedance Z_0 or (ii) characteristic impedance Z_0 and calculates the physical parameters. The two design methods are given here.

The characteristic impedance of a microstrip line is calculated for given/assumed physical dimensions of a microstrip line using

$$Z_0 = \begin{cases} \dfrac{60}{\sqrt{\varepsilon_e}} \ln\left(\dfrac{8d}{W} + \dfrac{W}{4d}\right) & \text{for } W/d \leq 1 \\[2ex] \dfrac{120\pi}{\sqrt{\varepsilon_e}[W/d + 1.393 + 0.667 \ln\left(W/d + 1.444\right)]} & \text{for } W/d > 1 \end{cases} \tag{3.126}$$

The physical dimensions of a microstrip line are calculated for a given characteristic impedance from

$$\frac{W}{d} = \begin{cases} \dfrac{8e^A}{e^{2A} - 2} & \text{for } W/d \leq 2 \\[2ex] \dfrac{2}{\pi}\left[B - 1 - \ln\left(2B - 1\right) + \dfrac{\varepsilon_r - 1}{2\varepsilon_r}\left\{\ln\left(B - 1\right) + 0.39 - \dfrac{0.61}{\varepsilon_r}\right\}\right] & \text{for } W/d > 2 \end{cases} \tag{3.127}$$

where

$$A = \frac{Z_0}{60}\sqrt{\frac{\varepsilon_r + 1}{2}} + \frac{\varepsilon_r - 1}{\varepsilon_r + 1}\left(0.23 + \frac{0.11}{\varepsilon_r}\right) \tag{3.128}$$

$$B = \frac{377\pi}{2Z_0\sqrt{\varepsilon_r}} \tag{3.129}$$

The wavelength on the microstrip line is

$$\lambda = \frac{\lambda_0}{\sqrt{\varepsilon_e}} \tag{3.130}$$

The phase shift due to a strip length, l, is calculated from

$$\beta = \frac{2\pi}{\lambda}l = \frac{2\pi}{\lambda_0}l\sqrt{\varepsilon_e} \tag{3.131}$$

The total loss in the microstrip structure is due to conductor and dielectric losses and it can be expressed as

$$\alpha = \alpha_c + \alpha_d \tag{3.132}$$

where

$$\alpha_d = \frac{k_0\varepsilon_r(\varepsilon_e - 1)\tan\delta}{2\sqrt{\varepsilon_e}(\varepsilon_r - 1)} \text{ Np/m} \tag{3.133}$$

$$\alpha_c = \frac{R_s}{Z_0 W} \text{ Np/m} \tag{3.134}$$

and

$$R_s = \sqrt{\frac{\omega\mu_0}{2\sigma}} \tag{3.135}$$

Example 3.9   Microstrip Line

If the substrate thickness is 0.158 cm, and dielectric constant is $\varepsilon_r = 2.20$, what would be the physical dimensions of a 100 Ω microstrip line? For this microstrip line, what is the guide wavelength if the frequency is 4.0 GHz?

Solution

In order to design microstrip dimensions based on characteristic impedance, we must first calculate A and B.

$$A = \frac{Z_0}{60\,\Omega} \sqrt{\frac{\varepsilon_r + 1}{2}} + \frac{\varepsilon_r - 1}{\varepsilon_r + 1} \left(0.23 + \frac{0.11}{\varepsilon_r}\right) = 2.213$$

$$B = \frac{377\,\Omega \cdot \pi}{2 Z_0 \sqrt{\varepsilon_r}} = 3.993$$

Now we also have an equation relating the ratio of the width to the dielectric thickness based on A and B.

$$\frac{W}{d} = \begin{cases} \dfrac{8e^A}{e^{2A} - 2} & \text{for } W/d \leq 2 \\[2mm] \dfrac{2}{\pi}\left[B - 1 - \ln(2B - 1) + \dfrac{\varepsilon_r - 1}{2\varepsilon_r}\left(\ln(B - 1) + 0.39 - \dfrac{0.61}{\varepsilon_r}\right)\right] & \text{for } W/d > 2 \end{cases}$$

Given that the thickness of 100 Ω trace is generally pretty thin, the top equation was chosen.

$$\frac{W}{d} = 0.896$$

Since the thickness is given, we can calculate the width as

$$W = 0.896d = 0.142 \text{ cm}$$

Now we must check the solution to verify the initial assumption that $W/d \leq 2$ was correct. But first, the effective dielectric constant must be found.

$$\varepsilon_e = \frac{\varepsilon_r + 1}{2} + \frac{\varepsilon_r - 1}{2} \frac{1}{\sqrt{1 + 12d/W}} = 1.758$$

This is the effective permittivity of a fictitious homogeneous dielectric containing the entire electric field of the microstrip. Now we calculate the impedance based on ε_e and the trace dimensions.

$$Z_0 = \begin{cases} \dfrac{60}{\sqrt{\varepsilon_e}} \ln\left(\dfrac{8d}{W} + \dfrac{W}{4d}\right) & \text{for } W/d \leq 1 \\[3mm] \dfrac{120\pi}{\sqrt{\varepsilon_e}[W/d + 1.393 + 0.667\ln(W/d + 1.444)]} & \text{for } W/d \geq 1 \end{cases}$$

We obviously choose the equation for $W/d \leq 1$. This leads to

$$Z_0 = 100.174 \,\Omega$$

This confirms the validity of the design. We can determine now the wavelength at 4.0 GHz from

$$\lambda = \frac{\lambda_0}{\sqrt{\varepsilon_e}} = \frac{c}{\sqrt{\varepsilon_e}f} = \frac{3 \times 10^8}{(\sqrt{1.758})4 \times 10^9} = 0.0565 \text{ m}$$

The Mathcad routine is given below for the calculation of microstrip parameters in this example.
Constant

$$f := 4.0 \text{ GHz} \quad d := 0.158 \text{ cm} \quad \varepsilon_r := 2.2 \quad Z_0 := 100 \,\Omega$$

Microstrip Dimension calculations

$$A := \frac{Z_0}{60\ \Omega} \cdot \sqrt{\frac{\varepsilon_r + 1}{2}} + \frac{\varepsilon_r - 1}{\varepsilon_r + 1} \cdot \left(0.23 + \frac{0.11}{\varepsilon_r} \right) \quad B := \frac{377\ \pi\Omega}{2Z_0 \cdot \sqrt{\varepsilon_r}}$$

$$A := 2.213 \quad B := 3.993 \quad W_{d\_guess} := 1$$

$$W_d := \left| \begin{array}{ll} \dfrac{8 \cdot e^A}{e^{2A} - 2} & \text{if } W_{d\_guess} < 2 \\[12pt] \dfrac{2}{\pi}\left[B - 1 - \ln(2 \cdot B - 1) + \dfrac{\varepsilon_r - 1}{2\varepsilon_r} \cdot \left(\ln(B-1) + 0.39 - \dfrac{0.61}{\varepsilon_r} \right) \right] & \text{if } W_{d\_guess} > 2 \end{array} \right.$$

$$W_d := 0.896 \quad W := W_d \cdot d \quad W = 0.142\ \text{cm}$$

Check calculation

$$\varepsilon_e := \frac{\varepsilon_r + 1}{2} + \frac{\varepsilon_r - 1}{2} \cdot \frac{1}{\sqrt{1 + 12 \cdot \frac{d}{W}}} \quad \varepsilon_e = 1.758$$

$$Z := \left| \begin{array}{ll} \left(\dfrac{60}{\sqrt{\varepsilon_e}} \cdot \ln\left(\dfrac{8d}{W} + \dfrac{W}{4 \cdot d} \right) \right) & \text{if } \dfrac{W}{d} \le 1 \\[18pt] \dfrac{120\pi}{\sqrt{\varepsilon_e} \cdot \left(\dfrac{W}{d} + 1.393 + 0.667 \ln\left(\dfrac{W}{d} + 1.444 \right) \right)} & \text{if } \dfrac{W}{d} > 1 \end{array} \right. \quad Z = 100.174$$

Determine the wavelength

$$v_p := \frac{c}{\sqrt{\varepsilon_e}}$$

$$k := 2\pi f \sqrt{\mu_0 \varepsilon_0 \varepsilon_e} \quad v_p = 2.261 \times 10^8\ \frac{m}{s} \quad \text{speed of wave on } \mu\text{strip}$$

$$k = 111.16\ \frac{1}{m} \quad \text{wave number on } \mu\text{strip}$$

$$\lambda := \frac{2\pi}{k} \quad \lambda = 5.652\ \text{cm}$$

Example 3.10 Microstrip Line Design
Design a microstrip transmission for a 50 Ω characteristic impedance and 90° phase shift. The operation frequency is 2.5 GHz. The substrate thickness is $d = 0.127$ cm with $\varepsilon_r = 2.20$. Validate your results with an electromagnetic simulator.

Solution
Analytical Derivation

1) Given that $Z_o = 50\ \Omega$, calculate W/d. Choose one of the equations for W/d and proceed.

$$\frac{W}{d} = \left\{ \frac{2}{\pi}\left[B - 1 - \ln(2B - 1) + \frac{\varepsilon_r - 1}{2\varepsilon_r}\left\{ \ln(B-1) + 0.39 - \frac{0.61}{\varepsilon_r} \right\} \right] \right\} \text{ for } W/d > 2$$

$$B = \frac{377\pi}{2 * Z_0 \sqrt{\varepsilon r}} = 7.985$$

$$\bar{d} = 3.081$$

2) Calculate W.

$$W = d*(3.081) = 0.391 \text{ cm}$$

3) Calculate l. This can be done by using the phase shift information.

$$\emptyset = \beta l = k_0 \sqrt{\varepsilon_e} l = 90°$$

$$k_0 = \frac{2\pi f}{c} = 52.35 \text{ m}^{-1}$$

$$\varepsilon_e = \frac{2.2 + 1}{2} + \frac{2.2 - 1}{2} \frac{1}{\sqrt{1 + 12(0.3245)}} = 1.871$$

$$l = \frac{90°\left(\pi/180\right)}{\sqrt{1.871}(52.35)} = 2.19 \text{ cm}$$

Verification with Simulation
The tutorial for this example is provided on the resources website.

Ansoft Designer
The three-dimensional (3D) layout of the simulation is given in Figure 3.31.

The characteristic impedance is plotted versus frequency and given in Figure 3.32.

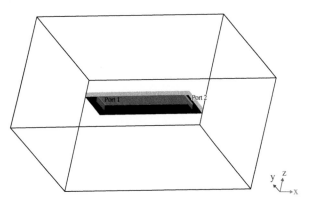

Figure 3.31 Three-dimensional view of the microstrip line created in Ansoft Designer.

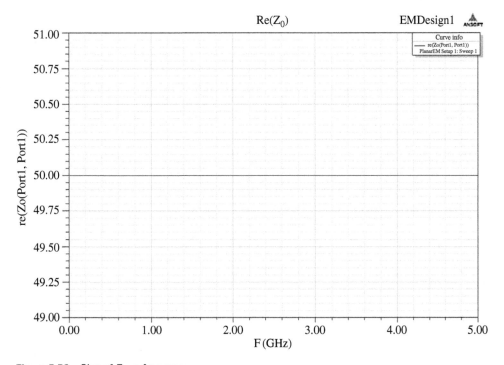

Figure 3.32 Plot of Z_0 vs frequency.

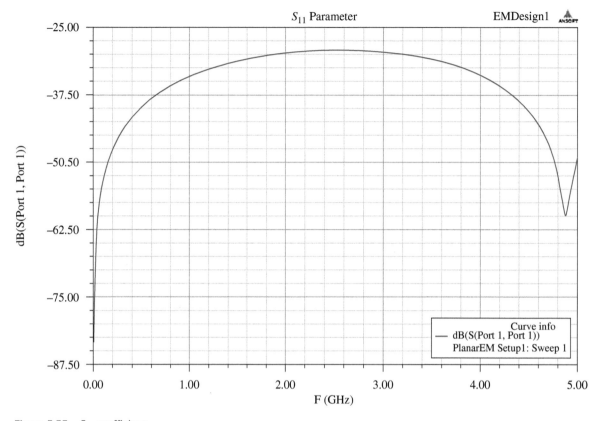

Figure 3.33 S_{11} coefficient.

The return loss is given in Figure 3.33.
The phase shift illustrating the 90° shift at 2.5 GHz is given in Figure 3.34.

Sonnet
The 3D layout of the simulation using Sonnet is given in Figure 3.35.
The characteristic impedance plot versus frequency is given in Figure 3.36.
The return loss is given in Figure 3.37.
The phase shift illustrating the 90° shift at 2.5 GHz is given in Figure 3.38.
Both simulations' results agree with the analytical results.

3.5 Striplines

Striplines are popular transmission lines that have two conductors and a homogeneous dielectric that can support a TEM wave. The stripline structure is able to support also higher order transverse magnetic (TM) and transverse electric (TE) modes. These modes can be avoided in practice with shorting screws and by making the ground plane spacing to less than $\lambda/4$. In addition, although the ends of the stripline structure are open, it can be considered a nonradiating structure. The configuration of the stripline is illustrated in Figure 3.39. One of the advantages of striplines is its broadband characteristics. It is electrically challenging to connect two conductor configurations

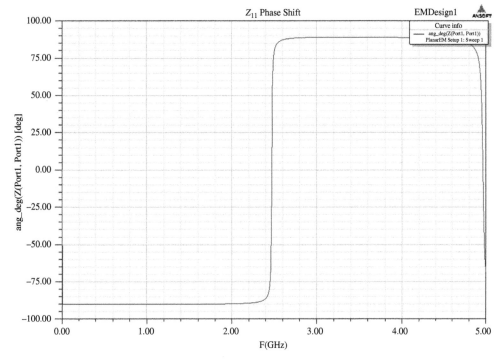

Figure 3.34 Z_{11} parameter plot illustrating a 90° phase shift.

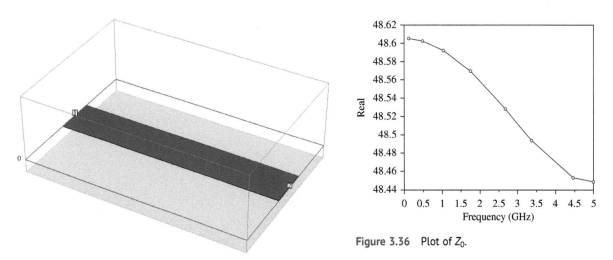

Figure 3.36 Plot of Z_0.

Figure 3.35 Three-dimensional view of the microstrip line created in Sonnet.

within a single structure. In striplines, the fields tend to concentrate next to the conductors, which limits the power handling capability. The field patterns of a stripline structure is shown in Figure 3.40.

Simple approximations developed by Howe [3] have been used for the formulation of striplines. The characteristic impedance is found from

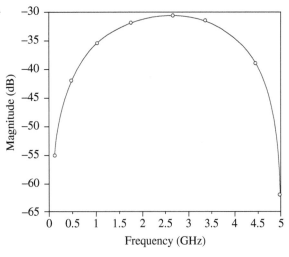

Figure 3.37 Plot of S11 coefficient.

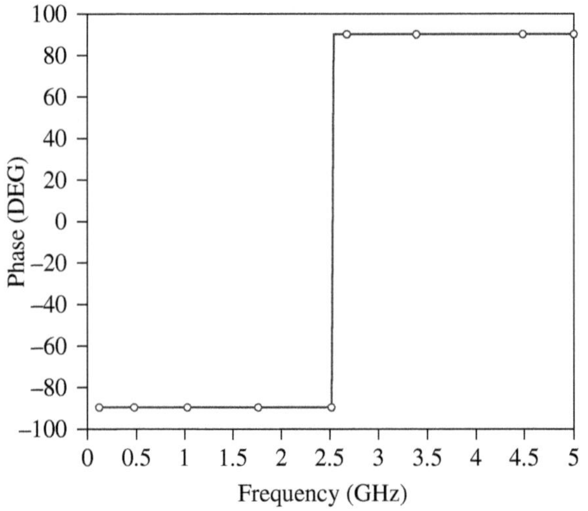

Figure 3.38 Z11 parameter plot illustrating a 90° phase shift.

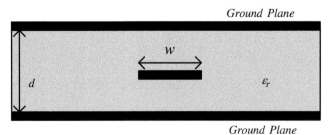

Figure 3.39 Stripline configuration.

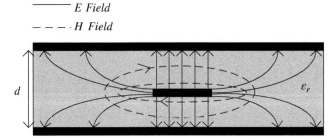

Figure 3.40 Field distribution for stripline configuration.

$$Z_0 = \frac{30\pi}{\sqrt{\varepsilon_r}} \frac{d}{W_e + 0.441d} \tag{3.136}$$

where W_e is the effective width of the center conductor and defined by

$$\frac{W_e}{d} = \frac{W}{d} - \begin{cases} 0 & \text{for } W/d > 0.35 \\ (0.35 - W/d)^2 & \text{for } W/d < 0.35 \end{cases} \tag{3.137}$$

In Eqs. (3.136) and (3.137), it is assumed that the strip thickness is zero. This gives around 1% of accuracy [4]. If the characteristic impedance is given, then the physical dimension of the stripline is found from

$$\frac{W}{d} = \begin{cases} x & \text{for } \sqrt{\varepsilon_r}Z_0 > 120 \\ 0.85 - \sqrt{0.6 - x} & \text{for } \sqrt{\varepsilon_r}Z_0 < 120 \end{cases} \tag{3.138}$$

where

$$x = \frac{30\pi}{\sqrt{\varepsilon_r}Z_0} - 0.441 \tag{3.139}$$

The total loss in the stripline structure is due to conductor and dielectric losses and can be expressed as

$$\alpha = \alpha_c + \alpha_d \tag{3.140}$$

where

$$\alpha_d = \frac{k\tan\delta}{2} \tag{3.141}$$

$$\alpha_c = \begin{cases} \dfrac{2.7 \times 10^{-3} R_s \varepsilon_r Z_0}{30\pi(d-t)} A & \text{for } \sqrt{\varepsilon_r}Z_0 > 120 \\ \dfrac{0.16 R_s}{Z_0 d} B & \text{for } \sqrt{\varepsilon_r}Z_0 < 120 \end{cases} \tag{3.142}$$

In (3.142)

$$A = 1 + \frac{2W}{d-t} + \frac{1}{\pi}\frac{d+t}{d-t}\ln\left(\frac{2d-t}{t}\right) \tag{3.143}$$

$$B = 1 + \frac{d}{(0.5W + 0.7t)}\left(0.5 + \frac{0.414t}{W} + \frac{1}{2\pi}\ln\frac{4\pi W}{t}\right) \tag{3.144}$$

3.6 Engineering Application Examples

In this section, the design of a microstrip device using microstrip transmission line principles is discussed. The specific microstrip device design that is detailed is a microstrip directional coupler. Directional couplers are used to sample signals which can then be used for power measurement. This is accomplished by coupling the signal through the main line, as illustrated in Figure 3.41. In Figure 3.41, the strip width is shown to be w, the spacing between the strips is s. The thickness of the dielectric is h. The microstrip implementation of the directional couplers is attractive due to its low profile, low cost, and ease of implementation.

Microstrip Coupler Design

Design, simulate, and implement a microstrip coupler with -15 dB coupling level at 300 MHz using TMM10 material which has $\varepsilon_r = 9.8$ and thickness of 120 mil for a 50 Ω system.

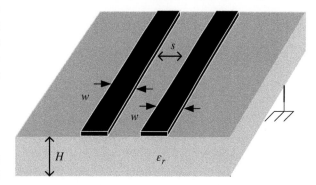

Figure 3.41 Typical microstrip directional coupler configuration.

Solution

The design method for a conventional microstrip coupler that reflects the practice is presented and reported in [1]. The design process in [1] requires only information about coupling level, port impedances, and operational frequency. It eliminates the use of design charts for odd and even mode impedances and provides a practical way of designing microstrip type couplers. The design procedure for the two-line microstrip coupler shown in Figure 3.41 is based on three steps, as outlined in [1]. The first step involves finding the even and odd mode impedances from

$$Z_{0e} = Z_0 \sqrt{\frac{1 + 10^{C/20}}{1 - 10^{C/20}}} \tag{3.145}$$

$$Z_{00} = Z_0 \sqrt{\frac{1 - 10^{C/20}}{1 + 10^{C/20}}} \tag{3.146}$$

which gives us the desired coupling value. In (3.145) and (3.146), C is the desired coupling value in dB. Then, spacing and shape ratios are found from

$$s/h = \frac{2}{\pi} \cosh^{-1} \left[\frac{\cosh\left[\frac{\pi}{2}\left(\frac{w}{h}\right)_{se}\right] + \cosh\left[\frac{\pi}{2}\left(\frac{w}{h}\right)'_{so}\right] - 2}{\cosh\left[\frac{\pi}{2}\left(\frac{w}{h}\right)'_{so}\right] - \cosh\left[\frac{\pi}{2}\left(\frac{w}{h}\right)_{se}\right]} \right] \tag{3.147}$$

and

$$\frac{w}{h} = \frac{8\sqrt{\left[\exp\left(\frac{R}{42.4}\sqrt{(\varepsilon_r + 1)}\right) - 1\right] \frac{7 + (4/\varepsilon_r)}{11} + \frac{1 + (1/\varepsilon_r)}{0.81}}}{\left[\exp\left(\frac{R}{42.4}\sqrt{\varepsilon_r + 1}\right) - 1\right]} \tag{3.148}$$

where

$$R = \frac{Z_{0e}}{2} \text{ or } R = \frac{Z_{00}}{2} \tag{3.149}$$

$(w/h)_{se}$ and $(w/h)_{so}$ are the shape ratios for the equivalent single case corresponding to even-mode and odd-mode geometry, respectively. $(w/h)'_{so}$ is the modified term for the shape ratio. The length of the coupler is then found from

$$l = \frac{\lambda}{4} = \frac{c}{4f\sqrt{\varepsilon_{eff}}} \tag{3.150}$$

where ε_{eff} is the effective permittivity constant of the coupled structure and defined as

$$\varepsilon_{eff} = \left[\frac{\sqrt{\varepsilon_{effe}} + \sqrt{\varepsilon_{effo}}}{2}\right]^2 \tag{3.151}$$

ε_{effe} and ε_{effo} are odd- and even-mode effective permittivity constants. All terms and their related equations are defined in [1]. The MATLAB GUI shown in Figure 3.42 is used to generate the physical dimensions to simulate a directional coupler using the formulations given above. Several other parameters, including spacing and shape ratios versus relative dielectric constants, odd- and even-mode capacitance variation, length, etc., are calculated and illustrated in this GUI.

The physical dimensions illustrated in Figure 3.42 are simulated using Ansoft Designer when the thickness of the substrate is 120 mi. The simulated results show that the coupling at 300 MHz is −17.26 dB, as illustrated in Figure 3.43. The isolation is −27.45 dB.

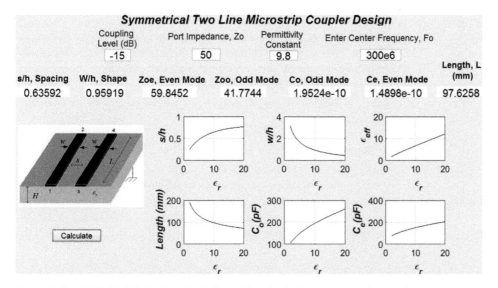

Figure 3.42 MATLAB GUI for the calculation of the physical parameters of a coupler.

Figure 3.43 Simulation results of a microstrip coupler at 300 MHz.

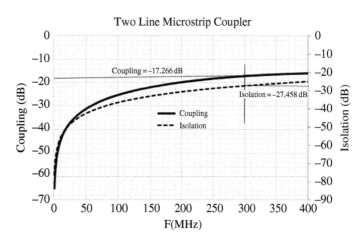

References

1 Eroglu, A. (2013). *RF Circuit Design Techniques for MF-UHF Applications*, 1e. CRC Press. ISBN: 978-1-4398-6165-3.
2 Gupta, K.C., Garg, R., and Bahl, I.J. (1979). *Microstrip Lines and Slotlines*. Norwood, MA: Artech House.
3 Howe, H. (1974). *Stripline Circuit Design*. Burlington, MA: Artech House.
4 Pozar, D. (2004). *Microwave Engineering*, 3e. Hoboken, NJ: Wiley.

Problems

Problem 3.1

The characteristic impedance of a lossy transmission line at 1 GHz is given to be $Z_0 = 100 + j15\,\Omega$. If the propagation constant is $\gamma = 0.2 + j0.15$/m, what are the equivalent lumped element values for R, G, L, and C for this transmission line?

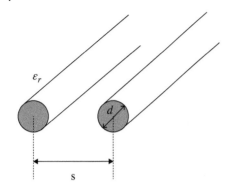

Figure 3.44 Twisted wire configuration.

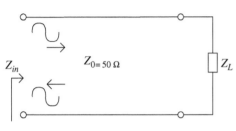

Figure 3.45 Problem 3.4.

Problem 3.2

Design the twisted wire transmission line configuration shown in Figure 3.44 with $30\,\Omega$ characteristic impedance that will be able to handle 6A current flow. Assume that the insulation thickness is 0.21 in. Polytetrafluoroethylene (PTFE) with $\varepsilon_r = 2.25$ and wall thickness $t = 0.045$ in. is used as insulation material to provide high voltage breakdown.

Problem 3.3

It is given that the inner conductor diameter of coaxial cable is 1.5 mm, and the inner diameter of the outer conductor is 4 mm. The dielectric constant of the material is 2.2 and its tangent loss is $\tan \delta = 0.0004$ and conductors are copper. (a) Calculate the R, L, G, and C at 2 GHz. (b) What are the characteristic impedance and attenuation of the line at 2 GHz?

Problem 3.4

The transmission line with characteristic impedance $Z_0 = 50\,\Omega$ shown in Figure 3.45 is connected to a $Z_L = 250 + j50\,\Omega$ Calculate Γ_L, SWR, l_{min}, l_{max}, V_{max}, and V_{min} on the transmission line if the incident voltage is $V^+ = 0.5 \angle 45^\circ \text{V}$.

Problem 3.5

The transmission line is measured at its input terminals for open circuit and short circuit conditions. If the transmission line length is 3 m, and the open circuit and short circuit measured impedance values are $Z_{0c} = 50 \angle 30^\circ$ and $Z_{sc} = 10 \angle 60^\circ$, calculate (a)

Z_0, attenuation constant, α, and phase constant, β, of the transmission line and (b) lumped element equivalent values, R, and L of the transmission line.

Problem 3.6

Locate the following normalized impedances on the Smith Chart, plot SWRs, and calculate reflection coefficients. (a) $z = 0.15 + j1.25$ (b) $z = 1.5 + j0.5$ (c) $z = 0.75 + j1.2$

Problem 3.7

What is the input impedance, Γ_L, and SWR using Smith Chart for a transmission line shown in (Figure 3.46) with characteristic impedance $Z_0 = 50\,\Omega$ length of $l = 0.33\,\lambda$ when it is connected to a load impedance of (a) $Z_L = 50 - j25\,\Omega$ (b) $Z_L = 100 + j75\,\Omega$.

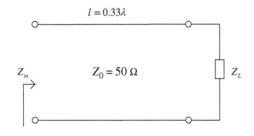

Figure 3.46 Problem 3.7.

Problem 3.8

Consider the lossless transmission line shown in Figure 3.47, which is terminated with $Z_L = 25 + j20$. Find the input impedance of the transmission line if the shunt open stub is placed at $d = 0.25\,\lambda$ away from the load and has a length of $l = 0.15\,\lambda$ (a) analytically, and then (b) confirm your solution using a Smith chart.

Figure 3.47 Problem 3.8.

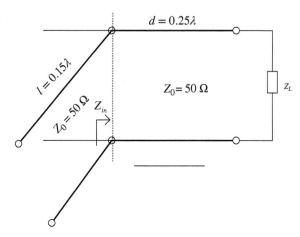

Problem 3.9
Consider the lossless transmission line shown in Figure 3.48, which is terminated with $Z_L = 60 - j80$. Find the input impedance of the transmission line if the series short stub is placed at $d = 0.25\,\lambda$ away from the load and has a length of $l = 0.11\,\lambda$ (a) analytically, and then (b) confirm your solution using a Smith chart.

Problem 3.10
Find the input impedance of the network shown in Figure 3.49 at $f = 1$ GHz (a) analytically and (b) using a Smith chart and obtain the network's frequency response up to 3 GHz.

Figure 3.48 Problem 3.9.

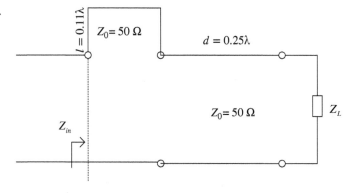

Figure 3.49 Problem 3.11.

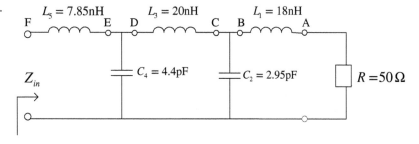

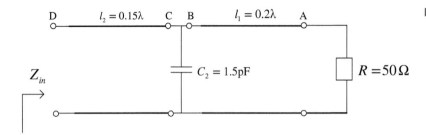

Figure 3.50 Problem 3.12.

Problem 3.11

Find the input impedance of the network which has lossless transmission lines and lumped elements shown in Figure 3.50 at $f = 2\,\text{GHz}$ (a) analytically (b) using a Smith chart and obtain the network's frequency response up to 3 GHz.

Problem 3.12

It is required to design a microstrip transmission line with 50 Ω characteristic impedance using a substrate thickness and dielectric constant 0.2 cm, with $\varepsilon_r = 4.4$ at $f = 2\,\text{GHz}$. (a) What is the width of the microstrip transmission line? (b) What is the guide wavelength, λ_g?

Problem 3.13

Obtain a $3:5$ voltage ratio using transmission line transformers (TLTs). Follow the synthesis technique and use Guanella's $1:1$ transformer as a basic block. Use PSpice to confirm your results.

Problem 3.14

Design a TLT to obtain a $4:1$ impedance ratio using a series type connection. Verify your results using PSpice and obtain voltage and current waveforms. Source voltage, frequency, and impedance are given to be 10 V, 2 MHz, and 50 Ω.

Design Challenge 3.1 Design of Two-wire Transmission Line

Design and build a twisted wire transmission line configuration with 25 Ω characteristic impedance that will be able to handle a 6 A current flow. Assume that the insulation thickness is 0.21 in. PTFE with $\varepsilon_r = 2.25$ and wall thickness $t = 0.045$ in. is used as insulation material to provide high voltage breakdown. Calculate, measure, and confirm the characteristic impedance. Compare your calculated and measured values.

4

Network Parameters

4.1 Introduction

Network parameters are mathematical tools for designers to model and characterize the critical parameters of devices by establishing relations between voltages and currents. Important transistor parameters, such as voltage and current gains, can be obtained with the application of these parameters. They can also be applied in small signal power amplifier design to obtain parameters, such as overall system gain, and loss, and several other responses.

Network parameters are analyzed and studied using two-port networks in [1]. A two-port network is shown in Figure 4.1. It is a set of four independent parameters, which can be related to voltage and current at any ports of the network. As a result a two-port network can be treated as a black box modeled by the relationships between the four variables. There exist six different ways to describe the relationships between these variables, depending on which two of the four variables are given, while the other two can always be derived. All voltages and currents are complex variables and represented by phasors containing both magnitude and phase. Two-port networks are characterized by using two-port network parameters such as Z impedance, Y admittance, h hybrid, and $ABCD$. High frequency (HF) networks are characterized by S parameters. They are usually expressed in matrix notation and establish relations between the following parameters: input voltage V_1, output voltage V_2, input current I_1, and output current I_2.

4.2 Impedance Parameters – Z Parameters

The voltages are represented in terms of currents through Z parameters as follows.

$$V_1 = Z_{11}I_1 + Z_{12}I_2 \tag{4.1}$$

$$V_2 = Z_{21}I_1 + Z_{22}I_2 \tag{4.2}$$

In matrix form Eqs. (4.1) and (4.2) can be combined and written as

$$\begin{bmatrix} V_1 \\ V_2 \end{bmatrix} = \begin{bmatrix} Z_{11} & Z_{12} \\ Z_{21} & Z_{22} \end{bmatrix} \begin{bmatrix} I_1 \\ I_2 \end{bmatrix} \tag{4.3}$$

RF/Microwave Engineering and Applications in Energy Systems, First Edition. Abdullah Eroglu.
© 2022 John Wiley & Sons Ltd. Published 2022 by John Wiley & Sons Ltd.
Companion website: www.wiley.com/go/eroglu/rfmicrowave

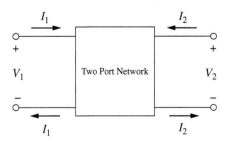

Figure 4.1 Two-port network representation.

The Z parameters for a two-port network cane be defined as

$$Z_{11} = \frac{V_1}{I_1}\bigg|_{I_2 = 0} \qquad Z_{12} = \frac{V_1}{I_2}\bigg|_{I_1 = 0}$$

$$Z_{21} = \frac{V_2}{I_1}\bigg|_{I_2 = 0} \qquad Z_{22} = \frac{V_2}{I_2}\bigg|_{I_1 = 0}$$

(4.4)

The formulation in (4.4) can be generalized for an n port network as

$$Z_{nm} = \frac{V_n}{I_m}\bigg|_{I_k = 0(k \neq m)}$$

(4.5)

Z_{nm} is the input impedance seen looking into port n, when all other ports are open-circuited. In other words, Z_{nm} is the transfer impedance between ports n and m when all other ports are open. It can be shown that for reciprocal networks

$$Z_{nm} = Z_{mn}$$

(4.6)

Example 4.1 Impedance Parameters

Find the impedance parameters of the circuit shown in Figure 4.2.

Solution

Consider the circuit in Figure 4.3 to calculate Z_{11} and Z_{21}. Apply nodal voltage method as illustrated in Figure 4.2.

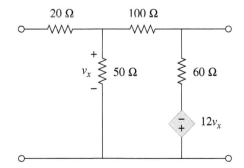

Figure 4.2 Circuit for Example 4.1.

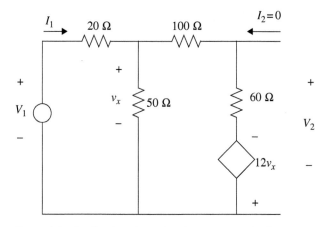

Figure 4.3 Application of nodal voltage method for Z_{11} and Z_{21}.

$$\frac{V_1-V_x}{20} = \frac{V_x}{50} + \frac{V_x+12V_x}{160} \longrightarrow V_x = \frac{40}{121}V_1$$

$$I_1 = \frac{V_1-V_x}{20} = \frac{81}{121}\left(\frac{V_1}{20}\right) \longrightarrow z_{11} = \frac{V_1}{I_1} = 29.88$$

$$V_2 = 60\left(\frac{13V_x}{160}\right) - 12V_x = -\frac{57}{8}V_x = -\frac{57}{8}\left(\frac{40}{121}\right)V_1 = -\frac{57}{8}\left(\frac{40}{121}\right)\frac{20x121}{81}I_1$$

$$= -70.37I_1 \longrightarrow z_{21} = \frac{V_2}{I_1} = -70.37$$

Z_{12} and Z_{22} can be calculated using Figure 4.4 as follows.

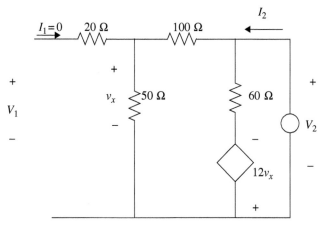

Figure 4.4 Application of nodal voltage method for Z_{11} and Z_{22}

$$V_x = \frac{50}{100+50}V_2 = \frac{1}{3}V_2, \quad I_2 = \frac{V_2}{150} + \frac{V_2+12V_x}{60} = 0.09V_2$$

$$z_{22} = \frac{V_2}{I_2} = 1/0.09 = 11.11$$

$$V_1 = V_x = \frac{1}{3}V_2 = \frac{11.11}{3}I_2 = 3.704I_2 \quad \longrightarrow \quad z_{12} = \frac{V_1}{I_2} = 3.704$$

Thus,

$$[z] = \begin{bmatrix} 29.88 & 3.704 \\ -70.37 & 11.11 \end{bmatrix} \Omega$$

4.3 Y Admittance Parameters

The currents are related to voltages through Y parameters as follows.

$$I_1 = Y_{11}V_1 + Y_{12}V_2 \tag{4.7}$$

$$I_2 = Y_{21}V_1 + Y_{22}V_2 \tag{4.8}$$

In matrix form, (4.7) and (4.8) can be written as

$$\begin{bmatrix} I_1 \\ I_2 \end{bmatrix} = \begin{bmatrix} Y_{11} & Y_{12} \\ Y_{21} & Y_{22} \end{bmatrix} \begin{bmatrix} V_1 \\ V_2 \end{bmatrix} \tag{4.9}$$

The Y parameters in (4.9) can be defined as

$$Y_{11} = \frac{I_1}{V_1}\bigg|_{V_2=0} \qquad Y_{12} = \frac{I_1}{V_2}\bigg|_{V_1=0}$$
$$Y_{21} = \frac{I_2}{V_1}\bigg|_{V_2=0} \qquad Y_{22} = \frac{I_2}{V_2}\bigg|_{V_1=0} \tag{4.10}$$

The formulation in (4.10) can be generalized for an n port network as

$$Y_{nm} = \frac{I_n}{V_m}\bigg|_{V_k=0(k\neq m)} \tag{4.11}$$

Y_{nm} is the input impedance seen looking into port n, when all other ports are short-circuited. In other words, Y_{nm} is the transfer admittance between ports n and m when all other ports are short. It can be shown that

$$Y_{nm} = Y_{mn} \tag{4.12}$$

In addition, it can be further proved that the impedance and admittance matrices are related through

$$[Z] = [Y]^{-1} \tag{4.13}$$

or

$$[Y] = [Z]^{-1} \tag{4.14}$$

4.4 *ABCD* Parameters

ABCD parameters relate the voltages to current in the following form of two-port networks.

$$V_1 = AV_1 - BI_2 \tag{4.15}$$

$$I_1 = CV_1 - DI_2 \tag{4.16}$$

which can be put in matrix form as

$$\begin{bmatrix} V_1 \\ I_1 \end{bmatrix} = \begin{bmatrix} A & B \\ C & D \end{bmatrix} \begin{bmatrix} V_1 \\ -I_2 \end{bmatrix} \tag{4.17}$$

ABCD parameters in (4.17) are defined as

$$A = \left.\frac{V_1}{V_2}\right|_{I_2=0} \quad B = \left.\frac{V_1}{-I_2}\right|_{V_2=0}$$
$$C = \left.\frac{I_1}{V_2}\right|_{I_2=0} \quad D = \left.\frac{I_1}{-I_2}\right|_{V_2=0} \tag{4.18}$$

When a network is reciprocal, it can be shown that

$$AD - BC = 1 \tag{4.19}$$

for a reciprocal network and $A = D$ for a symmetrical network. An *ABCD* network are useful in finding voltage or current gain of component or overall gain of a network. One of the great advantages of *ABCD* parameters is their use in cascaded networks or components. When this condition exists, the overall *ABCD* parameter of the network simply becomes the matrix product of individual network and component. This can be generalized for the N port network shown in Figure 4.5 as

$$\begin{Bmatrix} v_1 \\ i_1 \end{Bmatrix} = \left(\begin{bmatrix} A_1 & B_1 \\ C_1 & D_1 \end{bmatrix} \cdots\cdots \begin{bmatrix} A_n & B_n \\ C_n & D_n \end{bmatrix} \right) \begin{Bmatrix} v_2 \\ -i_2 \end{Bmatrix} \tag{4.20}$$

4.5 *h* Hybrid Parameters

Hybrid parameters relate voltage and current in a two-port network as

$$V_1 = h_{11}I_1 + h_{12}V_2 \tag{4.21}$$

$$I_2 = h_{21}I_1 + h_{22}V_2 \tag{4.22}$$

(4.21) and (4.22) can be put in matrix form as

$$\begin{bmatrix} V_1 \\ I_2 \end{bmatrix} = \begin{bmatrix} h_{11} & h_{12} \\ h_{21} & h_{22} \end{bmatrix} \begin{bmatrix} I_1 \\ V_2 \end{bmatrix} \tag{4.23}$$

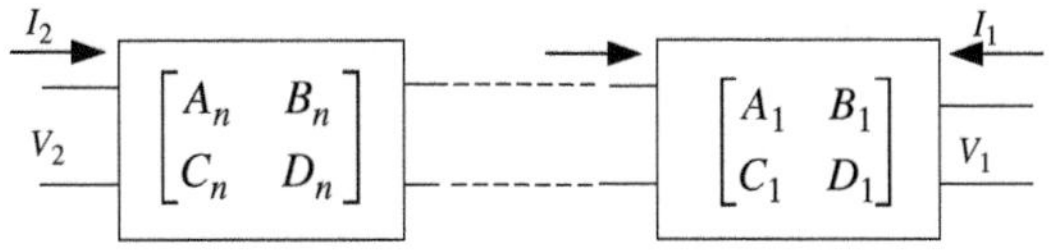

Figure 4.5 *ABCD* parameter of cascaded networks.

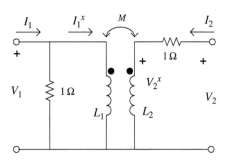

Figure 4.6 Coupling transformer example.

The hybrid parameters in (4.23) can be found from

$$h_{11} = \frac{V_1}{I_1}\bigg|_{V_2 = 0} \qquad h_{12} = \frac{V_1}{V_2}\bigg|_{I_1 = 0}$$

$$h_{21} = \frac{I_2}{I_1}\bigg|_{V_2 = 0} \qquad h_{22} = \frac{I_2}{V_2}\bigg|_{I_1 = 0}$$
(4.24)

Hybrid parameters are preferred for components such as transistors and transformers since they can be measured with ease in practice.

Example 4.2   Hybrid Parameters
Obtain the h parameter of the circuit shown in Figure 4.6 if $L_1 = L_2 = M = 1$ H.

Solution
There are two methods to solve this example.

- **First method**. From KVL on the primary side

$$V_1 = sL_1I_1^x + sMI_2$$
(4.25)

Application of KCL gives

$$I_1^x = I_1 - V_1$$
(4.26)

Substitution of I_1^x into the above equation gives

$$(1 + sL_1)V_1 - sMI_2 = sL_1I_1$$
(4.27)

From KVL on the secondary side

$$V_2 = V_2^x + I_2$$
(4.28)

Also

$$V_2^x = sL_2I_2 + sMI_1^x$$
(4.29)

Substitution of I_1^x into the above leads to

$$V_2^x = sL_2I_2 + sM(I_1 - V_1)$$
(4.30)

When V_2^x is inserted in V_2, we obtain

$$V_2 = (1 + sL_2)I_2 + sM(I_1 - V_1)$$
(4.31)

Figure 4.7 Conversion of transformer coupling circuit to equivalent circuit.

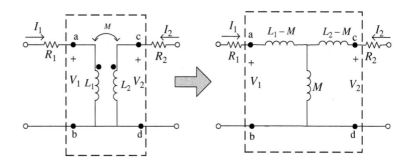

or

$$sMV_1 - (1 + sL_2)I_2 = sMI_1 - V_2 \qquad (4.32)$$

Equations (4.27) and (4.28) can be shown in matrix form as

$$\begin{bmatrix} 1 + sL_1 & -sM \\ sM & -(1 + sL_2) \end{bmatrix} \begin{bmatrix} V_1 \\ I_2 \end{bmatrix} = \begin{bmatrix} sL_1 & 0 \\ sM & -1 \end{bmatrix} \begin{bmatrix} I_1 \\ V_2 \end{bmatrix} \qquad (4.33)$$

or

$$\begin{bmatrix} V_1 \\ I_2 \end{bmatrix} = \begin{bmatrix} 1 + sL_1 & -sM \\ sM & -(1 + sL_2) \end{bmatrix}^{-1} \begin{bmatrix} sL_1 & 0 \\ sM & -1 \end{bmatrix} \begin{bmatrix} I_1 \\ V_2 \end{bmatrix} \qquad (4.34)$$

When $L_1 = L_2 = M = 1$ H, Eq. (4.34) becomes

$$\begin{bmatrix} V_1 \\ I_2 \end{bmatrix} = \begin{bmatrix} 1 + s & -s \\ s & -(1 + s) \end{bmatrix}^{-1} \begin{bmatrix} s & 0 \\ s & -1 \end{bmatrix} \begin{bmatrix} I_1 \\ V_2 \end{bmatrix} \qquad (4.35)$$

or

$$\begin{bmatrix} V_1 \\ I_2 \end{bmatrix} = \frac{1}{(2s + 1)} \begin{bmatrix} s & s \\ -s & s + 1 \end{bmatrix} \begin{bmatrix} I_1 \\ V_2 \end{bmatrix} \qquad (4.36)$$

Hence

$$[h] = \begin{bmatrix} h_{11} & h_{12} \\ h_{21} & h_{22} \end{bmatrix} = \begin{bmatrix} \dfrac{s}{2s + 1} & \dfrac{s}{2s + 1} \\ -\dfrac{s}{2s + 1} & \dfrac{s + 1}{2s + 1} \end{bmatrix} \qquad (4.37)$$

- **Second method**. The mutual inductance equivalent circuit can be converted into an equivalent circuit with the transformation parameters shown in Figure 4.7.

  So, the original circuit can then be translated to one shown in (Figure 4.8).

  ABCD parameter of Network 1, N_1, is

$$ABCD_{N1} = \begin{bmatrix} 1 & 0 \\ Y & 1 \end{bmatrix} = \begin{bmatrix} 1 & 0 \\ 1 & 1 \end{bmatrix} \qquad (4.38)$$

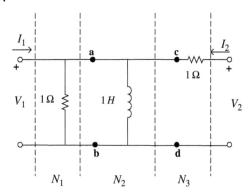

Figure 4.8 Transformer coupling circuit equivalent circuit.

ABCD parameter of Network 2, N_2, is

$$ABCD_{N2} = \begin{bmatrix} 1 & 0 \\ Y & 1 \end{bmatrix} = \begin{bmatrix} 1 & 0 \\ \dfrac{1}{j\omega} & 1 \end{bmatrix} \tag{4.39}$$

ABCD parameter of Network 3, N_3, is

$$ABCD_{N3} = \begin{bmatrix} 1 & Z \\ 0 & 1 \end{bmatrix} = \begin{bmatrix} 1 & 1 \\ 0 & 1 \end{bmatrix} \tag{4.40}$$

ABCD of overall network is

$$ABCD = (ABCD_{N1})(ABCD_{N2})(ABCD_{N3}) = \begin{bmatrix} 1 & 0 \\ 1 & 1 \end{bmatrix}\begin{bmatrix} 1 & 0 \\ \dfrac{1}{j\omega} & 1 \end{bmatrix}\begin{bmatrix} 1 & 1 \\ 0 & 1 \end{bmatrix} = \begin{bmatrix} 1 & 1 \\ 1 + \dfrac{1}{j\omega} & 2 + \dfrac{1}{j\omega} \end{bmatrix} \tag{4.41}$$

So, the hybrid parameters are

$$h = \begin{bmatrix} \dfrac{B}{D} & \dfrac{\Delta}{D} \\ -\dfrac{1}{D} & \dfrac{C}{D} \end{bmatrix} = \begin{bmatrix} \dfrac{j\omega}{2j\omega + 1} & \dfrac{j\omega}{2j\omega + 1} \\ -\dfrac{j\omega}{2j\omega + 1} & \dfrac{j\omega + 1}{2j\omega + 1} \end{bmatrix} \tag{4.42}$$

where $\Delta = 1$. The same result is obtained.

Example 4.3   *T* Network Parameters

Find (a) impedance, (b) admittance, (c) *ABCD*, and (d) hybrid parameters of the *T* network given in Figure 4.9.

Solution

a) *Z* parameters are found with an application of (4.4) by opening all the other ports except the measurement port. This leads to

$$Z_{11} = \frac{V_1}{I_1}\bigg|_{I_2 = 0} = Z_A + Z_C, Z_{21} = \frac{V_2}{I_1}\bigg|_{I_2 = 0} = Z_C$$

Figure 4.9 T network configuration.

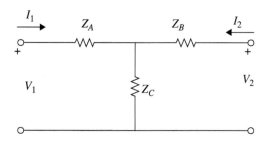

$$Z_{12} = \frac{V_1}{I_2}\bigg|_{I_1 = 0} = \frac{V_2}{I_2}\frac{Z_C}{Z_B + Z_C} = (Z_B + Z_C)\frac{Z_C}{Z_B + Z_C} = Z_C, Z_{22} = \frac{V_2}{I_2}\bigg|_{I_1 = 0} = Z_B + Z_C$$

The Z matrix is then constructed as

$$Z = \begin{bmatrix} Z_A + Z_C & (Z_B + Z_C)\dfrac{Z_C}{Z_B + Z_C} = Z_C \\ (Z_B + Z_C)\dfrac{Z_C}{Z_B + Z_C} = Z_C & Z_B + Z_C \end{bmatrix}$$

b) Y parameters are found from (4.10) by shorting all the other ports except the measurement port. Y_{11} and Y_{21} are found when port 2 is shorted as

$$Y_{11} = \frac{I_1}{V_1}\bigg|_{V_2 = 0} \rightarrow I_1 = \frac{V_1}{Z_A + (Z_B//Z_C)} = V_1\left(\frac{Z_B + Z_C}{Z_AZ_B + Z_AZ_C + Z_BZ_C}\right) \rightarrow Y_{11} = \left(\frac{Z_B + Z_C}{Z_AZ_B + Z_AZ_C + Z_BZ_C}\right)$$

$$Y_{21} = \frac{I_2}{V_1}\bigg|_{V_2 = 0} \rightarrow I_2 = \frac{-V_1}{(Z_A + (Z_B//Z_C))}\frac{Z_C}{(Z_C + Z_B)} \rightarrow Y_{21} = \left(\frac{-Z_C}{Z_AZ_B + Z_AZ_C + Z_BZ_C}\right)$$

Similarly, Y_{12} and Y_{22} are found when port 1 is shorted as

$$Y_{12} = \frac{I_1}{V_2}\bigg|_{V_1 = 0} \rightarrow I_1 = \frac{-V_2}{(Z_B + (Z_A//Z_C))}\frac{Z_C}{(Z_A + Z_C)} \rightarrow Y_{12} = \left(\frac{-Z_C}{Z_AZ_B + Z_AZ_C + Z_BZ_C}\right)$$

$$Y_{22} = \frac{I_2}{V_2}\bigg|_{V_1 = 0} \rightarrow I_2 = \frac{V_2}{Z_B + (Z_A//Z_C)} = V_1\left(\frac{Z_A + Z_C}{Z_AZ_B + Z_AZ_C + Z_BZ_C}\right) \rightarrow Y_{22} = \left(\frac{Z_A + Z_C}{Z_AZ_B + Z_AZ_C + Z_BZ_C}\right)$$

Y parameters can also be found by simply inverting the Z matrix as given by Eq. (4.14) as

$$[Y] = [Z]^{-1} = \frac{1}{(Z_AZ_B + Z_AZ_C + Z_BZ_C)}\begin{bmatrix} Z_B + Z_C & -Z_C \\ -Z_C & Z_A + Z_C \end{bmatrix}$$

So, the Y matrix for the T network is then

$$Y = \begin{bmatrix} \left(\dfrac{Z_B + Z_C}{Z_AZ_B + Z_AZ_C + Z_BZ_C}\right) & \left(\dfrac{-Z_C}{Z_AZ_B + Z_AZ_C + Z_BZ_C}\right) \\ \left(\dfrac{-Z_C}{Z_AZ_B + Z_AZ_C + Z_BZ_C}\right) & \left(\dfrac{Z_A + Z_C}{Z_AZ_B + Z_AZ_C + Z_BZ_C}\right) \end{bmatrix}$$

As seen from the results of part (a) and (b), the network is reciprocal since

$$Z_{12} = Z_{21} \text{ and } Y_{12} = Y_{21}$$

c) Hybrid parameters are found using Eq. (4.24). Parameters h_{11} and h_{21} are obtained when port 2 is shorted as

$$h_{11} = \frac{V_1}{I_1}\bigg|_{V_2 = 0} \rightarrow V_1 = I_1(Z_A + (Z_B//Z_C)) = I_1 \left(\frac{Z_A Z_B + Z_A Z_C + Z_B Z_C}{Z_B + Z_C} \right) \rightarrow h_{11} = \left(\frac{Z_A Z_B + Z_A Z_C + Z_B Z_C}{Z_B + Z_C} \right)$$

and

$$h_{21} = \frac{I_2}{I_1}\bigg|_{V_2 = 0} \rightarrow I_2 = -I_1 \left(\frac{Z_C}{Z_B + Z_C} \right) \rightarrow h_{21} = - \left(\frac{Z_C}{Z_B + Z_C} \right)$$

Parameters h_{12} and h_{22} are obtained when port 1 is open-circuited as

$$h_{12} = \frac{V_1}{V_1}\bigg|_{I_1 = 0} \rightarrow V_1 = V_2 \left(\frac{Z_C}{Z_B + Z_C} \right) \rightarrow h_{12} = \left(\frac{Z_C}{Z_B + Z_C} \right)$$

and

$$h_{22} = \frac{I_2}{V_2}\bigg|_{I_1 = 0} \rightarrow I_2 = V_2 \left(\frac{1}{Z_B + Z_C} \right) \rightarrow h_{22} = \left(\frac{1}{Z_B + Z_C} \right)$$

The hybrid matrix for the T network can now be constructed as

$$h = \begin{bmatrix} \left(\dfrac{Z_A Z_B + Z_A Z_C + Z_B Z_C}{Z_B + Z_C} \right) & \left(\dfrac{Z_C}{Z_B + Z_C} \right) \\ - \left(\dfrac{Z_C}{Z_B + Z_C} \right) & \left(\dfrac{1}{Z_B + Z_C} \right) \end{bmatrix}$$

d) *ABCD* parameters are found using Eqs. (4.1)–(4.18). Parameters A and C are determined when port 2 is open-circuited as

$$A = \frac{V_1}{V_2}\bigg|_{I_2 = 0} \rightarrow V_2 = \frac{Z_C}{Z_C + Z_A} V_1 \rightarrow A = \left(\frac{Z_C + Z_A}{Z_C} \right)$$

and

$$C = \frac{I_1}{V_2}\bigg|_{I_2 = 0} \rightarrow I_1 = V_2 \left(\frac{1}{Z_C} \right) \rightarrow C = \left(\frac{1}{Z_C} \right)$$

Parameters B and D are determined when port 2 is short-circuited as

$$B = \frac{V_1}{-I_2}\bigg|_{V_2 = 0} \rightarrow I_2 = \frac{-V_1}{Z_A + (Z_B//Z_C)} \frac{Z_C}{(Z_B + Z_C)} \rightarrow B = \left(\frac{Z_A Z_B + Z_A Z_C + Z_B Z_C}{Z_C} \right)$$

and

$$D = \frac{-I_1}{I_2}\bigg|_{V_2 = 0} \rightarrow I_2 = -I_1 \left(\frac{Z_C}{Z_B + Z_C} \right) \rightarrow D = \left(\frac{Z_B + Z_C}{Z_C} \right)$$

So, the *ABCD* matrix is

$$ABCD = \begin{bmatrix} \left(\dfrac{Z_C + Z_A}{Z_C} \right) & \left(\dfrac{Z_A Z_B + Z_A Z_C + Z_B Z_C}{Z_C} \right) \\ \left(\dfrac{1}{Z_C} \right) & \left(\dfrac{Z_B + Z_C}{Z_C} \right) \end{bmatrix}$$

It can be proven that the *Z, Y, h,* and *ABCD* parameters are related using the relations given in Table 4.1.

Table 4.1 Network parameter conversion table.

	[Z]	**[Y]**	**[ABCD]**	**[h]**
[Z]	$\begin{bmatrix} Z_{11} & Z_{12} \\ Z_{21} & Z_{22} \end{bmatrix}$	$\begin{bmatrix} \dfrac{Y_{22}}{\Delta Y} & -\dfrac{Y_{12}}{\Delta Y} \\ -\dfrac{Y_{21}}{\Delta Y} & \dfrac{Y_{11}}{\Delta Y} \end{bmatrix}$	$\begin{bmatrix} \dfrac{A}{C} & -\dfrac{\Delta_{ABCD}}{C} \\ \dfrac{1}{C} & \dfrac{D}{C} \end{bmatrix}$	$\begin{bmatrix} \dfrac{\Delta_h}{h_{22}} & \dfrac{h_{12}}{h_{22}} \\ -\dfrac{h_{21}}{h_{22}} & \dfrac{1}{h_{22}} \end{bmatrix}$
[Y]	$\begin{bmatrix} \dfrac{Z_{22}}{\Delta Z} & -\dfrac{Z_{12}}{\Delta Z} \\ -\dfrac{Z_{21}}{\Delta Z} & \dfrac{Z_{11}}{\Delta Z} \end{bmatrix}$	$\begin{bmatrix} Y_{11} & Y_{12} \\ Y_{21} & Y_{22} \end{bmatrix}$	$\begin{bmatrix} \dfrac{D}{B} & -\dfrac{\Delta_{ABCD}}{B} \\ -\dfrac{1}{B} & \dfrac{A}{B} \end{bmatrix}$	$\begin{bmatrix} \dfrac{1}{h_{11}} & -\dfrac{h_{12}}{h_{11}} \\ \dfrac{h_{21}}{h_{11}} & \dfrac{\Delta_h}{h_{11}} \end{bmatrix}$
[ABCD]	$\begin{bmatrix} \dfrac{Z_{11}}{Z_{21}} & \dfrac{\Delta Z}{Z_{21}} \\ \dfrac{1}{Z_{21}} & \dfrac{Z_{22}}{Z_{21}} \end{bmatrix}$	$\begin{bmatrix} -\dfrac{Y_{22}}{Y_{21}} & -\dfrac{1}{Y_{21}} \\ -\dfrac{\Delta Y}{Y_{21}} & -\dfrac{Y_{11}}{Y_{21}} \end{bmatrix}$	$\begin{bmatrix} A & B \\ C & D \end{bmatrix}$	$\begin{bmatrix} -\dfrac{\Delta_h}{h_{21}} & -\dfrac{h_{11}}{h_{21}} \\ -\dfrac{h_{22}}{h_{21}} & -\dfrac{1}{h_{21}} \end{bmatrix}$
[h]	$\begin{bmatrix} \dfrac{\Delta_Z}{Z_{22}} & \dfrac{Z_{12}}{Z_{22}} \\ -\dfrac{Z_{21}}{Z_{22}} & \dfrac{1}{Z_{22}} \end{bmatrix}$	$\begin{bmatrix} \dfrac{1}{Y_{11}} & -\dfrac{Y_{12}}{Y_{11}} \\ \dfrac{Y_{21}}{Y_{11}} & \dfrac{\Delta_Y}{Y_{11}} \end{bmatrix}$	$\begin{bmatrix} \dfrac{B}{D} & \dfrac{\Delta_{ABCD}}{D} \\ -\dfrac{1}{D} & \dfrac{C}{D} \end{bmatrix}$	$\begin{bmatrix} h_{11} & h_{12} \\ h_{21} & h_{22} \end{bmatrix}$

4.6 Network Connections

Networks and components in engineering applications can be connected in different ways to perform certain tasks. Commonly used network connection methods are series, parallel, and cascade connections. A series connection of two networks is shown in Figure 4.10. Since the network is connected in series, currents are the same and voltages are added across the network ports to find the overall voltage at the ports of the combined network. This can be represented by impedance matrices as

$$[Z] = [Z^x] + [Z^y] = \begin{bmatrix} Z_{11}^x & Z_{12}^x \\ Z_{21}^x & Z_{22}^x \end{bmatrix} + \begin{bmatrix} Z_{11}^y & Z_{12}^y \\ Z_{21}^y & Z_{22}^y \end{bmatrix} = \begin{bmatrix} Z_{11}^x + Z_{11}^y & Z_{12}^x + Z_{12}^y \\ Z_{21}^x + Z_{21}^y & Z_{22}^x + Z_{22}^y \end{bmatrix} \tag{4.43}$$

So

$$\begin{bmatrix} V_1 \\ V_2 \end{bmatrix} = \begin{bmatrix} Z_{11}^x + Z_{11}^y & Z_{12}^x + Z_{12}^y \\ Z_{21}^x + Z_{21}^y & Z_{22}^x + Z_{22}^y \end{bmatrix} \begin{bmatrix} I_1 \\ I_2 \end{bmatrix} \tag{4.44}$$

Figure 4.10 Series connection of two-port networks.

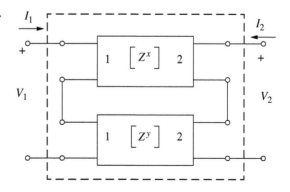

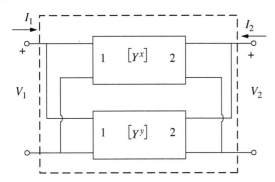

Figure 4.11 Parallel connection of two-port networks.

The parallel connection of a two-port network is illustrated in Figure 4.11. In parallel connected networks, voltages are the same across ports and currents are added to find the overall current flowing at the ports of the combined network. This can be represented by Y matrices as

$$[Y] = [Y^x] + [Y^y] = \begin{bmatrix} Y_{11}^x & Y_{12}^x \\ Y_{21}^x & Y_{22}^x \end{bmatrix} + \begin{bmatrix} Y_{11}^y & Y_{12}^y \\ Z_{21}^y & Y_{22}^y \end{bmatrix} = \begin{bmatrix} Y_{11}^x + Y_{11}^y & Y_{12}^x + Y_{12}^y \\ Y_{21}^x + Y_{21}^y & Y_{22}^x + Y_{22}^y \end{bmatrix} \tag{4.45}$$

As a result

$$\begin{bmatrix} I_1 \\ I_2 \end{bmatrix} = \begin{bmatrix} Y_{11}^x + Y_{11}^y & Y_{12}^x + Y_{12}^y \\ Y_{21}^x + Y_{21}^y & Y_{22}^x + Y_{22}^y \end{bmatrix} \begin{bmatrix} V_1 \\ V_2 \end{bmatrix} \tag{4.46}$$

The cascade connection of a two-port network is shown in Figure 4.12. In cascade connection, the magnitude of the current flowing at the output of the first network is equal to the current at the input port of the second network. The voltages at the output of the first network are also equal to the voltage across the input of the second network. This can be represented by using $ABCD$ matrices as

$$[ABCD] = [ABCD^x][ABCD^y] = \begin{bmatrix} A^x & B^x \\ C^x & D^x \end{bmatrix} \begin{bmatrix} A^y & B^y \\ C^y & D^y \end{bmatrix} = \begin{bmatrix} A^x A^y + B^x C^y & A^x B^y + B^x B^y \\ C^x A^y + D^x C^y & C^x B^y + D^x D^y \end{bmatrix} \tag{4.47}$$

Example 4.4   Network Parameters of RF Amplifier

Consider the radiofrequency (RF) amplifier given in Figure 4.12. It has a feedback network for stability, and input and output matching networks. The transistor used is an NPN bipolar junction transistor (BJT) and its characteristic parameters are given by $r_{BE} = 400\,\Omega$, $r_{CE} = 70\,\text{k}\Omega$, $C_{BE} = 15\,\text{pF}$, and $C_{BC} = 2\,\text{pF}$, and $g_m = 0.2\,\text{S}$. Find the voltage and current gain of this amplifier when $L = 2\,\text{nH}$, $C = 12\,\text{pF}$, $l = 5\,\text{cm}$, $v_p = 0.65\,c$.

Solution

The HF characteristics of the transistor are modeled using the hybrid parameters given by

$$h_{11} = h_{ie} = \frac{r_{BE}}{1 + j\omega(C_{BE} + C_{BC})r_{BE}} \tag{4.48}$$

$$h_{12} = h_{re} = \frac{j\omega C_{BC} r_{BE}}{1 + j\omega(C_{BE} + C_{BC})r_{BE}} \tag{4.49}$$

$$h_{21} = h_{fe} = \frac{r_{BE}(g_m - j\omega C_{BC})}{1 + j\omega(C_{BE} + C_{BC})r_{BE}} \tag{4.50}$$

$$h_{22} = h_{oe} = \frac{[1 + j\omega(C_{BE} + C_{BC})r_{BE}] + [(1 + r_{BE}g_m + j\omega C_{BE}r_{BE})]r_{CE}}{1 + j\omega(C_{BE} + C_{BC})r_{BE}} \tag{4.51}$$

Figure 4.12 Cascade connection of two-port networks.

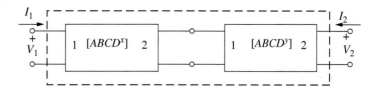

Figure 4.13 RF amplifier analysis by network parameters.

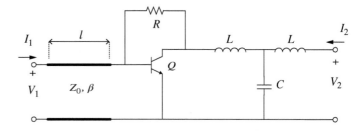

The amplifier network shown in Figure 4.12 is a combination of four networks that are connected in parallel and cascade. The overall network must first be partitioned. This is illustrated in Figure 4.13.

In the partitioned amplifier circuit, networks N_2 and N_3, are connected in parallel, as shown in Figure 4.14. Then, the parallel connected network, Y, can be represented by admittance matrix. The admittance matrix of Network 3 is

$$Y^y = \begin{bmatrix} \dfrac{1}{R} & -\dfrac{1}{R} \\ -\dfrac{1}{R} & \dfrac{1}{R} \end{bmatrix} \tag{4.52}$$

The admittance matrix for the transistor can be obtained by using the conversion table given in Table 4.1 since hybrid parameters for it are available. This can be done by using

$$Y^x = \begin{bmatrix} \dfrac{1}{h_{11}} & -\dfrac{h_{12}}{h_{11}} \\ \dfrac{h_{21}}{h_{11}} & \dfrac{\Delta h}{h_{11}} \end{bmatrix} \tag{4.53}$$

Figure 4.14 Partition of amplifier circuit for network analysis.

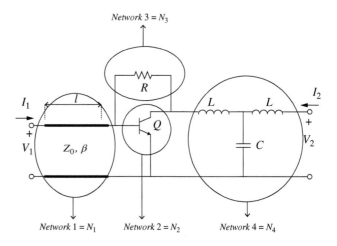

Then, the overall admittance matrix is found as

$$[Y] = [Y^x] + [Y^y] = \begin{bmatrix} \dfrac{1}{R} + \dfrac{1}{h_{11}} & -\dfrac{1}{R} - \dfrac{h_{12}}{h_{11}} \\ -\dfrac{1}{R} + \dfrac{h_{21}}{h_{11}} & \dfrac{1}{R} + \dfrac{\Delta h}{h_{11}} \end{bmatrix} \tag{4.54}$$

where Δ is for the determinant of the corresponding matrix. At this point, it is now clearer that networks 1, Y, and 4 are cascaded. We need to determine the $ABCD$ matrix of each network in this connection, as shown in Figure 4.15. The first step is then to convert the admittance matrix in (4.54) to an $ABCD$ parameter using the conversion table. The conversion table gives the relation as

$$ABCD^Y = \begin{bmatrix} \dfrac{Y_{22}}{Y_{21}} & -\dfrac{1}{Y_{21}} \\ \dfrac{\Delta Y}{Y_{21}} & \dfrac{Y_{11}}{Y_{21}} \end{bmatrix} \tag{4.55}$$

The $ABCD$ matrices for Network 1 and 4 are obtained as

$$ABCD^{N_1} = \begin{bmatrix} \cos(\beta l) & jZ_0 \sin(\beta l) \\ \dfrac{j \sin(\beta l)}{Z_0} & \cos(\beta l) \end{bmatrix} \tag{4.56}$$

$$ABCD^{N_4} = \begin{bmatrix} 1 - \omega^2 LC & j\omega L(2 - \omega^2 LC) \\ j\omega C & 1 - \omega^2 LC \end{bmatrix} \tag{4.57}$$

$ABCD$ parameter of the combined network shown in Figure 4.16 is

$$ABCD = ABCD^{N_1}\left(ABCD^Y\right)ABCD^{N_4} \tag{4.58}$$

$$ABCD = \begin{bmatrix} \cos(\beta l) & jZ_0 \sin(\beta l) \\ \dfrac{j \sin(\beta l)}{Z_0} & \cos(\beta l) \end{bmatrix} \begin{bmatrix} \dfrac{Y_{22}}{Y_{21}} & -\dfrac{1}{Y_{21}} \\ \dfrac{\Delta Y}{Y_{21}} & \dfrac{Y_{11}}{Y_{21}} \end{bmatrix} \begin{bmatrix} 1 - \omega^2 LC & j\omega L(2 - \omega^2 LC) \\ j\omega C & 1 - \omega^2 LC \end{bmatrix} \tag{4.59}$$

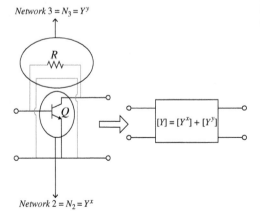

Network 3 = $N_3 = Y^y$

Network 2 = $N_2 = Y^x$

Figure 4.15 Illustration of parallel connection between Networks 2 and 3.

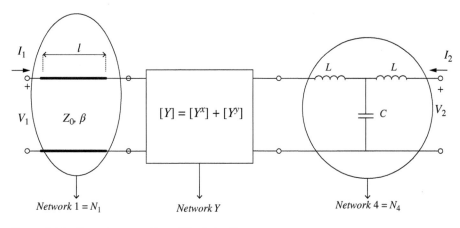

Figure 4.16 Cascade connection of final circuit.

Voltage and current gains from *ABCD* parameters are found using

$$V_{Gain} = 20 \log \left(\left| \frac{1}{A} \right| \right) \text{ (dB)} \tag{4.60}$$

$$I_{Gain} = 20 \log \left(\left| \frac{1}{D} \right| \right) \text{ (dB)} \tag{4.61}$$

MATLAB script has been written to obtain the voltage and current gains. The script that can be used for the analysis of any other amplifier network is given for reference. The voltage and current gains are plotted against frequency for various feedback resistors values and shown in Figures 4.17 and 4.18. This type of analysis gives the designer the effect on output response of several parameters in an amplifier circuit, including feedback, matching networks, and the parameters of the transistor.

Figure 4.17 Current gain of RF amplifier versus feedback resistor values and frequency.

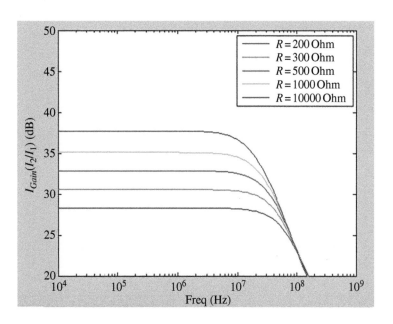

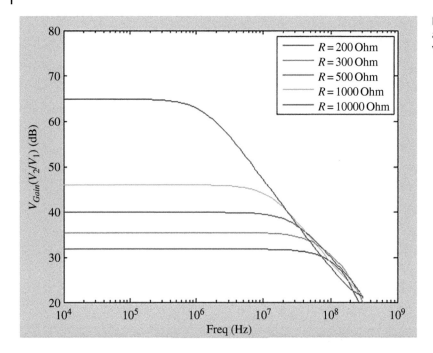

Figure 4.18 Voltage gain of RF amplifier versus feedback resistor values and frequency.

MATLAB script for the network analysis of an RF amplifier would be

```
Zo=50;
l=0.05;
L=2e-9;
C=12e-12;
rbe=400;
rce=70e3;
Cbe=15e-12;
Cbc=2e-12;
gm=0.2;
VGain=zeros(5,150);
IGain=zeros(5,150);
freq=zeros(1,150);
R=[200 300 500 1000 10000];

for i=1:5
for t=1:150;

f=10^((t+20)/20);
freq(t)=f;
lambda=0.65*3e8/(f);
bet=(2*pi)/lambda;
w=2*pi*f;
```

```
N1=[cos(bet*l) 1j*Zo*sin(bet*l);1j*(1/Zo)*sin(bet*l) cos(bet*l)];
Y1=[1/R(i) -1/R(i);-1/R(i) 1/R(i)];
k=(1+1j*w*rbe*(Cbc+Cbe));
h=[(rbe/k) (1j*w*rbe*Cbc)/k;(rbe.*(gm-1j*w*Cbc))/k
((1/rce)+(1j*w*Cbc*(1+gm*rbe+1j*w*Cbe*rbe)/k))];
Y2=[1/h(1,1) -h(1,2)/h(1,1);h(2,1)/h(1,1) det(h)/h(1,1)];
Y=Y1+Y2;
N23=[-Y(2,2)/Y(2,1) -1/Y(2,1);(det(Y)/Y(2,1)) -Y(1,1)/Y(2,1)];
N4=[(1-(w^2)*L*C) (2j*w*L-1j*(w^3)*L^2*C);1j*(w*C) (1-(w^2)*L*C)];
NT=N1*N23*N4;
VGain(i,t)=20*log10(abs(1/NT(1,1)));
IGain(i,t)=20*log10(abs(-1/NT(2,2)));

end
end

figure
semilogx(freq,(IGain))
axis([10^4 10^9 20 50]);
ylabel('I_{Gain} (I_2/I_1) (dB)';
xlabel('Freq (Hz)');
legend('R=20Ohm','R=30Ohm','R=50Ohm','R=100Ohm','R=1000Ohm')
figure
semilogx(freq,(VGain))
axis([10^4 10^9 20 80]);
ylabel('V_{Gain} (V_2/V_1) (dB)');
xlabel('Freq (Hz)');
legend('R=20Ohm','R=30Ohm','R=50Ohm','R=100Ohm','R=1000Ohm')
```

4.7 MATLAB Implementation of Network Parameters

Network parameters can be easily implemented by MATLAB, as illustrated elsewhere in this chapter, and the amount of computational time can be reduced. MATLAB scripts and functions can then be used to calculate two-port parameters, and conversion between them, for the networks. The MATLAB programs below are given to systematize these techniques. This is illustrated in the following examples.

```
%%%%%%%%%%%%%%%%%%%%%%%%%%%%%%%%%%%%%%%%%%%%%%%%%%%%%%%%%%%%%%%%%%%%%%%
% This m-file is function program to add two series connected network %
%                                  %
%%%%%%%%%%%%%%%%%%%%%%%%%%%%%%%%%%%%%%%%%%%%%%%%%%%%%%%%%%%%%%%%%%%%%%%

function [z11,z12,z21,z22]=SERIES(zx11,zx12,zx21,zx22,zw11,zw12,zw21,zw22)
z11= zx11 + zw11;
z12= zx12 + zw12;
```

```
z21= zx21 + zw21;
z22= zx22 + zw22;
Z=[z11 z12;z21 z22]
end


%%%%%%%%%%%%%%%%%%%%%%%%%%%%%%%%%%%%%%%%%%%%%%%%%%%%%%%%%%%%%%%%%%%%%%%%%
% This m-file is function program to add two parallel connected network %
%                                                      %
%%%%%%%%%%%%%%%%%%%%%%%%%%%%%%%%%%%%%%%%%%%%%%%%%%%%%%%%%%%%%%%%%%%%%%%%%
function [y11,y12,y21,y22]=PARALLEL(yx11,yx12,yx21,yx22,yw11,yw12,yw21,yw22)
y11= yx11 + yw11 ;
y12= yx12 + yw12 ;
y21= yx21 + yw21 ;
y22= yx22 + yw22 ;
Y=[y11 y12;y21 y22]
end



%%%%%%%%%%%%%%%%%%%%%%%%%%%%%%%%%%%%%%%%%%%%%%%%%%%%%%%%%%%%%%%%%%%%%%%%%
% This m-file is function program to add two cascaded connected network %
%                                                      %
%%%%%%%%%%%%%%%%%%%%%%%%%%%%%%%%%%%%%%%%%%%%%%%%%%%%%%%%%%%%%%%%%%%%%%%%%
function [a11,a12,a21,a22]=CASCADE(ax11,ax12,ax21,ax22,aw11,aw12,aw21,aw22)
a11=ax11.*aw11 + ax12.*aw21;
a12=ax11.*aw12 + ax12.*aw22;
a21=ax21.*aw11 + ax22.*aw21;
a22=ax21.*aw12 + ax22.*aw22;
A=[a11 a12;a21 a22]
end



%%%%%%%%%%%%%%%%%%%%%%%%%%%%%%%%%%%%%%%%%%%%%%%%%%%%%%%%%%%%%%%%%%%%%%%%%
% This m-file is function program to convert Z Parameters to Y Parameters%
%                                                      %
%%%%%%%%%%%%%%%%%%%%%%%%%%%%%%%%%%%%%%%%%%%%%%%%%%%%%%%%%%%%%%%%%%%%%%%%%

function [y11,y12,y21,y22]=Z2Y(z11,z12,z21,z22)
DET=z11.*z22-z21.*z12;
y11=z22./DET;
y12=-z12./DET;
y21=-z21./DET;
y22=z11./DET;
end
```

```
%%%%%%%%%%%%%%%%%%%%%%%%%%%%%%%%%%%%%%%%%%%%%%%%%%%%%%%%%%%%%%%%%%%%%%%%
% This m-file is function program to convert Y Parameters to Z Parameters%
%                                                                        %
%%%%%%%%%%%%%%%%%%%%%%%%%%%%%%%%%%%%%%%%%%%%%%%%%%%%%%%%%%%%%%%%%%%%%%%%

function [z11,z12,z21,z22]=Y2Z(y11,y12,y21,y22)
DET=y11.*y22-y21.*y12;
z11=y22./DET;
z12=-y12./DET;
z21=-y21./DET;
z22=y11./DET;
end
```

```
%%%%%%%%%%%%%%%%%%%%%%%%%%%%%%%%%%%%%%%%%%%%%%%%%%%%%%%%%%%%%%%%%%%%%%%%
% This m-file is function program to convert Z Parameters to A Parameters%
%                                                                        %
%%%%%%%%%%%%%%%%%%%%%%%%%%%%%%%%%%%%%%%%%%%%%%%%%%%%%%%%%%%%%%%%%%%%%%%%

function [a11,a12,a21,a22]=Z2A(z11,z12,z21,z22)
DET=z11.*z22-z21.*z12;
a11=z11./z21;
a12=DET./z21;
a21=1./z21;
a22=z22./z21;
end
```

```
%%%%%%%%%%%%%%%%%%%%%%%%%%%%%%%%%%%%%%%%%%%%%%%%%%%%%%%%%%%%%%%%%%%%%%%%
% This m-file is function program to convert Y Parameters to A Parameters%
%%%%%%%%%%%%%%%%%%%%%%%%%%%%%%%%%%%%%%%%%%%%%%%%%%%%%%%%%%%%%%%%%%%%%%%%

function [a11,a12,a21,a22]=Y2A(y11,y12,y21,y22)
DET=y11.*y22-y21.*y12;
a11=-y22./y21;
a12=-1./y21;
a21=-DET./y21;
a22=-y11./y21;
end
```

The following MATLAB program uses menu options and asks users to enter the two-port network parameters when networks are connected in series, parallel, or cascaded and then outputs the desired results.

```
%%%%%%%%%%%%%%%%%%%%%%%%%%%%%%%%%%%%%%%%%%%%%%%%%%%%%%%%%%%%%%%%%%%%%%%%%%%
% This m-file is a script  program that calculates the final 2-port% %parameters
when networks are connected in series, parallel or cascaded %
%%%%%%%%%%%%%%%%%%%%%%%%%%%%%%%%%%%%%%%%%%%%%%%%%%%%%%%%%%%%%%%%%%%%%%%%%%%
clear;
M = menu('Network Analysis','2 Network in Series', '2 Network in Parallel','2-
Network in Cascade');
switch M
    case 1
      zx11 = input('enter Z1_11: ');
      zx12 = input('enter Z1_12: ');
      zx21 = input('enter Z1_21: ');
      zx22 = input('enter Z1_22: ');
      zw11 = input('enter Z2_11: ');
      zw12 = input('enter Z2_12: ');
      zw21 = input('enter Z2_21: ');
      zw22 = input('enter Z2_22: ');
      SERIES(zx11,zx12,zx21,zx22,zw11,zw12,zw21,zw22);
    case 2
      yx11 = input('enter Y1_11: ');
      yx12 = input('enter Y1_12: ');
      yx21 = input('enter Y1_21: ');
      yx22 = input('enter Y1_22: ');
      yw11 = input('enter Y2_11: ');
      yw12 = input('enter Y2_12: ');
      yw21 = input('enter Y2_21: ');
      yw22 = input('enter Y2_22: ');
      PARALLEL(yx11,yx12,yx21,yx22,yw11,yw12,yw21,yw22);
    case 3
      ax11 = input('enter A1_11: ');
      ax12 = input('enter A1_12: ');
      ax21 = input('enter A1_21: ');
      ax22 = input('enter A1_22: ');
      aw11 = input('enter A2_11: ');
      aw12 = input('enter A2_12: ');
      aw21 = input('enter A2_21: ');
      aw22 = input('enter A2_22: ');
      CASCADE(ax11,ax12,ax21,ax22,aw11,aw12,aw21,aw22);

    otherwise
        disp('ERROR: invalid entry')
end
```

The following MATLAB program uses menu options and asks users to enter the two-port network parameters when networks are connected in series, parallel, or cascaded and then outputs the desired results.

When the program is run, the following window appears for the user. The user then specifies how the networks are to be connected. Then, the network parameters of the two networks can be manually entered from the MATLAB Command Window for execution.

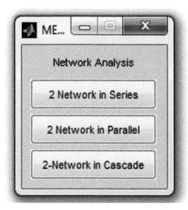

Example 4.5 Network Connections

Z parameters of the two-port network N in Figure 4.19a are $Z_{11} = 4s$, $Z_{12} = Z_{21} = 3s$, and $Z_{22} = 9s$ where $s = j\omega$. (a) Replace network N by its T-equivalent. (b) Use part (a) to find and input current i_1 (t) for $v_s = \cos 1000t$ (V) and write a MATLAB script to compute the equivalent network parameters of the circuit for (a) and (b). Your program should make the conversion from the two-port network to the T equivalent network by checking if the two-port network is reciprocal. Execute your program and plot $i(t)$ and confirm your results.

Solution

a) Any two-port reciprocal network can be converted to an equivalent T network, as shown here.

The transformation of the network to a T network shown in Figure 4.19b is valid with the following relations

$$Z_a = Z_{11} - Z_{12}$$
$$Z_b = Z_{22} - Z_{21}$$
$$Z_c = Z_{12} = Z_{21}$$

Based on the given information, the network is reciprocal because $Z_{12} = Z_{21}$. So, we can convert the network to a T network equivalent and obtain

$$Z_a = Z_{11} - Z_{12} = 4s - 3s = s$$
$$Z_b = Z_{22} - Z_{21} = 9s - 3s = 6s$$
$$Z_c = Z_{12} = Z_{21} = 3s$$

Hence, this simplified network can now be analyzed by establishing the relations between voltage and current.

$$V_1 = Z_{11}I_1 + Z_{12}I_2$$
$$V_2 = Z_{21}I_1 + Z_{22}I_2$$

b) From the final circuit, we obtain Z_{in} as

$$Z_{in}(s) = V_s/I_{in} = s + \frac{(3s + 6)(6s + 12)}{9s + 18} = 3s + 4 = 3j + 5 = 5\angle 36.9°$$

(a)

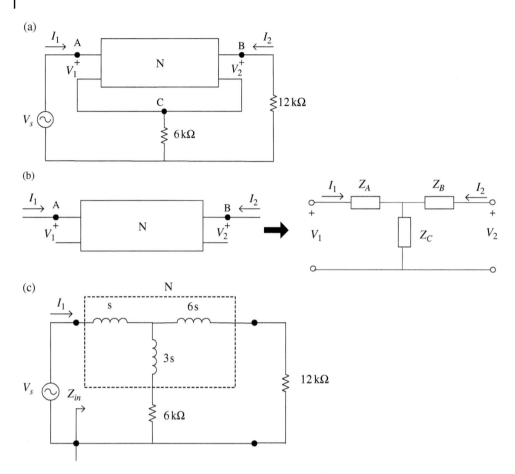

(b)

(c)

Figure 4.19 (a) *N* network for analysis; (b) network transformation to *T* network; (c) equivalent *T* network.

So, the current is

$$i(t) = 0.2\cos\left(1000t - 36.9°\right)\,\text{mA}$$

This operation can be implemented by MATLAB simply as

```
%%%%%%%%%%%%%%%%%%%%%%%%%%%%%%%%%%%%%%%%%%%%%%%%%%%%%%%%%%%%%%%%%%%%%%%%%%%%%%
% This m-file is a script  checks if a network is reciprocal network and then
converts it to its equivalent T-network and calculates current   %
%%%%%%%%%%%%%%%%%%%%%%%%%%%%%%%%%%%%%%%%%%%%%%%%%%%%%%%%%%%%%%%%%%%%%%%%%%%%%%
clear;
w = input('Circuit frequency: ');
s=sqrt(-1)*w;
'Enter Z parameters for network N in figure 2:'
z11 = input('Enter Z_11: ');
z12 = input('Enter Z_12: ');
z21 = input('Enter Z_21: ');
z22 = input('Enter Z_22: ');
```

```
if (z12==z21)
   za=z11-z12;
   zb=z22-z21;
   zc=z12;
   'The T equivalent of network N is:'
   za
   zb
   zc
   zeq=((6000+zc)*(12000+zb))/((6000+zc)+(12000+zb))+za;
   t=0:0.0001:0.1;
   v=cos(1000*t);
   i=1/(abs(zeq))*cos(w*t-angle(zeq));
   plot(t,i)
   title('i(t) for problem 2b');
   xlabel('t');
   ylabel('Amplitude');
else
    'Network N is not reciprocal'
end
```

Example 4.6 Network Parameters for Small Signal Model

(a) Obtain a small signal model of the metal oxide semiconductor field-effect transistor (MOSFET) using Y parameters for the equivalent shown in Figure 4.20. (b) Use MATLAB to compute the voltage gain and phase of the voltage gain of the model when $R_g = 5\ \Omega$, $C_{gs} = 10e{-}12$, $R_{gs} = 0.5\ \Omega$, $C_{gd} = 100e{-}12F$, $C_{ds} = 2e{-}12F$, $g_m = 20e{-}3S$, $R_{ds} = 70e3\ \Omega$, $R_S = 3\Omega$, $R_{highL} = 10e3\ \Omega$, and connected load is $R_L = 10e3\ \Omega$.

Solution

a) The equivalent circuit in Figure 4.20 is simplified and shown in Figure 4.21. When Figures 4.20 and 4.21 are compared, the following can be written.

Figure 4.20 Small signal MOSFET model.

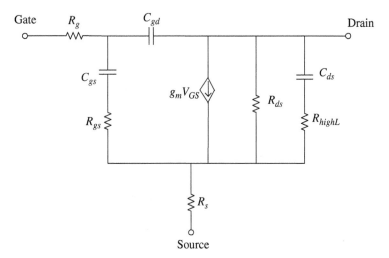

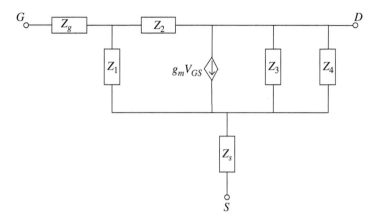

Figure 4.21 The simplified equivalent circuit for the MOSFET small signal model.

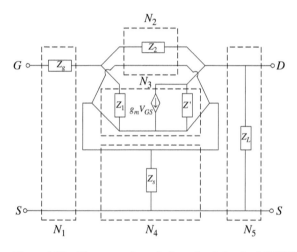

Figure 4.22 The network equivalent circuit for the MOSFET small signal model.

$$Z_g = R_g, Z_1 = R_{gs} - j\frac{1}{\omega C_{gs}}, Z_2 = -j\frac{1}{\omega C_{gd}},$$

$$Z_3 = R_{ds}, Z_4 = R_{highL} - j\frac{1}{\omega C_{ds}}, Z_s = R_s, Z_L = R_L, Z' = \frac{Z_3 Z_4}{Z_3 + Z_4}$$

The small signal MOSFET model circuit given in Figure 4.20 can be analyzed when each of the components is represented as a network with the connection that they are introduced in the circuit, as shown in Figure 4.22. From Figure 4.22, it can be seen that Network 2, N_2, and Network 3, N_3, are connected in parallel. The overall parallel connected network, N_{23}, can be found from

$$Y_{23} = Y_2 + Y_3 \rightarrow N_{23}$$

Figure 4.23 The equivalent circuit for Network 3.

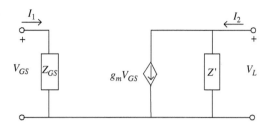

Now, Network 4, N_4, and the resultant parallel network, N_{23}, are connected in series. The combination of these two networks can be obtained from

$$Z_{234} = Z_{23} + Z_4 \rightarrow N_{234}$$

where $Z_{23} = (Y_{23})^{-1}$. From Figure 4.22, it can also be observed that networks N_1, N_{234}, and N_5 are cascaded. Hence, the overall network parameters can be found from $ABCD$ parameters.

$$[ABCD]_{Network} = [ABCD]_{N_1}[ABCD]_{N_{234}}[ABCD]_{N_5}$$

- $ABCD$ parameters for Network 1, 5, and Y parameters for Network 2 are

$$[ABCD]_{N_1} = \begin{bmatrix} 1 & Z_g \\ 0 & 1 \end{bmatrix}, [Y]_{N_2} = \begin{bmatrix} Y_{C_{gd}} & -Y_{C_{gd}} \\ -Y_{C_{gd}} & Y_{C_{gd}} \end{bmatrix}, \text{and } [ABCD]_{N_5} = \begin{bmatrix} 1 & 0 \\ 1/Z_L & 1 \end{bmatrix}$$

- Y parameters for Network 3 are found using Figure 4.23 Y parameters are found using Figure 4.23 as

$$Y_{11} = \left.\frac{I_1}{V_1}\right|_{V_2 = 0} = Y_{GS} \quad Y_{12} = \left.\frac{I_1}{V_2}\right|_{V_1 = 0} = 0$$

$$Y_{21} = \left.\frac{I_2}{V_1}\right|_{V_2 = 0} = g_m \quad Y_{22} = \left.\frac{I_2}{V_2}\right|_{V_1 = 0} = \frac{1}{Z_3}$$

So

$$[Y]_{N_3} = \begin{bmatrix} Y_{GS} & 0 \\ g_m & 1/Z' \end{bmatrix}$$

- The parallel connected network, N_{23}, can now be found from

$$[Y]_{N_{23}} = [Y]_{N_3} + [Y]_{N_2} = \begin{bmatrix} Y_{C_{gd}} & -Y_{C_{gd}} \\ -Y_{C_{gd}} & Y_{C_{gd}} \end{bmatrix} + \begin{bmatrix} Y_{GS} & 0 \\ g_m & 1/Z' \end{bmatrix}$$

Hence

$$[Y]_{N_{23}} = \begin{bmatrix} Y_{C_{gd}} + Y_{GS} & -Y_{C_{gd}} \\ -Y_{C_{gd}} + g_m & Y_{C_{gd}} + 1/Z' \end{bmatrix}$$

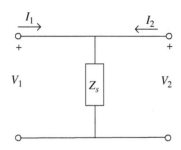

Figure 4.24 The equivalent circuit for Network 4.

- Z parameters for Network 4 are calculated using the circuit shown in Figure 4.24 as

$$Z_{11} = \frac{V_1}{I_1}\bigg|_{I_2 = 0} = Z_s \quad Z_{12} = \frac{V_1}{I_2}\bigg|_{I_1 = 0} = Z_s$$

$$Z_{21} = \frac{V_2}{I_1}\bigg|_{I_2 = 0} = Z_s \quad Z_{22} = \frac{V_2}{I_2}\bigg|_{I_1 = 0} = Z_s$$

So

$$[Z]_{N_4} = \begin{bmatrix} Z_S & Z_S \\ Z_S & Z_S \end{bmatrix}$$

- The port parameters for series connected networks N_4 and N_{23} are found by first converting Y parameters for N_{23} to Z parameters by

$$[Z]_{N_{23}} = \left([Y]_{N_{23}}\right)^{-1} = \begin{bmatrix} \dfrac{Y_{C_{gd}} + 1/Z'}{\Delta Y_{N_{23}}} & \dfrac{Y_{C_{gd}}}{\Delta Y_{N_{23}}} \\ \dfrac{Y_{C_{gd}} - g_m}{\Delta Y_{N_{23}}} & \dfrac{Y_{C_{gd}} + Y_{GS}}{\Delta Y_{N_{23}}} \end{bmatrix}$$

As a result, two-port parameters for the series connected networks are found from

$$[Z]_{N_{234}} = [Z]_{N_{23}} + [Z]_{N_4} = \begin{bmatrix} \dfrac{Y_{C_{gd}} + 1/Z'}{\Delta Y_{N_{23}}} & \dfrac{Y_{C_{gd}}}{\Delta Y_{N_{23}}} \\ \dfrac{Y_{C_{gd}} - g_m}{\Delta Y_{N_{23}}} & \dfrac{Y_{C_{gd}} + Y_{GS}}{\Delta Y_{N_{23}}} \end{bmatrix} + \begin{bmatrix} Z_S & Z_S \\ Z_S & Z_S \end{bmatrix}$$

Hence

$$[Z]_{N_{234}} = \begin{bmatrix} \dfrac{Y_{C_{gd}} + 1/Z'}{\Delta Y_{N_{23}}} + Z_S & \dfrac{Y_{C_{gd}}}{\Delta Y_{N_{23}}} + Z_S \\ \dfrac{Y_{C_{gd}} - g_m}{\Delta Y_{N_{23}}} + Z_S & \dfrac{Y_{C_{gd}} + Y_{GS}}{\Delta Y_{N_{23}}} + Z_S \end{bmatrix}$$

- We need to convert the resultant central Z parameters to $ABCD$ parameters using

$$[ABCD]_{N_{234}} = \begin{bmatrix} \dfrac{(Z_{11})_{[Z]_{N_{234}}}}{(Z_{21})_{[Z]_{N_{234}}}} & \dfrac{|[Z]_{N_{234}}|}{(Z_{21})_{[Z]_{N_{234}}}} \\ \dfrac{1}{(Z_{21})_{[Z]_{N_{234}}}} & \dfrac{(Z_{22})_{[Z]_{N_{234}}}}{(Z_{21})_{[Z]_{N_{234}}}} \end{bmatrix}$$

- The complete network parameters of the small signal model of MOSFET are found from $ABCD$ parameters since now N_1, N_{234}, and N_5 are all cascaded as

$$[ABCD] = [ABCD]_{N_1}[ABCD]_{N_{234}}[ABCD]_{N_5}$$

or

$$[ABCD] = \begin{bmatrix} 1 & Z_g \\ 0 & 1 \end{bmatrix} \begin{bmatrix} \dfrac{(Z_{11})_{[Z]_{N_{234}}}}{(Z_{21})_{[Z]_{N_{234}}}} & \dfrac{\left|[Z]_{N_{234}}\right|}{(Z_{21})_{[Z]_{N_{234}}}} \\ \dfrac{1}{(Z_{21})_{[Z]_{N_{234}}}} & \dfrac{(Z_{22})_{[Z]_{N_{234}}}}{(Z_{21})_{[Z]_{N_{234}}}} \end{bmatrix} \begin{bmatrix} 1 & 0 \\ 1/Z_L & 1 \end{bmatrix}$$

```
%%%%%%%%%%%%%%%%%%%%%%%%%%%%%%%%%%%%%%%%%%%%%%%%%%%%%%%%%%%%%%%%%%%%%%
% This m-file is developed to calculate the response of small signal MOSFET model  %
%%%%%%%%%%%%%%%%%%%%%%%%%%%%%%%%%%%%%%%%%%%%%%%%%%%%%%%%%%%%%%%%%%%%%%

clear
Rg=5;
Cgs=10e-12;
rgs=0.5;
Cgd=100e-12;
Cds=2e-12;
Gm =20e-3;
Rds=70e3;
RL=10e3;
RS =3;
Rhigh=10e3;
f=logspace(3,10);
omega=2*pi.*f;
Z3=Rds;
Z4=1./(1i.*omega.*Cds)+Rhigh;
Zp=(Z3.*Z4)./(Z3+Z4);


%++++++++++++++++++++++++++++++++VCCS (y-parameters)+++++++++++++++++++++++
VCCS_11=1./(rgs.*ones(size(f ))+1./(1i.*omega.*Cgs));
VCCS_12=zeros(size(f ));
VCCS_21=Gm.*ones(size(f ));
VCCS_22=1./(Zp);
%++++++++++++++++++++++++++++++++ CGD (y-parameters)+++++++++++++++++++++++
CGD_11= 1i.*omega.*Cgd;
CGD_12=-1i.*omega.*Cgd;
CGD_21=-1i.*omega.*Cgd;
CGD_22= 1i.*omega.*Cgd;
%++++++++++++++++++++++++++++++++++ RS (z-parameters)+++++++++++++++++++++++
RS_11=RS.*ones(size(f ));
RS_12=RS.*ones(size(f ));
RS_21=RS.*ones(size(f ));
RS_22=RS.*ones(size(f ));
%+++++++++++++++++++++++++++++++++ RL (ABCD-parameters)+++++++++++++++++++++++
```

```
RL_11=1.*ones(size(f));
RL_12=0.*ones(size(f));
RL_21=1./(RL);
RL_22=1.*ones(size(f));
%+++++++++++++++++++++++++++ Ci (ABCD-parameters)+++++++++++++++++++++++++++

Ci_11=1.*ones(size(f));
Ci_12=(Rg.*ones(size(f)));
Ci_21=0.*ones(size(f));
Ci_22=1.*ones(size(f));
```

- The MATLAB script is written to obtain the characteristics of the network by finding the voltage gain and phase of the voltage gain using the MATLAB functions developed and discussed elsewhere in this section.

```
%+++++++++++++++++++++++++++++++++++++++++++Xa = VCCS and CGD +++++++++++++++++++
[Xa_11,Xa_12,Xa_21,Xa_22]=PARALLEL(VCCS_11,VCCS_12,VCCS_21,VCCS_22,CGD_11,
CGD_12,CGD_21,CGD_22);
%++++++++++++++++++++++Convert Admittance to Impedance ++++++++++++++++++++++++++++++
[z11,z12,z21,z22]=Y2Z(Xa_11,Xa_12,Xa_21,Xa_22);
%++++++++++++++++++++++++++Xb = VCCS, CGD and RS +++++++++++++++++++++++++++++++
[Xb_11,Xb_12,Xb_21,Xb_22]=SERIES(z11,z12,z21,z22,RS_11,RS_12,RS_21,RS_22);
%++++++++++++++++++++++++Convert Impedance to ABCD  ++++++++++++++++++++++++++++++
[a11,a12,a21,a22]=Z2A(Xb_11,Xb_12,Xb_21,Xb_22);
%+++++++++++++++++++++++Xc = VCCS, CGD, RS and RL ++++++++++++++++++++++++++++
[Xc_11,Xc_12,Xc_21,Xc_22]=CASCADE(a11,a12,a21,a22,RL_11,RL_12,RL_21,RL_22);
%++++++++++++++++++++++++++Xd = VCCS, CGD, RS, RL and Ci++++++++++++++++++++++
[Xd_11,Xd_12,Xd_21,Xd_22]=CASCADE(Xc_11,Xc_12,Xc_21,Xc_22,Ci_11,Ci_12,Ci_21,Ci_22);
subplot(211)
semilogx(f,20*log10(abs(1./Xd_11)))
xlabel('Frequency (Hz)')
ylabel('Amplitude (dB)')
title('Magnitude of the voltage gain in dB')
subplot(212)
semilogx(f,((angle(1./Xd_11)).*180)./pi)
xlabel('Frequency (Hz)')
ylabel('Angle (deg)')
title('Phase of the voltage gain')
%==============================================================================
```

The voltage gain and phase responses when the program is run is given in Figure 4.25.

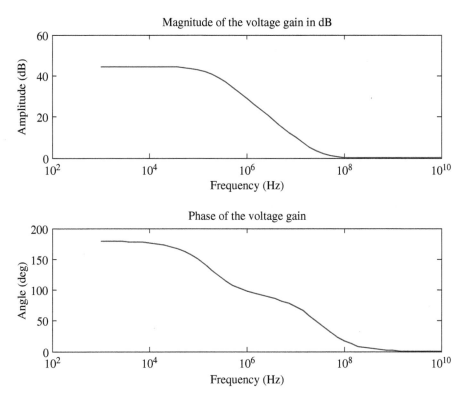

Figure 4.25 Voltage gain and phase responses.

4.8 *S*-Scattering Parameters

Scattering parameters are used to characterize RF/microwave devices and components at high frequencies [2]. Specifically, they are used to define the return loss and insertion loss of a component or device.

4.8.1 One-port Network

Consider the circuit given in Figure 4.26. The relationship between the current and the voltage can be written as

$$I = \frac{V_g}{Z_g + Z_L} \tag{4.62}$$

and

$$V = \frac{V_g Z_L}{Z_g + Z_L} \tag{4.63}$$

where Z_g is the generator impedance. The incident waves for voltage and current can be obtained when the generator is matched as

$$I_i = \frac{V_g}{Z_g + Z_g^*} = \frac{V_g}{2 \, \text{Re} \left\{ Z_g \right\}} \tag{4.64}$$

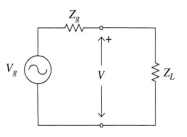

Figure 4.26 One-port network for scattering parameter analysis.

and

$$V_i = \frac{V_g Z_g^*}{Z_g + Z_g^*} = \frac{V_g Z_g^*}{2 \operatorname{Re}\{Z_g\}} \tag{4.65}$$

Then, the reflected waves are found from

$$I = I_i - I_r \tag{4.66}$$

and

$$V = V_i - V_r \tag{4.67}$$

Substituting (4.62) and (4.64) into (4.66) gives the reflected wave as

$$I_r = I_i - I = \left(\frac{Z_L - Z_g^*}{Z_L + Z_g^*}\right) I_i \tag{4.68}$$

or

$$I_r = S^I I_i \tag{4.69}$$

where

$$S^I = \left(\frac{Z_L - Z_g^*}{Z_L + Z_g^*}\right) \tag{4.70}$$

is the scattering matrix for the current. Similar analysis can be done to find the reflected voltage wave by substituting (4.63) and (4.65) into (4.67) as

$$V_r = V_i - V = \frac{Z_g}{Z_g^*}\left(\frac{Z_L - Z_g^*}{Z_L + Z_g^*}\right) V_i \tag{4.71}$$

or

$$V_r = \frac{Z_g}{Z_g^*} S^I V_i = S^V V_i \tag{4.72}$$

where

$$S^V = \frac{Z_g}{Z_g^*} S^I \tag{4.73}$$

is the scattering matrix for the voltage. It can also be shown that

$$V_i = Z_g^* I_i \tag{4.74}$$

$$V_r = Z_g I_r \tag{4.75}$$

When generator impedance is real, $Z_g = R_g$, then

$$S^I = S^V = \left(\frac{Z_L - R_g}{Z_L + R_g}\right)_i \tag{4.76}$$

4.8.2 N-port Network

The analysis described in Section 0 can be extended to the N-port network shown in Figure 4.27. The analysis is based on the assumption that generators are independent of each other. Hence, a Z generator matrix has no cross-coupling terms and can be expressed as a diagonal matrix.

$$[Z_g] = \begin{bmatrix} Z_{g1} & 0 & \cdots & 0 \\ 0 & Z_{g2} & \cdots & 0 \\ \vdots & \vdots & & \vdots \\ 0 & 0 & \cdots & Z_{gn} \end{bmatrix} \tag{4.77}$$

From Eqs. (4.66) to (4.67), the incident and reflected wave are related to the actual voltage and current values as

$$[I] = [I_i] - [I_r] \tag{4.78}$$

$$[V] = [V_i] + [V_r] \tag{4.79}$$

From (4.74) to (4.75), the incident and reflected components can be related through

$$[V_i] = \left[Z_g^*\right][I_i] \tag{4.80}$$

$$[V_r] = \left[Z_g\right][I_r] \tag{4.81}$$

similar to a one-port case as derived before. For an N-port network, Z parameters can be obtained as

$$[V] = [Z][I] \tag{4.82}$$

Figure 4.27 *N*-port network for scattering analysis.

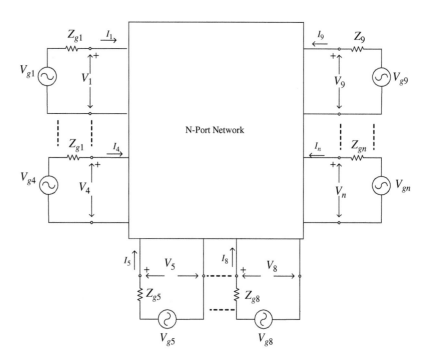

Using (4.77)–(4.82), we can obtain

$$[V_r] = [V] - [V_i] = [Z][I] - \left[Z_g^*\right][I_i] \tag{4.83}$$

Equation (4.83) can also be expressed as

$$\left[Z_g\right][I_r] = [Z][I] - \left[Z_g^*\right][I_i] = [Z]([I_i] - [I_r]) - \left[Z_g^*\right][I_i] \tag{4.84}$$

and simplified to

$$([Z] + [Z_g])[I_r] = \left([Z] - \left[Z_g^*\right]\right)[I_i] \tag{4.85}$$

Equation (4.85) can be put in the following form

$$[I_r] = ([Z] + [Z_g])^{-1}\left([Z] - \left[Z_g^*\right]\right)[I_i] \tag{4.86}$$

Since from (4.70) the scattering matrix for current for the *N*-port network is equal to

$$S^I = ([Z] + [Z_g])^{-1}\left([Z] - \left[Z_g^*\right]\right) \tag{4.87}$$

Then, (4.86) can be expressed as

$$[I_r] = \left[S^I\right][I_i] \tag{4.88}$$

For the *N*-port network, *Y* parameters for a short circuit case can be obtained similarly as

$$[I] = [Y][V] \tag{4.89}$$

It can also be shown that

$$[V_r] = -([Y] + [Y_g])^{-1}\left([Y] - \left[Y_g^*\right]\right)[V_i] \tag{4.90}$$

or

$$[V_r] = \left[S^V\right][V_i] \tag{4.91}$$

where

$$S^V = -([Y] + [Y_g])^{-1}\left([Y] - \left[Y_g^*\right]\right) \tag{4.92}$$

Example 4.7 Scattering Parameters

Consider a transistor network that is represented as a two-port network and connected between the source and the load. It is assumed that the generator or source and load impedances are equal, and are given as R_g. The transistor is represented by the *Z* parameters shown in Figure 4.28. Find current scattering matrix, S^I.

$$[Z] = \begin{bmatrix} Z_i & Z_r \\ Z_f & Z_o \end{bmatrix}$$

Solution

From Eq. (4.87), the scattering matrix for the current is

$$S^I = ([Z] + [Z_g])^{-1}\left([Z] - \left[Z_g^*\right]\right) \tag{4.93}$$

Figure 4.28 Two-port transistor network.

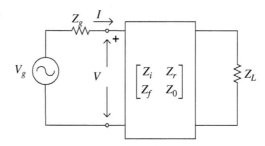

The generator Z_g matrix is

$$\left[Z_g\right] = \left[Z_g^*\right] = \begin{bmatrix} R_g & 0 \\ 0 & R_g \end{bmatrix} \tag{4.94}$$

Then

$$[Z] + \left[Z_g^*\right] = \begin{bmatrix} Z_i & Z_r \\ Z_f & Z_o \end{bmatrix} + \begin{bmatrix} R_g & 0 \\ 0 & R_g \end{bmatrix} = \begin{bmatrix} Z_i + R_g & 0 \\ 0 & Z_o + R_g \end{bmatrix} \tag{4.95}$$

The inverse of the matrix in (4.95) is

$$\left[\left([Z] + \left[Z_g^*\right]\right)\right]^{-1} = \frac{1}{\left|[Z] + \left[Z_g^*\right]\right|} \left(\left[[Z] + \left[Z_g^*\right]\right]^C\right)^T \tag{4.96}$$

$\left|[Z] + \left[Z_g^*\right]\right|$ is the determinant of $[Z] + \left[Z_g^*\right]$ and is calculated as

$$\left|[Z] + \left[Z_g^*\right]\right| = (Z_i + R_g)(Z_o + R_g) - Z_r Z_f \tag{4.97}$$

$\left[[Z] + \left[Z_g^*\right]\right]^C$ is the cofactor matrix for $[Z] + \left[Z_g^*\right]$ and is calculated as

$$\left([Z] + \left[Z_g^*\right]\right)^C = \begin{bmatrix} Z_o + R_g & -Z_f \\ -Z_r & Z_i + R_g \end{bmatrix} \tag{4.98}$$

Then

$$\left[\left([Z] + \left[Z_g^*\right]\right)^C\right]^T = \begin{bmatrix} Z_o + R_g & -Z_r \\ -Z_f & Z_i + R_g \end{bmatrix} \tag{4.99}$$

Hence, the inverse of the matrix from (4.97) to (4.99) is equal to

$$\left[\left([Z] + \left[Z_g^*\right]\right)\right]^{-1} = \frac{1}{\left((Z_i + R_g)(Z_o + R_g) - Z_r Z_f\right)} \begin{bmatrix} Z_o + R_g & -Z_r \\ -Z_f & Z_i + R_g \end{bmatrix} \tag{4.100}$$

Then, from (4.93)

$$S^I = \left([Z] + [Z_g]\right)^{-1}\left([Z] - [Z_g]\right) = \left(\left[\frac{1}{\left((Z_i + R_g)(Z_o + R_g) - Z_r Z_f\right)} \begin{bmatrix} Z_o + R_g & -Z_r \\ -Z_f & Z_i + R_g \end{bmatrix}\right] - \begin{bmatrix} Z_i - R_g & Z_r \\ Z_f & Z_o - R_g \end{bmatrix}\right) \tag{4.101}$$

which can be simplified to

$$S^I = \left([Z] + [Z_g]\right)^{-1}\left([Z] - [Z_g]\right) = \begin{pmatrix} \left(Z_o + R_g\right)\left(Z_i - R_g\right) - Z_r Z_f & 2Z_r R_g \\ 2Z_f R_g & \left(Z_i + R_g\right)\left(Z_o - R_g\right) - Z_r Z_f \end{pmatrix} \quad (4.102)$$

4.8.3 Normalized Scattering Parameters

Normalized scattering parameters can be introduced by a and b for the incident and reflected waves as follows.

$$[a] = \frac{1}{\sqrt{2}}\sqrt{\left([Z_g] + [Z_g^*]\right)}[I_i] \quad (4.103)$$

$$[b] = \frac{1}{\sqrt{2}}\sqrt{\left([Z_g] + [Z_g^*]\right)}[I_r] \quad (4.104)$$

where

$$\frac{1}{\sqrt{2}}\sqrt{\left([Z_g] + [Z_g^*]\right)} = \sqrt{\mathrm{Re}\left\{Z_g\right\}} = \begin{bmatrix} \sqrt{\mathrm{Re}\left\{Z_{g1}\right\}} & 0 & \cdots & 0 \\ 0 & \sqrt{\mathrm{Re}\left\{Z_{g2}\right\}} & \cdots & 0 \\ \vdots & \vdots & & \vdots \\ 0 & 0 & \cdots & \sqrt{\mathrm{Re}\left\{Z_{gn}\right\}} \end{bmatrix} \quad (4.105)$$

Substituting (4.86) into (4.103) and (4.104) gives

$$\frac{[b]}{\sqrt{\mathrm{Re}\left\{Z_g\right\}}} = [I_r] = [S^I][I_i] \quad (4.106)$$

or

$$\frac{[b]}{\sqrt{\mathrm{Re}\left\{Z_g\right\}}} = [S^I]\frac{[a]}{\sqrt{\mathrm{Re}\left\{Z_g\right\}}} \quad (4.107)$$

Then, from (4.106) to (4.107)

$$[b] = \sqrt{\mathrm{Re}\left\{Z_g\right\}}[S^I]\left[\mathrm{Re}\left\{Z_g\right\}\right]^{-1/2}[a] \quad (4.108)$$

Equation (4.108) can be simplified to

$$[b] = [S][a] \quad (4.109)$$

where

$$[S] = \sqrt{\mathrm{Re}\left\{Z_g\right\}}[S^I]\left[\mathrm{Re}\left\{Z_g\right\}\right]^{-1/2} \quad (4.110)$$

and

$$\left[\text{Re}\left\{Z_g\right\}\right]^{-1/2} = \begin{bmatrix} \dfrac{1}{\sqrt{\text{Re}\left\{Z_{g1}\right\}}} & 0 & \cdots & 0 \\ 0 & \dfrac{1}{\sqrt{\text{Re}\left\{Z_{g2}\right\}}} & \cdots & 0 \\ \vdots & \vdots & & \vdots \\ 0 & 0 & \cdots & \dfrac{1}{\sqrt{\text{Re}\left\{Z_{gn}\right\}}} \end{bmatrix}$$

(4.111)

The S matrix in (4.109) is called a normalized scattering matrix. It can be proven that

$$\left[S^I\right] = \left[Z_g\right]^{-1}\left[S^V\right]\left[Z_g^*\right]$$

(4.112)

When the generator or source impedance is real, $Z_g = R_g$, then, from (4.112), we obtain

$$\left[S^I\right] = \left[S^V\right]$$

(4.113)

In addition, Eqs. (4.87) and (4.110) take the following form

$$S^I = \left([Z] + [R_g]\right)^{-1}\left([Z] - [R_g]\right)$$

(4.114)

$$[S] = \sqrt{R_g}[S^I]\left[Z_g\right]^{-1/2}$$

(4.115)

From (4.113) to (4.115), we obtain

$$[S] = \left[S^I\right] = \left[S^V\right]$$

(4.116)

S parameters can be calculated using the two-port network shown in Figure 4.29. In Figure 4.29, the source or generator impedances are given as R_{g1} and R_{g2}. When Eq. (4.109) is expanded

$$\begin{bmatrix} b_1 \\ b_2 \end{bmatrix} = \begin{bmatrix} S_{11} & S_{12} \\ S_{21} & S_{22} \end{bmatrix}\begin{bmatrix} a_1 \\ a_2 \end{bmatrix}$$

(4.117)

From (4.117)

$$b_1 = S_{11}a_1 + S_{12}a_2$$

(4.118)

$$b_2 = S_{21}a_1 + S_{22}a_2$$

(4.119)

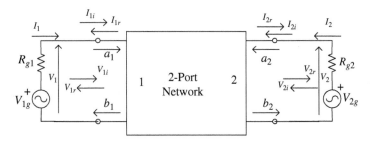

Figure 4.29 S parameters for two-port networks.

Hence, S parameters can be defined from (4.118) and (4.119) as

$$S_{11} = \left.\frac{b_1}{a_1}\right|_{a_2=0} \qquad S_{12} = \left.\frac{b_1}{a_2}\right|_{a_1=0}$$

$$S_{21} = \left.\frac{b_2}{a_1}\right|_{a_2=0} \qquad S_{22} = \left.\frac{b_2}{a_2}\right|_{a_1=0} \tag{4.120}$$

From Eq. (4.120), the scattering parameters are calculated when $a_1 = 0$ or $a_2 = 0$. a represents the incident waves. If Eq. (4.103) is reviewed again

$$[a] = \frac{1}{\sqrt{2}}\sqrt{\left([Z_g] + [Z_g^*]\right)}[I_i] \tag{4.121}$$

a_2 becomes zero when $I_{2i} = 0$. This can be obtained when there is no source connected to port 2, i.e. $V_{2g} = 0$ with the existence of source impedance R_{2g}. From KLV for the second port, we obtain

$$V_2 = -I_2 R_{g2} \quad \text{or} \quad V_2 + I_2 R_{g2} = 0 \tag{4.122}$$

Substituting (4.78) and (4.79) into (4.122) gives

$$V_2 + I_2 R_{g2} = V_{2i} + V_{2r} + R_{g2}(I_{2i} - I_{2r}) \tag{4.123}$$

which leads to

$$V_2 + I_2 R_{g2} = I_{2i} R_{g2} + R_{g2} I_{2r} + R_{g2} I_{2i} - I_{2r} R_{g2} \tag{4.124}$$

or

$$V_2 + I_2 R_{g2} = 2 R_{g2} I_{2i} \tag{4.125}$$

From (4.121), when $Z_g = R_g$

$$[a] = \sqrt{R_g}[I_i] \tag{4.126}$$

Substituting (4.124) into (4.125) gives

$$V_2 + I_2 R_{g2} = 2\sqrt{R_{g2}}a_2 \tag{4.127}$$

Then

$$a_2 = \frac{V_2 + I_2 R_{g2}}{2\sqrt{R_{g2}}} \tag{4.128}$$

It is then proven that, when (4.122) is substituted into (4.128), $a_2 = 0$ as expected. This also requires $I_i = 0$ from (4.127). Then, this shows that there is no reflected current which is the incident current, I_{2i}, at port 2 due to source generator incident wave from port 1.

Similar analysis can be done at port 1 when $a_1 = 0$. The same steps can be followed, and it can be shown that

$$a_1 = \frac{V_1 + I_1 R_{g1}}{2\sqrt{R_{g1}}} \tag{4.129}$$

Reflected waves b_1 and b_2 can be analyzed the same way using the analysis just presented for the incident waves a_1 and a_2. When no source voltage is connected at port 1, $a_1 = 0$ in the existence of source voltage R_{g1}, and we can write

$$V_1 = -I_1 R_{g1} \quad \text{or} \quad V_1 + I_1 R_{g1} = 0 \tag{4.130}$$

In terms of the reflected and incident voltage and current, we get

$$V_1 - I_1 R_{g1} = V_{1i} + V_{1r} - R_{g1}(I_{1i} - I_{1r}) \tag{4.131}$$

which leads to

$$V_1 - I_1 R_{g1} = I_{1i} R_{g1} + I_{1r} R_{g1} - I_{1i} R_{g1} + I_{1r} R_{g1} \tag{4.132}$$

or

$$V_1 - I_1 R_{g1} = 2 R_{g1} I_{1r} \tag{4.133}$$

From (4.104), when $Z_g = R_g$

$$[b] = \sqrt{R_g}[I_r] \tag{4.134}$$

Hence, (4.133) can be written as

$$V_1 - I_1 R_{g1} = 2\sqrt{R_{g1}}\, b_1 \tag{4.135}$$

Then

$$b_1 = \frac{V_1 - I_1 R_{g1}}{2\sqrt{R_{g1}}} \tag{4.136}$$

We can also show that when $a_2 = 0$

$$b_2 = \frac{V_2 - I_2 R_{g2}}{2\sqrt{R_{g2}}} \tag{4.137}$$

Incident and reflected parameters a and b for an N-port network can be written using the results given in (4.128), (4.129), (4.136) and (4.137) for real generator impedance R_g as

$$[a] = \frac{1}{2}[R_g]^{-1/2}([V] + [R_g][I]) \tag{4.138}$$

$$[b] = \frac{1}{2}[R_g]^{-1/2}([V] - [R_g][I]) \tag{4.139}$$

For an arbitrary impedance, (4.138) and (4.139) can be written as

$$[a] = \frac{1}{2}[\operatorname{Re}\{Z_g\}]^{-1/2}([V] + [Z_g][I]) \tag{4.140}$$

$$[b] = \frac{1}{2}[\operatorname{Re}\{Z_g\}]^{-1/2}\left([V] - [Z_g^*][I]\right) \tag{4.141}$$

Now, since we derived the conditions when a_1 and a_2 are zero, we can expand equations given by (4.120). When $a_2 = 0$, we can calculate S_{11} and S_{21}. From Eqs. (4.129), (4.136), (4.126), and (4.124), S_{11} can be expressed as

$$S_{11} = \frac{b_1}{a_1}\bigg|_{a_2=0} = \frac{\left(\dfrac{V_1 - I_1 R_{g1}}{2\sqrt{R_{g1}}}\right)}{\left(\dfrac{V_1 + I_1 R_{g1}}{2\sqrt{R_{g1}}}\right)}\bigg|_{I_{2i}=0} = \frac{V_1 - I_1 R_{g1}}{V_1 + I_1 R_{g1}} = \frac{V_{1r}}{V_{1i}} = \frac{\sqrt{R_{1g}}I_{1r}}{\sqrt{R_{1g}}I_{1i}} = \frac{I_{1r}}{I_{1i}} \tag{4.142}$$

or

$$S_{11} = \frac{Z_{11} - R_{g1}}{Z_{11} + R_{g1}} \tag{4.143}$$

In (4.143), S_{11} is the reflection coefficient at port 1 when port 2 is terminated with generator impedance R_{g2}. We express S_{21} using (4.129), (4.137), (4.126), and (4.134) as

$$S_{21} = \left.\frac{b_2}{a_1}\right|_{a_2 = 0} = \frac{\left(\dfrac{V_2 - I_2 R_{g2}}{2\sqrt{R_{g2}}}\right)}{\left(\dfrac{V_1 + I_1 R_{g1}}{2\sqrt{R_{g1}}}\right)}\Bigg|_{I_{2i} = 0} = \frac{(V_2 - I_2 R_{g2})\sqrt{R_{g1}}}{(V_1 + I_1 R_{g1})\sqrt{R_{g2}}} = \frac{\sqrt{R_{2g}}I_{2r}}{\sqrt{R_{1g}}I_{1i}} \tag{4.144}$$

When $a_2 = 0$, $V_{2g} = 0$ and that results in $V_2 = -I_2 R_{2g}$ and $V_{1g} = 2I_{1i}R_{1g}$, then (4.144) can be written as

$$S_{21} = \left.\frac{b_2}{a_1}\right|_{a_2 = 0} = -\frac{\sqrt{R_{2g}}I_2}{\sqrt{R_{1g}}(V_{1g}/2R_{1g})} = -2\sqrt{R_{1g}}\sqrt{R_{2g}}\frac{I_2}{V_{1g}} = 2\sqrt{\frac{R_{1g}}{R_{2g}}}\frac{V_2}{V_{1g}} = \frac{(V_2/\sqrt{R_{2g}})}{\left(\frac{1}{2}V_{1g}/\sqrt{R_{1g}}\right)} \tag{4.145}$$

As shown from (4.145), S_{21} is the forward transmission gain of the network from port 1 to port 2. A similar procedure can be repeated to derive S_{22} and S_{12} when $a_1 = 0$. Hence, it can be shown that

$$S_{22} = \left.\frac{b_2}{a_2}\right|_{a_1 = 0} = \frac{\left(\dfrac{V_2 - I_2 R_{g2}}{2\sqrt{R_{g2}}}\right)}{\left(\dfrac{V_2 + I_2 R_{g2}}{2\sqrt{R_{g2}}}\right)}\Bigg|_{I_{1i} = 0} = \frac{V_2 - I_2 R_{g2}}{V_2 + I_2 R_{g2}} = \frac{V_{2r}}{V_{2i}} = \frac{\sqrt{R_{2g}}I_{2r}}{\sqrt{R_{2g}}I_{2i}} = \frac{I_{2r}}{I_{2i}} \tag{4.146}$$

or

$$S_{22} = \frac{Z_{22} - R_{g2}}{Z_{22} + R_{g2}} \tag{4.147}$$

S_{22} is the reflection coefficient of the output. S_{12} can be obtained as

$$S_{12} = \left.\frac{b_1}{a_2}\right|_{a_1 = 0} = \frac{\left(\dfrac{V_1 - I_1 R_{g1}}{2\sqrt{R_{g1}}}\right)}{\left(\dfrac{V_2 + I_2 R_{g2}}{2\sqrt{R_{g2}}}\right)}\Bigg|_{I_{1i} = 0} = \frac{(V_1 - I_1 R_{g1})\sqrt{R_{g2}}}{(V_2 + I_2 R_{g2})\sqrt{R_{g1}}} = \frac{\sqrt{R_{1g}}I_{1r}}{\sqrt{R_{2g}}I_{2i}} \tag{4.148}$$

which can be put in the following form

$$S_{12} = \left.\frac{b_1}{a_2}\right|_{a_1 = 0} = -2\sqrt{R_{1g}}\sqrt{R_{2g}}\frac{I_1}{V_{2g}} = 2\sqrt{\frac{R_{2g}}{R_{1g}}}\frac{V_1}{V_{2g}} = \frac{(V_1/\sqrt{R_{1g}})}{\left(\frac{1}{2}V_{2g}/\sqrt{R_{2g}}\right)} \tag{4.149}$$

S_{12} is the reverse transmission gain of the network from port 2 to port 1. Overall, S parameters are found when $a_n = 0$, which means that there is no reflection at that port. This is only possible by matching all the ports except the measurement port. Insertion loss and return loss in terms of S parameters are defined as

$$\text{Insertion loss (dB)} = IL(\text{dB}) = 20\log\left(|S_{ij}|\right), i \neq j \tag{4.150}$$

$$\text{Re turn loss (dB)} = RL(\text{dB}) = 20\log\left(|S_{ii}|\right) \tag{4.151}$$

Another important parameter that can be defined using S parameters is the voltage standing wave ratio, (VSWR). For instance, the VSWR at port 1 is found from

$$\text{VSWR} = \frac{1 - |S_{11}|}{1 + |S_{11}|} \tag{4.152}$$

The two-port network is reciprocal if

$$S_{21} = S_{12} \tag{4.153}$$

It can be shown that a network is reciprocal if it is equal to its transpose. This is represented for two-port networks as

$$[S] = [S]^t \tag{4.154}$$

or

$$\begin{bmatrix} S_{11} & S_{12} \\ S_{21} & S_{22} \end{bmatrix}^t = \begin{bmatrix} S_{11} & S_{21} \\ S_{12} & S_{22} \end{bmatrix} \tag{4.155}$$

When a network is lossless, S parameters can be used to characterize this feature as

$$[S]^t [S]^* = [U] \tag{4.156}$$

where $*$ defines the complex conjugate of a matrix and U is the unitary matrix and defined by

$$[U] = \begin{bmatrix} 1 & 0 \\ 0 & 1 \end{bmatrix} \tag{4.157}$$

Equation (4.156) can be applied for a two-port network as

$$[S]^t [S]^* = \begin{bmatrix} \left(|S_{11}|^2 + |S_{21}|^2 \right) & \left(S_{11}S_{12}^* + S_{21}S_{22}^* \right) \\ \left(S_{12}S_{11}^* + S_{22}S_{21}^* \right) & \left(|S_{12}|^2 + |S_{22}|^2 \right) \end{bmatrix} = \begin{bmatrix} 1 & 0 \\ 0 & 1 \end{bmatrix} \tag{4.158}$$

We can further show that if a network is lossless and reciprocal it satisfies

$$|S_{11}|^2 + |S_{21}|^2 = 1 \tag{4.159}$$
$$S_{11}S_{12}^* + S_{21}S_{22}^* = 0 \tag{4.160}$$

Example 4.8   Scattering Parameters for T Network
Find the characteristic impedance of the T network given in Figure 4.30 to have no return loss at the input port.

Solution
The scattering parameter for the T network can be found from (4.120). From (4.120), S_{11} is equal to

$$S_{11} = \frac{b_1}{a_1}\bigg|_{a_2 = 0} = \frac{Z_{in} - Z_o}{Z_{in} + Z_o} \tag{4.161}$$

where

$$Z_{in} = Z_A + \left[\frac{Z_C(Z_B + Z_o)}{Z_C + (Z_B + Z_o)} \right] \tag{4.162}$$

Figure 4.30 T network configuration.

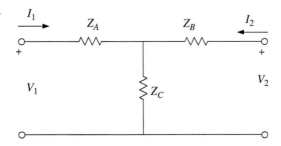

No return loss is possible when $S_{11} = 0$. This can be satisfied from (4.161) to (4.162) when

$$Z_o = Z_{in} = Z_A + \left[\frac{Z_C(Z_B + Z_o)}{Z_C + (Z_B + Z_o)} \right] \tag{4.163}$$

Example 4.9   Scattering Parameters for Transformers

Consider the typical transformer coupling circuit used for RF power amplifiers given in Figure 4.31. The primary and secondary sides of the transformer circuit become resonant at the frequency of operation. The coupling factor M is also set for maximum power transfer. Derive S parameters of the circuit.

Solution

The scattering parameters for S_{11} and S_{21} are found by connecting the source and generator impedance only at port 1 and the generator impedance at port 2, as shown in Figure 4.32.

Application of KVL on the left and right sides of the circuit gives

$$V_{1g} - I_1 R_{g1} + jI_1 X_{C1} - jI_1 X_{L1} - j\omega I_2 M = 0 \tag{4.164}$$

$$I_2 R_{g2} - jI_2 X_{C2} + jI_2 X_{L2} + j\omega I_1 M = 0 \tag{4.165}$$

From 4.164 to 4.165, we obtain

$$V_{g1} = I_1 \left(R_{g1} - jX_{C1} + jX_{L1} + \frac{\omega^2 M}{R_{g2} - jX_{C2} + jX_{L2}} \right) \tag{4.166}$$

$\dfrac{\omega^2 M}{R_{g2} - jX_{C2} + jX_{L2}}$ represented the secondary impedance referred to the primary side. It is communicated in the problem that the circuit is at resonance with the operational frequency. Then, 4.166 is simplified to

$$V_{g1} = I_1 \left(R_{g1} + \frac{\omega^2 M}{R_{g2}} \right) \tag{4.167}$$

From 4.167, it is seen that the maximum power transfer occurs when

$$R_{g1} = \frac{\omega^2 M}{R_{g2}} \tag{4.168}$$

Hence, the coupling factor, M, is found from 4.168 as

$$M = \omega \sqrt{R_{g1} R_{g2}} \tag{4.169}$$

As a result of 4.168 and 4.169, Z_{11} is equal to

$$Z_{11} = R_{g1} \tag{4.170}$$

From Eq. (4.143), S_{11} is found as

$$S_{11} = 0 \tag{4.171}$$

From (4.145)

$$S_{21} = -2\sqrt{R_{1g}}\sqrt{R_{2g}}\frac{I_2}{V_{1g}} \tag{4.172}$$

When 4.169 is substituted into 4.164 and 4.165, they are simplified to

$$V_{1g} - I_1 R_{g1} - jI_2 \sqrt{R_{g1} R_{g2}} = 0 \tag{4.173}$$

$$I_2 R_{g2} + jI_1 \sqrt{R_{g1} R_{g2}} = 0 \tag{4.174}$$

Solving 4.173 and 4.174 for ratio of (I_2/V_{1g}) gives

$$\frac{I_2}{V_{1g}} = \frac{-j}{2\sqrt{R_{1g}R_{2g}}} \tag{4.175}$$

Substitution of 4.175 into 4.172 gives S_{21} as

$$S_{21} = -2\sqrt{R_{1g}}\sqrt{R_{2g}}\left(\frac{-j}{2\sqrt{R_{1g}R_{2g}}}\right) \tag{4.176}$$

Then

$$S_{21} = j = 1\angle 90° \tag{4.177}$$

The scattering parameters for S_{22} and S_{12} are found by connecting the source and generator impedance only at port 2 and the generator impedance at port 1, as shown in Figure 4.33.

We can apply KVL for both sides of the circuit with 4.169 and obtain

$$I_1 R_{g1} + jI_2\sqrt{R_{g1}R_{g2}} = 0 \tag{4.178}$$

$$V_{g2} - I_2 R_{g2} - jI_1\sqrt{R_{g1}R_{g2}} = 0 \tag{4.179}$$

Solving 4.178 and 4.179 for V_{g2} gives

$$V_{g2} = I_2\left(R_{g1} + R_{g2}\right) \tag{4.180}$$

Since from 4.168 to 4.169

$$R_{g2} = \frac{\omega^2 M}{R_1} \tag{4.181}$$

Then

$$Z_{22} = R_{g2} \tag{4.182}$$

Substitution of 4.182 into (4.147) leads to

$$S_{22} = 0 \tag{4.183}$$

S_{12} is calculated by finding the ratio of (I_1/V_{g2}) from Eqs. 4.178 and 4.179 as

$$\frac{I_1}{V_{2g}} = \frac{-j}{2\sqrt{R_{1g}R_{2g}}} \tag{4.184}$$

From (4.149)

$$S_{12} = -2\sqrt{R_{1g}}\sqrt{R_{2g}}\frac{I_1}{V_{2g}} = -2\sqrt{R_{1g}}\sqrt{R_{2g}}\left(\frac{-j}{2\sqrt{R_{1g}R_{2g}}}\right) \tag{4.185}$$

Hence, S_{12} is equal to

$$S_{12} = j = 1\angle 90° \tag{4.186}$$

Then, the S matrix for a coupling transformer can be written as

$$S = \begin{bmatrix} 0 & j \\ j & 0 \end{bmatrix} \tag{4.187}$$

or

$$S = \begin{bmatrix} 0 & 1\angle 90° \\ 1\angle 90° & 0 \end{bmatrix} \tag{4.188}$$

The table that is illustrating the *ABCD* and *S* parameters for some of the basic RF components is given in Table 4.2.

4.9 Measurement of *S* Parameters

In this section, measurement of scattering parameters for 2-port and 3-port networks will be discussed. In addition, design of test fixture to measure scattering parameters will also be given (Figures 4.31–4.33 and Table 4.2).

4.9.1 Measurement of *S* Parameters for Two-port Network

Two-port scattering parameters can be measured by expressing the incident and reflected waves in terms of circuit parameters. From (4.118) to (4.119)

$$b_1 = S_{11}a_1 + S_{12}a_2 \tag{4.189}$$

$$b_2 = S_{21}a_1 + S_{22}a_2 \tag{4.190}$$

It is given before that by (4.80) (4.81) and (4.126) and (4.134) that

$$[V_i] = \left[Z_g^*\right][I_i] \tag{4.191}$$

$$[V_r] = \left[Z_g\right][I_r] \tag{4.192}$$

$$[a] = \sqrt{R_g}[I_i] \tag{4.193}$$

$$[b] = \sqrt{R_g}[I_r] \tag{4.194}$$

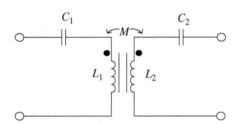

Figure 4.31 Transformer circuit.

Then, we can when we have real generator impedances we can write the following equations

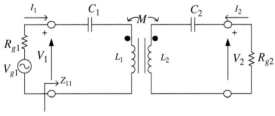

Figure 4.32 Transformer coupling circuit for S_{11} and S_{21}.

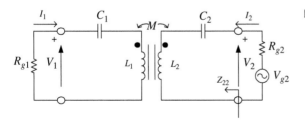

Figure 4.33 Transformer coupling circuit for S_{22} and S_{12}.

Table 4.2 *ABCD* and *S* parameters of basic network configurations.

			$N_1 : N_2$	ℓ
$z = \dfrac{Z}{Z_0}$	$y = \dfrac{Y}{Y_0}$		$n = N_1/N_2$	$\gamma = \alpha + j\beta$
1	1	**A**	$n = N_1/N_2$	$\cosh(\gamma\ell)$
Z	0	**B**	0	$Z_0 \sinh(\gamma\ell)$
0	Y	**C**	0	$\dfrac{\sinh(\gamma\ell)}{Z_0}$
1	1	**D**	$\dfrac{1}{n} = \dfrac{N_2}{N_1}$	$\cosh(\gamma\ell)$
$\dfrac{z}{z+2}$	$\dfrac{-y}{y+2}$	S_{11}	$\dfrac{n^2-1}{n^2+1}$	0
$\dfrac{2}{z+2}$	$\dfrac{2}{y+2}$	S_{12}	$\dfrac{2n}{n^2+1}$	$e^{-\gamma\ell}$
$\dfrac{2}{z+2}$	$\dfrac{2}{y+2}$	S_{21}	$\dfrac{2n}{n^2+1}$	$e^{-\gamma\ell}$
$\dfrac{z}{z+2}$	$\dfrac{-y}{y+2}$	S_{22}	$-\dfrac{n^2-1}{n^2+1}$	0

$$V_{i1} = R_{g1}I_{i1} \tag{4.195}$$

$$V_{r1} = R_{g1}I_{r1} \tag{4.196}$$

$$V_{i2} = R_{g2}I_{i2} \tag{4.197}$$

$$V_{r2} = R_{g2}I_{r2} \tag{4.198}$$

and

$$a_1 = \sqrt{R_{g1}}I_{i1} \tag{4.199}$$

$$b_1 = \sqrt{R_{g1}}I_{r1} \tag{4.200}$$

$$a_2 = \sqrt{R_{g2}}I_{i2} \tag{4.201}$$

$$b_2 = \sqrt{R_{g2}}I_{r2} \tag{4.202}$$

When (4.199)–(4.202) are substituted into (4.189) and (4.190), we obtain

$$\sqrt{R_{g1}}I_{r1} = S_{11}\sqrt{R_{g1}}I_{i1} + S_{12}\sqrt{R_{g2}}I_{i2} \tag{4.203}$$

$$\sqrt{R_{g2}}I_{r2} = S_{21}\sqrt{R_{g2}}I_{i1} + S_{22}\sqrt{R_{g2}}I_{i2} \tag{4.204}$$

When, the generator impedances at port 1 and 2 are equal, i.e. $R_{g1} = R_{g2} = R$, then (4.203) and (4.204) simplifies to

$$I_{r1} = S_{11}I_{i1} + S_{12}I_{i2} \tag{4.205}$$

$$I_{r2} = S_{21}I_{i1} + S_{22}I_{i2} \tag{4.206}$$

Similarly

$$V_{r1} = S_{11}V_{i1} + S_{12}V_{i2} \tag{4.207}$$

$$V_{r2} = S_{21}V_{i1} + S_{22}V_{i2} \tag{4.208}$$

Hence, from (4.207) to (4.208), the scattering parameters can be measured as

$$S_{11} = \left.\frac{V_{r1}}{V_{i1}}\right|_{V_{i2}=0} \qquad S_{12} = \left.\frac{V_{r1}}{V_{i2}}\right|_{V_{i1}=0}$$

$$S_{21} = \left.\frac{V_{r2}}{V_{i1}}\right|_{V_{i2}=0} \qquad S_{22} = \left.\frac{V_{r2}}{V_{i2}}\right|_{V_{i1}=0} \tag{4.209}$$

As can be seen from (4.209), the measurement of incident and reflected voltages at each port while the other port is terminated by a matched port will give the scattering parameters in (4.209). The incident and reflected voltage waves can be simply measured by directional couplers in practical applications.

4.9.2 Measurement of $\underline{S}$ Parameters for a Three-port Network

We obtain the following results by applying the same analysis for a three-port network. The incident and reflected voltage and current in terms of scattering parameters for a three-port network are obtained as

$$I_{r1} = S_{11}I_{i1} + S_{12}I_{i2} + S_{13}I_{i3} \tag{4.210}$$

$$I_{r2} = S_{21}I_{i1} + S_{22}I_{i2} + S_{23}I_{i3} \tag{4.211}$$

$$I_{r3} = S_{31}I_{i1} + S_{32}I_{i2} + S_{33}I_{i3} \tag{4.212}$$

Similarly

$$V_{r1} = S_{11}V_{i1} + S_{12}V_{i2} + S_{13}V_{i3} \tag{4.213}$$

$$V_{r2} = S_{21}V_{i1} + S_{22}V_{i2} + S_{23}V_{i3} \tag{4.214}$$

$$V_{r3} = S_{31}V_{i1} + S_{32}V_{i2} + S_{33}V_{i3} \tag{4.215}$$

Hence, the scattering parameters from (4.210) to (4.215) are found as

$$S_{11} = \left.\frac{V_{r1}}{V_{i1}}\right|_{V_{i2}=0,V_{i3}=0} \qquad S_{12} = \left.\frac{V_{r1}}{V_{i2}}\right|_{V_{i1}=0,V_{i3}=0}$$

$$S_{31} = \left.\frac{V_{r3}}{V_{i1}}\right|_{V_{i2}=0,V_{i3}=0} \qquad S_{33} = \left.\frac{V_{r3}}{V_{i3}}\right|_{V_{i1}=0,V_{i3}=0} \tag{4.216}$$

The complete conversion chart between S parameters and two-port parameters is given in Table 4.3.

MATLAB conversion codes to convert S parameters to Z parameters and $ABCD$ parameters to S parameters are given here.

```
%%%%%%%%%%%%%%%%%%%%%%%%%%%%%%%%%%%%%%%%%%%%%%%%%%%%%%%%%%%%%%%%%%%%%%%%%
% This m-file is function program to convert S Parameters to Z Parameters%
%                                                                       %
%%%%%%%%%%%%%%%%%%%%%%%%%%%%%%%%%%%%%%%%%%%%%%%%%%%%%%%%%%%%%%%%%%%%%%%%%

function [z11,z12,z21,z22]=S2Z(s11,s12,s21,s22,Zo)
zhi=(1-s11)*(1-s22)-s12*s21;
z11=Zo*(((1+s11)*(1-s22)+s12*s21)/zhi);
z12=Zo*(2*s12/zhi);
z21=Zo*(2*s21/zhi);
z22=Zo*(((1-s11)*(1+s22)+s12*s21)/zhi);
end
```

Table 4.3 Conversion chart between S parameters and two-port parameters.

S	Z	Y	ABCD
S_{11}	$\dfrac{(Z_{11}-Z_0)(Z_{22}+Z_0)-Z_{12}Z_{21}}{\Delta Z}$	$\dfrac{(Y_0-Y_{11})(Y_0+Y_{22})+Y_{12}Y_{21}}{\Delta Y}$	$\dfrac{A+B/Z_0-CZ_0-D}{A+B/Z_0+CZ_0+D}$
S_{12}	$\dfrac{2Z_{12}Z_0}{\Delta Z}$	$\dfrac{-2Y_{12}Y_0}{\Delta Y}$	$\dfrac{2(AD-BC)}{A+B/Z_0+CZ_0+D}$
S_{21}	$\dfrac{2Z_{21}Z_0}{\Delta Z}$	$\dfrac{-2Y_{21}Y_0}{\Delta Y}$	$\dfrac{2}{A+B/Z_0+CZ_0+D}$
S_{22}	$\dfrac{(Z_{11}+Z_0)(Z_{22}-Z_0)-Z_{12}Z_{21}}{\Delta Z}$	$\dfrac{(Y_0+Y_{11})(Y_0-Y_{22})+Y_{12}Y_{21}}{\Delta Y}$	$\dfrac{-A+B/Z_0-CZ_0+D}{A+B/Z_0+CZ_0+D}$
Z_{11}	$Z_0\dfrac{(1+S_{11})(1-S_{22})+S_{12}S_{21}}{(1-S_{11})(1-S_{22})-S_{12}S_{21}}$	Z_{11}	$\dfrac{Y_{22}}{|Y|}$	$\dfrac{A}{C}$
Z_{12}	$Z_0\dfrac{2S_{12}}{(1-S_{11})(1-S_{22})-S_{12}S_{21}}$	Z_{12}	$\dfrac{-Y_{12}}{|Y|}$	$\dfrac{AD-BC}{C}$
Z_{21}	$Z_0\dfrac{2S_{21}}{(1-S_{11})(1-S_{22})-S_{12}S_{21}}$	Z_{21}	$\dfrac{-Y_{21}}{|Y|}$	$\dfrac{1}{C}$
Z_{22}	$Z_0\dfrac{(1-S_{11})(1+S_{22})-S_{12}S_{21}}{(1-S_{11})(1-S_{22})-S_{12}S_{21}}$	Z_{22}	$\dfrac{Y_{11}}{|Y|}$	$\dfrac{D}{C}$
Y_{11}	$Y_0\dfrac{(1-S_{11})(1+S_{22})+S_{12}S_{21}}{(1+S_{11})(1+S_{22})-S_{12}S_{21}}$	$\dfrac{Z_{22}}{|Z|}$	Y_{11}	$\dfrac{D}{B}$
Y_{12}	$Y_0\dfrac{-2S_{12}}{(1+S_{11})(1+S_{22})-S_{12}S_{21}}$	$\dfrac{-Z_{12}}{|Z|}$	Y_{12}	$\dfrac{BC-AD}{B}$
Y_{21}	$Y_0\dfrac{-2S_{21}}{(1+S_{11})(1+S_{22})-S_{12}S_{21}}$	$\dfrac{-Z_{21}}{|Z|}$	Y_{21}	$\dfrac{-1}{B}$
Y_{22}	$Y_0\dfrac{(1+S_{11})(1-S_{22})+S_{12}S_{21}}{(1+S_{11})(1+S_{22})-S_{12}S_{21}}$	$\dfrac{Z_{11}}{|Z|}$	Y_{22}	$\dfrac{A}{B}$
A	$\dfrac{(1+S_{11})(1-S_{22})+S_{12}S_{21}}{2S_{21}}$	$\dfrac{Z_{11}}{Z_{21}}$	$\dfrac{-Y_{22}}{Y_{21}}$	A
B	$Z_0\dfrac{(1+S_{11})(1+S_{22})-S_{12}S_{21}}{2S_{21}}$	$\dfrac{|Z|}{Z_{21}}$	$\dfrac{-1}{Y_{21}}$	B
C	$\dfrac{1}{Z_0}\dfrac{(1-S_{11})(1-S_{22})-S_{12}S_{21}}{2S_{21}}$	$\dfrac{1}{Z_{21}}$	$\dfrac{-|Y|}{Y_{21}}$	C
D	$\dfrac{(1-S_{11})(1+S_{22})-S_{12}S_{21}}{2S_{21}}$	$\dfrac{Z_{22}}{Z_{21}}$	$\dfrac{-Y_{11}}{Y_{21}}$	D

$$|Z| = Z_{11}Z_{22} - Z_{12}Z_{21}; |Y| = Y_{11}Y_{22} - Y_{12}Y_{21}; \Delta Y = (Y_{11} + Y_0)(Y_{22} + Y_0) - Y_{12}Y_{21};$$

$$\Delta Z = (Z_{11} + Z_0)(Z_{22} + Z_0) - Z_{12}Z_{21}; Y_0 = 1/Z_0$$

and

```
%%%%%%%%%%%%%%%%%%%%%%%%%%%%%%%%%%%%%%%%%%%%%%%%%%%%%%%%%%%%%%%%%%%%%%%%
% This m-file is function program to convert ABCD Parameters to S Parameters%
%                                                                          %
%%%%%%%%%%%%%%%%%%%%%%%%%%%%%%%%%%%%%%%%%%%%%%%%%%%%%%%%%%%%%%%%%%%%%%%%

function [s11,s12,s21,s22]=A2S(a11,a12,a21,a22)
%Assume Zo=50
Zo=50;
DET=(a11+a12/Zo+a21*Zo+a22);
s11=(a11+a12/Zo-a21*Zo-a22)/DET;
s12=2*(a11*a22-a12*a21)/DET;
s21=2/DET;
s22=(-a11+a12/Zo-a21*Zo+a22)/DET;
end
```

4.10 Chain Scattering Parameters

When amplifier networks are cascaded, as shown in Figure 4.34, and the relation between the incident and reflected waves using scattering parameters is requested to be established, mathematically it is more efficient to use a direct matrix multiplication similar to *ABCD* matrices. We introduce a chain scattering matrix, *T*, to fulfill this requirement. The chain scattering matrix, *T*, can be expressed in terms of incident and reflected waves as

$$\begin{bmatrix} a_1 \\ b_1 \end{bmatrix} = \begin{bmatrix} T_{11} & T_{12} \\ T_{21} & T_{22} \end{bmatrix} \begin{bmatrix} b_2 \\ a_2 \end{bmatrix} \tag{4.217}$$

Hence, we can write chain scattering matrices for network A and B as

$$\begin{bmatrix} a_1^A \\ b_1^A \end{bmatrix} = \begin{bmatrix} T_{11}^A & T_{12}^A \\ T_{21}^A & T_{22}^A \end{bmatrix} \begin{bmatrix} b_2^A \\ a_2^A \end{bmatrix} \tag{4.218}$$

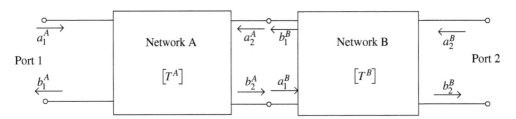

Figure 4.34 Illustration of chain scattering matrix for cascaded networks.

$$\begin{bmatrix} a_1^B \\ b_1^B \end{bmatrix} = \begin{bmatrix} T_{11}^B & T_{12}^B \\ T_{21}^B & T_{22}^B \end{bmatrix} \begin{bmatrix} b_2^B \\ a_2^B \end{bmatrix} \tag{4.219}$$

It is also seen that from Figure 4.34

$$\begin{bmatrix} a_2^A \\ b_2^A \end{bmatrix} = \begin{bmatrix} b_1^B \\ a_1^B \end{bmatrix} \tag{4.220}$$

Then, using (4.220), we can (4.219) into (4.218) and obtain

$$\begin{bmatrix} a_1^A \\ b_1^A \end{bmatrix} = \begin{bmatrix} T_{11}^A & T_{12}^A \\ T_{21}^A & T_{22}^A \end{bmatrix} \begin{bmatrix} T_{11}^B & T_{12}^B \\ T_{21}^B & T_{22}^B \end{bmatrix} \begin{bmatrix} b_2^B \\ a_2^B \end{bmatrix} \tag{4.221}$$

or

$$\begin{bmatrix} a_1^A \\ b_1^A \end{bmatrix} = [T] \begin{bmatrix} b_2^B \\ a_2^B \end{bmatrix} \tag{4.222}$$

where

$$[T] = \begin{bmatrix} T^A \end{bmatrix} \begin{bmatrix} T^B \end{bmatrix} \tag{4.223}$$

and

$$\begin{bmatrix} T^A \end{bmatrix} = \begin{bmatrix} T_{11}^A & T_{12}^A \\ T_{21}^A & T_{22}^A \end{bmatrix} \tag{4.224}$$

and

$$\begin{bmatrix} T^B \end{bmatrix} = \begin{bmatrix} T_{11}^B & T_{12}^B \\ T_{21}^B & T_{22}^B \end{bmatrix} \tag{4.225}$$

The chain scattering parameters are found from scattering parameters and defined by

$$\begin{aligned} T_{11} &= \frac{1}{S_{21}} \\ T_{21} &= \frac{S_{11}}{S_{21}} \end{aligned} \tag{4.226}$$

$$\begin{aligned} T_{12} &= -\frac{S_{22}}{S_{21}} \\ T_{22} &= \frac{-(S_{11}S_{22} - S_{12}S_{21})}{S_{21}} = -\frac{\Delta S}{S_{21}} \end{aligned} \tag{4.227a}$$

So

$$\begin{bmatrix} T_{11} & T_{12} \\ T_{21} & T_{22} \end{bmatrix} = \begin{bmatrix} \dfrac{1}{S_{21}} & -\dfrac{S_{22}}{S_{21}} \\ \dfrac{S_{11}}{S_{21}} & S_{12} - \dfrac{S_{11}S_{22}}{S_{21}} \end{bmatrix} \tag{4.227b}$$

4.11 Engineering Application Examples

In this section, the design of a test fixture to extract the *S* parameters of a transistor is given with analytical, simulation, and measurement results. Prior to characterizing a component with a given network analyzer, it is first necessary to calibrate the instrument for a given test setup. This is done in order to remove the effects of a test fixture in the measurement. Most of the modern network analyzers have integrated mathematical algorithms that can be utilized to calibrate out these effects seen by each port using standard network analyzer error models, and thus allowing the user to more easily obtain accurate measurements. Better accuracy in measurement of the device characteristics can be obtained using full two-port calibration with a method such as the short-open-load-thru (SOLT) method. In SOLT calibration, the analyzer is subjected to a series of known configuration setups, as shown in Figure 4.35. During these measurements, the network analyzer obtains the *S* parameters of the fixture used. Once these are known, the network analyzer can easily remove the effects of the fixturing through the utilization of an error matrix generated during calibration.

Test fixtures for the SOLT calibration method are designed using a grounded coplanar waveguides (GCPW) structure, as illustrated in Figure 4.36. A GCPW structure consists of a center conductor of width *W*, with a gap of width *s* on either side, separating it from a ground plane. Owing to the presence of the center conductor, the transmission line can support both even and odd quasi-TEM modes, which is dependent upon the E-fields in the tow gaps that are in either of the opposite directions, or the same directions. This type of transmission line is therefore considered a good design choice for active devices due to both the center conductor, and the close proximity of surrounding ground planes. The GCPW shown in Figure 4.35 with finite thickness dielectric,*h*, a finite

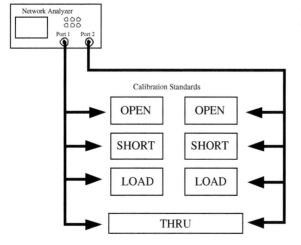

Figure 4.35 Implementation of SOLT calibration for network analyzer.

Figure 4.36 GCPW for SOLT calibration fixture implementation.

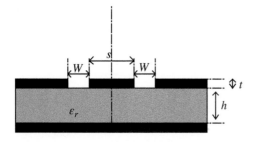

trace thickness, t, a center conductor of width W, a gap of width s, and an infinite ground plane is analyzed using the quasistatic approach given in [3–4] (Figure 4.36).

Design Example 4.1   Test Fixture Design for Measuring Scattering Parameters

Design and build a test fixture and then measure the S parameters of the BFR92 transistor at 700 MHz, and compare your results with published data.

Solution

The design details and measurement results are given below.

Coplanar Waveguide (CPW) Design

In order to provide an efficient, repeatable solution, the design equations are modeled with MATLAB. The model functions as a CPW calculator. It allows the user to enter specific parameters and requirements into the model, such as width, gap, relative permittivity of the dielectric, trace thickness, frequency, and length. The model then generates the resulting characteristic impedance and electrical length based on the input parameters. The user may continue to tune the input parameters until the characteristic impedance and electrical length meet the defined requirements. The analysis of the transistor is requested to be done using a CPW with a characteristic impedance of 50 Ω at 700 MHz. In order to easily verify the solution, a quarter-wave (90°) was targeted as the electrical length of the line. Using the MATLAB model the values detailed in Table 4.4 were calculated for the width, gap, and length of the CPW.

After using the model to determine the dimensions, Ansoft Designer's TRL calculator was used to verify the calculations. The TRL calculator in Ansoft Designer has more accurate modeling properties when it comes to the board and substrate materials. Plugging in the resulting values from the MATLAB model revealed the values detailed in Table 4.5.

When Tables 4.4 and 4.5 are compared, the MATLAB model results match the Ansoft Designer TRL calculator results.

Quarter-wave Stub Design

The stubs were to be designed using a parallel coaxial line and attached to both the input and output ports. The stubs are "invisible" at the design frequency of 700 MHz. A frequency of 100 MHz was chosen as the target frequency for the quarter-wave stub design. Ansoft Designer's coaxial line TRL calculator was used to design the quarter-wave 50 Ω coaxial stub lines at 100 MHz. From the calculator, a length of 29 507 mil is calculated for the stub.

SOLT Test Fixtures

SOLT test fixtures are designed repeating the method described above. They are implemented as shown in Figures 4.37–4.40. Rogers 4003 with a relative permittivity of 3.38 is used as a dielectric substrate. The board material has a substrate thickness of 0.060 in. with 1 oz copper plating on both sides. The gap size of 10 mi is used, as discussed and illustrated in Figure 4.36. In addition, several vias were placed through the board in order to provide a good electrical connection between the top and bottom layer ground planes to further improve the grounding of the fixture. Each board layout for the various fixtures incorporated a 0.25 in. section of transmission line at the output of each connector. This consistent length of transmission line is important in order to ensure that the ports of the network analyzer are subjected to the same additional line impedance regardless of the fixture connected. By doing so, the error matrix generated by the analyzer will be appropriate for this selected length. Two SMA (subminiature version A) connectors were mounted on port 1 and port 2 of each fixture, which were utilized for connecting to the network analyzer.

Once the network analyzer had been calibrated by means of a two-port calibration using the SOLT method with the designed calibration standards, as shown in Figures 4.37–4.40, the biasing circuit depicted in Figure 4.41 is simulated with a nonlinear circuit simulator of Ansoft Designer and implemented to characterize the active device, BFR92, for various biasing conditions. Simulation of the biasing circuit is required to determine the best direct

Figure 4.37 "Thru" calibration test fixture – top view.

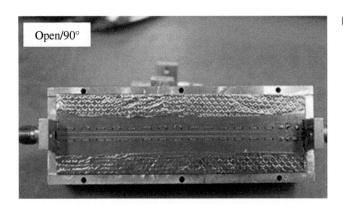

Figure 4.38 "Open" calibration test fixture.

Figure 4.39 "Load" calibration test fixture.

Figure 4.40 "Short" calibration test fixture.

Figure 4.41 BFR92 bias simulation test setup.

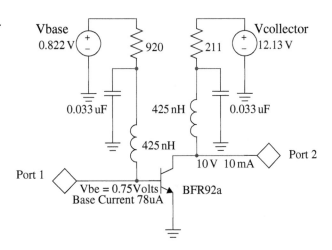

current (DC) biasing conditions. DC biasing conditions can be determined by adjusting the voltage sources V_{base} and $V_{collector}$ while maintaining constant current limiting resistors R_{base} and $R_{collector}$. A 425 nH inductor was placed in series with each current limiting resistor with their respective transistor pins in order to isolate these ports from the DC biases at microwave frequencies. The inductor selected (PN: CC21T36K2406S) was a conical inductor whose resonance frequency was greater than 1 GHz. In addition, a 0.033 µF capacitor was added in order to form a low pass filter with the current limiting resistors. The sole purpose of this RC was to provide a clean DC supply at the base and collector of the transistor under test. Utilizing the test setup shown in Figure 4.42, the BFR92 transistor could be easily tested under various bias conditions with the help of simulation. The circuit shown in Figure 4.42 is constructed, and characterization of the device has been performed under bias, as shown in Figure 4.43. The complete network analyzer measurement test setup used for characterization is illustrated in Figure 4.44.

Quarter-wave CPW Open and Short Circuit Simulation Results
The following plots provide the simulation results for the CPW design using the MATLAB model. A quarter-wave line was simulated using both short and open loads. Figure 4.44 shows the circuit constructed in Ansoft Designer used for the simulations. Simulation responses for both a quarter-wave short circuit and quarter-wave

Figure 4.42 Biasing test fixture for BRF92.

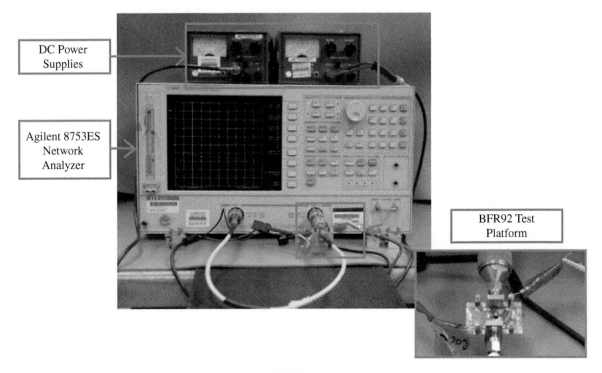

DC Power
Supplies

Agilent 8753ES
Network
Analyzer

BFR92 Test
Platform

Figure 4.43 Network analyzer measurement setup of BRF 92.

open circuit are illustrated in Figures 4.45 and 4.46, respectively. From the plots, it is clear that, for a quarter-wave short circuit, the input impedance appears as an open circuit at the design frequency of 700 MHz. Likewise,

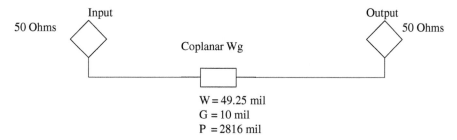

W = 49.25 mil
G = 10 mil
P = 2816 mil

Figure 4.44 Quarter-wave coplanar waveguide transmission line used for simulations.

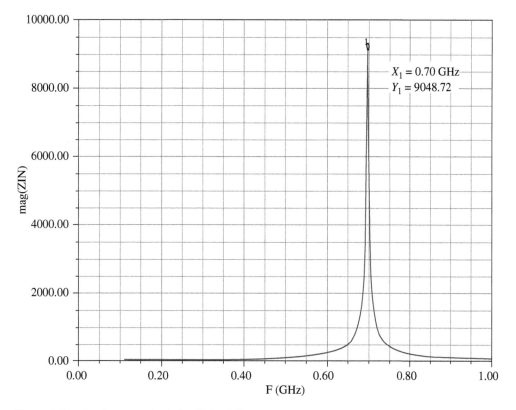

Figure 4.45 Quarter-wave short circuit simulation.

for a quarter-wave open circuit, the input impedance appears as a short circuit at 700 MHz. These plots verify that the CPW design used in the test fixture meets the specifications mentioned in the problem.

Investigation of the circuit when quarter-wave stubs were attached to the main transmission line's input and output ports has been performed. Both a short circuit stub and a stub with a capacitor on the end were introduced to the circuit and analyzed in Ansoft Designer. Before attaching the stubs, the S parameters of the CPW transmission line were captured to serve as a reference when comparing the results with the stubs attached. Figure 4.47 provides the plot of the S parameters with no stubs attached. It is shown in Figure 3.35 that $S_{11} = S_{22}$ and $S_{12} = S_{21}$. This is consistent with what would be expected. There should be no forward (S_{21}) or reverse gain (S_{12}) and very

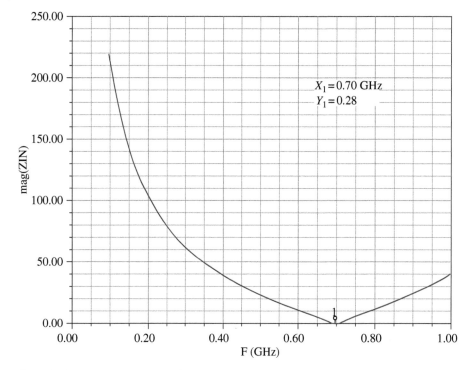

Figure 4.46 Quarter-wave open circuit simulation.

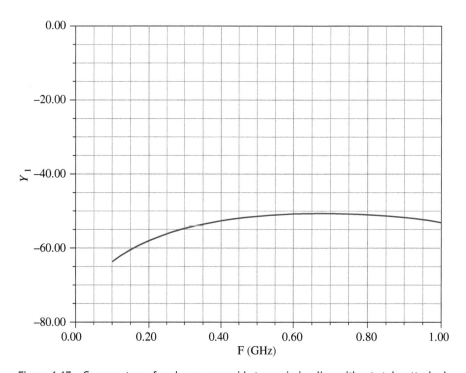

Figure 4.47 *S* parameters of coplanar waveguide transmission line without stubs attached.

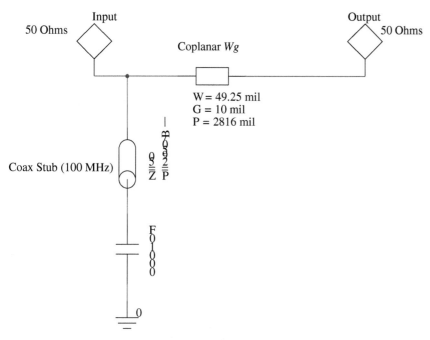

Figure 4.48 Quarter-wave stub attached to input port.

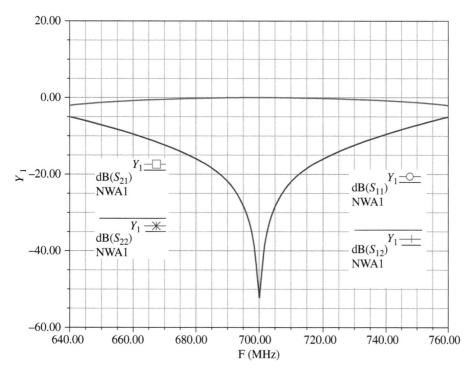

Figure 4.49 Quarter-wave short circuit stub attached at input port.

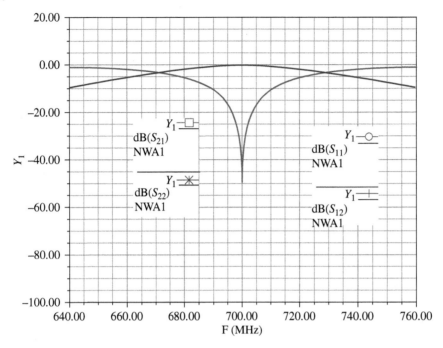

Figure 4.50 Quarter-wave stub with capacitor attached at input port.

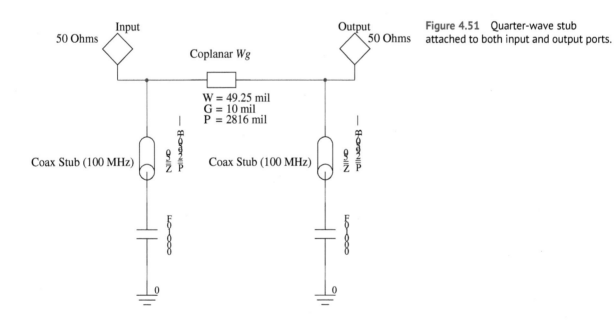

Figure 4.51 Quarter-wave stub attached to both input and output ports.

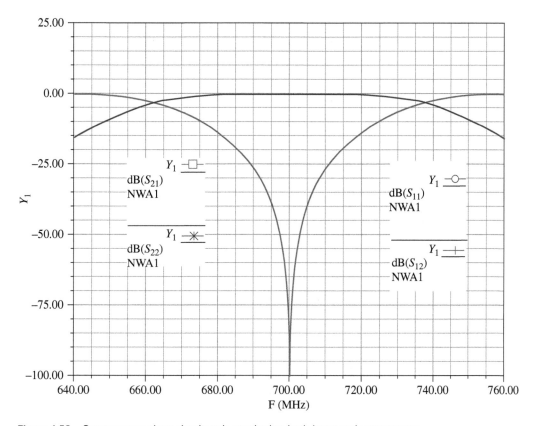

Figure 4.52 Quarter-wave short circuit stub attached to both input and output ports.

little reflection at both ends (S_{11} and S_{22}) due to the 50 Ω transmission line which provides a good impedance match to the ports.

Figure 4.48 shows the circuit in Ansoft Designer used for the analysis to verify that the short circuit stub attached to the input port is "invisible" at the design frequency of 700 MHz. Ansoft Designer is also used to do same analysis only replacing the short end of the stub by a small capacitor, which results in an open circuit stub at high frequencies Simulations were run and the S parameters were captured and plotted. Figures 4.49 and 4.50 provide the plots of the S parameters for the short circuit and capacitor attached cases respectively. Note that the capacitor was removed for the short circuit stub analysis.

From the results in Figures 4.49 and 4.50, it is clear that neither the short nor the open circuit quarter-wave stubs had any negative effects on the circuit's performance at the design frequency of 700 MHz. In both cases, the forward and reverse gains were unaffected, as were the reflection coefficients at both ports. Ansoft Designer is also used to attach the same short and open stub to the output port and verify that the stub is again "invisible" at the design frequency of 700 MHz, as shown in Figure 4.51. Simulations were run and the S parameters were captured and plotted. Figures 4.52 and 4.53 provide the plots of the S parameters for the short circuit and capacitor attached cases, respectively. Note that the capacitor was removed for the short circuit stub analysis.

From the results in Figures 4.47 and 4.48, it is clear that neither the short nor the open circuit quarter-wave stubs had any negative effects on the circuit's performance at our design frequency of 700 MHz. Both the forward and reverse gains were unaffected in both cases. However, the reflection coefficients in the short circuit case actually provided better performance (20–30 dB) than the original circuit.

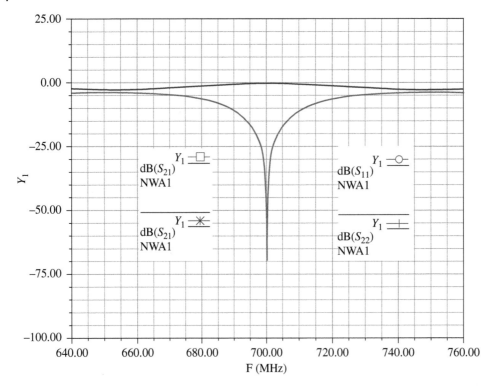

Figure 4.53 Quarter-wave stub with capacitor attached at both input and output ports.

Table 4.4 MATLAB CPW calculations.

MATLAB model calculations			
Input parameters		**Resulting calculations**	
Width (W)	49.25 mil		
Gap (s)	10 mil		
Length (l)	2816 mil		
Relative permittivity (er)	3.38		
Trace thickness (t)	1.4 mil		
Dielectric thickness (h)	32 mil		
Frequency (f)	700 MHz		
		Impedance (Z_0)	50.0105 Ω
		Electrical length (el)	90.0535°

Measurement Results

S parameter characterization data are obtained with the manufactured calibration set shown in Figure 4.42 and test fixtures Figures 4.37–4.40. The BFR92 transistor was measured under the six different bias conditions detailed in Table 4.6. The measured DC current gain under each biasing condition is summarized in the column title "β" in

Table 4.5   Ansoft designer CPW calculations.

	Ansoft designer TRL calculations		
Input parameters		**Resulting calculations**	
Width (W)	49.25 mil		
Gap (s)	10 mil		
Length (l)	2816 mil		
Relative permittivity (ε_r)	3.38		
Trace thickness (t)	1.4 mil		
Dielectric thickness (h)	32 mil		
Frequency (f)	700 MHz		
		Impedance (Z_0)	50.1467 Ω
		Electrical length (el)	90.5856º

Table 4.6   DC bias conditions.

V_{CE} (V)	I_C (mA)	V_{BE} (V)	V_{base} (V)	$V_{collector}$ (V)	I_B (μa)	β
10	5	0.769	0.823	11.04	59	85
10	10	0.765	0.865	12.11	109	91
10	15	0.751	0.889	13.16	150	100
5	5	0.785	0.845	6.05	65	77
5	10	0.794	0.909	7.10	125	80
5	15	0.797	0.966	8.15	184	82

Table 4.6. The measured S parameters were then plotted with the vendor-supplied spice model under the same biasing conditions, as well as the catalog vendor data for the given biases. These results are detailed in Figure 4.54–4.57 when $V_{CE} = 10$ V and $I_c = 5$ mA. As detailed in S parameter plots, the data measured using the BFR92 test fixture aligned themselves closely with both the vendor-supplied spice model and the vendor catalog data. The comparison of the measured and simulated data has also been done for all other conditions shown in Table 4.6. The agreement was again seen on all of them.

Design Example 4.2   Extrinsic and Intrinsic Parameters of Transistors

A manufacturer gives the following measured S parameters for the high power TO-247 MOSFET. It is communicated that at 3 MHz, low frequency measurement

$$S_{11} = 0.09 - j0.31; S_{12} = 0.89 + j0.01$$
$$S_{21} = 0.89 + j0.01; S_{22} = 0.02 - j0.34$$

at 300 MHz, HF measurement

$$S_{11} = -0.35 + j0.76; S_{12} = 0.2 + j0.1$$

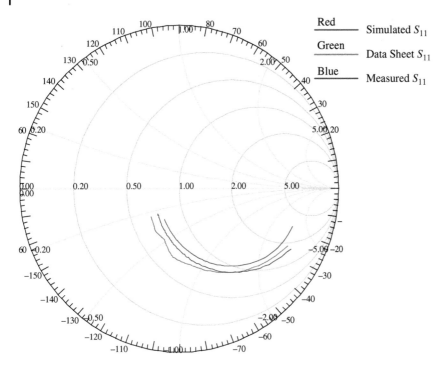

Figure 4.54　Input return loss comparison (S_{11}); V_{CE} = 10 V, I_C = 5 mA.

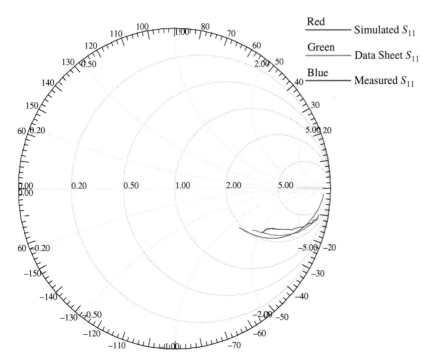

Figure 4.55　Output return loss comparison (S_{22}); V_{CE} = 10 V, I_C = 5 mA.

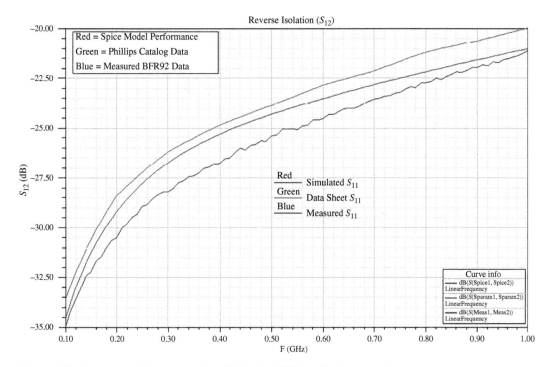

Figure 4.56 Reverse isolation comparison (S_{12}); V_{CE} = 10 V, I_C = 5 mA.

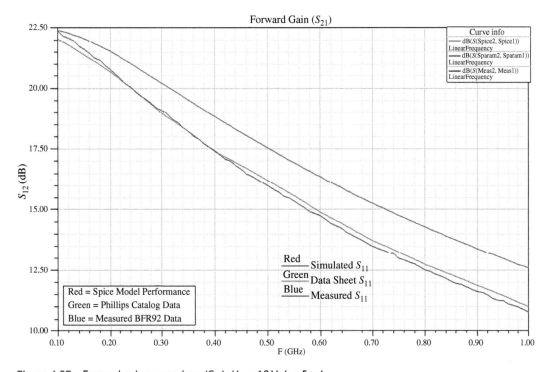

Figure 4.57 Forward gain comparison (S_{21}); V_{CE} = 10 V, I_C = 5 mA.

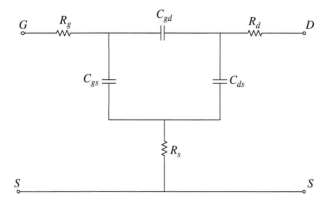

Figure 4.58 Zero biased small signal two-port network at low frequencies.

$$S_{21} = 0.2 + j0.1; S_{22} = -0.4 + j0.85$$

Calculate the extrinsic and intrinsic parameters of this device by ignoring test fixture effects. Compare your results with exact given high power TO-247 MOSFET extrinsic and intrinsic values which are

- $C_{iss} = C_{gs} + C_{gd} = 2700\,\text{pF}$, $C_{gs} = 2625\,\text{pF}$
- $C_{rss} = C_{gd} = 75\,\text{pF}$
- $C_{oss} = C_{ds} + C_{gd} = 350\,\text{pF}$, $C_{ds} = 275\,\text{pF}$
- $Lg = 13\,\text{nH}$, $Ls = 13\,\text{nH}$, $Ld = 5\,\text{nH}$
- $Rs = 0.95\,\Omega$, $Rd = 0.5\,\Omega$, $Rg = 5\,\Omega$

Use the zero bias MOSFET model given in Figure 4.58.

Solution

At zero bias, the network measured S parameters are dominated by capacitances and hence the small signal two-port network reduces to the circuit in Figure 4.58. The package parasitic resistances can be found by converting S parameters to Z parameters. The real part of the Z parameters are equal to the package inductance resistance values as given by

$$\text{Re}\{Z_{11}\} = R_g + R_s \tag{4.228}$$

$$\text{Re}\{Z_{22}\} = R_d + R_s \tag{4.229}$$

$$\text{Re}\{Z_{21}\} = \text{Re}\{Z_{21}\} = R_s \tag{4.230}$$

Frequency change does not affect the parasitic resistance values obtained in (4.228)–(4.230). Z parameters, Z^{DUT}, are the parameters measured with zero bias, as described elsewhere in this chapter. After parasitic resistances are identified using the Eqs. (4.228)–(4.230), device intrinsic parameters Z^i are found from

$$Z^i_{11} = Z^{DUT}_{11} - (R_g + R_s) \tag{4.231}$$

$$Z^i_{22} = Z^{DUT}_{22} - (R_d + R_s) \tag{4.232}$$

$$Z^i_{12} = Z^{DUT}_{12} - R_s \tag{4.233}$$

$$Z^i_{21} = Z^{DUT}_{21} - R_s \tag{4.234}$$

Y parameters of the intrinsic MOSFET components, Y^i, can be obtained from

$$Y^i = \frac{1}{Z^i} \tag{4.235}$$

or using the conversion parameters given previously. Y parameters can be obtained as

$$Y_{11}^i = j\omega(C_{gs} + C_{gd}) \tag{4.236}$$

$$Y_{22}^i = j\omega(C_{gd} + C_{ds}) + \frac{1}{R_{ds}} \tag{4.237}$$

$$Y_{12}^i = -j\omega(C_{gd}) \tag{4.238}$$

$$Y_{21}^i = g_m - j\omega C_{gd} \tag{4.239}$$

Hence, MOSFET intrinsic parameters are

$$C_{gd} = \frac{-\operatorname{Im}(Y_{12}^i)}{2\pi f} \tag{4.240}$$

$$C_{gs} = \frac{\operatorname{Im}(Y_{11}^i) + \operatorname{Im}(Y_{12}^i)}{2\pi f} \tag{4.241}$$

$$C_{ds} = \frac{\operatorname{Im}(Y_{22}^i) + \operatorname{Im}(Y_{12}^i)}{2\pi f} \tag{4.242}$$

$$R_{ds} = \frac{1}{\operatorname{Re}(Y_{22}^i)} \tag{4.243}$$

$$g_m e^{-j\omega\tau} = Y_{21}^i - Y_{12}^i \tag{4.244}$$

where

$$g_m = |Y_{21}^i - Y_{12}^i| \tag{4.245}$$

$$\tau = \frac{\tan^{-1}\left(\frac{\operatorname{Im}\{Y_{21}^i - Y_{12}^i\}}{\operatorname{Re}\{Y_{21}^i - Y_{12}^i\}}\right)}{2\pi f} \tag{4.246}$$

Using (4.236)–(4.238), Z^i parameters can be obtained as

$$Z_{11}^i = \frac{Y_{22}^i}{|Y^i|} = \frac{R_{ds} + j\omega(C_{gd} + C_{ds})}{Y_{11}^i Y_{22}^i - Y_{12}^i Y_{21}^i} \tag{4.247}$$

$$Z_{12}^i = -\frac{Y_{12}^i}{|Y^i|} = \frac{j\omega C_{gd}}{Y_{11}^i Y_{22}^i - Y_{12}^i Y_{21}^i} \tag{4.248}$$

$$Z_{21}^i = -\frac{Y_{21}^i}{|Y^i|} = \frac{-g_m + j\omega C_{gd}}{Y_{11}^i Y_{22}^i - Y_{12}^i Y_{21}^i} \tag{4.249}$$

$$Z_{22}^i = \frac{Y_{11}^i}{|Y^i|} = \frac{j\omega(C_{gd} + C_{gs})}{Y_{11}^i Y_{22}^i - Y_{12}^i Y_{21}^i} \tag{4.250}$$

When the intrinsic and extrinsic parameters of the device are combined, we obtain Z parameters of the device Z^{DUT} from (4.231) to (4.234) as

$$Z_{11}^{DUT} = [(R_g + R_s) + j\omega(L_g + L_s)] + Z_{11}^i = [(R_g + R_s) + j\omega(L_g + L_s)] + \frac{g_{ds} + j\omega(C_{gd} + C_{ds})}{Y_{11}^i Y_{22}^i - Y_{12}^i Y_{21}^i} \tag{4.251}$$

$$Z_{12}^{DUT} = [(R_s) + j\omega(L_s)] + Z_{12}^i = [R_s + j\omega L_s] + \frac{j\omega C_{gd}}{Y_{11}^i Y_{22}^i - Y_{12}^i Y_{21}^i} \tag{4.252}$$

$$Z_{21}^{DUT} = [(R_s) + j\omega(L_s)] + Z_{21}^i = [R_s + j\omega L_s] + \frac{-g_m + j\omega C_{gd}}{Y_{11}^i Y_{22}^i - Y_{12}^i Y_{21}^i} \tag{4.253}$$

$$Z_{22}^{DUT} = [(R_d + R_s) + j\omega(L_d + L_s)] + Z_{22}^i = [(R_d + R_s) + j\omega(L_d + L_s)] + \frac{j\omega(C_{gd} + C_{gs})}{Y_{11}^i Y_{22}^i - Y_{12}^i Y_{21}^i} \tag{4.254}$$

A zero bias MOSFET model is given in Figure 4.58. MATLAB script is written to extract the intrinsic and extrinsic values of MOSFET using the formulation given by Eqs. (4.228)–(4.254).

References

1 Eroglu, A. (2015). *Introduction to RF Power Amplifiers Design and Simulation*, 1e. CRC Press. ISBN: 978-1-4822-3164-9.
2 Matthaei, G., Jones, E.M.T., and Young, L. (1980). *Microwave Filters, Impedance-Matching Networks, and Coupling Structures*. Artech House.
3 Wadell, B.C. (1991). *Transmission Line Design Handbook*. Artech House.
4 Wolf, I. (2006). *Coplanar Microwave Integrated Circuits*. Wiley.

Problems

Problem 4.1
Obtain the Y parameters of the circuits in Figure 4.59.

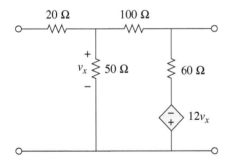

Figure 4.59 Problem 4.1.

Problem 4.2
Find the Z and Y parameters for the resistive π circuit shown in Figure 4.60.

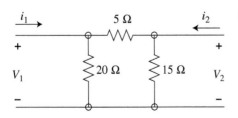

Figure 4.60 Problem 4.2.

Problem 4.3
Find *ABCD* parameters of the transformer in Figure 4.61

Figure 4.61   Problem 4.3.

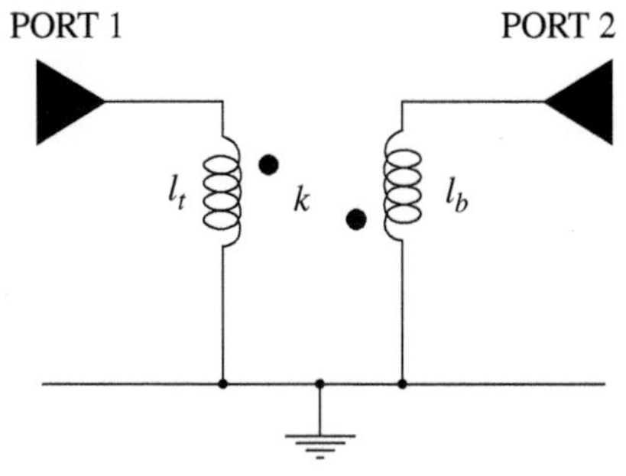

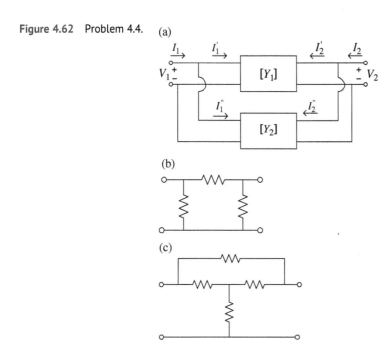

Problem 4.4
Consider the parallel connected network which is given in Figure 4.62a. Assume each network in this configuration is the π network shown in Figure 4.62b. (a) Prove that the overall Y parameters of the parallel connected network would be obtained by just adding the Y parameters of each network. (b) Use your results for the network given in Figure 4.62b.

Figure 4.62   Problem 4.4.

Problem 4.5
Find the return loss at the input port, port 1, for the series and shunt loads shown in Figure 4.63 when $Z_0 = 50\,\Omega$ and $Z = 100\,\Omega$.

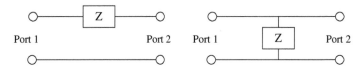

Figure 4.63 Problem 4.5.

Problem 4.6

Consider the amplifier circuit in Figure 4.64. In the circuit, $C_1 = 475.47\,\text{pF}$, $C_2 = 639\,\text{pF}$, $C_3 = 106.1\,\text{pF}$, $L_1 = 8.45\,\text{nH}$, $L_2 = 23.8\,\text{nH}$, and impedance transformer ratio $n_2/n_1 = 2.12$. The admittances of transistor are given as

$$Y_{\text{ie}} = (1.2 + j16)\,\text{mmho}, Y_{\text{fe}} = (40 - j150)\,\text{mmho}, Y_{\text{re}} = (0 - j1.5)\,\text{mmho}, Y_{\text{oe}} = (0.75 + j2.5)\,\text{mmho}$$

Calculate the *ABCD* parameters of the complete amplifier network and calculate the voltage gain.

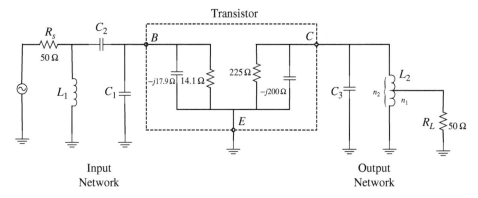

Figure 4.64 Problem 4.6.

Problem 4.7

Consider the network given in Figure 4.65. Calculate the *ABCD* parameters of the network and then find the load voltage, across Z_L. Show that the equivalent network obtained via *ABCD* parameters is a reciprocal network.

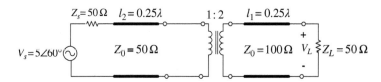

Figure 4.65 Problem 4.7.

Problem 4.8

If the voltage and current measurements of a two-port network are given as

$$\begin{bmatrix} V_1 \\ I_1 \end{bmatrix} = \begin{bmatrix} 10\angle 30^0 \\ 1\angle -60^0 \end{bmatrix} \text{ and } \begin{bmatrix} V_2 \\ I_2 \end{bmatrix} = \begin{bmatrix} 4\angle 60^0 \\ 0.5\angle -30^0 \end{bmatrix}$$

Calculate the incident voltages at the ports. Assume $Z_o = 50$.

Problem 4.9

Compare the 3 dB bandwidth and Q of the π network shown in Figure 4.66a and when it is cascaded in Figure 4.66b using scattering parameters.

Figure 4.66 Problem 4.9.

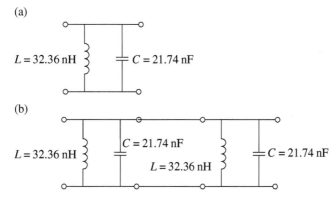

(a)

$L = 32.36\,\text{nH}$ $C = 21.74\,\text{nF}$

(b)

$L = 32.36\,\text{nH}$ $C = 21.74\,\text{nF}$ $L = 32.36\,\text{nH}$ $C = 21.74\,\text{nF}$

Problem 4.10

Calculate the hybrid parameters of a complementary metal oxide semiconductor (CMOS) amplifier cell given in Figure 4.67.

Figure 4.67 Problem 4.10.

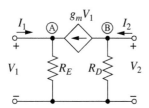

I_1 Ⓐ $g_m V_1$ Ⓑ I_2

V_1 R_E R_D V_2

Design Challenge 4.1 Calculation of Extrinsic and Intrinsic Parameters of Transistors with Test Fixture Effects

The test fixture shown in Figure 4.68 is used to measure the MOSFET with TO-247 package considered in Design Example 4.2.

It is communicated that the whole setup with the fixture gives the following measured S parameters.

At 3 MHz, low frequency measurement

$$S_{11} = 0.09 - j0.31; S_{12} = 0.89 + j0.01$$
$$S_{21} = 0.89 + j0.01; S_{22} = 0.02 - j0.34$$

(Continued)

Design Challenge 4.1 (Continued)

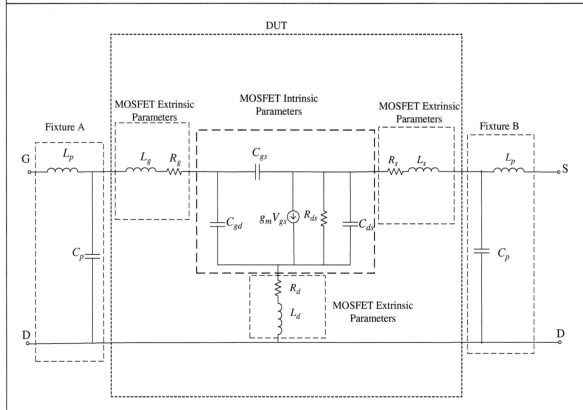

Figure 4.68 TO-247 MOSFET package measurement setup.

At 300 MHz, HF measurement

$$S_{11} = -0.31 + j0.78; S_{12} = 0.2 + j0.1$$
$$S_{21} = 0.2 + j0.1; S_{22} = -0.35 + j0.87$$

Calculate the extrinsic and intrinsic parameters of this device by including the test fixture effects. The test fixture is symmetric and can be represented by an LC network which has C_p = 0.1 pF and Lp = 1 nH.

Compare again your results with typical high power TO-247 MOSFET extrinsic and intrinsic values, which are

- $C_{iss} = C_{gs} + C_{gd}$ = 2700 pF, C_{gs} = 2625 pF
- $C_{rss} = C_{gd}$ = 75 pF
- $C_{oss} = C_{ds} + C_{gd}$ = 350 pF, C_{ds} = 275 pF
- L_g = 13 nH, Ls = 13 nH, Ld = 5 nH
- R_s = 0.95 Ω
- L_d = 5 nH

5

Impedance Matching

5.1 Introduction

Impedance matching provides the desired performance for high frequency (HF) circuits by minimizing the reflections. This provides the transfer of the maximum power to the load. This can be represented by

$$P_t = P_i + P_r = P_i\left(1 - |\Gamma|^2\right) \tag{5.1}$$

and

$$SWR = \frac{\left(1 + |\Gamma|^2\right)}{\left(1 - |\Gamma|^2\right)} \tag{5.2}$$

When impedance matching is accomplished, $|\Gamma| = 0$ and P_t maximizes and $SWR = 1$ from Eqs. (5.1) to (5.2) to provide optimum operation. Impedance matching can be done for a specific frequency of operation using narrow band matching techniques or it can be implemented to be effective over larger frequency bandwidths with broadband matching techniques. An impedance matching network is usually implemented between transmission and load, as illustrated in Figure 5.1. The matching networks can be implemented with lumped elements, distributed elements, or combination of both based on the application and frequency. In this chapter, design and implementation of matching networks are detailed using lumped and distributed elements for several cases using analytical and graphical methods with Smith Chart.

5.2 Impedance Matching Network with Lumped Elements

There are eight possible L matching networks that are shown in Figure 5.2 that can be used for matching source to load or load to source. These networks can be illustrated by generic two circuits, as shown in Figure 5.3 [1–2].

In either of the configurations of Figure 5.3, the reactive elements may be either inductors or capacitors. As a result, there are eight distinct possibilities, as shown in Figure 5.2 for the matching circuit for various load impedances. If the normalized load impedance, $z_L = Z_L/Z_0$, is inside the $1 + jx$ circle on the Smith chart, then the circuit of Figure 5.3a should be used. If the normalized load impedance is outside the $1 + jx$ circle on the Smith chart, the circuit of Figure 5.3b should be used. A $1 + jx$ circle is the resistance circle on the impedance Smith chart for which $r = 1$.

Consider first the circuit given in Figure 5.3a with $Z_L = R_L + jX_L$. It is assumed that $R_L > Z_0$ and $z_L = Z_L/Z_0$ maps inside the $1 + jx$ circle on the Smith chart. For a matched condition, the impedance seen looking into the matching network followed by the load impedance is then equal to Z_0 and can be written as

RF/Microwave Engineering and Applications in Energy Systems, First Edition. Abdullah Eroglu.
© 2022 John Wiley & Sons Ltd. Published 2022 by John Wiley & Sons Ltd.
Companion website: www.wiley.com/go/eroglu/rfmicrowave

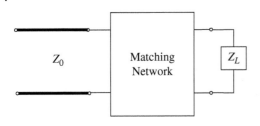

Figure 5.1 Implementation of matching network between t-line and load.

$$Z_0 = jX + \cfrac{1}{jB + \left(\cfrac{1}{R_L} + jX_L\right)} \qquad (5.3)$$

Separating Eq. (5.3) into real and imaginary parts gives two equations with two unknowns, X and B, as

$$B(XR_L - X_L Z_0) = R_L - Z_0 \qquad (5.4)$$

$$X(1 - BX_L) = BZ_0 R_L - X_L \qquad (5.5)$$

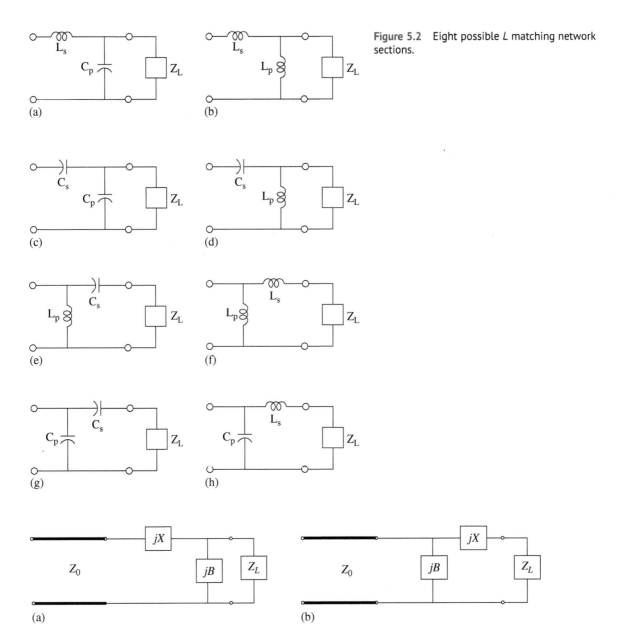

Figure 5.2 Eight possible L matching network sections.

Figure 5.3 Generic L matching network sections to represent eight L sections.

The solutions of (5.4) and (5.5) lead to

$$B = \frac{X_L \pm \sqrt{\frac{R_L}{Z_0}}\sqrt{R_L^2 + X_L^2 - Z_0 R_L}}{R_L^2 + X_L^2} \tag{5.6}$$

and

$$X = \frac{1}{B} + \frac{X_L Z_0}{R_L} - \frac{Z_0}{B R_L} \tag{5.7}$$

From (5.6), there exists two possible solutions for B and consequently X. Both of these solutions are physically realizable, and constitute all the values of B and X. Positive value of X gives an inductor, negative value of X gives a capacitor. Similarly, positive value of B gives a capacitor and negative value of B gives an inductor.

We can repeat the same procedure for the generic L matching network shown in Figure 5.3b. This circuit is used when $z_L = Z_L/Z_0$ and it maps outside the $1 + jx$ circle on the Smith chart since it is assumed that $R_L < Z_0$. For a matched condition, the admittance seen looking into the matching network followed by the load impedance $Z_L = R_L + jX_L$ is then equal to $1/Z_0$ and can be written as

$$\frac{1}{Z_0} = jB + \frac{1}{R_L + j(X + X_L)} \tag{5.8}$$

Separating Eq. (5.8) into real and imaginary parts gives the following two equations with two unknowns, X and B, as

$$BZ_0(X + X_L) = Z_0 - R_L \tag{5.9}$$

$$(X + X_L) = BZ_0 R_L \tag{5.10}$$

The solutions for (5.9) and (5.10) lead to

$$X = \sqrt{R_L(Z_0 - R_L)} - X_L \tag{5.11}$$

$$B = \pm \frac{\sqrt{(Z_0 - R_L)/R_L}}{Z_0} \tag{5.12}$$

Equation (5.12) has two possible solutions for B. In order to match an arbitrary complex load to a line of characteristic impedance Z_0, the real part of the input impedance to the matching network must be Z_0, while the imaginary part must be zero. This implies that a general matching network must have at least two degrees of freedom; in the L section matching circuit these two degrees of freedom are provided by the values of the two reactive components.

Example 5.1   L Matching Network
Apply impedance matching using L network in Figure 5.2h to match the output impedance of an amplifier, $Z_S = (150 + j75)\Omega$, at a frequency of 2 GHz to a terminated load whose impedance is $Z_{Load} = (75 + j15)\Omega$ so that the maximum power is delivered.

Solution
A maximum power transfer from source to load requires the source impedance to be equal to a complex conjugate of the load impedance.

$$Z_M = Z_{Load}^* = (75 - j15)\Omega$$

Z_M can be found as

$$Z_M = \frac{1}{Z_S^{-1} + jB_C} + jX_L = Z_{Load}^*$$

where

$$B_C = \omega C \text{ and } X_L = \omega L$$

Since

$$Z_S = R_S + jX_S \quad \text{and} \quad Z_{Load} = R_{Load} + jX_{Load}$$

and substituting them into

$$Z_M = \frac{1}{Z_S^{-1} + jB_C} + jX_L = Z_{Load}^*$$

gives

$$\frac{R_S + jX_S}{1 + jB_C(R_S + jX_S)} + jX_L = R_{Load} - jX_{Load}$$

Separating real and imaginary parts gives

$$R_S = R_{Load}(1 - B_C X_S) + (X_{Load} + X_L)B_C R_S$$

$$X_S = R_S R_{Load} B_C - (1 - B_C X_S)(X_{Load} + X_L)$$

Solving the equations above for B_C and X_L gives

$$B_C = \frac{X_S + \sqrt{\frac{R_S}{R_{Load}}\left(R_S^2 + X_S^2\right) - R_S^2}}{\left(R_S^2 + X_S^2\right)}$$

and

$$X_S = \frac{1}{B_C} - \frac{R_{Load}(1 - B_C X_S)}{B_C R_S} - X_{Load}$$

Since, $Z_S = (150 + j75)\Omega$ and $Z_{Load} = (75 + j15)\Omega$, then

$$B_C = 9.2 \text{ mS} \rightarrow C = B_C/\omega = 0.73 \text{ pF}$$

$$X_L = 76.9 \,\Omega \rightarrow L = X_L/\omega = 6.1 \text{ nH}$$

5.3 Impedance Matching with a Smith Chart – Graphical Method

As can be seen: the analytical calculation of the impedance matching is tedious. Another method is to use a *ZY* Smith chart as a graphical tool for the task illustrated in Figure 5.4. For this, there is a standard procedure that needs to be followed. The design procedure for matching source impedance to a load impedance using a Smith chart instead of an analytical method can be outlined as follows.

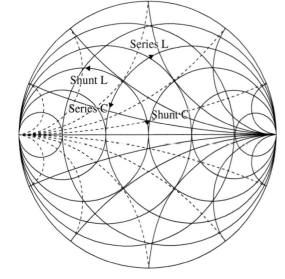

Figure 5.4 *ZY* Smith chart as a graphical tool for impedance matching.

- Normalize the given source and complex conjugate load impedances and locate them on the Smith chart.
- Plot constant resistance and conductance circles for the impedances located.
- Identify the intersection points between the constant resistance and conductance circle for the impedances located.
- The number of the intersection point correspond to the number of possible L matching networks.
- By following the paths that go through intersection points calculate normalized reactances and susceptances.
- Calculate the actual values of the inductors and capacitors by denormalizing at the given frequency.

Example 5.2   Smith Chart

Find the impedance matching network in Example 5.1 using a Smith chart. Assume $Z_0 = 75\,\Omega$.

Solution

In Example 5.1, $Z_S = (150 + j75)\Omega$, and $Z_{Load} = (75 + j15)\Omega$. When they are normalized

$$z_s = \frac{150 + j75}{75} = 2 + j1 \quad z_L = \frac{75 + j15}{75} = 1 + j0.2$$

In this problem, it is required to use the circuit in Figure 5.2h, as shown in Figure 5.5. We begin from Z_s, then add shunt C, and then we move on the circle shown until it intersects on the circle that Z_L^* lies. This point is illustrated as on the Smith chart. Since the movement is on the admittance circle, we also find the admittance value of the source impedance as

$$z_s = 2 + j1 \ \text{ or } \ y_s = 0.4 - j0.2$$

Z_1 is found on the Smith chart. Its impedance and admittance values are

$$z_1 = 1 - j1.22 \ \text{ or } \ y_1 = 0.4 + j0.49$$

The value of the added capacitance is calculated from the susceptance value as

$$jb_c = y_1 - y_s = j0.49 + j0.2 = j0.69$$

Then

$$C = \frac{b_c}{\omega Z_0} = \frac{0.69}{2\pi[2 \times 10^9]75} = 0.73 \text{ pF}$$

Now, we move from Z_1 to Z_s on the Smith chart. This movement is on the resistance circle, and it involves the addition of an inductor.

The value of the inductance is calculated from

$$jx_L = z_L^* - z_1 = -j0.2 + j1.22 = j1.02$$

The value of the inductance is found from

$$L = \frac{x_L Z_0}{\omega} = \frac{(1.02)75}{2\pi[2 \times 10^9]} = 6.09 \text{ nH}$$

Hence, the results confirm the analytical results.

Example 5.3   Two-element Matching Networks

Using a Smith chart, design all possible configurations of two-element matching networks that match source impedance

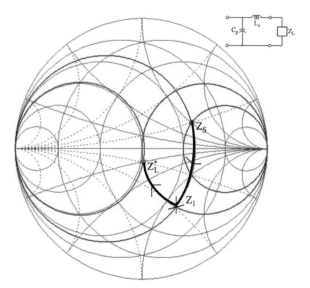

Figure 5.5   Smith chart solution for Example 5.2.

$Z_S = (25 + j70)\Omega$ to the load $Z_L = (10 - j10)\Omega$. Assume characteristic impedance of $Z_0 = 50\,\Omega$ and an operating frequency of $f = 2\,\text{GHz}$.

Solution

Given that $Z_L = 10 - j10$ and so normalized $z_{Ln} = \frac{Z_L}{Z_0} = \frac{10 - j10}{50} = 0.2 - j0.2$

So, the conjugate of normalized load impedance is $z_{Ln}^* = 0.2 + j0.2$

Source impedance, $Z_s = 25 + j70$ and so normalized $z_{sn} = \frac{Z_s}{Z_0} = \frac{25 + j70}{50} = 0.5 + j1.4$

Normalized source admittance $y_{sn} = \dfrac{1}{Z_{sn}} = 0.23 - j0.63$

We place these points on the Smith chart and draw resistance and conductance circles. Blue circles are for source impedance and red circles are for load impedance.

There are two intersections between the source and load at point A and B. So, there are two combinations of circuits.

First combination, $Z_S \rightarrow Z_A \rightarrow Z_{Ln}^*$

There will be shunt L and series C.

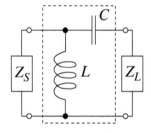

$$jB_A = y_A - y_S = 0.23 - j1.1 - (0.23 - j0.63) = -j0.47$$

$$0.47 = \frac{Z_0}{2\pi f \times L} = \frac{50}{2\pi \times 2 \times 10^9\,\text{Hz} \times L} = 8.465\,\text{nH}$$

$$jX_A = Z_{Ln}^* - Z_A = 0.2 + j0.2 - (0.2 + j0.9) = -j0.7$$

$$0.7 = \frac{1}{2\pi f \times C \times Z_0} = \frac{1}{2\pi \times 2 \times 10^9\,\text{Hz} \times C \times 50} = 2.2736\,\text{pF}$$

Second combination, $Z_S \rightarrow Z_B \rightarrow Z_{Ln}^*$

There will be shunt C and series L.

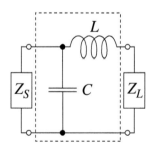

$$jB_B = y_B - y_S = 0.23 + j1.0 - (0.23 - j0.63) = j1.63$$

$$1.63 = 2\pi f \times C \times Z_0 = 2\pi \times 2 \times 10^9\,\text{Hz} \times C \times 50 = 2.59\,\text{pF}$$

$$jX_B = Z_{Ln}^* - Z_B = 0.2 + j0.2 - (0.2 - j0.95) = j1.15$$

$$1.15 = \frac{2\pi f \times L}{Z_0} = \frac{2\pi \times 2 \times 10^9 \text{Hz} \times L}{50} = 4.576 \text{ nH}$$

An illustration of the Smith chart solution is given in Figure 5.6.

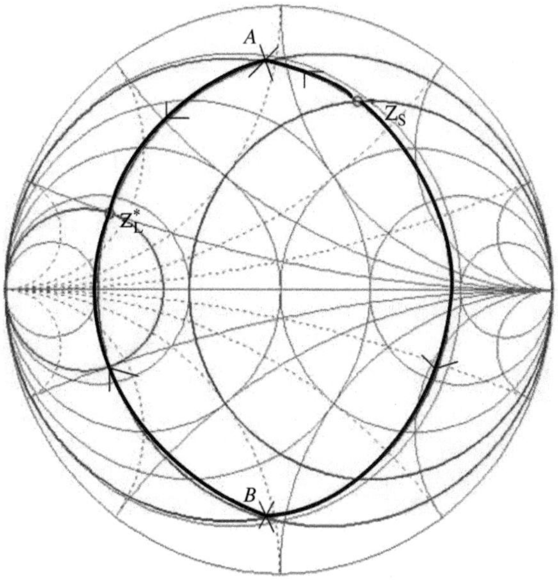

5.4 Impedance Matching Network with Transmission Lines

Transmission line matching becomes a necessity specifically for HF applications due to component parasitics. In this section, impedance matching using transmission lines such as quarter-wave transformers, single and double stub tuning is detailed.

Figure 5.6 Smith chart solution for Example 5.3.

5.4.1 Quarter-wave Transformers

Quarter-wave transformers can be used to match a resistive load impedance to another resistive impedance via quarter wavelength, $\lambda/4$, transmission line. This can be best understood using the relation given by Eq. (5.13). For a lossless case, the input impedance can be expressed as

$$Z_0 \frac{Z_L + jZ_0 \tan \beta l}{Z_0 + jZ_L \tan \beta l} \tag{5.13}$$

When the length of the transmission line is $\lambda/4$, (5.13) simplifies to

$$Z_{in} = \frac{Z_1^2}{R_L} \tag{5.14}$$

as illustrated in Figure 5.7.

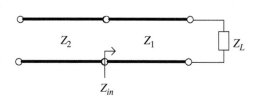

Figure 5.7 Input impedance using quarter-wave transformer.

Example 5.4 Quarter-wave Transmission Lines

Find the characteristic impedance of the transmission lines, Z_{01} and Z_{02}, if the load impedances are given as, $R_{L1} = 25\,\Omega$ and $R_{L2} = 75\,\Omega$ when they are matched with quarter wavelength transmission lines, as shown in Figure 5.8.

Solution

The characteristic impedance of each transmission line is found using (5.14) as

$$Z_{01} = \sqrt{Z'_{in} R_{L1}} = \sqrt{100(25)} = 50$$

$$Z_{02} = \sqrt{Z''_{in} R_{L1}} = \sqrt{100(75)} = 86.60$$

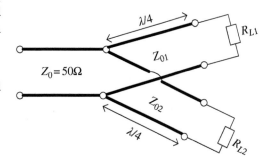

Figure 5.8 Illustration of Example 5.4.

5.4.2 Single Stub Tuning

At high frequencies, it may be desirable to match the given load to the transmission line using transmission lines instead of lumped element components discussed in Section 5.4.1. Impedance matching can then be done using a single open- or short-circuited length of transmission line called a "stub." It is connected either in parallel or in series with the transmission feed line at a certain distance from the load, as shown in Figure 5.9.

In single stub tuning, there are two design parameters: the distance, d, from the load to the stub position and the value of susceptance or reactance provided by the shunt or series stub.

5.4.2.1 Shunt Single Stub Tuning

When it is a shunt stub case, as shown in Figure 5.9a, we select d so that the admittance, Y, seen looking into the line at distance d from the load is equal to $Y_0 + jB$. Then, the matching is done by choosing the stub susceptance as $-jB$.

To obtain the relations for d and l, we write the input impedance, $Z_L = 1/Y_L = R_L + jX_L$ at a distance d from the load as

$$Z = Z_0 \frac{(R_L + jX_L) + jZ_0 \tan \beta d}{Z_0 + j(R_L + jX_L) \tan \beta d} \tag{5.15}$$

We then find the admittance from (5.15) as

$$Y = G + jB = \frac{1}{Z} \tag{5.16}$$

where

$$G = \frac{R_L(1 + \tan^2 \beta d)}{R_L^2 + (X_L + Z_0 \tan \beta d)^2} \tag{5.17}$$

$$B = \frac{R_L^2 \tan \beta d - (Z_0 - X_L \tan \beta d)(X_L + Z_0 \tan \beta d)}{Z_0 [R_L^2 + (X_L + Z_0 \tan \beta d)^2]} \tag{5.18}$$

To have the matching condition, we need to set G in Eq. (5.17) to $G = Y_0 = 1/Z_0$. Hence

$$Z_0(R_L - Z_0) \tan^2 \beta d - 2X_L Z_0 \tan \beta d + (R_L Z_0 - R_L^2 - X_L^2) = 0 \tag{5.19}$$

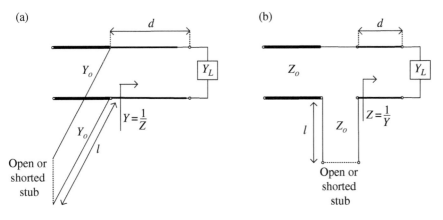

Figure 5.9 Single stub matching: (a) parallel; (b) series.

which leads to two solutions for $\tan \beta d$ as

$$\tan \beta d = \frac{X_L \pm \sqrt{R_L \left[(Z_0 - R_L)^2 + X_L^2\right]/Z_0}}{R_L - Z_0}, \quad \text{for } R_L \neq Z_0 \tag{5.20}$$

If $R_L = Z_0$ then $\tan \beta d = -X_L/2Z_0$. As a result, we have solutions for d as

$$\frac{d}{\lambda} = \begin{cases} \dfrac{1}{2\pi} \tan^{-1}\left(-\dfrac{X_L}{2Z_0}\right) & \text{for } -\dfrac{X_L}{2Z_0} \geq 0 \\[3mm] \dfrac{1}{2\pi}\left(\pi + \tan^{-1}\left(-\dfrac{X_L}{2Z_0}\right)\right) & \text{for } -\dfrac{X_L}{2Z_0} < 0 \end{cases} \tag{5.21}$$

To find the required stub lengths, we first set $B_s = -B$. This leads to the final solutions for open and shorted stubs shown in Figure 5.9a as

$$\frac{l}{\lambda} = \frac{1}{2\pi} \tan^{-1}\left(\frac{B_S}{Y_0}\right) = -\frac{1}{2\pi} \tan^{-1}\left(\frac{B}{Y_0}\right) \quad \text{for the open stub} \tag{5.22}$$

$$\frac{l}{\lambda} = -\frac{1}{2\pi} \tan^{-1}\left(\frac{Y_0}{B_S}\right) = \frac{1}{2\pi} \tan^{-1}\left(\frac{Y_0}{B}\right) \quad \text{for the shorted stub} \tag{5.23}$$

A Smith chart solution for the matching with open stub is practical and can be described as follows.

- Normalize load impedance and locate corresponding admittance on the Z Smith chart.
- Rotate clockwise around the Smith chart from y_L until it intersects the $g = 1$ circle. It intersects the $g = 1$ circle at two points. The "length" of this rotation determines the value of d. There are two possible solutions.
- Rotate clockwise from the short/open circuit point around the $g = 0$ circle, until stub b equals $-b$. The "length" of this rotation determines the stub length l.

5.4.2.2 Series Single Stub Tuning

For the series stub case shown in Figure 5.9b, d is chosen so that the impedance looking into the line at a distance d from the load is equal to $Z_0 + jX$. Then, we select the stub reactance to be $-jX$ to match the line.

To obtain the relations for d and l, we write the input impedance and then we write the input admittance, $Y_L = 1/Z_L = G_L + jB_L$ at a distance d from the load as

$$Y = Y_0 \frac{(G_L + jB_L) + jY_0 \tan \beta d}{Y_0 + j(G_L + jB_L)\tan \beta d} \tag{5.24}$$

We then wire the impedance from (5.24) as

$$Z = R + jX = \frac{1}{Y} \tag{5.25}$$

where

$$R = \frac{G_L(1 + \tan^2 \beta d)}{G_L^2 + (B_L + Y_0 \tan \beta d)^2} \tag{5.26}$$

$$X = \frac{G_L^2 \tan \beta d - (Y_0 - B_L \tan \beta d)(B_L + Y_0 \tan \beta d)}{Y_0\left[G_L^2 + (B_L + Y_0 \tan \beta d)^2\right]} \tag{5.27}$$

To have the matching condition, we need to set G in Eq. (5.26) to $R = Z_0 = 1/Y_0$. Hence

$$Y_0(G_L - Y_0) \tan^2 \beta d - 2B_L Y_0 \tan \beta d + (G_L Y_0 - G_L^2 - B_L^2) = 0 \tag{5.28}$$

which lead to two solution for $\tan \beta d$ as

$$\tan \beta d = \frac{B_L \pm \sqrt{G_L \left[(Y_0 - G_L)^2 + B_L^2 \right] / Y_0}}{G_L - Y_0}, \quad \text{for } G_L \neq Y_0 \tag{5.29}$$

If $G_L = Y_0$ then $\tan \beta d = -B_L/2Y_0$. As a result, we have solutions for d as

$$\frac{d}{\lambda} = \begin{cases} \dfrac{1}{2\pi} \tan^{-1}\left(-\dfrac{B_L}{2Y_0} \right) & \text{for } -\dfrac{B_L}{2Y_0} \geq 0 \\[3mm] \dfrac{1}{2\pi} \left(\pi + \tan^{-1}\left(-\dfrac{B_L}{2Y_0} \right) \right) & \text{for } -\dfrac{B_L}{2Y_0} < 0 \end{cases} \tag{5.30}$$

To find the required stub lengths, we first set $X_s = -X$. This leads to the final solutions for open and shorted stubs shown in Figure 5.9b as

$$\frac{l}{\lambda} = \frac{1}{2\pi} \tan^{-1}\left(\frac{X_S}{Z_0} \right) = -\frac{1}{2\pi} \tan^{-1}\left(\frac{X}{Z_0} \right) \tag{5.31}$$

$$\frac{l}{\lambda} = -\frac{1}{2\pi} \tan^{-1}\left(\frac{Z_0}{X_S} \right) = \frac{1}{2\pi} \tan^{-1}\left(\frac{Z_0}{X} \right) \tag{5.32}$$

The Smith chart solution for the matching with series stub is practical and can be described as follows.

- Normalize load impedance and locate it on the Z Smith chart.
- Rotate clockwise around the Smith chart from Z_L until it intersects the $r = 1$ circle. It intersects the $r = 1$ circle at two points. The "length" of this rotation determines the value of d. There are two possible solutions.
- Rotate clockwise from the short/open circuit point around the $r = 0$ circle, until stub X equals in $-X$. The "length" of this rotation determines the stub length l.

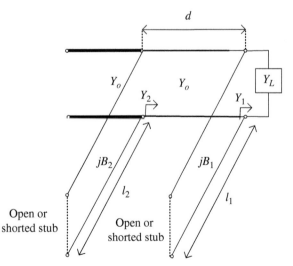

Figure 5.10 Illustration of double stub tuner circuit.

5.4.3 Double Stub Tuning

Single stub tuners can be used to match any load with real and reactive parts using a variable length transmission line between stub and load. This will pose a challenge for loads which are variable. This challenge can be overcome by using double stub tuners which have adjustable transmission line lengths. The disadvantage of a double stub tuner is that it cannot match all loads. Let's consider the double stub tuner shown in Figure 5.10.

From Figure 5.10, the admittance, Y_1, can be found from

$$Y_1 = G_L + j(B_L + B_1) \tag{5.33}$$

In (5.33), $Y_L = G_L + jB_L$ is the load admittance and B_1 is the susceptance of the first stub. The equivalent admittance, Y_2, after the transmission line of length d in the circuit just before the second stub is calculated using

$$Y_2 = Y_0 \frac{G_L + j(B_L + B_1 + Y_0 \tan(\beta d))}{Y_0 + j \tan(\beta d)(G_L + jB_L + jB_1)} \tag{5.34}$$

We equate the real part of Y_2 to Y_0 as

$$\text{Re}\left\{ Y_0 \frac{G_L + j(B_L + B_1 + Y_0 \tan(\beta d))}{Y_0 + j \tan(\beta d)(G_L + jB_L + jB_1)} \right\} = Y_0 \tag{5.35}$$

or

$$\text{Re}\left\{ \frac{G_L + j(B_L + B_1 + Y_0 \tan(\beta d))}{Y_0 + j \tan(\beta d)(G_L + jB_L + jB_1)} \right\} = 1 \tag{5.36}$$

which gives

$$G_L^2 - G_L Y_0 \frac{1 + (\tan(\beta d))^2}{(\tan(\beta d))^2} + \frac{(Y_0 - B_L(\tan(\beta d)) - B_1(\tan(\beta d)))Y_0}{(\tan(\beta d))^2} = 0 \tag{5.37}$$

and

$$G_L = Y_0 \frac{1 + (\tan(\beta d))^2}{2(\tan(\beta d))^2}\left[1 \pm \sqrt{\frac{1 - 4(\tan(\beta d))^2(Y_0 - B_L(\tan(\beta d)) - B_1(\tan(\beta d)))^2}{Y_0(1 + (\tan(\beta d))^2)^2}} \right] \tag{5.38}$$

The constraint is for G_L to be real and this leads to

$$0 \le \frac{4(\tan(\beta d))^2(Y_0 - B_L(\tan(\beta d)) - B_1(\tan(\beta d)))^2}{Y_0(1 + (\tan(\beta d))^2)^2} \le 1 \tag{5.39}$$

Hence

$$0 \le G_L \le \frac{Y_0}{\sin^2 \beta d} \tag{5.40}$$

Equation (5.40) gives the range for G_L that can be matched for a given stub with length d. After d is determined, the susceptance of the first stub, B_1, is calculated from (5.37)

$$B_1 = -B_L + \frac{Y_0 \pm \sqrt{(1 + (\tan(\beta d))^2)G_L Y_0 - G_L^2(\tan(\beta d))^2}}{(\tan(\beta d))} \tag{5.41}$$

The susceptance of the second stub is found by equating the susceptance to the negative of the imaginary part of (5.34) as

$$B_2 = \frac{\pm Y_0 \sqrt{Y_0 G_L (1 + (\tan(\beta d))^2) - G_L^2(\tan(\beta d))^2 + G_L Y_0}}{G_L(\tan(\beta d))} \tag{5.42}$$

The length of the open-circuited stub length is then equal to

$$\frac{\ell_0}{\lambda} = \frac{1}{2\pi} \tan^{-1}\left(\frac{B_i}{Y_0}\right) \tag{5.43}$$

whereas the length of the short-circuited stub length is found from

$$\frac{\ell_S}{\lambda} = \frac{1}{2\pi} \tan^{-1}\left(\frac{Y_0}{B_i}\right) \tag{5.44}$$

where $i = 1,2$.

In a Smith chart solution for the double stub tuner, d is assumed to be fixed based on the constraints as discussed. It is usually assumed to be equal to $\lambda/16$, $\lambda/8$, $3\lambda/16$, etc. A Smith chart solution for the matching with double stub is practical and can described as follows for the short circuit stubs.

a) Draw a $g = 1$ circle on the Smith chart.
b) Rotate the $g = 1$ circle toward load (counterclockwise) by d/λ. This is where Y_2 will be located.
c) Normalize the load impedance and convert it to admittance, $y_L = g_L + jb_L$ and mark that on the Smith chart.
d) Now draw a g_L circle and identify the intersections with the rotated $g = 1$ circle. The intersection points will be the admittances, $y_1 = g_1 + jb_1$.
e) Use a compass and identify the corresponding admittances, $y_2 = g_2 + jb_2$, on the $g = 1$ circle.
f) Determine the stub length, l_1, needed to move from y_1 to y_L.
g) Finally, determine the stub length, l_2, needed to move from $-jb_2$ to y_{sc}.

Example 5.5 Double Stub Tuner

A double stub shunt tuner is used to match a load impedance $Z_L = 60 + j80\ \Omega$ to a 50 Ω transmission line at $f = 2\,\text{GHz}$. The stubs are separated by $\lambda/8$ and short-circuited. Design the matching circuits and calculate the lengths.

Solution

We follow the steps outlined for the double tuning matching network.

a) Draw a $g = 1$ circle on the Smith chart, as shown in Figure 5.11.
b) Rotate the $g = 1$ circle toward load (counterclockwise) by $\lambda/8$. This is where Y_2 will be located.
c) Normalize the load impedance and convert it to admittance, $y_L = g_L + jb_L = 0.3 - j0.4$ and mark that on the Smith chart as P_L.

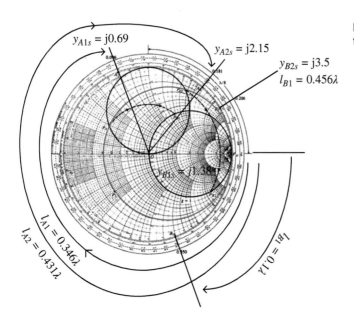

Figure 5.11 Smith chart solution for double tuner for Example 5.5.

d) Now draw a $g_L = 0.3$ circle and identify the intersections with the rotated $g = 1$ circle. The intersection points will be the admittances and are marked on the Smith chart as P_{A1} and P_{A2}. The admittance values at these points are

e)

$$y_{A1} = 0.3 + j0.29 \text{ at } P_{A1}$$

$$y_{A2} = 0.3 + j1.75 \text{ at } P_{A2}$$

f) Use a compass from the origin for points P_{A1} and P_{A2} and move until intersecting the corresponding admittances on the $g = 1$ circle as

$$Y_{B1} = 1 + j1.38 \text{ at } P_{B1}$$

$$Y_{B2} = 1 - j3.5 \text{ at } P_{B2}$$

g) Determine the stub length, l_1, needed to move from y_A to y_L as

$$y_{A1s} = y_{A1} - y_L = j0.29 + j0.4 = j0.69, l_{A1} = 0.096\lambda + 0.25\lambda = 0.346\lambda \text{ at Point } A_1$$

$$y_{A1s} = y_{A2} - y_L = j1.75 + j0.4 = j2.15, l_{A1} = 0.181\lambda + 0.25\lambda = 0.431\lambda \text{ at Point } A_2$$

The stub length l_{A1} is found from P_{sc} to y_{A1s} and stub length l_{A2} is found from P_{sc} to y_{A12s}.

h) Finally, determine the stub length, l_2, needed to move from $-jb_2$ to y_{sc}.

$$y_{B1s} = -j1.38, l_{B1} = 0.350\lambda - 0.25\lambda = 0.1\lambda \text{ at Point } B_1$$

$$y_{B2s} = +j3.5, l_{B2} = 0.206\lambda + 0.25\lambda = 0.456\lambda \text{ at Point } B_2$$

5.5 Impedance Transformation and Matching between Source and Load Impedances

Consider a matching network between source and load, as shown in Figure 5.12. As discussed in Section 5.2, there are eight possible matching networks, as shown in Figure 5.2, which can be represented by the two generic L matching networks shown in Figure 5.13. We will first derive the analytical equations as we did before in Section 5.4. This time, consider first the generic L type matching network shown in Figure 5.13b.

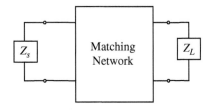

Figure 5.12 Matching network between load and transmission line.

Since source is matched to load impedance, the complex conjugate impedance of the load should be equal to the overall impedance connected to the load impedance. This can be expressed by

$$Z_L^* = \frac{1}{Z_s^{-1} + jB} + jX \tag{5.45}$$

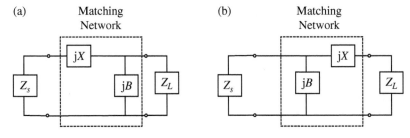

Figure 5.13 Generic two L matching networks between source and load impedances.

Express

$$Z_s = R_s + jX_s \quad \text{and} \quad Z_L = R_L + jX_L \tag{5.46}$$

Then

$$Z^*_{Load} = \frac{1}{Z_s^{-1} + jB} + jX = \frac{R_s + jX_s}{1 + jB(R_s + jX_s)} + jX = R_L - jX_L \tag{5.47}$$

Separate real and imaginary parts

$$R_s = R_L(1 - BX_s) + (X_L + X)BR_s \tag{5.48}$$

$$X_s = R_sR_LB - (1 - B_CX_L)(X_L + X) \tag{5.49}$$

Solving for B and X gives

$$X = -X_L \pm \sqrt{R_L(R_s - R_L) + \frac{R_L}{R_s}X_s^2} \tag{5.50}$$

$$B = \frac{R_s - R_L}{R_sX + R_sX_L - R_LX_s} \tag{5.51}$$

The solution given by (5.50) and (5.51) is valid only when $R_s > R_L$. A similar procedure can be applied to the circuit shown in Figure 5.13a. We obtain the following equations for reactance and susceptance by assuming $R_s < R_L$.

$$B = \frac{R_sX_L \pm \sqrt{R_sR_L\left(R_L^2 + X_L^2 - R_sR_L\right)}}{R_s\left(R_L^2 + X_L^2\right)} \tag{5.52}$$

$$X = \frac{\left(R_L^2 + X_L^2\right)B - X_L + \frac{X_s}{R_s}R_L}{\left(R_L^2 + X_L^2\right)B^2 - 2X_LB + 1} \tag{5.53}$$

As can be seen, the analytical calculation of the impedance transformation and matching is tedious. Instead, we can use a Smith chart for the same task. For this there is a standard procedure that must be followed. The design procedure for matching a source impedance to a load impedance using a Smith chart instead of an analytical method can be outlined as follows.

- Normalize the given source and complex conjugate load impedances and locate them on the Smith chart.
- Plot constant resistance and conductance circles for the impedances located.
- Identify the intersection points between the constant resistance and conductance circle for the impedances located.
- The number of the intersection point corresponds to the number of possible L matching networks.
- By following the paths that go through intersection points calculate normalized reactances and susceptances.
- Calculate the actual values of the inductors and capacitors by denormalizing at the given frequency.

Example 5.6   *ZY* Smith Chart
Using a Smith chart, design all possible configurations of two-element matching networks that match source impedance $Z_s = (15 + j50)\Omega$ to the load $Z_L = (20 - j30)\Omega$. Assume characteristic impedance of $Z_0 = 50\,\Omega$ and an operating frequency of $f = 4\,\text{GHz}$.

Solution
The MATLAB program was developed to plot resistance and conductance circles for the source and complex conjugate of the load impedances. As shown in Figure 5.14, there are four possible L matching networks. These networks are illustrated in Figure 5.15.

5.6 Bandwidth of Matching Networks

The bandwidth of the network can be best understood with its quality factor, Q. If Q and the frequency of the network are known then the bandwidth can be found from

$$BW = \frac{f_o}{Q} \tag{5.54}$$

At each node of the matching network circuit, for an equivalent series input impedance the quality factor is defined as

$$Q_n = \frac{|X_s|}{R_s} \tag{5.55}$$

Similarly, at each node, for an equivalent parallel admittance the quality factor is defined as

$$Q_n = \frac{|B_p|}{G_p} \tag{5.56}$$

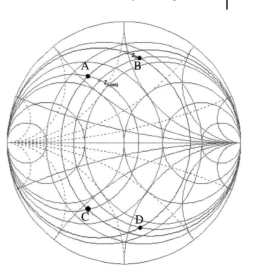

Figure 5.14 Number of possible L matching networks to match source and load impedance.

Consider the Smith chart solution given in Example 5.3, which is given in Figure 5.16 as a reference again. The impedances at node A and B are

$$Z_A = (0.2 + j0.9) \text{ and } Z_B = (0.2 - j0.95)$$

For instance, the quality factors at node A and B are

$$Q_A = \frac{|0.9|}{0.2} = 4.5 \text{ and } Q_B = \frac{|0.95|}{0.2} = 4.75$$

It is higher than the quality factor at node A giving higher bandwidth. For L-type networks, the node quality factor, Q_n, and load quality factor, Q_L, are related through

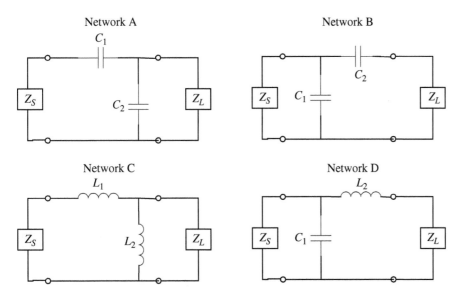

Figure 5.15 Possible L matching networks to match source and load impedance.

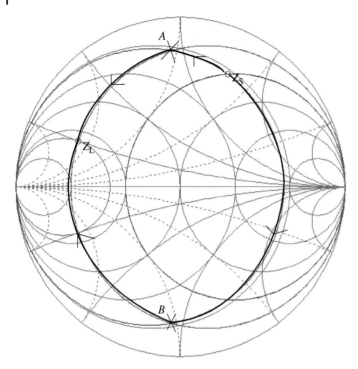

Figure 5.16 Node quality factor illustration.

$$Q_n = \frac{Q_L}{2} \tag{5.57}$$

When the matching network is complicated, the loaded quality factor is estimated using the maximum node quality factor. Constant Q_n contours on a Smith chart are given in Figure 5.17.

The process of designing a matching network based on a quality factor can be simplified for matching load impedance to source impedance as follows.

- Normalize the load and source impedances.
- Find the complex conjugate of the source impedance.
- Mark the load impedance and complex conjugate of source impedances on the Smith chart.
- Draw the quality factor line on the Smith chart.
- Find the possible impedance matching network without exceeding the quality factor lines.

Example 5.7   Quality Factor of Matching Network
Design an L-type matching network with a maximum quality factor of 1.95 to match a load with $10\,\Omega$ resistive element and 1.59 nH inductive element to a $50\,\Omega$ source impedance at $f = 1$ GHz using a Smith chart.

Solution
There are two possible L-type matching networks that can be used satisfying the quality factor of 1.95 requirement. The two possible matching networks meeting this criterion are given in Figure 5.18.

The Smith chart showing the solution is given in Figure 5.19.

In practice, the quality factor of the matching network is important. If a broadband circuit is desired, the matching networks with lower Q are used to increase the bandwidth. With L-type networks we have a limitation in the

Figure 5.17 *Q*-factor Smith chart.

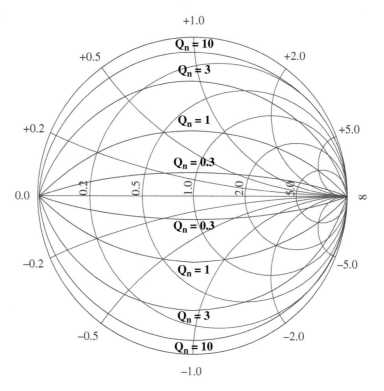

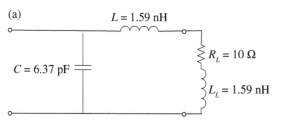

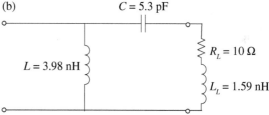

Figure 5.18 Two possible *L*-type matching networks.

value of *Q*. This can be overcome using *T*-type or *PI*-type matching networks. The *Q* of *PI* or *T* network is usually taken as the highest value of Q_n in the circuit.

The general topologies for *T* and *PI* networks are given in Figure 5.20.

5.7 Engineering Application Examples

In this section, single stub tuners are designed and simulated, a prototype built and the measurements results presented.

Design Example 5.1   Design and Implementation of Single Stub Tuning Network

Design a single stub matching system to match an unknown load to a 50 Ω system impedance at 500, 100, and 30 MHz. Then, simulate and prototype it at 30 MHz by considering the two coaxial cable sections, as shown in Figure 5.21. The RG58U coaxial cable has the following electrical characteristics.

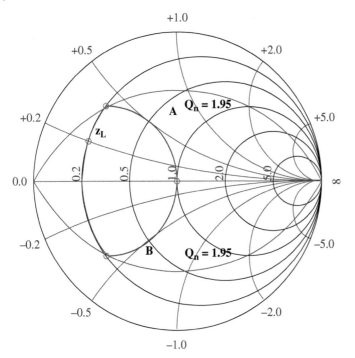

Figure 5.19 Smith chart solution for Example 5.7.

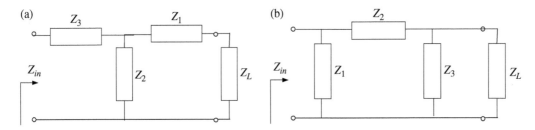

Figure 5.20 General topologies for (a) *T* network and (b) *PI* network.

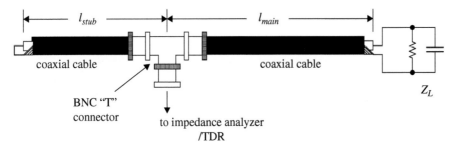

Figure 5.21 Dimensions of single stub tuner.

- Conductor diameter: 0.032 in.
- Core outer diameter: 0.116 in.
- Total outer diameter: 0.193 in.
- Dielectric: solid polyethylene ($\varepsilon_r = 2.26$, $\tan\delta = 0.0002$)
- Copper conductivity: 5.96×10^7 Sie/m
- Nominal velocity of propagation: 66%
- Nominal impedance (Z_0): 53.5 Ω

Hint: One cable will serve as the "main" line between the load formed of a 100 Ω resistor and 100 pF capacitor and the shunt stub, and the other will serve as the stub itself. Each cable has a BNC connector at one end. These cable ends will be attached to a BNC "Tee" connector to allow the tuning stub to be connected in shunt with the main transmission line. A diagram of the assembly is shown in Figure 5.21. One advantage of BNC connectors is that they are designed to have characteristic impedances close to 50 Ω.

If the stub has been designed correctly, the input impedance looking into the freeport (male connector) of the BNC Tee should be close to 52 Ω.

Solution

a) Analytical solution for 500 MHz

From the given information, we calculate Y_L as

$$Y_L = 0.01 + j0.3141592654$$

and

$$Z_L = \frac{1}{Y_L} = 0.1012186277 - j3.179876973$$

Now that we have Z_L, we can use the closed form quadratic equation to find $t = \tan(\beta d)$ with

$$t = \frac{X_L \pm \sqrt{R_L\left((Z_0 - R_L)^2 + X_L^2\right)/Z_0}}{R_L - Z_0}$$

From the equation above, we get $t_1 = 0.0186422505$ and $t_2 = 0.108810841$. Next, we calculate length for the constant conductance circle $Y_0 = \frac{1}{Z_0} = 0.02$. This can be from the input impedance transmission line equation as

$$Z' = Z_0 \frac{Z_L + jZ_0 t}{Z_0 + jZ_L t} = \begin{cases} 0.1010141375 - j2.245106459, & t = t_1 \\ 0.1010141375 + j2.245106459, & t = t_2 \end{cases}$$

We then find the "real" part of $\mathrm{Re}\{Y'\} = \mathrm{Re} = \{1/Z'\}$. If it is equal to Y_0, then the calculated $\tan(\beta d)$ is correct.

$$Y' = \frac{1'}{Z} = \begin{cases} 0.02 + j0.4445133162, & t = t_1 \\ 0.02 - j0.4445133162, & t = t_2 \end{cases}$$

As can be shown above $\mathfrak{R}(Y') = Y_0$ hence our calculation for $t = \tan(\beta d)$ is correct. The next step is to find $\frac{d}{\lambda}$ in order to have a practical length-to-wavelength ratio to build the coaxial line. Since $t > 0$ in both cases, we can use the following equation to find $\frac{d}{\lambda}$.

$$\frac{d}{\lambda} = \frac{\tan^{-1}(t)}{2\pi} = \begin{cases} 0.002966662673, & t = t_1 \\ 0.017249918302, & t = t_2 \end{cases}$$

We have now completed the first half of the design by getting the series transmission line length to match to Y_0. Next, we need to find the shunt stub values in order to match back to 50 Ω. It is known that $Y' = G + jB$.

Currently, we have two values for B: one positive, one negative. This gives us two options for matching: a shunt open stub or a shunt shorted stub. For $t = t_1$ we need a shunt shorted stub, since it will act more closely to an inductor. For $t = t_2$ we will need a shunt open stub, since it will act more closely to a capacitor. The equations we need are the following

$$\frac{l_0}{\lambda} = \frac{\tan^{-1}\left(\dfrac{Y_0}{B}\right)}{2\pi} \quad \text{for a shunt shorted stub}$$

$$\frac{l_0}{\lambda} = \frac{-\tan^{-1}\left(\dfrac{B}{Y_0}\right)}{2\pi} \quad \text{for a shunt open stub}$$

Thus for $t = t_1$ and $t = t_2$ we have the following results

$$\frac{l_0}{\lambda} = 0.007156036562 \text{ for } t = t_1, \text{short circuit stub}$$

$$\frac{l_0}{\lambda} = 0.242843963438 \text{ for } t = t_2, \text{open circuit stub}$$

So we now have derived equations for d and l_0 which corresponds to l_{main} and l_{stub}, respectively, as illustrated in Figure 5.21. We choose the smallest values that will give us the results we seek so that we minimize any losses and space the matching network takes up. This results in choosing $t = t_1$ where our corresponding results are

$$\frac{l_{main}}{\lambda} = 0.017249918302$$

$$\frac{l_{stub}}{\lambda} = 0.242843963438$$

So, the next goal is to get realistic lengths of these stubs so that we can implement them directly. In order to do this we need to find the effective wavelength, λ_{eff}, so that $\dfrac{l_{main}}{\lambda_{eff}}$ and $\dfrac{l_{stub}}{\lambda_{eff}}$ are equal to the above ratios. λ_{eff} can be estimated from

$$\lambda_{eff} = \frac{cv_p}{f}$$

where $c = 3 \times 10^8 \dfrac{m}{s}$, $v_p = 0.66$ (velocity factor), and $f = 5 \times 10^8$ MHz. Hence

$$\lambda_{eff} = 0.396 \text{ m} = 15.58 \text{ in.}$$

This leads to

$$l_{main} = 0.26875 \text{ in.}$$

$$l_{stub} = 3.78350895 \text{ in.}$$

b) Analytical solution for 100 MHz

The following equations are the same as those above except with the values changed.

1) Calculate the load impedance and admittance values at $f = 100$ MHz

$$Y_L = 0.01 + j0.0628318531$$

or

$$Z_L = \frac{1}{Y_L} = 2.470452301 - j15.52230961$$

2)

$$t = \frac{X_L \pm \sqrt{R_L\left((Z_0 - R_L)^2 + X_L^2\right)/Z_0}}{R_L - Z_0}$$

$$t_1 = 0.0927472906 \quad t_2 = 0.5604173349$$

3)

$$Z' = Z_0 \frac{Z_L + jZ_0t}{Z_0 + jZ_Lt} = \begin{cases} 2.354136655 - j10.59079191, & t = t_1 \\ 2.354136655 + j10.59079191, & t = t_2 \end{cases}$$

$$Y' = \frac{1}{Z'} = \begin{cases} 0.02 + j0.0899760164, & t = t_1 \\ 0.02 - j0.0899760164, & t = t_2 \end{cases}$$

4)

$$\frac{d}{\lambda} = \frac{\tan^{-1}(t)}{2\pi} = \begin{cases} 0.0147190813, & t = t_1 \\ 0.0812972949, & t = t_2 \end{cases}$$

$$\frac{l_0}{\lambda} = \frac{\tan^{-1}\left(\dfrac{Y_0}{B}\right)}{2\pi} = 0.0348112273 \text{ for the shunt shorted stub for } t = t_1, \text{ short circuit stub}$$

$$\frac{l_0}{\lambda} = \frac{-\tan^{-1}\left(\dfrac{B}{Y_0}\right)}{2\pi} = 0.2151887727 \text{ for shunt open stub for } t = t_2, \text{ open circuit stub}$$

$$\frac{l_{main}}{\lambda} = 0.0147190813$$

$$\frac{l_{stub}}{\lambda} = 0.0348112273$$

5)

$$\lambda_{eff} = \frac{cv_p}{f}$$

$$\lambda_{eff} = 1.98 \text{ m} = 77.9528 \text{ in.}$$

$$l_{main} = 1.1474 \text{ in.}$$

$$l_{stub} = 2.7136 \text{ in.}$$

c) Analytical solution for 30 MHz

1) Calculate the load impedance and admittance values at $f = 100$ MHz

$$Y_L = 0.01 + j0.0188 \text{ or } Z_L = \frac{1}{Y_L} = 21.9633 - j41.4000$$

2) $$t = \frac{X_L \pm \sqrt{R_L\left((Z_0 - R_L)^2 + X_L^2\right)/Z_0}}{R_L - Z_0}$$

$$t_1 = 0.2947 \quad t_2 = 2.6586$$

3) $$Z' = Z_0 \frac{Z_L + jZ_0t}{Z_0 + jZ_Lt} = \begin{cases} 15.2601 - j23.0246, & t = t_1 \\ 15.2601 + j23.0246, & t = t_2 \end{cases}$$

4) $$Y' = \frac{1}{Z'} = \begin{cases} 0.02 + j0.0302, & t = t_1 \\ 0.02 - j0.0302, & t = t_2 \end{cases}$$

$$\frac{d}{\lambda} = \frac{\tan^{-1}(t)}{2\pi} = \begin{cases} 0.0456, & t = t_1 \\ 0.1927, & t = t_2 \end{cases}$$

$$\frac{l_0}{\lambda} = \frac{\tan^{-1}\left(\dfrac{Y_0}{B}\right)}{2\pi} = 0.0931 \text{ for shunt shorted stub for } t = t_1, \text{ short circuit stub}$$

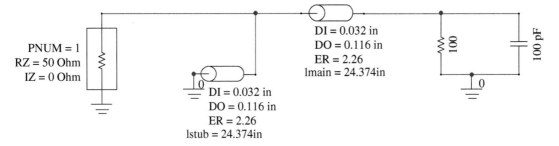

Figure 5.22 Ansoft simulation of single sub tuner.

$$\frac{l_0}{\lambda} = \frac{-\tan^{-1}\left(\dfrac{B}{Y_0}\right)}{2\pi} = 0.1569 \text{ for shunt open stub for } t = t_2, \text{ open circuit stub}$$

$$\frac{l_{main}}{\lambda} = 0.0456$$

$$\frac{l_{stub}}{\lambda} = 0.0931$$

5) $\lambda_{eff} = \dfrac{cv_p}{f}$

$\lambda_{eff} = 6.65 \text{ m} = 261.811 \text{ in.}$

$$l_{main} = 11.939 \text{ in.}$$

$$l_{stub} = 24.374 \text{ in.}$$

The structure can be simulated using Ansoft Designer, as shown in the circuit in Figure 5.22.

The electrical lengths in the simulation can be found using the TRL calculator within Ansoft Designer, as shown in Figure 5.23.

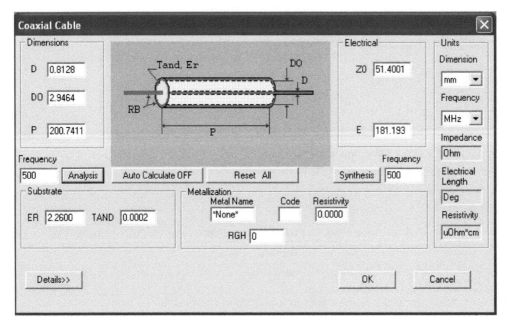

Figure 5.23 TRL calculator for electrical length in Ansoft Simulator.

The results of Ansoft Simulator for the single stub tuner is

	F (MHz)	Z_{11} (Ohm) NWA1 real, imaginary		Y_{11} (Sie) NWA1 real, imaginary		S_{11} NWA1 real, imaginary	
1	30.000 000	49.023 235	−1.702 617	0.020 374	0.000 708	0.019 820	−118.857

Since the simulation results are good, the prototype can now be built. The following parts are used to build the prototype.

- 6 ft. RG58U coaxial cable
- 50 Ω coaxial Tee connector
- 100 Ω resistor (SMT used)
- 100 pF capacitor (SMT used)
- Network analyzer
- Soldering iron and solder
- Wire cutters
- Measuring device (ruler)

The prototypes built are shown in Figures 5.24 and 5.25, respectively.

The prototypes are measured using a network analyzer and results and are illustrated in Figures 5.26 and 5.27.

The measured impedance at $f = 30$ MHz is 43.68 Ω with some small reactance. The logarithmic plot of the prototype gives a slight shift in the tuning frequency to around 27 MHz, which explains the small mismatch measured for 30 MHz.

Design Example 5.2 Design and Simulation of Microstrip Matching Networks

Design two matching networks shown in Figure 5.28 with the following configuration to match a load $Z_L = 200 - j100$ Ω with a 50 Ω transmission line. The operation frequency is 5 GHz. The shunt transmission line can be either an open or a short circuit stub. In addition, please consider the following in the design of the matching network.

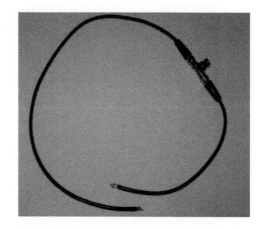

Figure 5.24 Single stub tuner prototype.

Figure 5.25 Prototype sections: (a) load; (b) shorted stub.

(a)

(b)

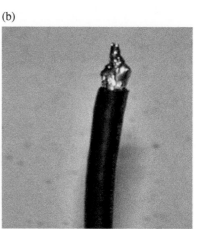

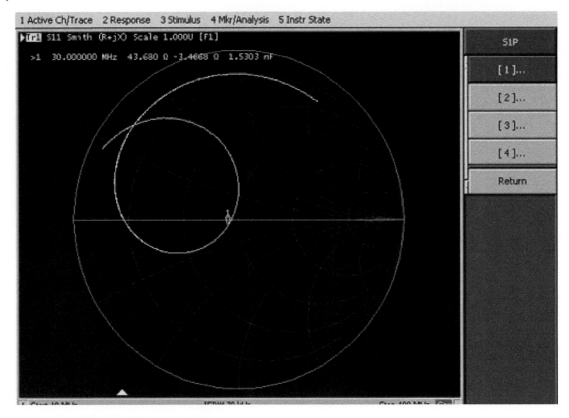

Figure 5.26 Measured S11 for the prototype.

1) Design the matching network using analytical equations.
2) Using an advanced design system (ADS), implement the two matching networks and simulate it to obtain the S_{11} and $VSWR_{in}$.
3) Implement the matching networks using microstrip. Use the following properties for substrate. You can use MTAPER or MTEE for discontinuities for the layout.
   a) $Er = 9.6$
   b) $H = 10$ mil
   c) $Hu = 3.9e34$ mil
   d) $T = 0.150$ mil
   e) $\sigma = 4.1e7$
   f) $Tan\delta = 0$
4) Confirm your simulation results with an electromagnetic (EM) simulator to verify your layout and optimize if necessary (Figure 5.28).

Solution

a) **Analytical Derivation of Transmission Line Matching Network**
   Series Transmission Line

   It is known that the input impedance of a lossless transmission line of length d and a characteristic impedance of Z_0 being terminated in a load designated Z_L is given by the equation

   $$Z_{IN} = Z_0 \frac{Z_L + jZ_0 \tan{(\beta d)}}{Z_0 + jZ_L \tan{(\beta d)}} \tag{5.58}$$

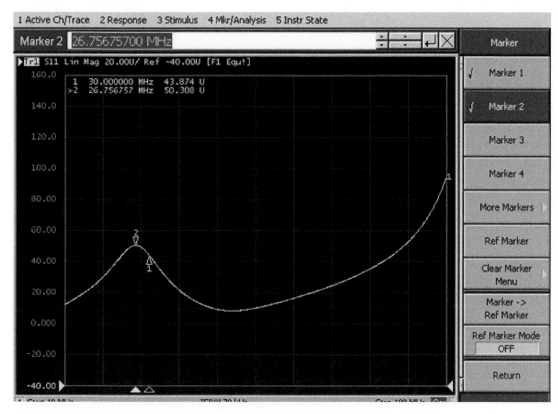

Figure 5.27 Illustration of the measured tuning frequency.

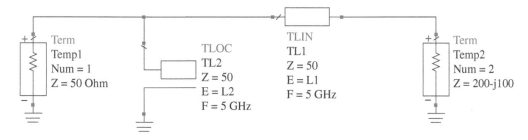

Figure 5.28 Illustration of the matching network for Design Example 5.2.

The input admittance of a series transmission line can be written as

$$Y_{IN} = \frac{1}{Z_{IN}} = \frac{R_S + jZ_L t}{R_S(Z_L + jR_S t)} = G_{IN} + jB_{IN} \tag{5.59}$$

where

$$Z_0 = R_S$$
$$t = \tan(\beta d) \tag{5.60}$$

We can now substitute the normalized load impedance into Eq. (5.59), and separating the real and imaginary parts gives

$$\frac{Z_L}{R_S} = r + jx \overset{\text{Subsitution}}{\Rightarrow} G_{IN} + jB_{IN} \tag{5.61}$$

$$G_{IN} = \frac{r(1 + t^2)}{R_S(r^2 + x^2 + t^2 + 2xt)} \tag{5.62}$$

$$B_{IN} = \frac{xt^2 + (r^2 + x^2 - 1)t + x}{R_S(r^2 + x^2 + t^2 + 2xt)} \tag{5.63}$$

As we are interested in obtaining the electrical length of the transmission line, we must isolate and solve for t. This can be achieved by setting the input conductance to be equal to the source conductance as

$$\frac{r(1 + t^2)}{R_S(r^2 + x^2 + t^2 + 2xt)} = \frac{1}{R_S} \tag{5.64}$$

$$(r - 1)t^2 - 2xt - (r^2 + x^2 - r) = 0 \tag{5.65}$$

$$t = \frac{x \pm \sqrt{r(r^2 + x^2 - 2r + 1)}}{r - 1} \tag{5.66}$$

Since $t = \tan(\beta d)$, and it is known that $\beta\lambda = 2\pi$, we can then find the electrical length of the stub from

$$d = \frac{\lambda}{2\pi} \tan^{-1}(t) \tag{5.67}$$

d in degrees is obtained from

$$d = \frac{360}{2\pi} \tan^{-1}(t) \tag{5.68}$$

As for every half wavelength the input impedance of the transmission line will repeat, this means there are an infinite number of transmission line lengths that will provide a match. Typically, the shorter length is selected to improve the matching bandwidth. Also, in the event that a solution for t happens to be negative, a half wavelength should be added to each line to obtain a positive result.

$$d = \frac{360(\pi + \tan^{-1}(t))}{2\pi} \tag{5.69}$$

From (5.66), (5.68), and (5.69) we obtain

$$t_1 = 1.737 \text{ and } t_2 = -3.07$$

$$d_1 = 60.07^\circ \text{ and } d_2 = 108.04^\circ$$

Shunt Stub Transmission Line

As t was previously obtained in determining the electrical length of the series transmission line component, we will now use this information to design the shunt stub. We will begin by substituting t into the following equation.

$$B = \frac{xt^2 \pm (r^2 + x^2 - 1)t + x}{R_S(r^2 + x^2 + t^2 + 2xt)} \tag{5.70}$$

The electrical length of an open circuit stub is found by setting the susceptance of the stubs equal to the negative value of the input susceptance as

$$S_O = \frac{-\lambda(\tan^{-1}(R_S B))}{2\pi} \tag{5.71}$$

The electrical length of a short circuit stub is found similarly by

$$S_S = \frac{\lambda\left(\tan^{-1}\left(\frac{1}{R_S B}\right)\right)}{2\pi} \tag{5.72}$$

The electrical length is calculated from (5.70) to (5.72) and given as

$$S_{o1} = 119.01° \text{ and } S_{o2} = 60.98°$$

$$S_{c1} = 150.98° \text{ and } S_{c2} = 29.01°$$

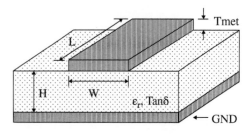

Figure 5.29 Representation of microstrip line.

Microstrip Line Implementation

The microstrip line design and implementation is discussed in Chapter 3, Section 3.4. A typical representation of a microstrip transmission line is given as reference again in Figure 5.29.

For a given line width, w, and substrate thickness, h, an approximation for ε_{eff} can be arrived at and given by (3.125) as

$$\varepsilon_{\text{eff}} = \frac{(\varepsilon_r + 1)}{2} + \frac{(\varepsilon_r - 1)}{2\sqrt{1 + 12\frac{h}{w}}} \tag{5.73}$$

From (3.126), the characteristic impedance of the transmission line is found as

$$Z_0 = \begin{cases} \dfrac{60}{\sqrt{\varepsilon_{\text{eff}}}} \ln\left(\dfrac{8h}{w} + \dfrac{w}{4h}\right), & \dfrac{w}{h} \leq 1 \\[2ex] \dfrac{1}{\sqrt{\varepsilon_{\text{eff}}}} \dfrac{120\pi}{\dfrac{w}{h} + 1.393 + 0.667\ln\left(\dfrac{w}{h} + 1.444\right)}, & \dfrac{w}{h} \geq 1 \end{cases} \tag{5.74}$$

The thickness of the substrate, h, is given in the problem. Thus, if ε_r and Z_0 are known, then the ratio of $\dfrac{w}{h}$

necessary to achieve Z_0 has been determined and given by (3.127) as

$$\frac{w}{h} = \begin{cases} \dfrac{8e^A}{e^{2A} - 2}, & \dfrac{w}{h} \leq 2 \\[2ex] \dfrac{2}{\pi}\left\{B - 1 - \ln(2B - 1) + \dfrac{\varepsilon_r - 1}{2\varepsilon_r}\left[\ln(B - 1) + 0.39 - \dfrac{0.61}{\varepsilon_r}\right]\right\}, & \dfrac{w}{h} > 2 \end{cases} \tag{5.75}$$

From (3.128) to (3.129)

$$A = \frac{Z_0}{60}\sqrt{\frac{\varepsilon_r + 1}{2}} + \frac{\varepsilon_r - 1}{\varepsilon_r + 1}\left(0.23 + \frac{0.11}{\varepsilon_r}\right)$$

$$B = \frac{60\pi^2}{Z_0\sqrt{\varepsilon_r}} \tag{5.76}$$

From the given information, $\varepsilon_r = 9.6$, $h = 10\,\text{mi}$, $Z_0 = 50\,\Omega$, we can find that

$$\frac{w}{h} = 0.9947 \rightarrow w = 9.947\,\text{mil} \tag{5.77}$$

Additionally, the length of the microstrip line can easily be calculated from the wavelength, λ, once the electrical length of the ideal line is known.

$$\lambda = \frac{u}{f} = \frac{c}{f\sqrt{\varepsilon_{\text{eff}}}} \tag{5.78}$$

Using the equations gives

$$\varepsilon_{eff} = 6.4897, \lambda = 927.27 \text{ mil}$$

Length of series stub $(d_1) = l_1 = 154.73$ mil, length of series stub $(d_2) = l_2 = 278.28$ mil

Length of shunt open stub $(S_{o1}) = l_{o1} = 306.56$ mil, length of shunt open stub $(S_{o2}) = l_{o2} = 157.08$ mil

Length of shunt short stub $(S_{c1}) = l_{s1} = 388.89$ mil, length of short stub $(S_{c2}) = l_{o2} = 74.74$ mil

b) **ADS Simulation of Transmission Line Matching Network**
   Open Stub Matching Network – Ideal Transmission Line

   The calculated values are used in ADS and the circuit is constructed, as shown in Figure 5.30. The simulation results are illustrated in Figure 5.31 and confirm the matching at $f = 5$ GHz.

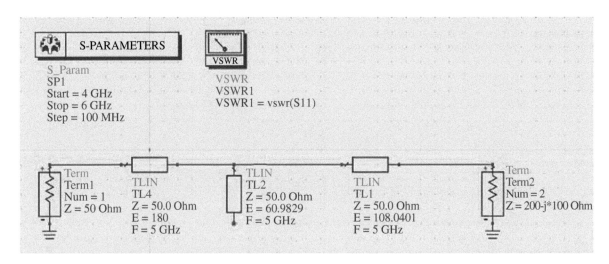

Figure 5.30 The simulated matching network with open stub.

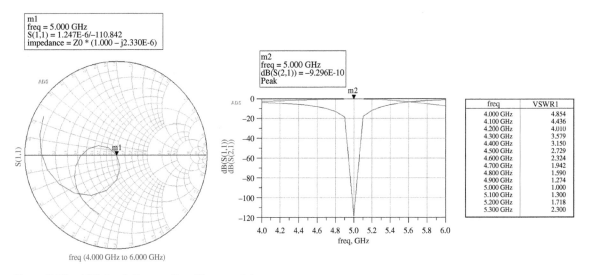

Figure 5.31 ADS simulation results with open stub.

Shorted Stub Matching Network – Ideal Transmission Line

The circuit shown in Figure 5.30 is reconstructed using the shorted stub in Figure 5.32. The simulation results are given in Figure 5.33. The simulation results show successful matching at $f = 5$ GHz when the open stub is replaced with a short stub.

c) Implementation of Microstrip Lines

Open Stub Matching Network – Ideal Microstrip

The implementation of the microstrip lines for open stub matching is illustrated in Figure 5.34. The simulation results are given in Figure 5.35. The microstrip implementation is successfully implemented as illustrated.

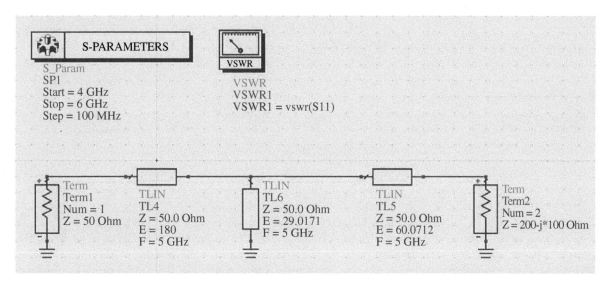

Figure 5.32 The simulated matching network with short stub.

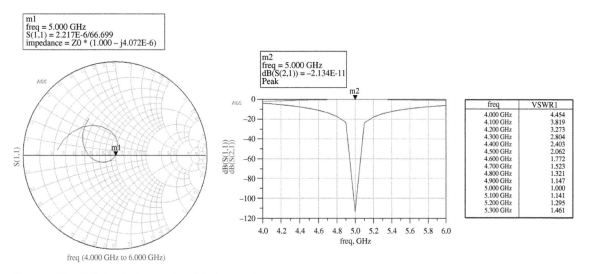

Figure 5.33 ADS simulation results with short stub.

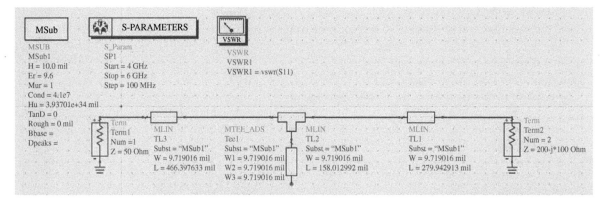

Figure 5.34 Microstrip line implementation with open stub.

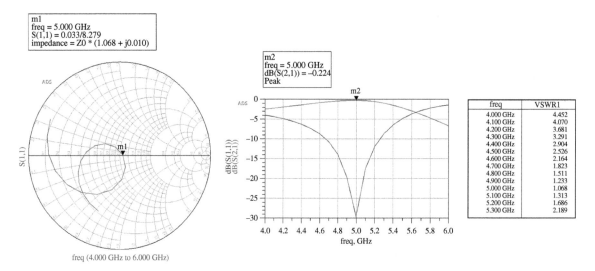

Figure 5.35 ADS simulation results of microstrip line implementation with open stub.

Shorted Stub Matching Network – Ideal Microstrip

The microstrip matching network is now implemented using a short stub, as shown in Figure 5.36. The simulation results are given in Figure 5.37. The results show that microstrip line implementation with the short stub also successfully matched load to source impedance as desired.

d) **EM Simulation of the Layout**

The electromagnetic simulator Momentum was used to simulate the matching network for verification for this section, as shown in Figure 5.38. The ideal microstrip schematics were copied and then modified to eliminate the simulation instances and terminations, replacing the latter with the pins "P1" and "P2." It should also be noted that, for shorted connections, it is a best practice to go ahead and define a via in the schematics to create a seamless transfer of the schematic to an accurate layout capture. Additionally, the ideal ground cannot

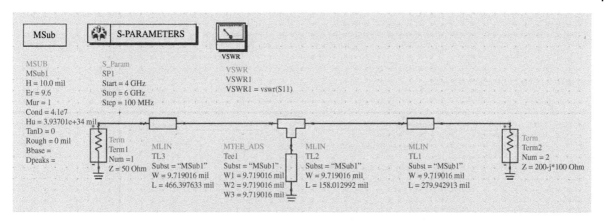

Figure 5.36 Microstrip line implementation with short stub.

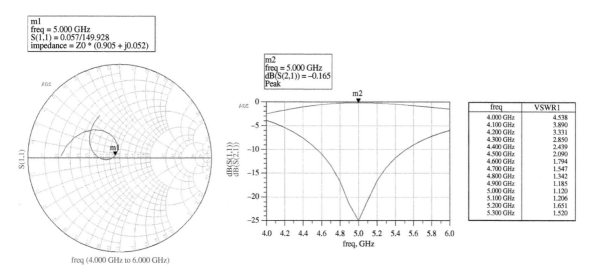

Figure 5.37 ADS simulation results of microstrip line implementation with short stub.

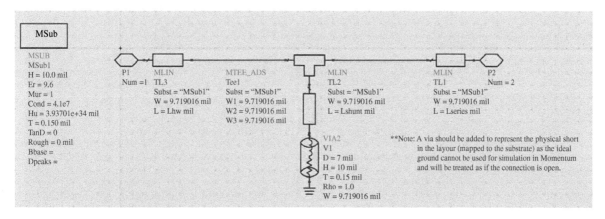

Figure 5.38 The layout to be used for EM simulation.

be used for EM simulation in Momentum and the connection will be treated as open. Also, to parameterize the design to allow for tuning or optimization, global design parameters were defined and used for the lengths of the microstrip transmission lines.

File- > Design Parameters

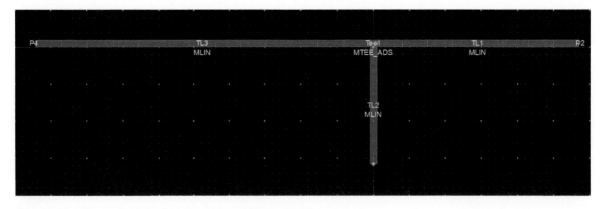

With all of these items in place, the user can now simply select *Layout- > Generate/Update Layout.* This will automatically generate the layout from the schematic capture. This will give the open stub and short stub matching networks shown in Figures 5.39 and 5.40.

From this point, the user will need to define the EM Simulation Settings. This mainly consists of setting up the following items:

Figure 5.39 Microstrip layout for open stub matching network.

Figure 5.40 Microstrip layout for short stub matching network.

- Setup type
- Substrate
- Ports
- Frequency plan.

For this given methodology, the user should select the setup type to be EM Simulation/Model and the EM Simulator to be Momentum Microwave, as shown here.

Setting up the Substrate
If a proper substrate has not already been created, then it will need to be created by selecting "New…" on the following window (EM Simulation Settings):

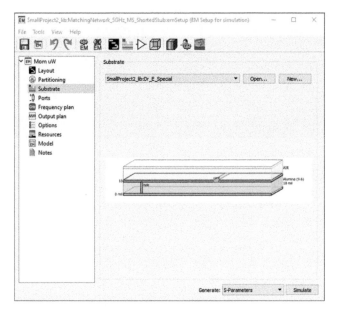

Name the substrate and select an appropriate template if available. In this case, the given *Er* is the same as Alumina, therefore the 25 mil Alumina was chosen as a starting point.

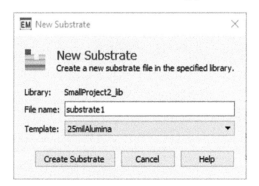

This will create the following substrate:

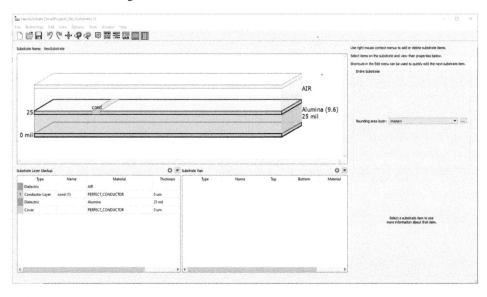

From this point, the substrate will be modified to represent the substrate defined in the schematic captures (MSub). First, the dielectric thickness will be changed from 25 to 10 mi. Next, the conductor layer should be selected so that its thickness can be defined as 0.150 mi. In addition to these changes, the user will also need to add a conductor via for use with the shorted stub by right clicking on the appropriate layer and "mapping the conductor via."

The final substrate should look as follows:

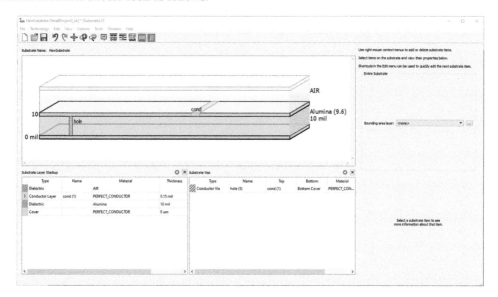

Next, the ports for the layout will need to be modified in the EM Simulation Settings:

It is likely that you will have to select the button "Open Layout Port Editor" and change the reference imped-ance for the port that the load is connected to its specified value, $200 - j^*100$.

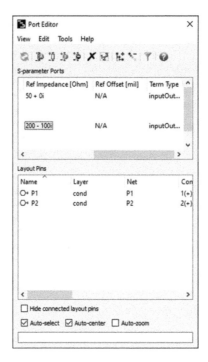

In this case, the last thing that needs to be set up is the "Frequency Plan" in the EM Simulation Settings:

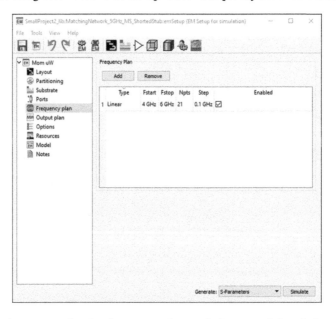

With reasonable points set up for the frequency plan and the rest of the vital settings entered, the user can now press "Simulate" to generate the results of the EM Simulation in Momentum. As this simulation is providing a closer "real world" approximation of the expected phenomena, the results will no longer be ideal. The simulation results are given in Figure 5.41.

However, as the design has been parameterized, the circuit can now be tuned to have near-perfect performance in the EM simulation. This is accomplished by creating an EM Model and Symbol that can be used in a test bench environment for the tuning. To do this, the user needs to select **EM- > Component- > Create**

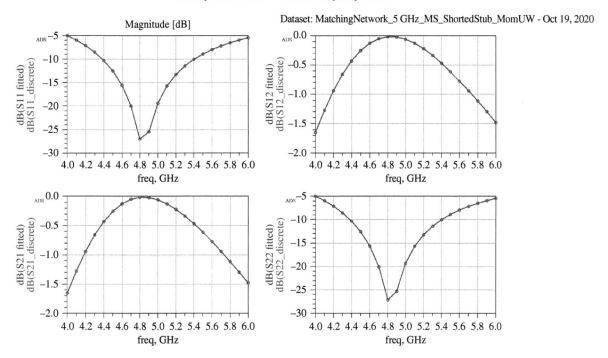

Figure 5.41 EM simulation results of the matching network.

EM Model and Symbol... from the layout capture window. These models and symbols will be created automatically at this point. Now, a co-simulation can be done in ADS for optimization/tuning using the EM model. The co-simulation circuits for open and short stubs are given in Figures 5.42 and 5.43, respectively.

Tuning can now be accomplished by selecting ***Simulate- > Tuning...***

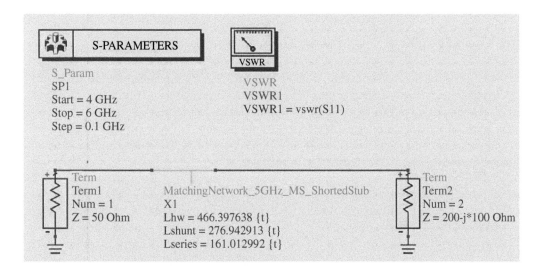

Figure 5.42 Co-simulation circuit for open stub.

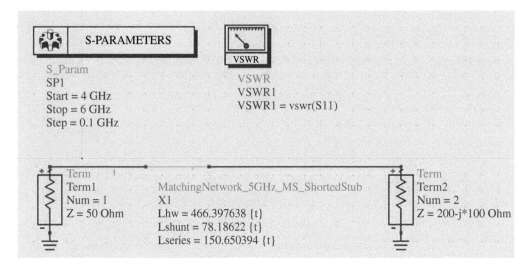

Figure 5.43 Co-simulation circuit for short stub.

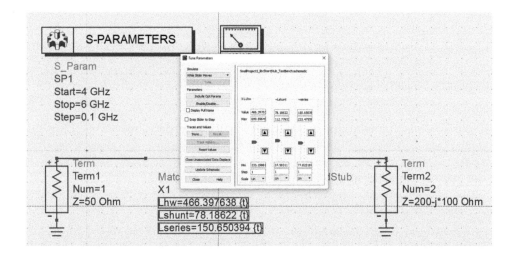

This will simultaneously open a simulation output window and the "Tune Parameters" dialog. From this dialog, you can select the parameters that you wish to tune from the top-level schematic. As the sliders are moved, the simulation results will be updated in real time. By this adjustment, the user should be able to easily arrive at near-perfect performance for the matching network layout. Figures 5.44 and 5.45 give the simulation results for open and short stubs using electromagnetic models when co-simulated with ADS. The results confirm excellent matching and confirm the accuracy of the final layout that is optimized.

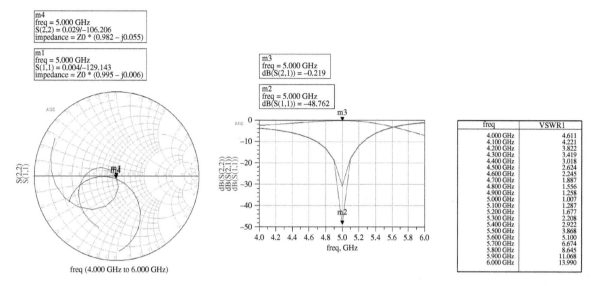

Figure 5.44 Simulation results for the final layout for co-simulated open stub.

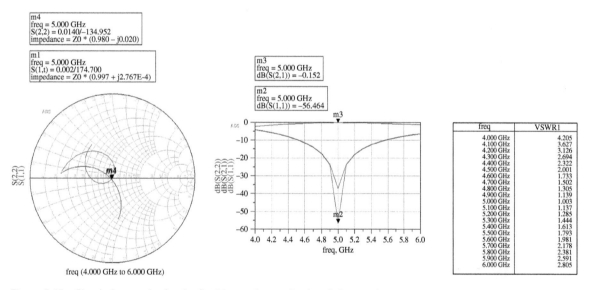

Figure 5.45 Simulation results for the final layout for co-simulated short stub.

References

1 Eroglu, A. (2013). *RF Circuit Design Techniques for MF-UHF Applications*, 1e. CRC Press. ISBN: 978-1-4398-6165-3.
2 Eroglu, A. (2015). *Introduction to RF Power Amplifiers Design and Simulation*, 1e. CRC Press. ISBN: 978-1-4822-3164-9.

Problems

Problem 5.1
Using a Smith chart, design all possible configurations of two-element matching networks that match source impedance $Zs = (25 + j100)\Omega$ to the load $Z_L = (30 - j10)\Omega$. Assume characteristic impedance of $Z_0 = 50\,\Omega$ and an operating frequency of $f = 2\,\text{GHz}$.

Problem 5.2
It is required to match a load impedance, $Z_{Load} = 10 + j10\,\Omega$, to a transmission line with 50 Ω characteristic impedance at $f = 500\,\text{MHz}$. Design L-type matching possible networks. Confirm your results with ADS simulation.

Problem 5.3
It is required to design a lumped element matching network to transform load $Z_{load} = (100 + j100)\,\Omega$ to input impedance $Z_{in} = (50 + j20)\,\Omega$ at $f = 500\,\text{MHz}$ and calculate component values.

Problem 5.4
Design a T network to transform $Z_L = 50\,\Omega$ to input impedance $Z_{in} = 20 - j30\,\Omega$ at $f = 500\,\text{MHz}$ using T-type matching network for $Q_n = 5$ and calculate component values using a Smith chart.

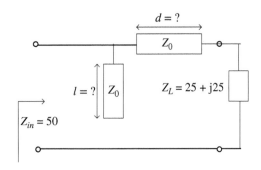

Figure 5.46   Problem 5.6.

Problem 5.5
Design a π network to transform $Z_L = 50\,\Omega$ to input impedance $Z_{in} = 20 - j30\,\Omega$ at $f = 500\,\text{MHz}$ using a T-type matching network for $Q_n = 5$ and calculate the component values using a Smith chart.

Problem 5.6
Design a matching network to match the load impedance $Z_{Load} = 25 + j25\,\Omega$ to input impedance $Z_{in} = 50\,\Omega$ at $f = 1.5\,\text{GHz}$ using only a transmission line and shunt open stub, as shown in Figure 5.46.

Problem 5.7
Design a matching network to match the load impedance $Z_{Load} = 25 + j75\,\Omega$ to input impedance $Z_{in} = 50\,\Omega$ at $f = 1.5\,\text{GHz}$ using only a transmission line and a series open stub, as shown in Figure 5.47.

Problem 5.8
Implement a matching network to transform $Z_{Load} = 75 + j25\,\Omega$ to input impedance $Z_{in} = 150\,\Omega$ at $f = 1\,\text{GHz}$. The desired matching network topology is illustrated in Figure 5.48.

Problem 5.9
Use a double stub tuner to match $Z_{Load} = 100 + j100\,\Omega$ to a transmission line with 300 Ω characteristic impedance. The desired matching network topology is illustrated in Figure 5.49.

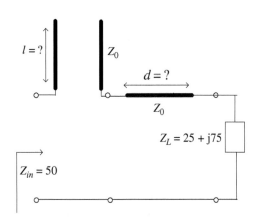

Figure 5.47   Problem 5.7.

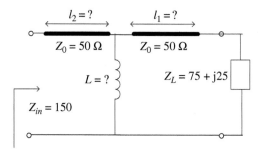

Figure 5.48 Problem 5.8.

Figure 5.49 Problem 5.9.

Problem 5.10

Match a transmission line with a characteristic impedance of $150\,\Omega$ with $100\,\Omega$ transmission line using a quarter-wave transformer. What should be the characteristic impedance of the quarter-wave transformer?

Problem 5.11

Interface $10\,\Omega$ differential output of the amplifier 1 (Amp 1) to $100\,\Omega$ input impedance of the second amplifier (Amp 2) using an unbalanced LC network at 100 MHz, as shown in Figure 5.50.

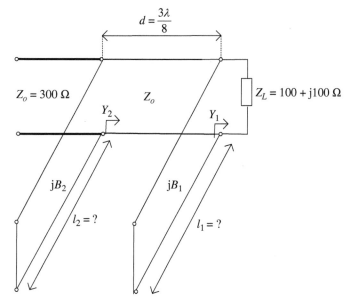

Figure 5.50 Amplifier impedance transformer design.

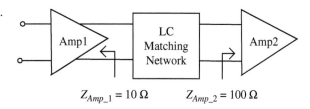

Design Challenge 5.1 Design of Impedance Matching Network for Amplifiers

Design an impedance matching network for the amplifier system shown in Figure 5.51 for the specified frequency range 1.4–1.6 GHz. Unlike single frequency amplifiers, the impedance of the devices must be matched across the range of frequency. Based on the transistor scattering parameters, the input and output impedances for the transistor are given as

$$Z_{in} = Z_s = Z_0 (1 + S_{11})/1 - S_{11} = 11.73 - j12.8$$
$$Z_{out} = Z_L = Z_0 (1 + S_{22})/1 - S_{22} = 41.4 - j17.6$$

a) Calculate the matching network using the analytical method.
b) Confirm your analytical results with ADS simulation.

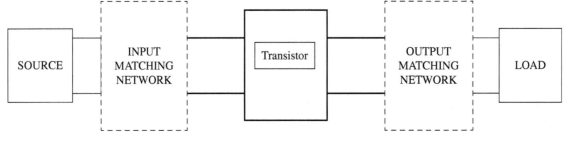

Figure 5.51 Design Challenge 5.1.

6

Resonator Circuits

6.1 Introduction

Resonators have frequency characteristics which give them the ability to present specific impedances, quality factors, and bandwidths. They can eliminate the reactive component effects and introduce only resistive portion of the impedance at a frequency called resonance frequency. A circuit which is capable of producing these effects is called a resonant circuit. The ideal resonant circuit acts like a filter and eliminates the unwanted signal content out of the frequency of interest, as shown in Figure 6.1. Resonant circuits can also be used as part of the impedance matching networks to transform one impedance at one point to another impedance [1, 2]. In RF amplifier circuits, it is a commonly used technique to present a matched impedance at one frequency and introduce high impedance levels at others. When the amplifier is matched at the input and output for maximum gain, it is possible to deliver the highest amount of power by keeping the circuit stable. Most of the time, the stability of the circuit is accomplished by using filters and resonators to eliminate the spurious contents and oscillations.

This chapter discusses resonant networks, transmission lines, Smith charts, and impedance matching networks.

6.2 Parallel and Series Resonant Networks

6.2.1 Parallel Resonance

Consider the parallel resonant circuit given in Figure 6.2. The response of the circuit can be obtained by finding the voltage with an application of KCL. An application of KCL gives the first order differential equation for voltage v as

$$\frac{v}{R} + \frac{1}{L}\int_0^t v d\tau + I_o + C\frac{dv}{dt} = 0 \tag{6.1}$$

where I_o is the initial charged current on the inductor. Equation (6.1) can be rewritten as

$$\frac{d^2v}{dt^2} + \frac{1}{RC}\frac{dv}{dt} + \frac{v}{LC} = 0 \tag{6.2}$$

The solution for the voltage in Eq. (6.2) will be in the following form

$$v = Ae^{st} \tag{6.3}$$

RF/Microwave Engineering and Applications in Energy Systems, First Edition. Abdullah Eroglu.
© 2022 John Wiley & Sons Ltd. Published 2022 by John Wiley & Sons Ltd.
Companion website: www.wiley.com/go/eroglu/rfmicrowave

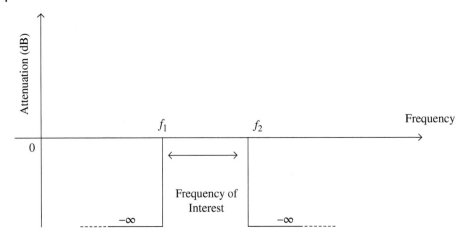

Figure 6.1 Ideal resonant network response.

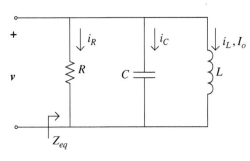

Figure 6.2 Parallel resonant circuit.

where A is constant and $s = j\omega$. Substitution of (6.3) into (6.2) gives

$$Ae^{st}\left(s^2 + \frac{s}{RC} + \frac{1}{LC}\right) = 0 \tag{6.4}$$

which can be simplified to

$$s^2 + \frac{s}{RC} + \frac{1}{LC} = 0 \tag{6.5}$$

Equation (6.5) is called the characteristic equation. The roots of the equation are

$$s_1 = -\frac{1}{2RC} + \sqrt{\left(\frac{1}{2RC}\right)^2 - \left(\frac{1}{LC}\right)} \tag{6.6}$$

$$s_2 = -\frac{1}{2RC} - \sqrt{\left(\frac{1}{2RC}\right)^2 - \left(\frac{1}{LC}\right)} \tag{6.7}$$

The complete solution for voltage v is then obtained as

$$v = v_1 + v_2 = A_1 e^{s_1 t} + A_2 e^{s_2 t} \tag{6.8}$$

We can express the roots given by (6.6) and (6.7) as

$$s_1 = -\alpha + \sqrt{\alpha^2 - \omega_0^2} \tag{6.9}$$

$$s_2 = -\alpha - \sqrt{\alpha^2 - \omega_0^2} \tag{6.10}$$

In (6.10), α is the damping coefficient, and ω_0 is the resonant frequency. At resonant frequency, the reactive components cancel each other. The damping coefficient and resonant frequency are given by the following equations

$$\alpha = \frac{1}{2RC} \tag{6.11}$$

and

$$\omega_0 = \frac{1}{\sqrt{LC}} \tag{6.12}$$

When

$$
\begin{aligned}
\omega_0^2 < \alpha^2, \quad & s_1 \text{ and } s_2 \text{ are real and distinct, voltage is overdamped} \\
\omega_0^2 > \alpha^2, \quad & s_1 \text{ and } s_2 \text{ are compelx, voltage is underdamped} \\
\omega_0^2 = \alpha^2, \quad & s_1 \text{ and } s_2 \text{ real and equal, voltage is critically damped}
\end{aligned}
\tag{6.13}
$$

The time domain representation of a parallel resonant network voltage response to illustrate underdamped and overdamped cases are illustrated in Figures 6.3 and 6.4, respectively. L and C values are taken to be 0.1 H and 0.001 F for an underdamped case, whereas for overdamped case L and C values are taken to be 50 mH and 0.2 μF. R values are varied to see its effect on voltage response for damping.

The quality factor Q and the bandwidth BW of the parallel resonant network are

$$Q = \frac{R}{\omega_0 L} = \omega_0 R C \tag{6.14}$$

$$BW = \frac{\omega_0}{Q} = \frac{1}{RC} \tag{6.15}$$

In terms of quality factor, the roots given by (6.6) and (6.7) can be expressed as

$$s_1 = \omega_0 \left[-\frac{1}{2Q} + \sqrt{\left(\frac{1}{2Q}\right)^2 - 1} \right] \tag{6.16}$$

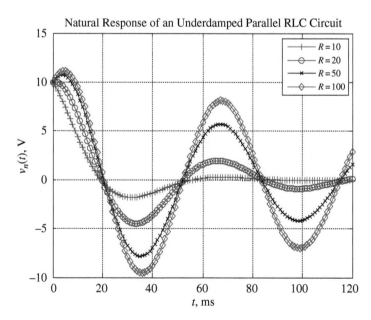

Figure 6.3 Parallel resonant network response for an underdamped case.

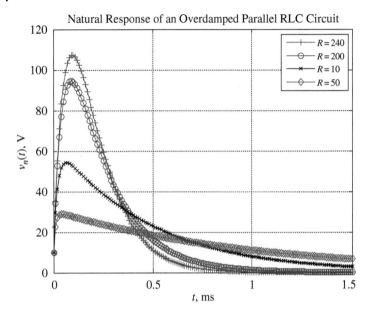

Figure 6.4 Parallel resonant network response for an overdamped case.

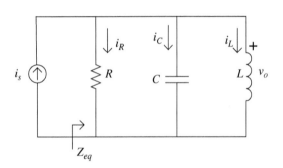

Figure 6.5 Parallel resonant circuit with a source current.

$$s_2 = \omega_0 \left[-\frac{1}{2Q} - \sqrt{\left(\frac{1}{2Q}\right)^2 - 1} \right] \qquad (6.17)$$

Assume now there is a source current connected to parallel resonant network in Figure 6.2 as illustrated in Figure 6.5.

The equivalent impedance of the parallel resonant network is found from Figure 6.5 as

$$Z_{eq}(s) = \frac{V_o(s)}{I_s(s)} = \frac{s/C}{s^2 + s(1/RC) + (1/LC)} = \frac{s/C}{(s - s_1)(s + s_2)} \qquad (6.18a)$$

which can be written as

$$Z_{eq}(j\omega) = \frac{1}{R} + j\frac{\omega L}{(1 - \omega^2 LC)} \qquad (6.18b)$$

In (6.18), s_1 and s_2 are now the poles of the impedance. When

$$\left(\frac{1}{2RC}\right)^2 \geq \left(\frac{1}{LC}\right) \quad \text{or} \quad R \leq \left(\frac{\omega_0 L}{2}\right) = \left(\frac{1}{2\omega_0 C}\right) \qquad (6.19)$$

poles of the impedance lie on the negative real axis. Hence, the value of R is small in comparison to the values of the reactances and as a result the resonant network has a broadband response. When

$$\left(\frac{1}{2RC}\right)^2 < \left(\frac{1}{LC}\right) \quad \text{or} \quad R > \left(\frac{\omega_0 L}{2}\right) \qquad (6.20)$$

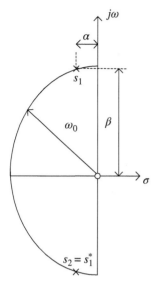

The poles become complex and they take the following form.

$$s_1 = -\alpha + j\sqrt{\omega_0^2 - \alpha^2} = -\alpha + j\beta \qquad (6.21)$$

$$s_2 = -\alpha - j\sqrt{\omega_0^2 - \alpha^2} = -\alpha - j\beta \qquad (6.22)$$

This can be illustrated on the pole-zero diagram shown in Figure 6.6.

The transfer function for the parallel resonant network is found using Figure 6.5 as

$$|H(\omega)| = \left|\frac{I_R}{I_S}\right| = \left|\frac{\omega(L/R)}{\sqrt{(1-\omega^2 LC)^2 + (\omega L/R)^2}}\right| \qquad (6.23)$$

At the resonant frequency, the transfer function will be real and equal to its maximum value as

$$|H(\omega = \omega_0)| = \frac{\omega_0(L/R)}{\sqrt{(1-\omega_0^2 LC)^2 + (\omega_0 L/R)^2}} = 1 = |H(\omega)|_{max} \qquad (6.24)$$

Figure 6.6 Pole-zero diagram for complex conjugate roots.

The network response is obtained using a transfer function given by (6.24) for different values of R, as shown in Figure 6.7. The values of the inductance and capacitance are taken to be 1.25 μH and 400 nF.

This gives the resonant frequency as 0.22508 MHz. The condition for a broadband network is accomplished when $R = 5$, as shown in Figure 6.7. As R increases, the quality factor of the network increases, which agrees with Eq. (6.14).

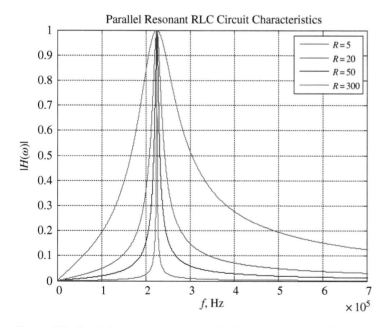

Figure 6.7 Parallel resonant circuit transfer function characteristics.

Quality factor, Q, is an important parameter in resonant network responses as it can be used as a measure for the loss and bandwidth of the circuit. The quality factor of the circuit defines the ratio of the peak energy stored to the energy dissipated per cycle as given by

$$Q = \frac{2\pi(\text{Peak Energy Stored})}{(\text{Energy Dissipated Per Cycle Stored})} = \frac{2\pi\left(\frac{1}{2}CV^2\right)}{(2\pi/\omega_0)(V^2/2R)} = \omega_0 CR \tag{6.25}$$

Example 6.1 Parallel Resonant Circuit

A parallel resonant circuit with a source resistance of 50 Ω and a load resistance of 25 Ω. The loaded Q must be equal to 12 at the resonant frequency of 60 MHz.

a) Design the resonant circuit.

b) Calculate the 3 dB bandwidth of the resonant circuit.

c) Obtain the frequency response of this circuit versus frequency, i.e. plot $20 \log(V_o/V_{in})$ vs frequency.

Solution

a) The effective parallel resistance across the parallel resonance circuit is

$$R_p = \frac{(50)25}{50 + 25} = 16.67 \ \Omega$$

Then

$$X_p = \frac{R_p}{Q} = \frac{16.67}{12} = 1.4$$

Since

$$X_p = \omega L = \frac{1}{\omega C}$$

Then, the resonance element values are

$$L = \frac{X_p}{\omega} = \frac{1.4}{2\pi(60 \times 10^6)} = 3.71 \ \text{nH}$$

and

$$C = \frac{1}{\omega X_p} = \frac{1}{2\pi(60 \times 10^6)(1.4)} = 1894.7 \ \text{pF}$$

b) BW is found from

$$Q = \frac{f_c}{BW_{3dB}} \rightarrow BW_{3dB} = \frac{f_c}{Q} = \frac{60 \times 10^6}{12} = 5 \times 10^6 \ \text{Hz}$$

c) The plot of the attenuation profile is shown in Figure 6.8.

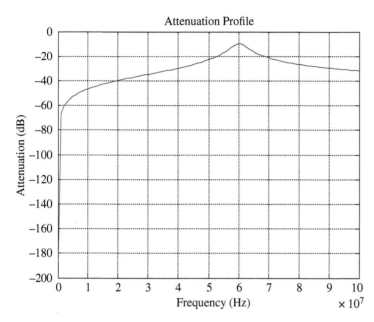

Figure 6.8 Attenuation profile.

6.2.2 Series Resonance

Consider the series resonant circuit given in Figure 6.9a. The response of the circuit can be obtained by application of KVL, which gives the first order differential equation for current as

$$Ri + L\frac{di}{dt} + \frac{1}{C}\int_0^t id\tau + V_0 = 0 \tag{6.26}$$

where V_0 is the initial charged voltage on capacitor. Equation (6.24) can be written as

$$\frac{d^2i}{dt^2} + \frac{R}{L}\frac{di}{dt} + \frac{i}{LC} = 0 \tag{6.27}$$

Following the same solution technique for parallel resonant network leads to the following equation in frequency domain.

$$s^2 + \frac{R}{L}s + \frac{1}{LC} = 0 \tag{6.28}$$

The roots of the equation are

$$s_1 = -\frac{R}{2L} + \sqrt{\left(\frac{R}{2L}\right)^2 - \left(\frac{1}{LC}\right)} \tag{6.29}$$

$$s_1 = -\frac{R}{2L} - \sqrt{\left(\frac{R}{2L}\right)^2 - \left(\frac{1}{LC}\right)} \tag{6.30}$$

(a)

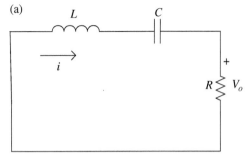

(b)

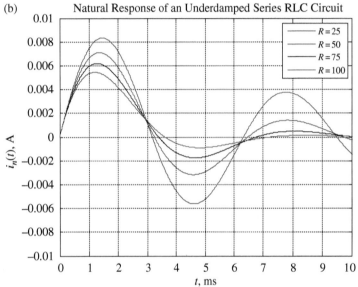

(c)

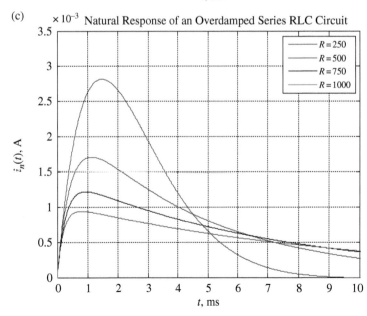

Figure 6.9 (a) Series resonant network. (b) Series resonant network response for underdamped case. (c) Series resonant network response for overdamped case.

Equations (6.25) and (6.26) as

$$s_1 = -\alpha + \sqrt{\alpha^2 - \omega_0^2} \tag{6.31}$$

$$s_2 = -\alpha - \sqrt{\alpha^2 - \omega_0^2} \tag{6.32}$$

where α and ω are defined for a series resonant network as

$$\alpha = \frac{R}{2L} \tag{6.33}$$

and

$$\omega_0 = \frac{1}{\sqrt{LC}} \tag{6.34}$$

The quality factor and the bandwidth of the series resonant network are

$$Q = \frac{\omega_0 L}{R} = \frac{1}{\omega_0 RC} \tag{6.35}$$

$$BW = \frac{\omega_0}{Q} = \frac{R}{L} \tag{6.36}$$

In terms of quality factor, the roots given by (6.29) and (6.30) can be obtained as

$$s_1 = \omega_0 \left[-\frac{1}{2Q} + \sqrt{\left(\frac{1}{2Q}\right)^2 - 1} \right] \tag{6.37}$$

$$s_2 = \omega_0 \left[-\frac{1}{2Q} - \sqrt{\left(\frac{1}{2Q}\right)^2 - 1} \right] \tag{6.38}$$

Equations (6.37) and (6.38) are identical to the ones obtained for the parallel resonant circuit. The damping characteristics of the series resonant network follow the conditions listed by Eq. (6.13). The time domain representation of a series resonant network current response illustrating underdamped and overdamped cases are illustrated in Figure 6.9b and c, respectively. L and C values are taken to be 100 mH and 10 μF for an underdamped case, whereas for an overdamped case L and C values are taken to be 200 mH and 10 μF. R values are varied to see their effect on the current response for damping.

The transfer function of this network can be found by connecting a source voltage, as shown in Figure 6.10 and obtained as

$$|H(j\omega)| = \left| \frac{V_o(s)}{V_s(s)} \right| = \frac{\omega \dfrac{R}{L}}{\sqrt{\left(\dfrac{1}{LC} - \omega^2\right)^2 + \left(\omega \dfrac{R}{L}\right)^2}} \tag{6.39}$$

The phase of the transfer function is found from

$$\theta(j\omega) = 90^0 - \tan^{-1} \left(\frac{\omega \dfrac{R}{L}}{\dfrac{1}{LC} - \omega^2} \right) \tag{6.40}$$

At resonant frequency, the transfer function is maximum and will be equal to

$$|H(j\omega)| = \frac{\sqrt{\dfrac{1}{LC}} \dfrac{R}{L}}{\sqrt{\left(\dfrac{1}{LC} - \dfrac{1}{LC}\right)^2 + \left(\sqrt{\dfrac{1}{LC}} \dfrac{R}{L}\right)^2}} = 1 = |H(j\omega)|_{\max} \tag{6.41}$$

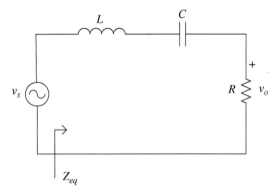

Figure 6.10 Series resonant network with source voltage.

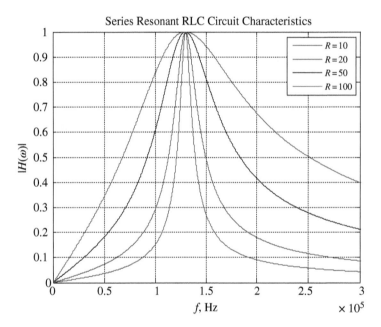

Figure 6.11 Series resonant circuit transfer function characteristics.

The resonant characteristics of the network can be obtained by plotting the transfer function given by (6.39) versus different values of R, as shown in Figure 6.11. The values of the inductance and capacitance are taken to be $150\,\mu$H and $10\,$nF.

6.3 Practical Resonances with Loss, Loading, and Coupling Effects

6.3.1 Component Resonances

RF components such as resistors, inductors, and capacitors in practice exhibit resonances at high frequencies due to their high frequency (HF) characteristics. The HF representation of an inductor and capacitor are given in Figure 6.12.

(a)   (b)

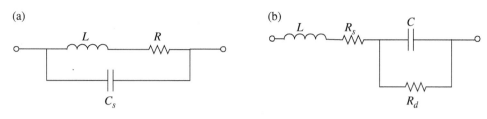

Figure 6.12 High frequency representation of (a) inductor and (b) capacitor.

As seen from the equivalent circuit in Figure 6.12, the inductor will act as an inductor until it reaches resonant frequency, then it gets into resonance and exhibits capacitive effects afterward. The expression that gives this characteristics for an inductor can be obtained as

$$Z = \frac{(j\omega L + R)\dfrac{1}{j\omega C_s}}{(j\omega L + R) + \dfrac{1}{j\omega C_s}} = \frac{R}{(1 - \omega^2 L C_s)^2 + (\omega R C_s)^2} + j\frac{\omega(L - R^2 C_s) - \omega^3 L^2 C_s}{(1 - \omega^2 L C_s)^2 + (\omega R C_s)^2} \tag{6.42}$$

The equation in (6.42) can also be written as

$$Z = R_s + jX_s \tag{6.43}$$

where

$$R_s = \frac{R}{(1 - \omega^2 L C_s)^2 + (\omega R C_s)^2} \tag{6.44}$$

and

$$X_s = \frac{\omega(L - R^2 C_s) - \omega^3 L^2 C_s}{(1 - \omega^2 L C_s)^2 + (\omega R C_s)^2} \tag{6.45}$$

We can now use Eqs. (6.44) and (6.45) and represent the circuit given in Figure 6.12a with an equivalent series circuit shown in Figure 6.13.

The resonance frequency is found when $X_s = 0$ as

$$f_r = \frac{1}{2\pi}\sqrt{\frac{L - R^2 C_s}{L^2 C_s}} \tag{6.46}$$

The quality factor is obtained from

$$Q = \frac{|X_s|}{R_s} \tag{6.47}$$

R_s in Figure 6.13 is the series resistance of the inductor and includes the distributed resistance effect of the wire. It is calculated from

$$R = \frac{l_W}{\sigma A} \tag{6.48}$$

Figure 6.13 Equivalent series circuit.

C_s in Figure 6.12a is the capacitance including the effects of distributed capacitance of the inductor and is given by

$$C_s = \frac{2\pi\varepsilon_0 daN^2}{l_W} \tag{6.49}$$

For an air core inductor, the value of L shown in Figure 6.12a and used in Eq. (6.45) is found from

$$L = \frac{d^2 N^2}{18d + 40l}\ \mu\mathrm{H} \tag{6.50}$$

In this equation, L is given as inductance in μH, d is the coil inner diameter in inches, l is the coil length in inches, and N is the number of turns in the coil. The formula given in (6.50) can be extended to include the spacing between each turn of the air coil inductor. Then, Eq. (6.50) can be modified as

$$L = \frac{d^2 N^2}{18d + 40(Na + (N-1)s)}\ \mu\mathrm{H} \tag{6.51}$$

In (6.51), a represents the wire diameter in inches and s represents the spacing in inches between each turn. When air is replaced with a magnetic material, such as that used for a toroidal core, the inductance of the formed inductor can be calculated using

$$L = \frac{4\pi N^2 \mu_i A_{Tc}}{l_e}\ \mathrm{nH} \tag{6.52}$$

In (6.52), L is the inductance in nanohenries (nH), N is the number of turns, μ_i is the initial permeability, A_{Tc} is the total cross-sectional area of the core in cm^2 and l_e is the effective length of the core in centimeters. The details of the derivation and implementation of inductor design and design tables are given in [1].

From Figure 6.12b, it is also clear that the nonideal capacitor has also resonances due to its HF characteristics. The HF model of the capacitor has parasitic components such as lead inductance, L, conductor loss, R_s, and dielectric loss, R_d, which only become relevant at high frequencies. The characteristics of the capacitor can be obtained by finding the equivalent impedance as

$$Z = (j\omega L_s + R_s) + \left(\frac{1}{G_d + j\omega C}\right) = \frac{R_s G_d^2 + (\omega C)^2 + G_d}{G_d^2 + (\omega C)^2} + j\frac{\omega L G_d^2 + L\omega(\omega C)^2 - \omega C}{G_d^2 + (\omega C)^2} \tag{6.53}$$

The equation in (6.53) can be expressed as

$$Z = R_s + jX_s \tag{6.54}$$

where

$$R_s = \frac{R_s G_d^2 + (\omega C)^2 + G_d}{G_d^2 + (\omega C)^2} \tag{6.55a}$$

$$X_s = \frac{\omega L G_d^2 + L\omega(\omega C)^2 - \omega C}{G_d^2 + (\omega C)^2} \tag{6.55b}$$

The impedance given in (6.54) can be converted to admittance as

$$Y = Z^{-1} = \frac{R_s}{R_s^2 + X_s^2} + j\frac{-X_s}{R_s^2 + X_s^2} = G + jB \tag{6.56}$$

Figure 6.14 Equivalent parallel circuit.

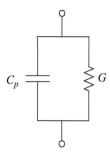

or

$$Y = \frac{\left(R_s G_d^2 + (\omega C)^2 + G_d\right)\left(G_d^2 + (\omega C)^2\right)}{\left(R_s G_d^2 + (\omega C)^2 + G_d\right)^2 + \left(\omega L G_d^2 + L\omega(\omega C)^2 - \omega C\right)^2} + j\frac{\left(\omega C - \omega L G_d^2 - L\omega(\omega C)^2\right)\left(G_d^2 + (\omega C)^2\right)}{\left(R_s G_d^2 + (\omega C)^2 + G_d\right)^2 + \left(\omega L G_d^2 + L\omega(\omega C)^2 - \omega C\right)^2} \tag{6.57}$$

Then, the capacitor can be represented by a parallel equivalent circuit, as shown in Figure 6.14, where

$$G = \frac{\left(R_s G_d^2 + (\omega C)^2 + G_d\right)\left(G_d^2 + (\omega C)^2\right)}{\left(R_s G_d^2 + (\omega C)^2 + G_d\right)^2 + \left(\omega L G_d^2 + L\omega(\omega C)^2 - \omega C\right)^2} \tag{6.58}$$

$$B = \frac{\left(\omega C - \omega L G_d^2 - L\omega(\omega C)^2\right)\left(G_d^2 + (\omega C)^2\right)}{\left(R_s G_d^2 + (\omega C)^2 + G_d\right)^2 + \left(\omega L G_d^2 + L\omega(\omega C)^2 - \omega C\right)^2} \tag{6.59}$$

The resonance frequency for the circuit shown in Figure 6.14 is found when $B = 0$ as

$$f_r = \frac{1}{2\pi}\sqrt{\frac{R^2 C - L}{R^2 C^2 L}} \tag{6.60}$$

The quality factor for the parallel network is then obtained from

$$Q = \frac{|B|}{G} = \frac{R_p}{|X_p|} \tag{6.61}$$

6.3.2 Parallel *LC* Networks

6.3.2.1 Parallel *LC* Networks with Ideal Components

When ideal components are used, the typical circuit would be represented as the one shown in Figure 6.15. At resonance, the magnitudes of the reactances of the *L* and *C* elements are equal. The reactances of two components have opposite signs so the net reactance is zero for a series circuit or infinity for a parallel circuit. Hence, we can obtain the resonance frequency from

$$\omega_0 L = \frac{1}{\omega_0 C} \rightarrow \omega_0 = \frac{1}{\sqrt{LC}} \rightarrow f_0 = \frac{1}{2\pi\sqrt{LC}} \tag{6.62}$$

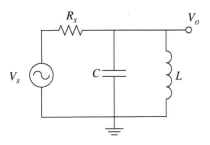

Figure 6.15 *LC* resonant network with ideal components and source.

The transfer function for the *LC* resonant network in Figure 6.16 is found from

$$|H(\omega)|_{dB} = \left|\frac{V_{out}}{V_{in}}\right|_{dB} = 20\log\left(\frac{X_{total}}{R_s + X_{total}}\right) \quad (6.63)$$

where

$$X_{total} = \frac{X_C X_L}{X_C + X_L} \quad (6.64)$$

Substitution of (6.64) into (6.63) gives

$$|H(\omega)|_{dB} = \left|\frac{V_{out}}{V_{in}}\right|_{dB} = 20\log\left|\frac{j\omega L/R_s}{(1 - \omega^2 LC) + j\omega L/R_s}\right| \quad (6.65)$$

The frequency characteristics of the circuit are plotted in Figure 6.16, when $L = 0.5\,\mu\text{H}$, $C = 2500\,\text{pF}$, and $R = 50\,\Omega$.

The quality factor of the circuit is found from its equivalent impedance as

$$Z_{eq} = R_s + \frac{X_C X_L}{X_C + X_L} = \frac{R_s}{(1 - \omega^2 LC)} + j\frac{\omega L}{(1 - \omega^2 LC)} \quad (6.66)$$

Then, the quality factor of the circuit is

$$Q = \frac{|X|}{R} = \frac{\omega L}{R_s} \quad (6.67)$$

6.3.2.2 Parallel *LC* Networks with Nonideal Components

Now, assume we have some additional loss for the inductor for the *LC* network, as shown in Figure 6.17.

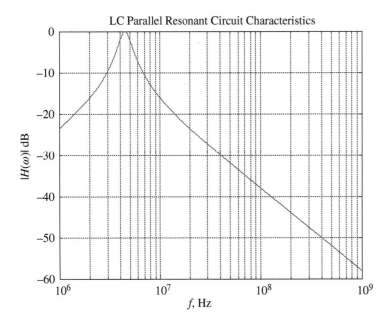

Figure 6.16 The frequency characteristics of an *LC* network with source.

The equivalent impedance of the network in Figure 6.17 takes the following form with the addition of loss component r.

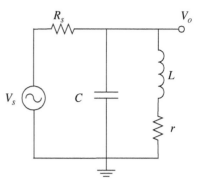

Figure 6.17 Addition of loss to parallel LC network.

$$Z_{eq} = R_s + \frac{r + j\omega L}{(j\omega rC - \omega^2 LC + 1)} = R_s + \frac{(r + j\omega L)(1 - \omega^2 LC - j\omega rC)}{(1 - \omega^2 LC)^2 + (\omega rC)^2} \qquad (6.68)$$

which can be simplified to

$$Z_{eq} = \frac{R_s\left[(1 - \omega^2 LC)^2 + (\omega rC)^2\right] + r}{(1 - \omega^2 LC)^2 + (\omega rC)^2} + j\frac{\omega\left[(L - Cr^2) - \omega^2 L^2 C\right]}{(1 - \omega^2 LC)^2 + (\omega rC)^2} \qquad (6.69)$$

The resonant frequency of the network is now equal to

$$\omega_0 = \sqrt{\frac{L - Cr^2}{L^2 C}} \rightarrow f_0 = \frac{1}{2\pi}\sqrt{\frac{L - Cr^2}{L^2 C}} \qquad (6.70)$$

The loaded quality factor of the circuit is obtained as

$$Q_L = \frac{\omega\left[(L - Cr^2) - \omega^2 L^2 C\right]}{R_s\left[(1 - \omega^2 LC)^2 + (\omega rC)^2\right]} \qquad (6.71)$$

when $r = 0$ for the lossless case, (6.70) and (6.71) reduce to the ones given in (6.62) and (6.67). The transfer function with the loss resistance changes to

$$|H(\omega)|_{dB} = \left|\frac{V_{out}}{V_{in}}\right|_{dB} = 20\log\left|\frac{r + j\omega\left[(L - Cr^2) - \omega^2 L^2 C\right]}{R_s\left[(1 - \omega^2 LC)^2 + (\omega rC)^2\right] + j\omega\left[(L - Cr^2) - \omega^2 L^2 C\right]}\right| \qquad (6.72)$$

The transfer function showing the attenuation profile with different loss resistance values when $L = 0.5\,\mu\text{H}$, $C = 2500\,\text{pF}$, and $R = 50\,\Omega$ is illustrated in Figure 6.18. The network bandwidth broadens as the value of r increases as expected.

The quality factor of the network with the addition of loss resistance significantly differs from the original LC parallel resonant network. Ideally, the original network has an infinite value for its quality factor. In agreement with this, a very large value of the quality factor for the original network is obtained at resonance frequency when $r = 0$ is also seen from Figure 6.19.

6.3.2.3 Loading Effects on Parallel LC Networks

A resonant circuit becomes loaded when it is connected to a load or fed by a source. The Q of the circuit under these conditions is called the loaded Q, or simply, Q_L. The loaded quality factor of the circuit then depends on source resistance, load resistance, and the individual Q of the reactive components.

When the reactive components are lossy, they impact the Q factor of the overall circuit. For instance, consider the resonant circuit with source resistance in Figure 6.17. The quality factor of the circuit versus various source resistance values when $L = 0.5\,\mu\text{H}$ and $C = 2500\,\text{pF}$ are illustrated in Figure 6.20. For the same frequency, the quality factor increases as the value of the source resistance, R_s, increases. As a result, the selectivity of the network can be adjusted by setting the value of the source resistance. The attenuation profile showing the response when the source resistance is changed is given in Figure 6.21.

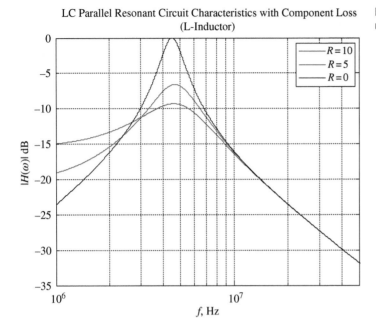

LC Parallel Resonant Circuit Characteristics with Component Loss (L-Inductor)

Figure 6.18 Attenuation profile of an *LC* network with loss resistor.

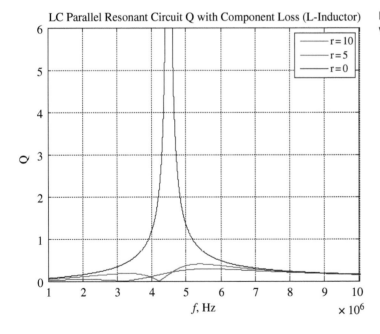

LC Parallel Resonant Circuit Q with Component Loss (L-Inductor)

Figure 6.19 Quality factor of an *LC* network with a loss resistor.

Figure 6.20 Quality factor of an *LC* network for different source resistance values.

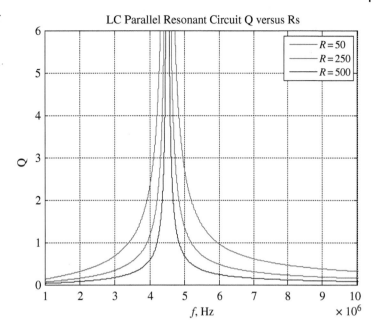

Figure 6.21 Attenuation profile of an *LC* network for different source resistance values.

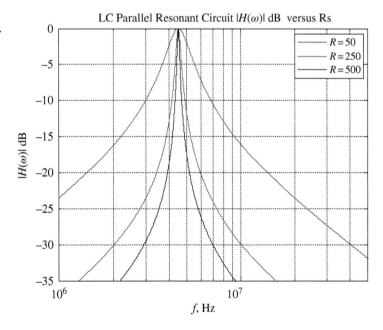

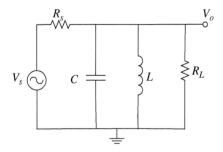

Figure 6.22 Loaded LC resonant circuit.

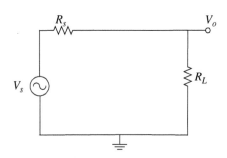

Figure 6.23 Equivalent loaded LC resonant circuit at resonance.

The circuit shown in Figure 6.22 has both source and load resistances. The impedance of the loaded resonant network can be obtained as

$$Z_{eq} = R_s + \frac{j\omega R_L L}{\omega L + j\omega R_L(\omega^2 LC - 1)} = R_s + \frac{\omega R_L L[\omega L - j\omega R_L(\omega^2 LC - 1)]}{(\omega L)^2 + [\omega R_L(\omega^2 LC - 1)]^2} \tag{6.73}$$

which can be simplified to

$$Z_{eq} = \frac{R_s\left((\omega L)^2 + [\omega R_L(\omega^2 LC - 1)]^2\right) + (\omega L)^2 R_L}{(\omega L)^2 + [\omega R_L(\omega^2 LC - 1)]^2}$$

$$+ j\frac{\omega R_L^2 L(1 - \omega^2 LC)}{(\omega L)^2 + [\omega R_L(\omega^2 LC - 1)]^2} \tag{6.74}$$

The loaded quality factor of the circuit is obtained as

$$Q_L = \frac{\omega R_L^2 L(1 - \omega^2 LC)}{R_s\left((\omega L)^2 + [\omega R_L(\omega^2 LC - 1)]^2\right) + (\omega L)^2 R_L} \tag{6.75}$$

At resonant frequency, this circuit is simplified to the one in Figure 6.23.

The equivalent impedance from (6.74) is equal to

$$Z_{eq} = R_s + R_L \tag{6.76}$$

It is important to note that at resonant frequency

$$\omega_0 L = \frac{1}{\omega_0 C} \tag{6.77}$$

6.3.2.4 *LC* Network Transformations
RL *Networks*

When any one of the reactive components is lossy, impedance transformation from parallel to series or series to parallel will greatly facilitate the analysis of the problem.

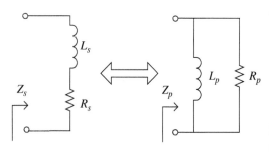

Figure 6.24 *RC* series to parallel *RC* network transformation.

The equivalent impedances of the series and parallel *RL* networks shown in Figure 6.24 are

$$Z_s = R_s + j\omega L_s \tag{6.78}$$

$$Z_p = \frac{\omega^2 L_p^2 R_p}{R_p^2 + \omega^2 L_p^2} + j\frac{\omega L_p R_p^2}{R_p^2 + \omega^2 L_p^2} \tag{6.79}$$

To transform a parallel network to a series network in Figure 6.24, we assume that their quality factors are the same, $Q_p = Q_s$, and we use the impedance equality as

$$Z_s = Z_p \tag{6.80}$$

This can be accomplished by equating the real and imaginary parts as

$$R_s = R_p \quad \text{or} \quad R_s = \frac{\omega^2 L_p^2 R_p}{R_p^2 + \omega^2 L_p^2} \tag{6.81}$$

which can be rewritten as

$$R_s = \frac{R_p}{\left(R_p/\omega L_p\right)^2 + 1} \tag{6.82}$$

The quality factor of the parallel network, Q_p, in Figure 6.24b is

$$Q = \frac{R_p}{\omega L_p} \tag{6.83}$$

Substitution of (6.83) into (6.82) gives the equation that relates the resistances of two networks via quality factor as

$$R_s = \frac{R_p}{Q_p^2 + 1} \tag{6.84}$$

When the imaginary parts are equated

$$\omega L_s = \frac{\omega L_p R_p^2}{R_p^2 + \omega^2 L_p^2} \quad \text{or} \quad L_s = \frac{L_p}{1 + \left(\omega L_p/R_p\right)^2} \tag{6.85}$$

which can be expressed as

$$L_s = L_p \left(\frac{Q_p^2}{1 + Q_p^2} \right) \tag{6.86}$$

The same procedure can be applied to transform a series network to a parallel RL network. The results are summarized in Table 6.1.

Table 6.1 *LC* parallel and series *RL* network transformation.

Series network (a)		Parallel network (b)	
$Q_s = \dfrac{\omega L_s}{R_s}$	(6.87a)	$Q_p = \dfrac{R_p}{\omega L_p}$	(6.88a)
$L_s = L_p \left(\dfrac{Q_p^2}{1 + Q_p^2} \right)$	(6.87b)	$L_p = L_s \left(\dfrac{1 + Q_s^2}{Q_s^2} \right)$	(6.88b)
$R_s = \dfrac{R_p}{1 + Q_p^2}$	(6.87c)	$R_p = R_s \left(1 + Q_s^2 \right)$	(6.88c)

Example 6.2 Network Transformation

A parallel resonant circuit with a 3 dB bandwidth of 5 MHz and a center frequency of 40 MHz. It is given that the resonant circuit has source and load impedances of 100 Ω. The Q of the inductor is given to be 120. The capacitor is assumed to be an ideal capacitor.

a) Design the resonant circuit.
b) What is the loaded Q of the resonant circuit?
c) What is the insertion loss of the network?
d) Obtain the frequency response of this circuit versus frequency, i.e. plot 20log(Vo/Vin) vs frequency.

Solution

a) $Q = \dfrac{f_c}{f_2 - f_1} = \dfrac{40}{5} = 8$

b) Since the inductor is lossy, we need to make conversion from the series to the parallel circuit, as shown in Figure 6.25
   For this

$$X_p = \frac{R_p}{Q_p} \rightarrow R_p = Q_p X_p = 120 X_p$$

The loaded Q of the resonant circuit is

$$Q = \frac{R_{total}}{X_p} \rightarrow 8 = \frac{R_{total}}{X_p}$$

R_{total} is found from

$$8 = \frac{R_{total}}{X_p} = \frac{\dfrac{R_p(50)}{R_p + (50)}}{X_p} \rightarrow 8 = \frac{120 X_p(50)}{(120 X_p + 50) X_p}$$

So

$$X_p = \frac{5600}{960} = 5.83 \, \Omega$$

Then, and

$$R_p = 120 X_p = 699.6 \, \Omega$$

The values of the L and C are found from

$$L = \frac{X_p}{\omega} = \frac{5.83}{2\pi(40 \times 10^6)} = 23.2 \, \text{nH}$$

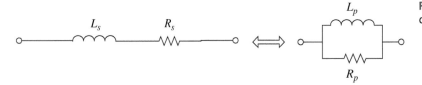

Figure 6.25 Series to parallel conversion.

Figure 6.26 Attenuation profile for parallel resonant network.

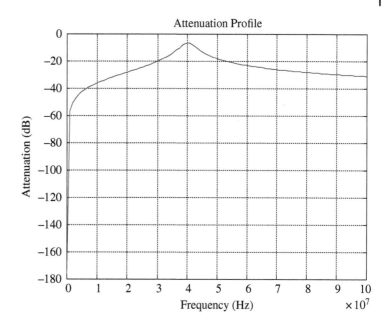

and

$$C = \frac{1}{\omega X_p} = \frac{1}{2\pi(40 \times 10^6)(5.83)} = 682.5 \, \text{pF}$$

c) The load voltage with the resonant circuit is found from

$$V_L = \frac{84.1}{84.1 + 100} V_s = 0.457 V_s$$

The insertion loss is then found from

$$IL = 20 \log \left(\frac{0.457 V_s}{0.5 V_s} \right) = 0.78 \, \text{dB}$$

d) The attenuation profile that is obtained with the program is given in Figure 6.26.

RC Networks

Consider the *RC* parallel circuits shown in Figure 6.27. To transform the parallel *RC* network to a series *RC* network so that both circuits will have the same quality factors, we use the same approach in Section 6.3.2.4.1 and equate impedances as

$$Z_s = Z_p \tag{6.89}$$

where

$$Z_p = \frac{R_p}{1 + (\omega R_p C_p)^2} - \frac{\omega R_p^2 C_p}{1 + (\omega R_p C_p)^2} \tag{6.90}$$

$$Z_s = R_s - \frac{j}{\omega C_s} \tag{6.91}$$

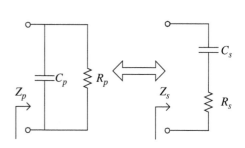

Figure 6.27 *LC* series to parallel *RL* network transformation.

Equating the real and imaginary parts give

$$R_s = \frac{R_p}{1 + (\omega R_p C_p)^2} = \frac{R_p}{1 + Q_p^2} \tag{6.92}$$

and

$$\frac{1}{\omega C_s} = \frac{\omega R_p^2 C_p}{1 + (\omega R_p C_p)^2} \tag{6.93}$$

which leads to

$$C_s = C_p \left(\frac{1 + (\omega R_p C_p)^2}{(\omega R_p C_p)^2} \right) = C_p \left(\frac{1 + Q_p^2}{Q_p^2} \right) \tag{6.94a}$$

The same procedure can be applied to transform a series network to a parallel *RL* network. The results are summarized in Table 6.2.

6.3.2.5 *LC* Network with Series Loss

When a parallel *LC* network has a component with a series loss, we can use the transformations discussed in Sections 0 and 6.3.2.4.2 to simplify the analysis of the circuit. Consider the parallel *LC* network with a series loss shown in Figure 6.28.

Table 6.2 Transform of series network to parallel network.

Series network (a)		Parallel network (b)	
$Q_s = \dfrac{1}{\omega R_s C_s}$	(6.94b)	$Q_p = \omega R_p C_p$	(6.95a)
$C_s = C_p \left(\dfrac{1 + Q_p^2}{Q_p^2} \right)$	(6.94c)	$C_p = C_s \left(\dfrac{Q_s^2}{1 + Q_s^2} \right)$	(6.95b)
$R_s = \dfrac{R_p}{1 + Q_p^2}$	(6.94d)	$R_p = R_s \left(1 + Q_s^2 \right)$	(6.95c)

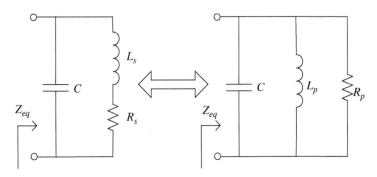

Figure 6.28 *LC* parallel network using *L* with series loss transformation.

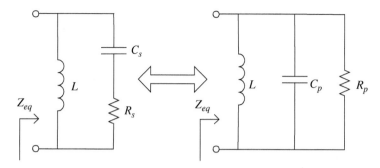

Figure 6.29 *LC* parallel network using *C* with series loss transformation.

The transformation of the circuit given in Figure 6.28a and b can be done using the relations given by

$$R_p = R_s\left(1 + Q_s^2\right) \tag{6.96}$$

$$L_p = L_s\left(\frac{1 + Q_s^2}{Q_s^2}\right) \tag{6.97}$$

$$Q_s = \frac{\omega L_s}{R_s} \tag{6.98}$$

The same procedure can be applied if a capacitor has a series loss, as shown in Figure 6.28a. The transformation from Figure 6.29a and b can then be performed using (6.89), (6.91), and

$$C_p = C_s\left(\frac{Q_s^2}{1 + Q_s^2}\right) \tag{6.99}$$

6.4 Coupling of Resonators

Single resonators can be coupled via capacitors or inductors to produce wide, flat passbands and steeper skirts. The conventional way of coupling single identical resonators via an inductor is shown in Figure 6.30. The circuit shown in Figure 6.31 will exhibit as a single shunt tapped inductance with a 6 dB/octave slope below resonance since each reactive element presents this amount for the slope. The circuit will be behaving like a three-element low pass filter above resonance, as shown in Figure 6.31 with an 18 dB/octave slope.

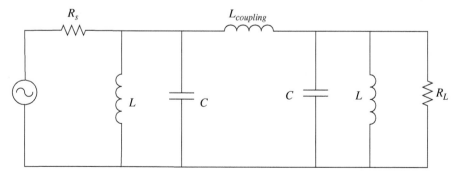

Figure 6.30 Inductively coupled resonators.

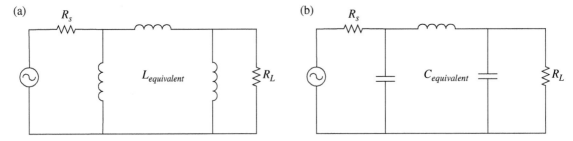

Figure 6.31 Inductively coupled resonators (a) below resonance and (b) above resonance.

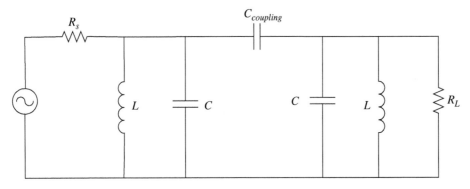

Figure 6.32 Capacitively coupled resonators.

The value of the inductor used to couple two identical resonant circuits is found from

$$L_{coupling} = Q_R L \tag{6.100}$$

where Q_R is the loaded quality factor of the single resonator. Coupling of single identical resonators via capacitor is shown in Figure 6.32. The circuit shown in Figure 6.32 presents a three-element low pass filter with an 18 dB/octave slope below resonance and effective single shunt tapped capacitance with a 6 dB/octave slope above resonance, as shown in Figure 6.33.

The value of the capacitor used to couple two identical resonant circuits is found from

$$C_{coupling} = \frac{C}{Q_R} \tag{6.101}$$

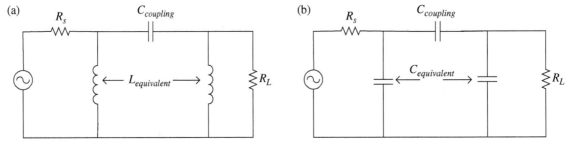

Figure 6.33 Inductively coupled resonators (a) below resonance and (b) above resonance.

The relation between the loaded quality factor of the single resonator and total loaded Q_t of the entire resonator circuit is obtained from

$$Q_R = \frac{Q_t}{0.707} \tag{6.102}$$

Example 6.3 Two-resonator Tuned Circuit

Design a two-resonator tuned circuit at a resonant frequency of 75 MHz, 3 dB bandwidth of 3.75 MHz, source and load impedances of 100 and 1000 Ω, respectively, using inductively coupled and capacitively coupled circuits. Assume that inductors Q of 85 are the frequency of interest. Use computer-aided design (CAD) to obtain the frequency response of the coupled resonator circuits.

Solution

The loaded quality factor of the overall resonant circuit is

$$Q_{total} = \frac{f_o}{BW} = \frac{75}{3.75} = 20$$

Then, the quality factor of the single resonator is found from

$$Q_R = \frac{Q_{total}}{0.707} = \frac{20}{0.707} = 28.3$$

The inductor is lossy. Its quality factor is found from

$$Q_p = \frac{R_p}{X_p} = 85 \quad \text{or} \quad R_p = 85X_p$$

The loaded quality factor of a single resonator found in (6.96) can also be found from

$$Q_R = \frac{R_{total}}{X_p}$$

where

$$R_{total} = \frac{R'_s R_p}{R'_s + R_p}$$

$$R'_s = 1000 \ \Omega$$

This leads to

$$Q_R = \frac{R'_s R_p}{(R'_s + R_p)X_p} = 28.3$$

So

$$X_p = \frac{R'_s R_p}{(R'_s + R_p)Q_R} = \frac{1000R_p}{(1000 + R_p)28.3}$$

or

$$X_p = \frac{1000(85)X_p}{(1000 + (85)X_p)28.3} = 23.57 \ \Omega$$

Then

$$R_p = 85X_p = 2003 \ \Omega$$

The components' values are

$$L_1 = L_2 = \frac{X_p}{\omega} = 50 \, \text{nH}$$

$$C_s = \frac{1}{X_p \omega} = 90 \, \text{pF}$$

The coupling inductance is found from

$$L_{12} = Q_R L = (28.3)50 \, \text{nH} = 1.415 \, \mu\text{H}$$

and the coupling capacitance is found from

$$C_{12} = \frac{C}{Q_R} = \frac{90 \times 10^{-12}}{28.3} = 3.18 \times 10^{-12}$$

The attenuation profile is obtained using Ansoft Designer with the circuits shown in Figures 6.34 and 6.35. The response for capacitive coupling is given in Figure 6.36 and inductive coupling is given in 6.36b.

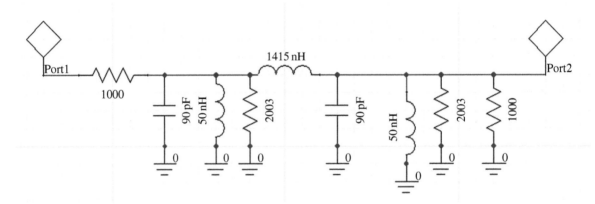

Figure 6.34 Inductively coupled resonators.

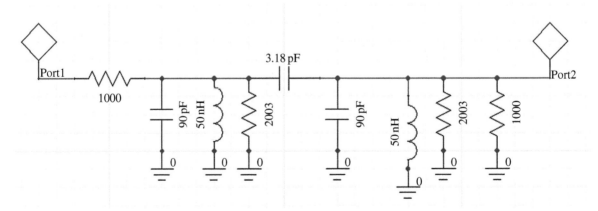

Figure 6.35 Capacitively coupled resonators.

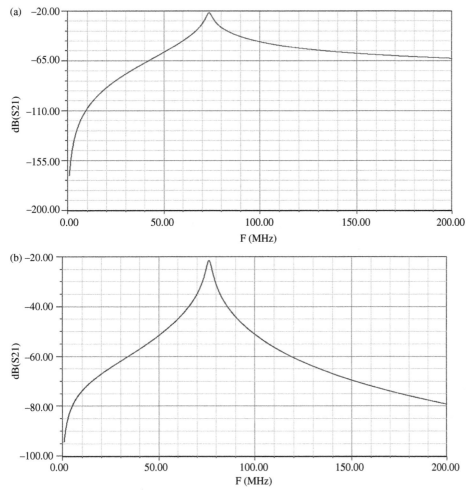

Figure 6.36 (a) Attenuation profile for capacitively coupled resonators. (b) Attenuation profile for inductively coupled resonators.

6.5 *LC* Resonators as Impedance Transformers

6.5.1 Inductive Load

Consider the *LC* parallel network shown in Figure 6.19 by ignoring the source resistance. This time assume the loss resistor is part of the load resistance, *R*. The new circuit is illustrated in Figure 6.37.

The equivalent impedance at the input for the circuit in Figure 6.37 can be written as

$$Z_{eq} = \frac{R}{(1 - \omega^2 LC)^2 + (\omega RC)^2} + j\frac{\omega[(L - CR^2) - \omega^2 L^2 C]}{(1 - \omega^2 LC)^2 + (\omega RC)^2} \qquad (6.103)$$

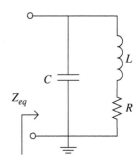

Figure 6.37 *LC* impedance transformer for inductive load.

The resonant frequency of the network is now equal to

$$\omega_0 = \sqrt{\frac{L - CR^2}{L^2C}} \rightarrow f_0 = \frac{1}{2\pi}\sqrt{\frac{L - CR^2}{L^2C}} \tag{6.104}$$

At resonance frequency, the equivalent impedance will be purely resistive, $Z_{eq} = R_{eq}$, and equal to

$$R_{eq} = \frac{R}{(1 - \omega^2 LC)^2 + (\omega RC)^2} = \frac{L}{RC} \tag{6.105}$$

Hence, the network at resonance converts the inductive load impedance to a resistive impedance. Since the quality factor, Q_{load}, of the load at resonance is

$$Q_{load} = \frac{\omega_0 L}{R} \tag{6.106}$$

The following relation can be written between the load quality factor and the equivalent impedance at resonance as

$$R_{eq} = \frac{L}{RC} = (Q_{load}^2 + 1)R \tag{6.107}$$

6.5.2 Capacitive Load

The same principle for the inductive load can be applied to convert the capacitive load to a resistive load at the resonant frequency using the LC resonant circuit shown in Figure 6.38.

The equivalent impedance at the input for the circuit in Figure 6.38 can be written as

$$Z_{eq} = \frac{\omega^4 RL^2 C^2}{(1 - \omega^2 LC)^2 + (\omega RC)^2} + j\frac{\omega L\left[\omega^2\left(R^2 C^2 - LC\right) + 1\right]}{(1 - \omega^2 LC)^2 + (\omega RC)^2} \tag{6.108}$$

The resonant frequency of the network is now equal to

$$\omega_0 = \sqrt{\frac{1}{LC - R^2 C^2}} \rightarrow f_0 = \frac{1}{2\pi}\sqrt{\frac{1}{LC - R^2 C^2}} \tag{6.109}$$

At resonance frequency, $Z_{eq} = R_{eq}$, and it can be expressed as

$$R_{eq} = \frac{L}{RC} \tag{6.110}$$

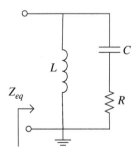

Figure 6.38 *LC* impedance transformer for capacitive load.

Hence, the network at resonance converts the inductive load impedance to a resistive impedance. Since the quality factor, Q_{load}, of the load at resonance is

$$Q_{load} = \frac{1}{\omega_0 RC} \tag{6.111}$$

Then, the following relation can be established

$$R_{eq} = \frac{L}{RC} = \omega_0 L Q_{load} \tag{6.112}$$

Example 6.4   Impedance Transformer

An amplifier output needs to be terminated with a load line resistance of 2000 Ω at 1.6 MHz. It is given in the datasheet that the transistor has 20 pF at 1.6 MHz. There is an inductive load connected to the output of the load line circuit of the amplifier with $RL = 5\,\Omega$. The configuration of this circuit is given in Figure 6.39.

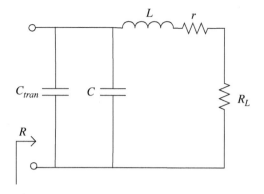

a) Calculate the values of L and C by assuming that load inductor has a negligible loss, i.e. $r = 0$.
b) The inductor is changed to a magnetic core inductor which has the quality factor of 50. Calculate loss resistance, r, for the reactive component values obtained in (a). What is the value of new load line resistance?
c) If the quality factor of the inductor is 50 and the load line resistor is required to be 2000 Ω as set in the problem, what are the values of L and C with $RL = 5\,\Omega$?

Figure 6.39   Amplifier output load line circuit.

Solution

It is given that $R_{eq} = 2000\,\Omega$, $RL = R = 5\,\Omega$, $C_{tran} = 20\,\text{pF}$, $f = 1.6\,\text{MHz}$, $\omega_0 = 10^7\,\text{rad/s}$ and $C_T = C_{tran} + C$.

a) When $r = 0\,\Omega$, we can use Eq. (6.107) as

$$R_{eq} = (Q_{load}^2 + 1)R \rightarrow \frac{R_{eq}}{R} - 1 = Q_{load}^2 \rightarrow Q_{load} = 19.98$$

From Eq. (6.106)

$$Q_{load} = \frac{\omega_0 L}{R} \rightarrow L = \frac{Q_{load}R}{\omega_0} \rightarrow L = 10\,\mu\text{H}$$

Using now Eq. (6.105)

$$R_{eq} = \frac{L}{RC_T} \rightarrow C_T = \frac{L}{R_{eq}R} \rightarrow C_T = 1\,\text{nF}$$

Since

$$C_T = C_{tran} + C \rightarrow C = C_T - C_{tran} \rightarrow C = 980\,\text{pF}$$

b) The Q of the inductor is given to be equal to 50. Then

$$Q_{inductor} = \frac{\omega_0 L}{r} \rightarrow r = \frac{\omega_0 L}{Q_{inductor}} \rightarrow r = 2\,\Omega$$

So, the new load resistance, R_{eq}, from (6.105) is

$$R_{eq} = \frac{L}{(R + r)C_T} = \frac{10 \times 10^{-6}}{(7)(1 \times 10^{-9})} = 1428.6\,\Omega$$

c) The Q of the inductor is given to be equal to 50. Then

$$Q_{inductor} = \frac{\omega_0 L}{r} \rightarrow r = 2 \times 10^5 L = aL$$

Since

$$Req = R\left(Q_{load}^2 + 1\right) \rightarrow R_{eq} = (R + r)\left(Q_{load}^2 + 1\right) = \frac{\omega_0^2 L^2 + (R + aL)^2}{R + aL}$$

which leads to the solution for L as

$$L^2 - L\frac{a\left(R_{eq} - 2R\right)}{\omega_0^2 + a^2} - \frac{R\left(R_{eq} - R\right)}{\omega_0^2 + a^2} = 0$$

So, the inductance value as $L = 12.2\,\mu\text{H}$. Substituting the value of L into the $Q_{inductor}$ gives the value of r as

$$r = 2 \times 10^5 L = 2.44\,\Omega \text{ where } L = 12.2\,\mu\text{H}$$

Using

$$R_{eq} = \frac{L}{RC_T} \rightarrow C_T = \frac{L}{R_{eq}R} \rightarrow C_T = 820\,\text{pF}$$

Since

$$C_T = C_{tran} + C \rightarrow C = 820 - 20 \rightarrow C = 800\,\text{pF}$$

6.6 Tapped Resonators as Impedance Transformers

6.6.1 Tapped-C Impedance Transformer

To understand the operation of tapped-C impedance transformers, consider the capacitive voltage divider circuit shown in Figure 6.40. The output voltage can be found from

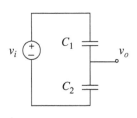

$$v_0 = v_i\frac{1/(j\omega C_2)}{1/(j\omega C_2) + 1/(j\omega C_1)} = v_i\frac{C_1}{C_1 + C_2} \tag{6.113}$$

which can be expressed as

$$v_0 = v_i n$$

where

Figure 6.40 Capacitive voltage divider.

$$n = \frac{C_1}{C_1 + C_2} \tag{6.114}$$

Now, assume there is a load resistor connected to the output of the capacitor and resonator inductor connected to the input of the divider circuit, as shown in Figure 6.41.

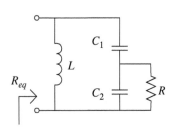

We then convert the output shunt connected circuit to a series connection by using the parallel series conversion shown in Figure 6.42.

The relation of the components in Figures 6.42 and 6.41 are

$$C_s = C_2\left(\frac{Q_p^2 + 1}{Q_p^2}\right) \tag{6.115}$$

Figure 6.41 Capacitive voltage divider with load resistor.

$$R_s = \frac{R}{Q_p^2 + 1} \tag{6.116}$$

$$R_s = \frac{R_{eq}}{Q_r^2 + 1} \qquad (6.117)$$

where

$$Q_p = \frac{R}{X_{C_2}} = \omega_0 R C_2 \qquad (6.118)$$

$$Q_r = \frac{R_{eq}}{\omega_0 L} = \frac{1}{\omega_0 R_s C} \qquad (6.119)$$

The equivalent capacitance can then be written as

$$C = \frac{C_1 C_s}{C_1 + C_s} \qquad (6.120)$$

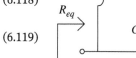

Figure 6.42 Capacitive voltage divider with parallel to series transformation.

Equating (6.116) and (6.117) gives

$$Q_p = \sqrt{\left[(Q_r^2 + 1) \frac{R}{R_{eq}} - 1 \right]} \qquad (6.121)$$

Overall, using the transformations given, the tapped-C circuit in Figure 6.42 can be simplified and transformed to the one in Figure 6.43 with the following relations as

$$R_s' = R_s \left(1 + \frac{C_1}{C_2} \right)^2 \qquad (6.122)$$

and

$$C_T = \frac{C_1 C_2}{C_1 + C_2} \qquad (6.123)$$

At resonance, the circuit can be simplified to the impedance transformer circuit shown in Figure 6.44 with

$$N^2 = \frac{R}{R_{eq}} \qquad (6.124)$$

Figure 6.43 Tapped equivalent circuit.

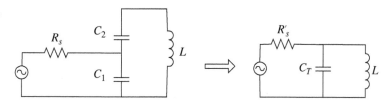

Figure 6.44 Equivalent tapped-C circuit representation using transformer.

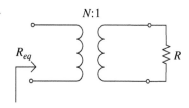

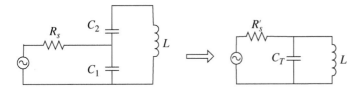

Figure 6.45 Parallel resonant circuit with tapped-C approach.

Substitution of (4.136) into (4.135) gives

$$Q_p = \sqrt{\left[\frac{(Q_r^2 + 1)}{N^2} - 1\right]}$$ (6.125)

Example 6.5 Tapped-C Resonator Circuit

Design a parallel resonant circuit with the tapped-C approach where 3 dB bandwidth is 3 MHz and there is a center frequency of 27.12 MHz. A resonant circuit will operate between a source resistance of 50 Ω and a load resistance of 100 Ω. Assume that Q of the inductor is 150 at 27.12 MHz.

a) Obtain element values of the circuit shown in Figure 6.45a.
b) Obtain element values of the equivalent circuit shown in Figure 6.45 b.
c) Obtain the frequency response of the circuits shown in Figure 6.45a and b.

Solution

The equivalent circuit shown in Figure 6.46b has

$$R_s' = 100 \,\Omega$$

Since

$$R_s' = R_s\left(1 + \frac{C_1}{C_2}\right)^2 \rightarrow \left(1 + \frac{C_1}{C_2}\right)^2 = 2 \rightarrow C_1 = 0.414C_2$$

Since the inductor is lossy

$$X_p = \frac{R_p}{Q_p} \rightarrow R_p = Q_p X_p = 150X_p$$

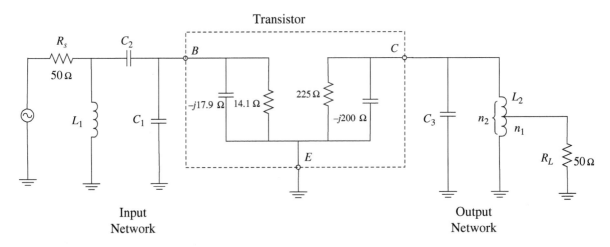

Transistor

Input Network

Output Network

Figure 6.46 Tapped C and L implementation for amplifiers.

The loaded Q of the resonant circuit is found from

$$Q = \frac{f_c}{f_2 - f_1} = \frac{27.12}{3} = 9.04$$

Since

$$Q = \frac{R_{total}}{X_p} \rightarrow 9.04 = \frac{R_{total}}{X_p} = \frac{50R_p}{(50 + R_p)X_p} \rightarrow 9.04 = \frac{(150X_p)50}{(50 + 150X_p)X_p}$$

Then

$$X_p = \frac{7048}{1356} = 5.2\ \Omega$$

So

$$R_p = 150X_p = 780\ \Omega$$

The values of the L and C are found from

$$L = \frac{X_p}{\omega} = \frac{5.2}{2\pi(27.12 \times 10^6)} = 30.5\ \text{nH}$$

$$C_T = \frac{1}{\omega X_p} = \frac{1}{2\pi(27.12 \times 10^6)(5.2)} = 1128\ \text{pF}$$

The capacitor values for the circuit in Figure 6.46a are found from

$$C_T = \frac{C_1 C_2}{C_1 + C_2} \rightarrow 1128 = \frac{0.414C_2}{1.414} \rightarrow C_2 = 3852.6\ \text{pF and } C_1 = 1595\ \text{pF}$$

The attenuation profile is shown in Figure 6.47.

Figure 6.47 Attenuation profile for Example 6.5.

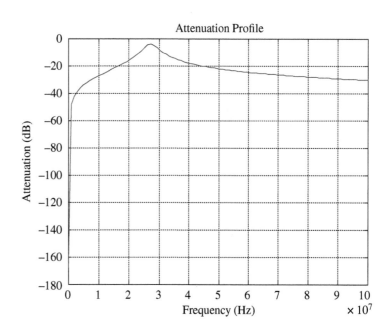

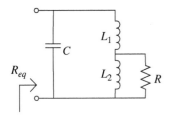

Figure 6.48 Tapped-L impedance transformer.

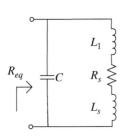

Figure 6.49 Tapped-L impedance transformer with parallel to series transformation.

6.6.2 Tapped-L Impedance Transformer

A typical tapped-L impedance transformer circuit is shown in Figure 6.48. We can follow the same procedure outlined in Section 6.6.1 and convert the circuit to its equivalent circuit, shown in Figure 6.49, by using parallel to series transformation relations as done in that section.

Overall, the tapped-L circuit in Figure 6.48 can be simplified to the one shown in Figure 6.50 using the transformations with the following relation as

$$R'_s = R_s \left(\frac{n}{n_1}\right)^2 \tag{6.126}$$

6.7 Engineering Application Examples

In this section, design examples for implementing impedance transformers using resonator networks for RF amplifiers are presented and detailed.

Design Example 6.1 Tapped-C and L Resonators for RF Amplifier Circuits

Consider the transistor amplifier circuit given in Figure 6.46. It is required to match the low transistor input impedance 14.1–50 Ω using a tapped-C circuit at the input and match the output of the transistor impedance 225 Ω down to 50 Ω using an L-tapped circuit at 100 MHz. A bandwidth of 3 dB of the amplifier circuit is given to be 10 MHz. The loaded Q of the input matching network is given to be 5 and the loaded Q of the output matching network is given be equal to 7.5. Calculate

a) C_1, C_2, C_3 L_1, L_2, and impedance transformer ratio n_1/n_2 for this amplifier circuit.

b) Develop MATLAB GUI (graphical user interface) to match the input and output impedances of the transistor to the given source impedance at the input using a tapped-C circuit and a load impedance at the output using an L-tapped circuit just like the amplifier circuit shown in Figure 6.46. Your program should take source impedance, load impedance, 3 dB bandwidth, center frequency, quality factors of the input, and output matching networks as input values and calculate and illustrate C_1, C_2, C_3 L_1, L_2, and impedance transformer ratio n_1/n_2 as output values. Test the accuracy of your program using the values in part (a).

c) Consider the amplifier circuit given in Figure 6.51 with the calculated values from part (a). Represent the complete circuit with ABCD network parameters and calculate the overall ABCD parameters of the network and gain. Check your analytical results with the results of your program.

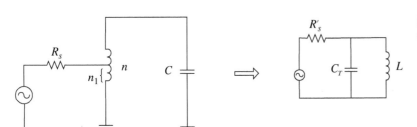

Figure 6.50 Tapped-L equivalent circuit.

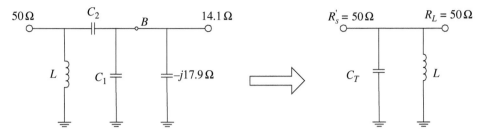

Figure 6.51 Input network transformation for tapped C transformer.

Solution

a) The input side of the network is given in Figure 6.51.
   From

$$R'_s = R_s \left(1 + \frac{C_b}{C_2}\right)^2 \rightarrow \frac{C_b}{C_2} = C_{ratio} = \sqrt{\frac{R'_s}{R_s} - 1} = \sqrt{\frac{50}{14.1} - 1} = 0.883$$

Since the quality factor of the input matching network is 5, $Q_{in} = 5$ pF

$$X_{C_b} = \frac{R_s}{Q_{in}} = \frac{14.1}{5} = 2.82 \rightarrow C_b = \frac{1}{\omega X_{C_b}} = 564.3 \, \text{pF}$$

Form the given information for the transistor

$$C_{tran\_in} = \frac{1}{\omega(17.9)} = 88.91 \, \text{pF}$$

Then

$$C_1 = C_b - C_{trans\_in} = 564.3 - 88.91 = 475.47 \, \text{pF}$$

Now, we can find C_2 from

$$\frac{C_b}{C_2} = C_{ratio} = 0.883 \rightarrow C_2 = \frac{C_b}{C_{ratio}} = \frac{564.3}{0.883} = 639 \, \text{pF}$$

Now, using

$$C_T = \frac{C_b C_2}{C_b + C_2} = 299.7 \, \text{pF}$$

Since

$$X_{C_t} = \frac{1}{\omega C_T} = 5.31$$

Then, inductance L_1 can be found from

$$L_1 = \frac{X_{C_T}}{\omega} = 8.45 \, \text{nH}$$

The transformation of the output network can be done, as shown in Figure 6.52, with the transformations obtained before.

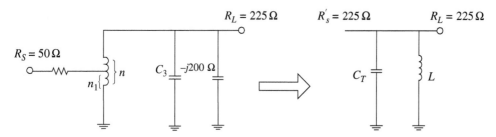

Figure 6.52 Output network transformation for tapped *L* transformer.

From

$$R'_s = R_s \left(\frac{n}{n_1}\right)^2 \rightarrow \left(\frac{n}{n_1}\right) = \sqrt{\frac{225}{50}} = 2.12$$

It is given that the quality of the output network is 7.5, $Q_{out} = 7.5$. So

$$X_p = \frac{R_p}{Q_p} = \frac{(225/2)}{7.5} = 15 \rightarrow C_p = \frac{1}{\omega X_p} = 106.1 \, \text{pF}$$

Since

$$C_p = C_3 + 7.95 \rightarrow C_3 = 106.1 - 7.95 = 98.15 \, \text{pF}$$

where 7.95 pF is obtained from the reactance given in the question, $-j200 \, \Omega$. We can then find the inductance, L_2, as

$$L_2 = \frac{X_p}{\omega} = 23.8 \, \text{nH}$$

b) (b) and (c).

The following MATLAB GUI is developed to match any transistor input and output impedances to the desired impedances via the tapped-*C* impedance transformer at the input and tapped-*L* impedance transformer at the output.

The last section of the program is not included, owing to its standard format for GUIs. When the program is executed the following MATLAB GUI window showing the results is illustrated in Figure 6.53.

```
function varargout = AmplifierGUI(varargin)
% AMPLIFIERGUI MATLAB code for AmplifierGUI.fig
%      AMPLIFIERGUI, by itself, creates a new AMPLIFIERGUI or raises the
%
%      AMPLIFIERGUI('CALLBACK',hObject,eventData,handles,...) calls the
%localfunction named CALLBACK in AMPLIFIERGUI.M with the given input
%arguments. AMPLIFIERGUI('Property','Value',...) creates a new
%AMPLIFIERGUI or raises the      existing singleton*. Starting from the
%left, property value pairs are
%      applied to the GUI before AmplifierGUI_OpeningFcn gets called.

% Begin initialization code - DO NOT EDIT
gui_Singleton = 1;
```

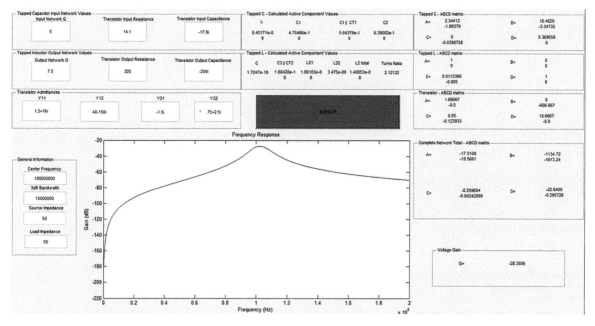

Figure 6.53 MATLAB GUI to design tapped *C* and tapped *L* impedance transformers for amplifiers.

```
gui_State = struct('gui_Name',        mfilename, ...
                   'gui_Singleton',  gui_Singleton, ...
                   'gui_OpeningFcn', @AmplifierGUI_OpeningFcn, ...
                   'gui_OutputFcn',  @AmplifierGUI_OutputFcn, ...
                   'gui_LayoutFcn',  [] , ...
                   'gui_Callback',   []);
if nargin && ischar(varargin{1})
    gui_State.gui_Callback = str2func(varargin{1});
end


if nargout
    [varargout{1:nargout}] = gui_mainfcn(gui_State, varargin{:});
else
    gui_mainfcn(gui_State, varargin{:});
end
% End initialization code - DO NOT EDIT



% - Executes just before AmplifierGUI is made visible.

% - Executes just before AmplifierGUI is made visible.
function AmplifierGUI_OpeningFcn(hObject, eventdata, handles, varargin)

% hObject      handle to figure
% eventdata    reserved - to be defined in a future version of MATLAB
% handles      structure with handles and user data (see GUIDATA)
% varargin     command line arguments to AmplifierGUI (see VARARGIN)
```

```matlab
% Choose default command line output for AmplifierGUI
handles.output = hObject;

% Update handles structure
guidata(hObject, handles);

% UIWAIT makes AmplifierGUI wait for user response (see UIRESUME)
% uiwait(handles.figure1);

% — Outputs from this function are returned to the command line.
function varargout = AmplifierGUI_OutputFcn(hObject, eventdata, handles)
% varargout   cell array for returning output args (see VARARGOUT);
% hObject     handle to figure
% eventdata   reserved - to be defined in a future version of MATLAB
% handles     structure with handles and user data (see GUIDATA)

% Get default command line output from handles structure
varargout{1} = handles.output;
% — Executes on button press in pushbutton1.
function pushbutton1_Callback(hObject, eventdata, handles)
% hObject     handle to pushbutton1 (see GCBO)
% eventdata   reserved - to be defined in a future version of MATLAB
% handles     structure with handles and user data (see GUIDATA)
Qin=str2num(get(handles.Qin,'String'));
Qout=str2num(get(handles.Qout,'String'));
RT1=str2num(get(handles.RT1,'String'));
TR2=str2num(get(handles.TR2,'String'));
TC1=str2double(get(handles.TC1,'String'));
TC2=str2double(get(handles.TC2,'String'));
Cw=str2num(get(handles.Cw,'String'));
w=str2num(get(handles.w,'String'));
Rsource=str2double(get(handles.Rsource,'String'));
Rload=str2num(get(handles.Rload,'String'));
Qp=w/Cw;
C1P=Qin/(2*pi*w*RT1);
set(handles.C1p,'String',C1P);
XTC1=1/(2*pi*w*1i*TC1);
c1=C1P-XTC1;
set(handles.C1,'String',c1);
c2=C1P/(sqrt(Rsource/RT1)-1);
set(handles.C2,'String',c2);
Ceq=(C1P*c2)/(C1P+c2);
Xp=1/(2*pi*w*Ceq);
L1=Xp/(2*pi*w);
set(handles.L1,'String',L1);
XCT2=1/(2*pi*w*1i*TC2);
N=sqrt(TR2/Rload);
```

```
set(handles.N,'String',N);
LRin=(Rload*(Qp*Qp+1))/(Qout*Qout+1);
C3p=Qp/(2*pi*w*LRin);
set(handles.C3p,'String',C3p);
C3=C3p-XCT2;
set(handles.C3,'String',C3);
L21=Rload/(2*pi*w*Qout);
L22=(L21*(Qp*Qout-(Qout*Qout)))/((Qout*Qout)+1);
L2=L22+L21;
set(handles.L2,'String',L2);
set(handles.L21,'String',L21);
set(handles.L22,'String',L22);
A_1=1;B_1=Rsource;C_1=0;D_1=1;
ABCD_1=[A_1 B_1;C_1 D_1];
YC1=1/(2*pi*w*c1*1i);
YC11=1/YC1;
YC2=1/(2*pi*w*c2*1i);
YC22=1/YC2;
YL1=(1i*L1*2*pi*w);
YL11=1/YL1;
A_2=1+(YC22/YC11);
B_2=1/YC11;
C_2=YL11+YC22+((YL11*YC22)/YC11);
D_2=1+(YL11/YC11);
ABCD_2=[A_2 B_2;C_2 D_2];
ABCDtotal_1=ABCD_1*ABCD_2;
set(handles.CA,'String',[real(ABCDtotal_1(1,1)), imag(ABCDtotal_1(1,1))].');
set(handles.CB,'String',[real(ABCDtotal_1(1,2)), imag(ABCDtotal_1(1,2))].');
set(handles.CC,'String',[real(ABCDtotal_1(2,1)), imag(ABCDtotal_1(2,1))].');
set(handles.CD,'String',[real(ABCDtotal_1(2,2)), imag(ABCDtotal_1(2,2))].');
TY11=str2double(get(handles.edit4,'String'));
TY12=str2double(get(handles.edit5,'String'));
TY21=str2double(get(handles.edit6,'String'));
TY22=str2double(get(handles.edit7,'String'));
Ytrans=(1/1000).*[TY11 TY12; TY21 TY22];
Ydet=det(Ytrans);
ABCDtrans=[-Ytrans(2,2)/Ytrans(2,1) -1./Ytrans(2,1);-Ydet/Ytrans(2,1)
-Ytrans(1,1)/Ytrans(2,1)];
set(handles.TA,'String',[real(ABCDtrans(1,1)), imag(ABCDtrans(1,1))].');
set(handles.TB,'String',[real(ABCDtrans(1,2)), imag(ABCDtrans(1,2))].');
set(handles.TC,'String',[real(ABCDtrans(2,1)), imag(ABCDtrans(2,1))].');
set(handles.TD,'String',[real(ABCDtrans(2,2)), imag(ABCDtrans(2,2))].');
ImpC3=1/(2*pi*w*C3*1i);
YL1=(2*pi*w*L22*1i);
YL2=(2*pi*w*L21*1i);
Series=((Rload*YL2)/(Rload+YL2))+YL1;
```

```
Outputsimple=(Series*ImpC3)/((Series+ImpC3));
A=1;B=0;C=1./Outputsimple;D=1;
ABCDOutput=[A B;C D];
set(handles.LA,'String',[real(ABCDOutput(1,1)), imag(ABCDOutput(1,1))].');
set(handles.LB,'String',[real(ABCDOutput(1,2)), imag(ABCDOutput(1,2))].');
set(handles.LC,'String',[real(ABCDOutput(2,1)), imag(ABCDOutput(2,1))].');
set(handles.LD,'String',[real(ABCDOutput(2,2)), imag(ABCDOutput(2,2))].');
ABCD=ABCDtotal_1*ABCDtrans*ABCDOutput;
set(handles.A,'String',[real(ABCD(1,1)), imag(ABCD(1,1))].');
set(handles.B,'String',[real(ABCD(1,2)), imag(ABCD(1,2))].');
set(handles.C,'String',[real(ABCD(2,1)), imag(ABCD(2,1))].');
set(handles.D,'String',[real(ABCD(2,2)), imag(ABCD(2,2))].');
Vg=20*log10(abs(1/ABCD(1,1)));
set(handles.Vg,'String',Vg);
%Frequency Response Plotting over range of Frequencies.
range=[1:100000:2*w];
n=1;
V=length(range);
for n=(1:1:V)
freq=n*100000;
Qp=freq/(Cw);
A_1=1;B_1=Rsource;C_1=0;D_1=1;
ABCD_1=[A_1 B_1;C_1 D_1];
YC1=1/(2*pi*freq*c1*1i);
YC11=1/YC1;
YC2=1/(2*pi*freq*c2*1i);
YC22=1/YC2;
YL1=(1i*L1*2*pi*freq);
YL11=1/YL1;
A_2=1+(YC22/YC11);
B_2=1/YC11;
C_2=YL11+YC22+((YL11*YC22)/YC11);
D_2=1+(YL11/YC11);
ABCD_2=[A_2 B_2;C_2 D_2];
ABCDtotal_1=ABCD_1*ABCD_2;
Ytrans=(1/1000).*[TY11 TY12; TY21 TY22];
Ydet=det(Ytrans);
ABCDtrans=[-Ytrans(2,2)/Ytrans(2,1) -1./Ytrans(2,1);-Ydet/Ytrans(2,1) -Ytrans
(1,1)/Ytrans(2,1)];
ImpC3=1/(2*pi*freq*C3*1i);
YL1=(2*pi*freq*L22*1i);
YL2=(2*pi*freq*L21*1i);
Series=((Rload*YL2)/(Rload+YL2))+YL1;
Outputsimple=(Series*ImpC3)/((Series+ImpC3));
A=1;B=0;C=1./Outputsimple;D=1;
ABCDOutput=[A B;C D];
```

```
ABCD=ABCDtotal_1*ABCDtrans*ABCDOutput;
gain=1/(ABCD(1,1));
VG(n)= 20*log10(abs(gain));
 end
VG;
plot(range,VG);
title('Frequency Response');
xlabel('Frequency (Hz)');
ylabel('Gain (dB)');
```

Design Example 6.2   Capacitively Coupled Amplifier Circuit

Consider the capacitively coupled amplifier circuit shown in Figure 6.54. Design the resonator tuned amplifier circuit at a resonant frequency of 100 MHz, 3 dB bandwidth of 5 MHz, and source and load impedances of 50 and 12 Ω, respectively. Assume that inductor Qs are 65 at the frequency of interest. Integrate a tapped-C transformer to your circuit to match load impedance to source impedance. Calculate resonator and tapped-C component values

Solution

Given parameters are

$$f_c = 100 \text{ MHz}, f_{3db} = 5 \text{ MHz}, R_s = 50 \ \Omega, R_L = 12 \ \Omega, Q_{ind} = 65$$

From the given parameters, the Q of the network is found as

$$Q_t = \frac{f_c}{f_{3dB}} = \frac{100}{5} = 20$$

We can the find the quality factor of the resonator from

$$Q_R = \frac{Q_t}{0.707} = \frac{20}{0.707} = 28.3$$

Since, the Q of the inductor is given, then

$$Q_{ind} = \frac{R_p}{X_p} = 65 \rightarrow R_p = 65X_p$$

Q of the single resonator can also be found from

$$Q_R = \frac{R_{tot}}{X_p} = 28.3$$

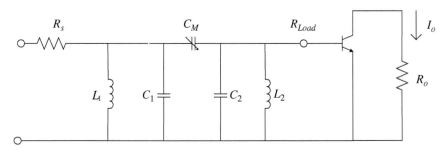

Figure 6.54   Capacitively coupled amplifier circuit.

where

$$R_{tot} = R'_s//R_p = R_s//R_p = \frac{50R_p}{50 + R_p}$$

Hence

$$Q_R = \left(\frac{50R_p}{50 + R_p}\right)\frac{1}{X_p} = 28.3$$

Hence

$$X_p = \frac{50(Q_{ind} - Q_R)}{Q_{ind}Q_R} = 1\,\Omega$$

Now using the relation

$$R_p = 65X_p = 65\,\Omega$$

The values of the inductors are obtained from

$$L_1 = L_2 = \frac{X_p}{\omega} = \frac{1}{2\pi(100 \times 10^6)} = 1.59\,\text{nH}$$

and the capacitor is found from

$$C_T = \frac{1}{\omega X_p} = \frac{1}{2\pi(100 \times 10^6)(1)} = 1590\,\text{pF}$$

Since

$$C_T = \frac{C'_1 C'_2}{C'_1 + C'_2} \rightarrow \frac{C'_1}{C'_2} = \sqrt{\frac{R'_s}{R_s} - 1}$$

Hence

$$\frac{C'_1}{C'_2} = \sqrt{\frac{R'_s}{R_s} - 1} = \sqrt{\frac{50}{12} - 1} = 1.04 \rightarrow C'_1 = 1.04C'_2$$

So

$$C_T = \frac{C'_1 C'_2}{C'_1 + C'_2} = \frac{1.04C'^2_2}{2.04C'_2} = 1590 \rightarrow C'_2 = 3119\,\text{pF}$$

So

$$C'_1 = 1.04C'_2 = 3244\,\text{pF}$$

The coupling capacitor is found as

$$C_M = \frac{C_T}{Q_R} = \frac{1590}{28.3} = 56.2\,\text{pF}$$

References

1 Eroglu, A. (2013). *RF Circuit Design Techniques for MF-UHF Applications*. CRC Press.
2 Eroglu, A. (2015). *Introduction to RF Power Amplifiers Design and Simulation*, ISBN: 978-1-4822-3164-9, 1e. CRC Press - Taylor and Francis.

Problems

Problem 6.1

A parallel resonant circuit with a source resistance of 50 Ω and a load resistance of 200 Ω. The loaded Q must be equal to 15 at the resonant frequency of 100 MHz.

a) Design the resonant circuit
b) Calculate the 3 dB bandwidth of the resonant circuit.
c) Obtain the frequency response of this circuit versus frequency
   Assume lossless components and no impedance matching.

Problem 6.2

A parallel resonant circuit with a 3 dB bandwidth of 6 MHz and a center frequency of 150 MHz. It is given that the resonant circuit has source and load impedances of 75 and 100 Ω, respectively. The Q of the inductor is given to be 200. Capacitor is assumed to be an ideal capacitor.

a) Design the resonant circuit
b) What is the loaded Q of the resonant circuit?
c) Obtain the frequency response of this circuit versus frequency

Problem 6.3

Design a parallel resonant circuit with the tapped-C approach where 3 dB bandwidth is 5 MHz and center frequency of 50 MHz. Resonant circuit will operate between a source resistance of 50 Ω and a load resistance of 75 Ω. Assume that Q of the inductor is 25 at 50 MHz.

a) Obtain element values of the circuit shown in Figure 6.55a
b) Obtain element values of the equivalent circuit shown in Figure 6.55b
c) Obtain the frequency response of the circuits shown in Figure 6.55b

Figure 6.55 Problem 6.3.

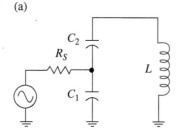 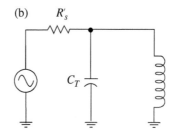

Problem 6.4

(a) Design two-resonator tuned circuit at resonant frequency of 125 MHz, 3 dB bandwidth of 5.75 MHz, source and load impedances of 250 and 2500 Ω, respectively using top C and top L coupling technique, as shown in Figure 6.56a,b. Assume that inductor Qs are 65 at the frequency of interest. (b) Finally, use a tapped-C transformer to present an effective source resistance (Rs) of 1000 Ω to the filter. Obtain the frequency response of the circuits shown in Figure 6.56a,b using ADS.

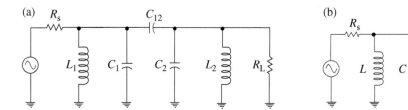

Figure 6.56 Problem 6.4.

Problem 6.5

An amplifier output needs to be terminated with a load line resistance of 4000 Ω at 2 MHz. It is given in the datasheet that the transistor has 40 pF at 2 MHz. There is an inductive load connected to the output of the load line circuit of the amplifier with $RL = 15\,\Omega$. The configuration of this circuit is given in Figure 6.57.

a) Calculate the values of L and C by assuming that load inductor has a negligible loss, i.e. $r = 0$.
b) The inductor is changed to a magnetic core inductor which has the quality factor of 50. Calculate loss resistance, r for the reactive component values obtained in (a). What is the value of new load line resistance.
c) If the quality factor of 50 the inductor is 50 and load line resistor is required to be 4000 Ω as set in the problem, what are the values of L and C.
d) Obtain frequency characteristics and confirm your results with ADS

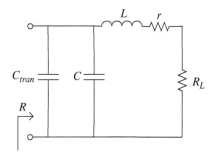

Figure 6.57 Problem 6.5.

Problem 6.6

Design and simulate a resonant LC circuit shown in Figure 6.58 which is terminated with a 2 kΩ and driven by a source with impedance 50 Ω at $f = 100$ MHz. The network Q should be 20. Use a capacitive transformer to match the load with the source for maximum power transfer. Assume lossless capacitors, inductor with Q of 100 and the amplitude of supply voltage 1 mV. Calculate the power transfer to the load.

Figure 6.58 Problem 6.6.

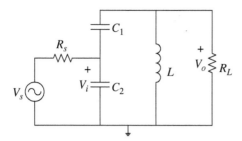

Problem 6.7

a) An interstage matching network is required to transform a load resistance $R_2 = 400\,\Omega$ into $R_1 = 1\,k\Omega$ at fo = 5 MHz with bandwidth BW = 50 kHz. Inductance coils with L $2\,\mu$, H are desired if feasible. Neglect the coil resistance. Use the circuit of Figure 6.59a and calculate the values of L, C_1, and C_2

b) For the problem in part Figure 6.59a, now use the circuit in Figure 6.59b and transform R2 into R1 and calculate values of L_1, L_2, C_1 and C_2

c) Verify your results in part (a) and (b) by ADS.

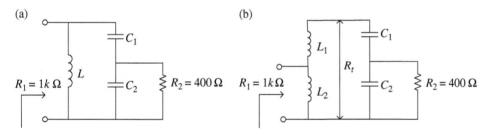

Figure 6.59 Problem 6.7.

Problem 6.8

Consider the transistor amplifier circuit given in Figure 6.60. It is required to match the low transistor input impedance 25–50 Ω using tapped-C circuit at the input and match the output of the transistor impedance 250 Ω down to 50 Ω using L-tapped circuit at 300 MHz. 3 dB bandwidth of the amplifier circuit is given to be 15 MHz. The loaded Q of the input matching network is given to be 5 and loaded Q of the output matching network is given be equal to 10. Calculate C_1, C_2, C_3 L_1, L_2 and impedance transformer ratio n_1/n_2 for this amplifier circuit.

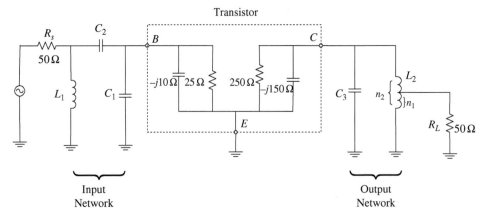

Figure 6.60 Problem 6.8.

Problem 6.9

Consider the capacitively coupled amplifier circuit shown in Figure 6.61. Design the resonator tuned amplifier circuit at a resonant frequency of 50 MHz, 3 dB bandwidth of 7.5 MHz, source and load impedances of 50 and 25 Ω, respectively. Assume that inductor Q_s are 85 at the frequency of interest. Integrate a tapped-C transformer to your circuit to match the load impedance to the source impedance. (a) Calculate resonator and tapped-C component values. (b) Obtain the frequency response of the circuits.

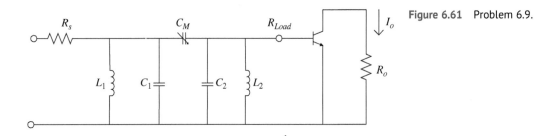

Figure 6.61 Problem 6.9.

Problem 6.10

Consider the series resonator circuit shown in Figure 6.62. The component values are given to be $L = 100\,\mu\text{H}$, $C = 100\,\text{pF}$, and $R = 5\,\Omega$. Calculate the resonant frequency, f_r, the quality factor of the circuit, Q, and the voltage across the capacitor, V_c, at resonance.

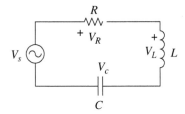

Figure 6.62 Problem 6.10.

Design Challenge 6.1 Transformation of Coupled Circuits for RF Amplifiers

Convert the capacitively coupled amplifier circuit discussed in Problem 6.9 to an inductively coupled circuit for the same requirements and calculate the value of L_M.

7

Couplers, Combiners, and Dividers

7.1 Introduction

An RF/microwave system will only work if all of its subcomponents are correctly designed and interfaced according to operational requirements. The integration of the subcomponents and assemblies in practice has been done by system engineers in coordination with RF design engineers. In this chapter, the design methods for couplers, reflectometers, combiners, and inverters are given.

7.2 Directional Couplers

Directional couplers are used widely as sampling devices for measuring forward and reflected power based on the magnitude of the reflection coefficient. Directional couplers can be implemented as a planar device using transmission lines, such as the microstrip and stripline, or lumped elements with transformers based on frequency of operation. Conventional directional couplers are four-port devices consisting of main and coupled lines, as shown in Figure 7.1.

Under the matched conditions, when the device is assumed to be lossless, the following relations are valid.

$$S_{11} = S_{22} = S_{33} = S_{44} = 0 \tag{7.1}$$

$$S^\dagger S = I \tag{7.2}$$

where $\dagger$ is used for the conjugate transpose of the matrix and I represents the unit matrix. From (7.2)

$$S_{14}^* \left(|S_{13}|^2 - |S_{24}|^2 \right) = 0 \tag{7.3}$$

$$S_{23} \left(|S_{12}|^2 - |S_{34}|^2 \right) = 0 \tag{7.4}$$

If the network is assumed to be a symmetrical device, then

$$S_{14} = S_{41} = S_{23} = S_{32} = 0 \tag{7.5}$$

Hence

$$|S_{12}|^2 + |S_{13}|^2 = 1 \tag{7.6}$$

$$|S_{12}|^2 + |S_{24}|^2 = 1 \tag{7.7}$$

$$|S_{13}|^2 + |S_{34}|^2 = 1 \tag{7.8}$$

$$|S_{24}|^2 + |S_{34}|^2 = 1 \tag{7.9}$$

RF/Microwave Engineering and Applications in Energy Systems, First Edition. Abdullah Eroglu.
© 2022 John Wiley & Sons Ltd. Published 2022 by John Wiley & Sons Ltd.
Companion website: www.wiley.com/go/eroglu/rfmicrowave

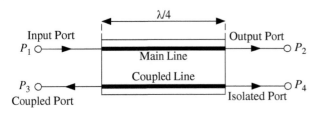

$\lambda/4$

Input Port
P_1 o→

Output Port
→o P_2

Main Line

Coupled Line

P_3 o←

→o P_4

Coupled Port

Isolated Port

Figure 7.1 Directional coupler as a four-port device.

which lead to

$$|S_{13}| = |S_{24}|, |S_{12}| = |S_{34}| \tag{7.10}$$

As a result, the scattering matrix for symmetrical, lossless directional couplers can be obtained as

$$S = \begin{bmatrix} 0 & \alpha & j\beta & 0 \\ \alpha & 0 & 0 & j\beta \\ j\beta & 0 & 0 & \alpha \\ 0 & j\beta & \alpha & 0 \end{bmatrix} \tag{7.11}$$

where

$$S_{12} = S_{34} = \alpha \tag{7.12}$$

$$S_{13} = S_{24} = j\beta \tag{7.13}$$

Important directional coupler performance parameters, the coupling level, isolation level, and directivity can be found from

$$\text{Coupling level (dB)} = 10\log\left(\frac{P_1}{P_3}\right) = -20\log(\beta) \tag{7.14}$$

$$\text{Isolation level (dB)} = 10\log\left(\frac{P_1}{P_4}\right) = -20\log(|S_{14}|) \tag{7.15}$$

$$\text{Directivity level (dB)} = 10\log\left(\frac{P_3}{P_4}\right) = 20\log\left(\frac{\beta}{|S_{14}|}\right) \tag{7.16}$$

7.2.1 Microstrip Directional Couplers

The design of microstrip directional couplers is discussed in [1–3]. This section discusses two-line, three-line, and multilayer planar directional coupler designs.

7.2.1.1 Two-line Microstrip Directional Couplers

Consider the geometry of a symmetrical microstrip directional coupler, as shown in Figure 7.2.

In practice, port termination impedances, coupling level and operational frequency are input design parameters that are being used to realize couplers. The matched system is accomplished when the characteristic impedance

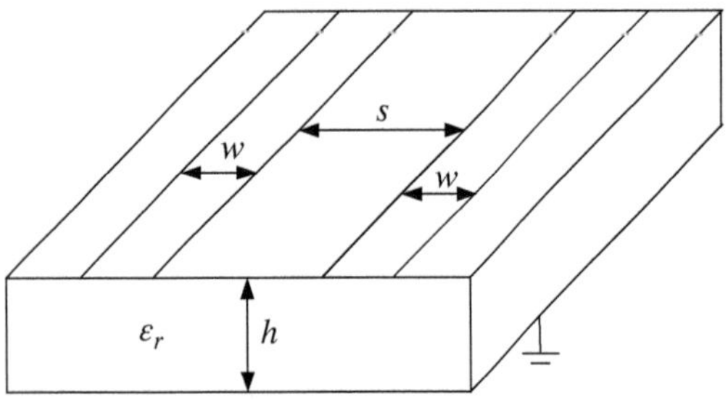

Figure 7.2 Symmetrical two-line microstrip directional coupler.

$$Z_0 = \sqrt{Z_{oe} Z_{oo}} \tag{7.17}$$

is equal to the port impedance. In (7.17), Z_{oe} and Z_{oo} are even and odd mode impedances, respectively. The even and odd impedances, Z_{oe} and Z_{oo}, of the microstrip coupler given in Figure 7.2 can be found from

$$Z_{oe} = Z_0 \sqrt{\frac{1 + 10^{C/20}}{1 - 10^{C/20}}} \tag{7.18}$$

$$Z_{oo} = Z_0 \sqrt{\frac{1 - 10^{C/20}}{1 + 10^{C/20}}} \tag{7.19}$$

where C is the forward coupling requirement and given in dB. The physical dimensions of the directional coupler are found using the synthesis method. Application of the synthesis method gives the spacing ratio s/h of the coupler in Figure 7.2 as

$$s/h = \frac{2}{\pi} \cosh^{-1} \left[\frac{\cosh\left[\frac{\pi}{2}\left(\frac{w}{h}\right)_{se}\right] + \cosh\left[\frac{\pi}{2}\left(\frac{w}{h}\right)_{so}'\right] - 2}{\cosh\left[\frac{\pi}{2}\left(\frac{w}{h}\right)_{so}'\right] - \cosh\left[\frac{\pi}{2}\left(\frac{w}{h}\right)_{se}\right]} \right] \tag{7.20}$$

$(w/h)_{se}$ and $(w/h)_{so}$ are the shape ratios for the equivalent single case corresponding to even-mode and odd-mode geometry, respectively. $(w/h)_{so}'$ is the second term for the shape ratio. (w/h) is the shape ratio for the single microstrip line and it is expressed as

$$\frac{w}{h} = \frac{8\sqrt{\left[\exp\left(\frac{R}{42.4}\sqrt{(\varepsilon_r + 1)}\right) - 1\right]\frac{7 + (4/\varepsilon_r)}{11} + \frac{1 + (1/\varepsilon_r)}{0.81}}}{\left[\exp\left(\frac{R}{42.4}\sqrt{\varepsilon_r + 1}\right) - 1\right]} \tag{7.21}$$

where

$$R = \frac{Z_{oe}}{2} \text{ or } R = \frac{Z_{oo}}{2} \tag{7.22}$$

Z_{ose} and Z_{oso} are the characteristic impedances corresponding to single microstrip shape ratios $(w/h)_{se}$ and $(w/h)_{so}$, respectively. They are given as

$$Z_{ose} = \frac{Z_{oe}}{2} \tag{7.23}$$

$$Z_{oso} = \frac{Z_{oo}}{2} \tag{7.24}$$

and

$$(w/h)_{se} = (w/h) \Big|_{R = Z_{ose}} \tag{7.25}$$

$$(w/h)_{so} = (w/h) \Big|_{R = Z_{oso}} \tag{7.26}$$

The term $(w/h)_{so}'$ in the Eq. (7.20) is given as

$$\left(\frac{w}{h}\right)_{so}' = 0.78\left(\frac{w}{h}\right)_{so} + 0.1\left(\frac{w}{h}\right)_{se} \tag{7.27}$$

After the spacing ratio s/h for the coupled lines is found, we can proceed to find w/h for the coupled lines. The shape ratio for the coupled lines is

$$\left(\frac{w}{h}\right) = \frac{1}{\pi} \cosh^{-1}(d) - \frac{1}{2}\left(\frac{s}{h}\right) \tag{7.28}$$

where

$$d = \frac{\cosh\left[\frac{\pi}{2}\left(\frac{w}{h}\right)_{se}\right](g+1)+g-1}{2} \tag{7.29}$$

$$g = \cosh\left[\frac{\pi}{2}\left(\frac{s}{h}\right)\right] \tag{7.30}$$

The physical length of the directional coupler is obtained using

$$l = \frac{\lambda}{4} = \frac{c}{4f\sqrt{\varepsilon_{eff}}} \tag{7.31}$$

where $c = 3*10^8 \text{m/sec}$ and f is operational frequency in Hz. Hence, the length of the directional coupler can be found if the effective permittivity constant ε_{eff} of the coupled structure shown in Figure 7.1 is known. ε_{eff} can be found from

$$\varepsilon_{eff} = \left[\frac{\sqrt{\varepsilon_{effe}} + \sqrt{\varepsilon_{effo}}}{2}\right]^2 \tag{7.32}$$

ε_{effe} and ε_{effo} are the effective permittivity constants of the coupled structure for odd and even modes, respectively. ε_{effe} and ε_{effo} depend on even- and odd-mode capacitances C_e and C_o as

$$\varepsilon_{effe} = \frac{C_e}{C_{e1}} \tag{7.33}$$

$$\varepsilon_{effo} = \frac{C_o}{C_{o1}} \tag{7.34}$$

$C_{e1, o1}$ is the capacitance with air as a dielectric. All the capacitances are given as capacitance per unit length. The even-mode capacitance C_e is

$$C_e = C_p + C_f + C_f' \tag{7.35}$$

The capacitances in even mode for the coupled lines and is shown in Figure 7.3a.

C_p is the parallel plate capacitance and defined as

$$C_p = \varepsilon_0 \varepsilon_r \frac{w}{h} \tag{7.36}$$

where w/h is found in the previous section. C_f is the fringing capacitance due to the microstrip's being taken alone as if it were a single strip. That is equal to

$$C_f = \frac{\sqrt{\varepsilon_{seff}}}{2cZ_0} - \frac{C_p}{2} \tag{7.37}$$

Here, ε_{seff} is the effective permittivity constant of a single strip microstrip. It can be expressed as

$$\varepsilon_{seff} = \frac{\varepsilon_r + 1}{2} - \frac{\varepsilon_r - 1}{2}F(w/h) \tag{7.38}$$

(a)

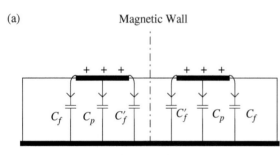

Magnetic Wall

Even Mode

(b)

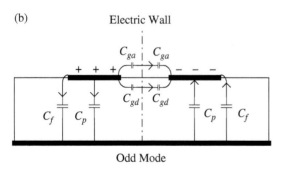

Electric Wall

Odd Mode

Figure 7.3 Coupled lines mode representation: (a) even mode; (b) odd mode.

where

$$F(w/h) = \begin{cases} (1 + 12h/w)^{-1/2} + 0.041(1 - w/h)^2 & \text{for} \quad \left(\dfrac{w}{h} \leq 1\right) \\ (1 + 12h/w)^{-1/2} & \text{for} \quad \left(\dfrac{w}{h} \geq 1\right) \end{cases} \tag{7.39}$$

C'_f is given by the following equation.

$$C'_f = \frac{C_f}{1 + A\left(\frac{h}{s}\right)\tanh\left(\frac{10s}{h}\right)}\left(\frac{\varepsilon_r}{\varepsilon_{seff}}\right)^{1/4} \tag{7.40}$$

and

$$A = \exp\left[-0.1\exp\left(2.33 - 1.5\frac{w}{h}\right)\right] \tag{7.41}$$

The odd-mode capacitance C_o is

$$C_o = C_p + C_f + C_{ga} + C_{gd} \tag{7.42}$$

The capacitances in odd mode for the coupled lines are shown in Figure 7.3b. C_{ga} is the capacitance term in odd mode for the fringing field across the gap in the air region. It can be written as

$$C_{ga} = \varepsilon_0 \frac{K(k')}{K(k)} \tag{7.43}$$

where

$$\frac{K(k')}{K(k)} = \begin{cases} \dfrac{1}{\pi}\ln\left[2\dfrac{1 + \sqrt{k'}}{1 - \sqrt{k'}}\right], & 0 \leq k^2 \leq 0.5 \\ \dfrac{\pi}{\ln\left[2\dfrac{1 + \sqrt{k'}}{1 - \sqrt{k'}}\right]}, & 0.5 \leq k^2 \leq 1 \end{cases} \tag{7.44}$$

and

$$k = \frac{\left(\frac{s}{h}\right)}{\left(\frac{s}{h}\right) + \left(\frac{2w}{h}\right)} \tag{7.45}$$

$$k' = \sqrt{1 - k^2} \tag{7.46}$$

C_{gd} represents the capacitance in odd mode for the fringing field across the gap in the dielectric region. It can be found using

$$C_{gd} = \frac{\varepsilon_0\varepsilon_r}{\pi}\ln\left\{\coth\left(\frac{\pi}{4}\frac{s}{h}\right)\right\} + 0.65C_f\left[\frac{0.02}{\left(\frac{s}{h}\right)}\sqrt{\varepsilon_r} + \left(1 - \frac{1}{\varepsilon_r^2}\right)\right] \tag{7.47}$$

Since

$$Z_{oe} = \frac{1}{c\sqrt{C_e C_{e1}}} \tag{7.48}$$

$$Z_{oo} = \frac{1}{c\sqrt{C_o C_{o1}}} \tag{7.49}$$

then we can write

$$C_{e1} = \frac{1}{c^2 C_e Z_{oe}^2} \tag{7.50}$$

$$C_{o1} = \frac{1}{c^2 C_o Z_{oo}^2} \tag{7.51}$$

Substituting (7.35), (7.32), (7.40), and, (7.51) into (7.34) and (7.35) gives the even- and odd-mode effective permittivities ε_{effe} and ε_{effo}. When (7.34) and (7.35) are substituted into (7.33), we can find the effective permittivity constant ε_{eff} of the coupled structure. Now, (7.31) can be used to calculate the physical length of the directional coupler at the operational frequency. Several application examples of two-line microstrip coupler designs with various commonly used RF materials are given in [1].

7.2.1.2 Three-line Microstrip Directional Couplers

Symmetrical three-line microstrip couplers, which are also known as six-port couplers, are shown in Figure 7.4 and can be used for several applications, including the detection of forward and reverse power simultaneously and accurately using phase velocity equalization techniques with lumped elements terminated at the ports of the coupling lines. The design procedure is described in [4] and repeated below.

Step 1 Generate the design specifications for a two-line coupler using equations in Section 7.2.1.1.

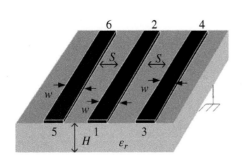

Figure 7.4 Three-line symmetrical microstrip coupler.

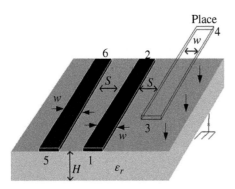

Figure 7.5 Generation of three-line coupler from the design parameters of a two-line coupler.

In this step, the required design parameters – coupling level, operational frequency, and port impedance – are used to create the design for the two-line symmetrical coupler shown in Figure 7.2. Spacing ratio, shape ratio, and length of the two-line coupler are then found using Eqs. (7.20), (7.21), and (7.31).

Step 2 Implementation of three-line couplers

A symmetrical three-line coupler, having the same coupling characteristics of two-line couplers, is then obtained by placing the second coupling line on the other side of the main line by keeping the same spacing with the main line, as shown in Figure 7.5.

Step 3 Verification of coupling levels via semi-empirical formulation

The coupling levels of a three-line microstrip coupler are given by the following equations

$$K_{13} = K_{15} = \frac{Z_{ee} - Z_{oo}}{Z_{ee} + Z_{oo}} \tag{7.52}$$

$$K_{53} = \frac{\sqrt{Z_{ee}Z_{oo}} - Z_{oe}}{\sqrt{Z_{ee}Z_{oo}} + Z_{oe}} \tag{7.53}$$

where $K_{13} = K_{15}$ (not in dB) represents the coupling from the side lines into the main line. K_{53} is the coupling level between the two coupled lines.

$K_{13} = K_{15}$ is equal to C, which is the coupling between the main line and the coupling line for two-line microstrip couplers [1, 2]. The semi-empirical formulation derived in this paper for K_{53} represents the coupling level between the two coupled lines through the center line and is found using the analysis given in [1, 2]. The formulation is obtained by finding the coupling between the two-coupled lines without the presence of the main line, as shown in Figure 7.6, with an introduced error by its removal. The error is found through simulations and integrated to the formulation obtained. After the formulation for K_{53} is obtained, the relations for three-line mode impedances can be obtained as

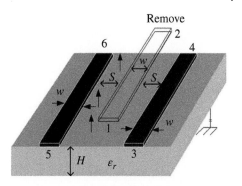

$$Z_{oo} = Z_o \left(\frac{1 + K_{53}}{1 - K_{53}}\right) \sqrt{\frac{1 - K_{13}}{1 + K_{13}}} \tag{7.54}$$

$$Z_{ee} = Z_o \left(\frac{1 + K_{53}}{1 - K_{53}}\right) \sqrt{\frac{1 + K_{13}}{1 - K_{13}}} \tag{7.55}$$

Figure 7.6 Removal of the main line in a three-line coupler for formulation for K_{13}.

with K_{53} given in (7.53) depends on the three-line coupler mode impedances Z_{ee}, Z_{oo}, and Z_{oe}. Since the coupling level, port impedance, and operational frequency are known, using the formulation in [1, 2], spacing ratio, s/h, and shape ratio, w/h, are obtained. This leads to

$$d_{new} = \cosh\left(pi\left[(w/h) + \frac{1}{2}(s/h)_{new}\right]\right) \tag{7.56}$$

where

$$\left(\frac{s}{h}\right)_{new} = 2\left(\frac{s}{h}\right) + \left(\frac{w}{h}\right) \tag{7.57}$$

In (7.57), s/h and w/h are the values that are obtained using the formulation in [1, 2]. $(w/h)_{se}$ is the shape ratio for the even modes for the new coupled system and found from

$$\left(\frac{w}{h}\right)_{se} = \frac{2}{\pi} \cosh^{-1}\left[\frac{(2d_{new} - g + 1)}{g + 1}\right] \tag{7.58}$$

where

$$g = \cosh\left[\frac{\pi}{2}\left(\frac{s}{h}\right)\right] \tag{7.59}$$

The characteristic impedances corresponding to single microstrip shape ratios $(w/h)_{se}$ can be obtained from

$$Z_{ose} = 42.4 \log\left[1 + \frac{1}{2}\left(\frac{64a}{(w/h)_{se}^2} + \Delta\right)\right] \frac{1}{\sqrt{\varepsilon_r + 1}} \tag{7.60}$$

Then, the characteristic impedance for oe propagating mode for the coupled lines is equal to

$$Z_{oe} = 2Z_{ose} \tag{7.61}$$

In (7.60)

$$\Delta = \sqrt{\left(\frac{64a}{(w/h)_{se}^2} + \Delta\right)^2 + \left(\frac{256b}{(w/h)_{se}^2}\right)} \tag{7.62}$$

where

$$a = \left(\frac{7}{11}\right) + \left(\frac{4}{11\varepsilon_r}\right) \tag{7.63a}$$

and

$$b = \left(\frac{1}{0.81}\right) + \left(\frac{1}{0.81\varepsilon_r}\right)$$ (7.63b)

The characteristic impedance of two-line couplers for oo mode is then obtained from

$$Z_{oo} = \frac{1}{(2lfC_o)} - \frac{C_e Z_{oe}}{C_o}$$ (7.64)

where C_e and C_o are the capacitances for the even and odd modes, respectively. They are expressed as

$$C_e = C_p + C_f + C'_f$$ (7.65)

$$C_o = C_p + C_f + C_{ga} + C_{gd}$$ (7.66)

The relations for the capacitances in (7.65) and (7.66) are given in [1, 2]. Then, K_{53} for three-line microstrip couplers can then be found from (7.61) and (7.64) with an error function as

$$K_{53}(dB) = 20 \log \left(\frac{Z_{oe} - Z_{oo}}{Z_{oe} + Z_{oo}}\right) - erf(\varepsilon_r)$$ (7.67)

where $erf(\varepsilon_r) = a\varepsilon_r^2 + b\varepsilon_r + c$ is defined as the error function and obtained from simulations. The error function is approximated to be a polynomial function of dielectric constant at order 2 and is given in Table 7.1 below at each coupling level.

Step 4 Calculation of the mode impedances for three-line couplers

Eqs. (7.54) and (7.55) can now be used to calculate the mode impedances of three-line couplers to confirm the results.

Example 7.1 Symmetrical Two-line Microstrip Coupler
Design 15 dB three-line coupler using Teflon at 300 MHz with the method introduced.

Solution
The design procedure for two-line conventional directional couplers and 3-line directional couplers for Teflon with relative permittivity constant 2.08 has been applied at 300 MHz to realize a 15 dB coupler. A two-line microstrip is first designed using MATLAB GUI developed with the formulation given in this chapter and the results showing the physical dimensions of the microstrip coupler are illustrated in Figure 7.7.

Based on the results obtained, Ansoft Designer is used to simulate the same coupler, as shown in Figure 7.8. The simulation results are illustrated in Figure 7.9.

Table 7.1 $erf(\varepsilon_r) = a\varepsilon_r^2 + b\varepsilon_r + c$ for different coupling levels.

Coupling level (dB)	a	b	c
−10	0.0121	−0.6817	12.766
−13	−0.0202	−0.2442	10.265
−15	−0.0941	0.1357	9.725
−18	0.0438	−0.2626	8.9564
−20	0.0543	−0.2873	9.2126

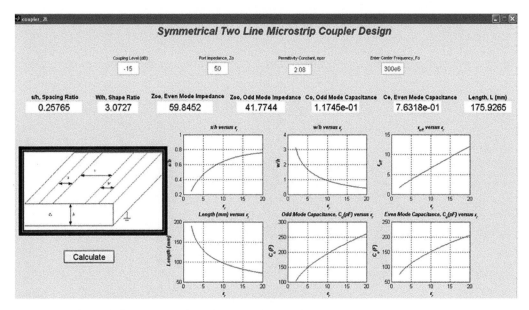

Figure 7.7 MATLAB GUI for a two-line microstrip directional coupler.

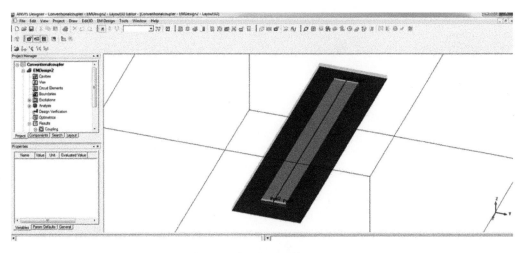

Figure 7.8 Simulated microstrip two-line directional coupler.

7.2.2 Multilayer and Multiline Planar Directional Couplers

There are several benefits to using multilayer directional couplers in RF applications, including improvement in directivity, better isolation, and providing better performance against arcing for high power applications. The design process given in this section can be used to implement both two- and three-line multilayer couplers in conjunction with the method given in Section 7.2.1.2. Illustrations of the two- and three-line multilayer couplers are given in Figures 7.10 and 7.11. In Figures 7.10 and 7.11, the dielectric thickness is H and equal to $H_1 + H_2$.

The step-by-step design procedure to realize two- and three-line multilayer planar directional couplers are given as follows.

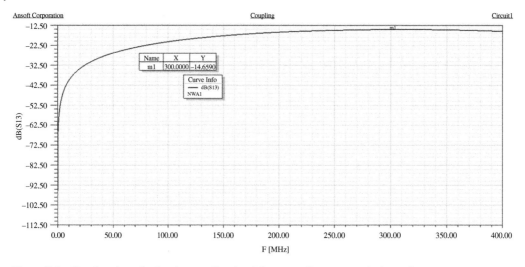

Figure 7.9 Simulated results for the coupling level for a two-line symmetrical coupler.

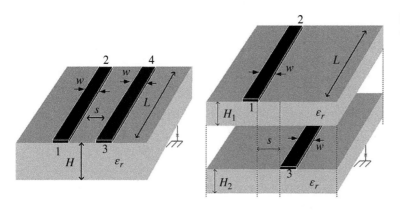

Figure 7.10 Two-line multilayer directional coupler.

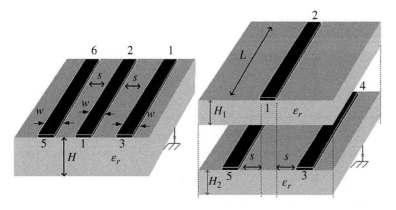

Figure 7.11 Three-line multilayer directional coupler.

Step 1 Generate Base Design

Generate the design parameters for a two-line coupler using Eqs. (7.1)–(7.7) for the base design and obtain spacing and shape ratios. The thickness and coupling levels can be chosen based on the application.

Step 2 Simulation for Parametric Analysis

Use simulation to conduct parametric analysis by moving the coupled line at a predetermined distance inside dielectric toward ground plane. Obtain the new coupling for the predetermined distance and repeat this until the full thickness of the dielectric is reached.

Step 3 Curve Fitting for Equation

Use the curve fitting method to obtain an equation for the material used for the desired coupling level to calculate the physical dimensions, including the height of the coupled line from the ground plane. This procedure is applied for commonly used RF materials, such as TMM10, FR4, RF60, RO4003, and Teflon, where the relative dielectric constant ranges from 2.08 to 9.8 to have the design equations for the two-line multilayer configuration shown in Figure 7.10. The three-line multilayer structure shown in Figure 7.11 is then formed by application of the procedure described in Section 7.2.1.2. The summary of the equations, which were obtained using the curve fitting technique for each material, is tabulated in Table 7.2.

The formulation and analytical results obtained are used to develop MATLAB GUI to design multilayer microstrip couplers and to obtain their physical dimensions. The designer is required to only enter the desired coupling level, port impedance, dielectric contact of the material used, and its thickness in the GUI program. The program then calculates all the physical dimensions, including the spacing between the coupled and main lines, and the width and height of the coupled line from the ground plane. The GUI program results can also be used to design three-line multilayer couplers using the same physical dimensions with an addition of the second coupling line, as described in Section 7.2.1.2.

Table 7.2 Multilayer coupler design equations.

Material	Dielectric constant (ε_r)	Equation
Teflon	2.08	$C\ (dB) = 42.85 H^{-0.225}$
RO 4003	3.38	$C\ (dB) = 43.897 H^{-0.246}$
FR4	4.4	$C\ (dB) = 42.387 H^{-0.245}$
RF60	6.15	$C\ (dB) = 45.873 H^{-0.276}$
TMM10	9.8	$C\ (dB) = 44.139 H^{-0.282}$

7.2.3 Transformer Coupled Directional Couplers

Couplers can be implemented using distributed elements or lumped elements, as discussed elsewhere in this section. The type of application, operational frequency, and power handling capability are among the important factors that dictate the type of directional coupler used in the RF system. Conventional directional couplers are designed as four-port couplers and have been studied extensively. However, better performance and more functionality from couplers can be obtained when they are implemented as six-port couplers.

It is possible to use six-port couplers for voltage standing wave ratio (VSWR) measurement. They also play a very important role in measuring the voltage, current, power, impedance, and phase, as discussed in [5]. A detailed study of a wide band impedance measurement using a six-port coupler is given in [6]. The design and analysis of a six-port stripline coupler with a high phase and amplitude balance has been studied in [7]. One of the important applications of six-port couplers is their implementation as reflectometers. The theory of a six-port reflectometer is detailed in [8, 9]. In [10, 11], a six-port reflectometer based on four-port coplanar-waveguide couplers has been modified to meet optimum design specifications. Similar studies to realize reflectometers using couplers are reported [12–14]. Theoretical analysis of the impedance measurement using six-port coupler as reflectometer has been introduced in [15, 16]. The design and performance of six-port reflectometers based on microstrip type couplers is analyzed in [17–19]. Hansson and Riblet [20] managed to realize a six-port network of an ideal q-point distribution by using a matched reciprocal lossless five-port, directional coupler. An improved complex reflection coefficient measurement device consisting of two six-port couplers is presented in [20]. Similarly, a six-port device has been designed for power measurement with two six-port directional couplers and is discussed in [21]. Six-port couplers can also be used in designing power splitting and combining networks [22]. Six-port devices have also been commonly used for source pull and load pull characterization of active devices and systems [23, 24]. As a result, the design of six-port couplers reported in the literature is based on the planar structures involving microstrips, striplines, or different waveguide structures. For high power and low cost applications, directional couplers can be implemented using RF transformers. Four-port directional coupler designs using transformer coupling are given in [25, 26].

In this section, a detailed analytical analysis of a four- and a six-port directional coupler using ideal RF transformers is presented. Closed form expressions at each port are obtained, and coupling, isolation, and directivity levels of six-port couplers using transformer coupling are given. S parameters for four- and six-port coupler are derived and coupler performance parameters are expressed in terms of S parameters. Based on the analytical model, MATLAB GUI has been developed and used for the design, simulation, and analysis of four- and six-port couplers using transformer coupling. The directional coupler is then simulated using frequency domain and time domain simulators, such as Ansoft Designer and PSpice, and simulation results are compared with the analytical results. The six-port coupler is then implemented and measured with network analyzer HP 8753ES. The proposed model can be used as a building block in various applications, such as reflectometers, high power impedance and power measurements, VSWR measurement, load pull or source pull of active devices, etc.

7.2.3.1 Four-port Directional Coupler Design and Implementation

The design, simulation, and implementation of a four-port coupler are given in [1] and are only briefly described here. S parameters of the four-port directional coupler shown in Figure 7.12 can be represented in matrix form as

$$S = \begin{bmatrix} S_{11} & S_{12} & S_{13} & S_{14} \\ S_{21} & S_{22} & S_{23} & S_{24} \\ S_{31} & S_{31} & S_{33} & S_{34} \\ S_{41} & S_{42} & S_{43} & S_{44} \end{bmatrix} \tag{7.68}$$

The performance of a four-port coupler can be calculated using S_{13} and S_{14} for coupling, isolation, and directivity levels when excitation is from port 1 on the main line. In Figure 7.12, T_1 is the transformer with turns ratio $N_1 : 1$ and T_2 is the transformer with turns ratio $N_2 : 1$. The transformers are assumed to be ideal and lossless. The relations between voltages and currents through turn ratios of the directional coupler at the ports can be obtained as

$$V_2 = N_2(V_4 - V_3) \tag{7.69a}$$

$$V_4 = N_1(V_2 - V_1) \tag{7.69b}$$

and

$$I_1 = N_1(I_3 + I_4) \tag{7.70a}$$

$$I_3 = N_1(I_1 + I_2) \tag{7.70b}$$

Scattering parameters of the coupler can be obtained by using the incident and reflected waves which are designated by a_i and b_i. Then, the voltages and currents can be expressed in terms of waves as

$$V_i = \sqrt{Z}(a_i + b_i) \tag{7.71}$$

$$I_i = \frac{1}{\sqrt{Z}}(a_i - b_i) \tag{7.72}$$

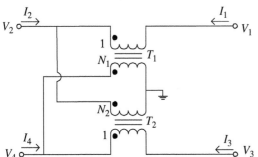

Figure 7.12 Four-port transformer directional coupler.

Z is the characteristic impedance at the ports of the directional coupler. The scattering parameters of the coupler are obtained by relating the incident and reflected waves using

$$S_{ij} = \frac{b_i}{a_j}\Big|_{a_k = 0 \text{ for } k \neq j} \tag{7.73}$$

The scattering parameters that are required to calculate the coupler performance parameters are

$$S_{13} = \frac{(-2N_1N_2)(N_1 + N_2)}{\left(4N_1^2N_2^2 + 1 + \left(N_1^2 - N_2^2\right)\right)} \tag{7.74}$$

$$S_{14} = \frac{(-2N_1)\left(-N_1N_2 + N_2^2 + 1\right)}{\left(4N_1^2N_2^2 + 1 + \left(N_1^2 - N_2^2\right)\right)} \tag{7.75}$$

The coupling, isolation, and directivity levels of four-port couplers are expressed using S parameters by Eqs. (7.14)–(7.16). Eqs. (7.64) and (7.65) facilitate directional coupler performance parameter calculations when only turns ratios are known, under the assumption that all ports are matched However, one other important aspect of the directional coupler in practical applications is real operating conditions including voltage, current, power ratings, and operational frequency. These parameters dictate the type of core, the winding and the wire, the coax line, and the insulation that will be used in the design. As a result, circuit analysis is required to determine the operating conditions on the coupler at each node. This analysis is detailed in [1] using the circuit analysis of the four-port coupler shown in Figure 7.13.

The application of nodal analysis for the coupler circuit in Figure 7.13 gives the performance parameters for the coupler as

$$\text{Coupling level (dB)} = 20\log\left(\frac{V_{coupled}}{\left(\frac{R_2}{R_1 + R_2}\right)V_1}\right) \tag{7.76}$$

$$\text{Isolation level (dB)} = 20\log\left(\frac{V_{isolated}}{\left(\frac{R_2}{R_1 + R_2}\right)V_1}\right) \tag{7.77}$$

Similarly, the directivity level is found from

$$\text{Directivty level (dB)} = \text{Isolation level (dB)} - \text{Coupling level (dB)} \tag{7.78}$$

Figure 7.13 Four-port transformer directional coupler for circuit analysis.

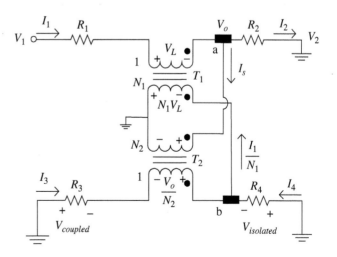

7.2.3.2 Six-port Directional Coupler Design

We shall use the four-port coupler introduced in Section 7.2.3.1 as a basic element to realize a six-port coupler using transformer coupling for high power RF applications. Six-port coupler design and analysis using transformer coupling have not been reported in the literature, according to the authors' knowledge. In Figure 7.14, T_1 is the transformer with turns ratio $N_1 : 1$ and T_2 is the transformer with turns ratio $N_2 : 1$. The transformers are assumed to be ideal and lossless. S parameters of the six-port directional coupler shown in Figure 7.14 can be obtained from

$$
\begin{pmatrix} b_1 \\ b_2 \\ \vdots \\ \\ \\ b_6 \end{pmatrix} = \begin{pmatrix} S_{11} & S_{12} & & & S_{16} \\ S_{21} & \ddots & & & \vdots \\ S_{31} & & \ddots & & \vdots \\ \\ S_{61} & \cdots & \cdots & & S_{66} \end{pmatrix} \begin{pmatrix} a_1 \\ a_2 \\ \vdots \\ \\ \\ a_6 \end{pmatrix}
\tag{7.79}
$$

The relations between voltages and currents through turn ratios of the directional coupler at each port can be obtained as

$$v = N_2(V_4 - V_3) \tag{7.80}$$

$$V_4 = N_1(v - V_1) \tag{7.81b}$$

$$V_2 = N_2(V_6 - V_5) \tag{7.81c}$$

$$V_6 = N_1(V_2 - v) \tag{7.81d}$$

and

$$I_1 = N_1(I_4 + I_3) \tag{7.82a}$$

$$I_3 = N_2(-i + I_1) \tag{7.82b}$$

$$i = N_1(I_5 + I_6) \tag{7.83a}$$

$$I_5 = N_2(I_2 + i) \tag{7.83b}$$

Then, the voltages and currents are expressed in terms of waves using the relations (7.61) and (7.62) and the scattering parameters for coupler performance are then obtained with an application of Eq. (7.63) as

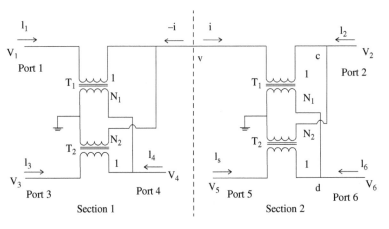

Figure 7.14 Six-port transformer directional coupler.

$$S_{62} = S_{31} = \frac{(-4N_1^4N_2^3 - 4N_1^3N_2^4 + 2N_1^3N_2^2 - 2N_1^2N_2^3 - 2N_1^2N_2)}{8N_1^4N_2^4 + 4N_1^4N_2^2 - 8N_1^3N_2^3 - 2N_1^3N_2 + 4N_1^2N_2^4 + 10N_1^2N_2^2 + N_1^2 + N_2^2 - 2N_1N_2^3 - 4N_1N_2 + 1} \quad (7.84a)$$

$$S_{42} = S_{51} = \frac{(-4N_1^4N_2^3 - 4N_1^3N_2^4 + 2N_1^3N_2^2 + 2N_1^2N_2^3)}{8N_1^4N_2^4 + 4N_1^4N_2^2 - 8N_1^3N_2^3 - 2N_1^3N_2 + 4N_1^2N_2^4 + 10N_1^2N_2^2 + N_1^2 + N_2^2 - 2N_1N_2^3 - 4N_1N_2 + 1} \quad (7.84b)$$

$$S_{52} = S_{41} = \frac{(4N_1^4N_2^3 - 4N_1^3N_2^4 - 10N_1^3N_2^2 + 2N_1^2N_2^3 + 6N_1^2N_2 - 2N_1N_2^2 - 2N_1)}{8N_1^4N_2^4 + 4N_1^4N_2^2 - 8N_1^3N_2^3 - 2N_1^3N_2 + 4N_1^2N_2^4 + 10N_1^2N_2^2 + N_1^2 + N_2^2 - 2N_1N_2^3 - 4N_1N_2 + 1} \quad (7.84c)$$

$$S_{32} = S_{61} = \frac{(4N_1^4N_2^3 - 4N_1^3N_2^4 - 6N_1^3N_2^2 + 2N_1^2N_2^3 + 2N_1^2N_2)}{8N_1^4N_2^4 + 4N_1^4N_2^2 - 8N_1^3N_2^3 - 2N_1^3N_2 + 4N_1^2N_2^4 + 10N_1^2N_2^2 + N_1^2 + N_2^2 - 2N_1N_2^3 - 4N_1N_2 + 1} \quad (7.84d)$$

The coupling and isolation levels of a six-port directional coupler when operating in forward and reverse modes are then expressed using S parameters as

$$\text{First coupling level (Port 3)} = 20\log(-S_{13}) \text{ dB} \quad (7.85a)$$

$$\text{Second coupling level (Port 5)} = 20\log(-S_{15}) \text{ dB} \quad (7.85b)$$

$$\text{First isolation level (Port 4)} = 20\log(-S_{14}) \text{ dB} \quad (7.85c)$$

$$\text{Second isolation level (Port 6)} = 20\log(-S_{16}) \text{ dB} \quad (7.85d)$$

The directivity level can be obtained again from Eq. (7.78) accordingly. Equation (7.83) with (7.78) lead to directional coupler performance parameter calculations through only knowledge of turns ratios under the assumption that all ports are matched. The real operating conditions require us to analyze the six-port directional coupler using circuit analysis techniques. A complete analysis of a six-port coupler has been performed using forward and reverse modes for the circuits shown in Figures 7.15 and 7.16, respectively.

Forward Mode Analysis

In the forward mode operation, V_1 is the excitation voltage with the other port voltages replaced by shorts, i.e. $V_2 = V_3 = V_4 = V_5 = V_6 = 0$. V_{o2} is the output voltage. Ports 3 and 5 are the first and second coupled ports,

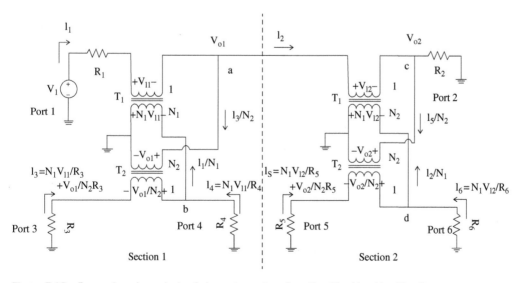

Figure 7.15 Forward mode analysis of six-port coupler when $V_2 = V_3 = V_4 = V_5 = V_6 = 0$.

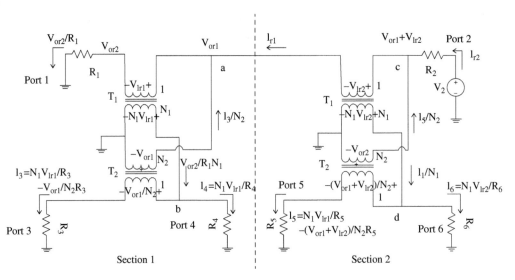

Figure 7.16 Reverse mode analysis of six-port coupler when $V_1 = V_3 = V_4 = V_5 = V_6 = 0$.

respectively, and ports 4 and 6 are the first and second isolated ports, respectively. The circuit analysis of a six-port coupler in forward mode then gives the following relations

$$I_1 = \left(\frac{N_1 V_{l1}}{R_3 N_2}\right) + \left(\frac{V_{o1}}{R_3 N_2^2}\right) + I_2 \text{ at the "node a"} \tag{7.86a}$$

$$\frac{I_1}{N_1} = \left(\frac{N_1 V_{l1}}{R_3}\right) + \left(\frac{V_{o1}}{R_3 N_2}\right) + \left(\frac{N_1 V_{l1}}{R_4}\right) \text{ at the "node b"} \tag{7.86b}$$

$$I_2 = \left(\frac{N_1 V_{l2}}{R_5 N_2}\right) + \left(\frac{V_{o2}}{R_5 N_2^2}\right) + \frac{V_{o2}}{R_2} \text{ at the "node c"} \tag{7.86c}$$

$$\frac{I_2}{N_1} = \left(\frac{N_1 V_{l2}}{R_5}\right) + \left(\frac{V_{o2}}{R_5 N_2}\right) + \left(\frac{N_1 V_{l2}}{R_6}\right) \text{ at the "node d"} \tag{7.86d}$$

Furthermore, the voltages can be related as

$$V_1 - I_1 R_1 - V_{l1} = V_{o1} \tag{7.87a}$$

$$V_{o1} - V_{l2} = V_{o2} \tag{7.87b}$$

In practical applications, the terminal resistances are assumed to be equal, i.e. $R_1 = R_2 = R_3 = R_4 = R_5 = R_6 = r$. This leads to two important equations as

$$V_{o2} = aV_{l2} \tag{7.87c}$$

$$bV_{l2} = cV_{o2} + V_1 \tag{7.87d}$$

which leads to

$$V_{o2} = \left(\frac{a}{b - ca}\right)V_1 \tag{7.87e}$$

where

$$a = \frac{N_1 N_2 (1 - 2N_1 N_2)}{\left(N_1 N_2 - N_1^2 - 1\right)},$$

(7.88a)

$$b = \left(\frac{N_1 + N_2}{N_2}\right) + 2N_1^2 + 1 - \left(\frac{2N_1^2 + 1}{N_1 N_2 - 2(N_1 N_2)^2}\right)$$

(7.88b)

and

$$c = \left(\frac{2N_1^2 + 1}{N_1 N_2 - 2(N_1 N_2)^2}\right) - \left(\frac{N_1 + N_2}{N_2}\right)$$

(7.88c)

The input resistance and the coupled port resistance can now be from

$$R_{in} = \frac{V_1}{I_1} - r = R_{coupled}$$

(7.89)

Reverse Mode Analysis

In the reverse mode operation, V_2 is the applied input voltage with the other port voltages replaced by shorts, i.e. $V_1 = V_3 = V_4 = V_5 = V_6 = 0$. V_{or2} is the output voltage. Ports 4 and 6 are the first and second reverse coupled ports and ports 3 and 5 are the first and second reverse isolated ports. The circuit analysis in reverse mode gives the following relations

$$I_{r1} + \left(\frac{N_1 V_{lr1}}{R_3 N_2}\right) - \left(\frac{V_{or1}}{R_3 N_2^2}\right) = \frac{V_{or2}}{R_1} \text{ at the "node a"}$$

(7.90)

$$\frac{V_{or2}}{R_1 N_1} = \left(\frac{N_1 V_{lr1}}{R_3}\right) - \left(\frac{V_{or1}}{R_3 N_2}\right) + \left(\frac{N_1 V_{lr1}}{R_4}\right) \text{ at the "node b"}$$

(7.91)

$$I_{r2} + \left(\frac{N_1 V_{lr2}}{R_5 N_2}\right) - \left(\frac{V_{or1} + V_{lr2}}{R_5 N_2^2}\right) = I_{r1} \text{ at the "node c"}$$

(7.92)

$$\frac{I_{r1}}{N_1} = \left(\frac{N_1 V_{lr2}}{R_6}\right) - \left(\frac{V_{or1} + V_{lr2}}{R_5 N_2}\right) + \left(\frac{N_1 V_{lr2}}{R_5}\right) \text{ at the "node d"}$$

(7.93)

Similar to forward mode analysis, the voltage relations for reverse mode can be written as

$$V_{or2} + V_{lr1} = V_{or1}$$

(7.94)

$$V_2 - I_{r2} R_2 = V_{or1} + V_{lr2}$$

(7.95)

When $R_1 = R_2 = R_3 = R_4 = R_5 = R_6 = r$, we obtain

$$a_r V_{lr2} + b_r V_{or1} = V_2$$

(7.96)

$$c_r V_{lr1} = V_{or1}$$

(7.97)

$$d_r V_{or1} - e_r V_{lr1} + b_r V_{lr2} = V_2$$

(7.98)

which leads to

$$V_{or1} = \left[\frac{c_r (a_r - b_r)}{a_r (c_r d_r - e_r) - c_r b_r^2}\right] V_2$$

(7.99)

where

$$a_r = \frac{N_2^2 + 1 - 2N_1N_2 + 2(N_1N_2)^2}{N_2^2}, \tag{7.100a}$$

$$b_r = \frac{N_2^2 + 1 - N_1N_2}{N_2^2} \tag{7.100b}$$

$$c_r = \frac{N_2(1 + 2N_1^2)}{N_1 + N_2}, \tag{7.101}$$

$$d_r = \frac{2(1 + N_2^2)}{N_2^2}, \tag{7.102}$$

and

$$e_r = \frac{(N_1 + N_2)}{N_2} \tag{7.103}$$

The output and the isolated port resistances can be calculated as

$$R_{out} = \frac{V_2}{I_{r2}} - r = R_{isolated} \tag{7.104}$$

The summary of the analytical results giving design parameters is illustrated in Table 7.3.

Table 7.3 Design equations for six-port transformer coupled directional coupler.

Forward mode		Reverse mode	
Output voltage	V_{o2}	Reverse output voltage	V_{or2}
First coupled voltage	$N_1V_{l1} + \frac{V_{o1}}{N_2}$	First reverse coupled voltage	N_1V_{lr2}
Second coupled voltage	$N_1V_{l2} + \frac{V_{o2}}{N_2}$	Second reverse coupled voltage	N_1V_{lr1}
First isolated voltage	N_1V_{l1}	First reverse isolated voltage	$-N_1V_{lr2} + \left(\frac{V_{or1} + V_{lr2}}{N_2}\right)$
Second isolated voltage	N_1V_{l2}	Second reverse isolated voltage	$-N_1V_{lr1} + \frac{V_{or1}}{N_2}$
Input return loss	$-20\log\left(\frac{r - R_{in}}{r + R_{in}}\right)$	Output return loss	$-20\log\left(\frac{r - R_{out}}{r + R_{out}}\right)$
Coupled port return loss	$-20\log\left(\frac{r - R_{coupled}}{r + R_{coupled}}\right)$	Isolated port return loss	$-20\log\left(\frac{r - R_{isolated}}{r + R_{isolated}}\right)$
Insertion loss	$-20\log\left(\frac{V_{o2}}{0.5*V_1}\right)$	Reverse insertion loss	$-20\log\left(\frac{V_{or2}}{0.5*V_2}\right)$
First coupled port loss	$-20\log\left(\frac{N_1V_{l1} + \frac{V_{o1}}{N_2}}{0.5*V_1}\right)$	First reverse coupled port loss	$-20\log\left(\frac{N_1V_{lr2}}{0.5*V_2}\right)$
Second coupled port loss	$-20\log\left(\frac{N_1V_{l2} + \frac{V_{o2}}{N_2}}{0.5*V_1}\right)$	Second reverse coupled port loss	$-20\log\left(\frac{N_1V_{lr1}}{0.5*V_2}\right)$
First isolated port loss	$-20\log\left(\frac{N_1V_{l1}}{0.5*V_1}\right)$	First reverse isolated port loss	$-20\log\left(\frac{-N_1V_{lr2} + \left(\frac{V_{or1} + V_{lr2}}{N_2}\right)}{0.5*V_2}\right)$
Second isolated port loss	$-20\log\left(\frac{N_1V_{l2}}{0.5*V_1}\right)$	Second reverse isolated port loss	$-20\log\left(\frac{-N_1V_{lr1} + \frac{V_{or1}}{N_2}}{0.5*V_2}\right)$

7.3   Multistate Reflectometers

The concept of six-port reflectometer theory was first introduced by Hoer and Engen [9, 10] and then became an attractive method for the measurement of voltage, current, impedance, phase, and power. Although six- and five-port reflectometers are attractive and low cost alternatives to network analyzers, it is not convenient to use them in practice since commercially available devices are mainly four-port directional couplers.

Four-port networks such as couplers can also be used to detect power and measure the magnitude of the reflection coefficient. However, it is not possible to measure the complex reflection coefficient with four-port couplers using conventional techniques. There has been research on the use of four-port couplers as multistate reflectometers to be implemented in an automated environment and used for the measurement of several important parameters, such as complex reflection coefficient. The analysis of multistate reflectometers is applicable in practice using a variable attenuator concept with four-port networks. However, no equations or solutions have been obtained or presented in [27] for the reflection coefficient calculation where power circles are constructed and intersection point is obtained and impedance is determined accurately.

In this section, the analysis of multistate reflectometers based on four-port networks and variable attenuators, shown in Figure 7.17, is extended to examine the more general theory of using scalar power measurements to determine the complex reflection coefficient. The explicit closed form relations and solutions for the system of the equations are derived and used to calculate the complex reflection coefficient with the concept of the radical center for three power circles. Analytical results based on the derivations are obtained and the general theory is verified and calibration of multistate reflectometers is discussed.

7.3.1   Multistate Reflectometer Based on Four-port Network and Variable Attenuator

The multistate reflectometer based on the four-port network and variable attenuator is shown in Figure 7.17. It consists of one arbitrary four-port network, a power source, the device under test (DUT), a variable attenuator, and two scalar power detectors. The terms a_i represent the complex amplitude of the voltage wave incident to port i, and the b_i terms represent the emergent voltage wave amplitude from port i. The general operating principles of the device shown in Figure 7.17 are based on those of the more well-known six-port reflectometer design developed by Engen [10]. The key difference is that in the four-port design shown in Figure 7.17, the system should be measured under two different attenuator settings to obtain the necessary number of equations to solve for the value of the complex reflection coefficient for the DUT.

The initial analysis of the multistate reflectometer system with the attenuator shown in Figure 7.17 is performed and the measured powers at ports 2 and 3 in terms of reflection coefficient Γ were given as

$$P_i = |b_i|^2 = q_i \left| \frac{1 + A_i\Gamma}{1 + A_0'\Gamma} \right|^2, \quad (i = 2, 3) \qquad (7.105)$$

Since, in the system illustrated in Figure 7.17, there are two different network statuses, (7.105) leads to two sets of equations as

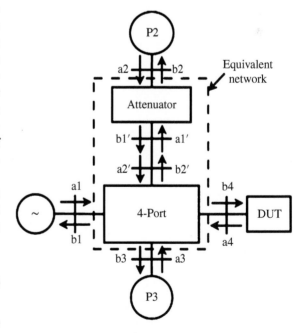

Figure 7.17   Multistate reflectometer based on four-port coupler and attenuator.

$$P_i' = |b_i'|^2 = q_i' \left| \frac{1 + A_i' \Gamma}{1 + A_0' \Gamma} \right|^2, \quad (i = 2, 3) \tag{7.106}$$

$$P_i'' = |b_i''|^2 = q_i'' \left| \frac{1 + A_i'' \Gamma}{1 + A_0'' \Gamma} \right|^2, \quad (i = 2, 3) \tag{7.107}$$

where q_i' and q_i'' for $i = 2, 3$ are real constants, A_i' and A_i'' for $i = 0, 2, 3$ are complex constants, and P_i' and P_i'' for $i = 2, 3$ are the power meter readings for each of the two network statuses. Eqs. (7.106) and (7.107) present a set of bilinear equations that need to be solved for the complex reflection coefficient, Γ. The procedure that is to find power ratio equations for five-port networks that can be implemented for the system shown in Figure 7.17. The analysis begins with separating (7.105) into its corresponding real and imaginary parts as

$$P_i = q_i \left| \frac{1 + (c_i + jd_i)(x + jy)}{1 + (c_0 + jd_0)(x + jy)} \right|^2 \tag{7.108}$$

or

$$P_i = q_i \frac{(c_i^2 + d_i^2)x^2 + (c_i^2 + d_i^2)y^2 + 2c_i x - 2d_i y + 1}{(c_0^2 + d_0^2)x^2 + (c_0^2 + d_0^2)y^2 + 2c_0 x - 2d_0 y + 1} \tag{7.109}$$

where

$$\Gamma = |\Gamma| \angle \psi^\circ = x + jy \tag{7.110}$$

and

$$A_i = \alpha_i \angle \phi_i^\circ = c_i + jd_i \tag{7.111}$$

Equation (7.109) can be expressed as

$$\left(P_i \alpha_0^2 - q_i \alpha_i^2 \right) (x^2 + y^2) + 2(P_i c_0 - q_i c_i)x + 2(q_i d_i - P_i d_0)y = q_i - P_i \tag{7.112}$$

Furthermore, we can put Eq. (7.112) in the following form

$$x^2 + 2u_i x + y^2 + 2v_i y = 2r_i, \quad (i = 2, 3) \tag{7.113}$$

where

$$\omega_i = P_i \alpha_0^2 - q_i \alpha_i^2 \tag{7.114}$$

$$u_i = \frac{P_i c_0 - q_i c_i}{\omega_i} \tag{7.115}$$

$$v_i = \frac{q_i d_i - P_i c_0}{\omega_i} \tag{7.116}$$

$$r_i = \frac{q_i - P_i}{2\omega_i} \tag{7.117}$$

It is now clear from Eq. (7.113) that it represents the general form of the equation for a circle with center $(-u_i, -v_i)$. Hence, Eq. (7.113) can be written in center radius form as

$$(x - (-u_i))^2 + (y - (-v_i))^2 = R_i^2, \quad (i = 2, 3) \tag{7.118}$$

where $R_i^2 = 2r_i + u_i^2 + v_i^2$. Equation (7.118) defines a set of circles in the complex plane indicating possible values of complex reflection coefficient, Γ. In order to solve this system for Γ, at least two independent circle equations must be solved for their intersection points. Most solution methods for six-port reflectometer designs utilize a ratio of power readings, as opposed to each independent power reading [10]. The resulting equation is of identical form to (7.105), and this approach yields many benefits. Additionally, if the power reading being normalized to is highly independent of a_2, it acts to stabilize the system against power fluctuations.

The reference port is coupled to forward power, while the variable-state port is connected to a phase-shifting network. The method is more complicated than that proposed in [27], but approaches the ideal behavior proposed by Engen in [10]. Three power ratios are measured by dividing the variable reading by the forward-coupled reading. Regardless of the specific calibration/measurement scheme being used, the general solution for Γ is described by the intersection of three circles. In reality, however, the circles will not intersect, owing to noise and inaccuracies, but this can be overcome quite simply by using the concept of the radical center. The radical center of the three circles is the unique point, which possesses equal power with respect to all three circles. In other words, it is the point where tangent lines to all circles are of equal length. For three overlapping circles, the radical center is given by the intersection point of the three common chords between all three circles [10]. Additionally, the radical center is still defined when no circle intersections occur as the intersection point of the three radical axes. If the measured location of Γ is interpreted as the radical center of three power or power ratio circles, the bilinear equations of the form of (7.113) may be reduced to a simple system of linear equations by subtraction

$$2(u_i - u_j)x + 2(v_i - v_j)y = 2(r_i - r_j) \qquad (7.119)$$

where the equation of circle j is subtracted from circle i. Equation (7.118) is the equation of the radical axis between circles i and j. In a three-circle system two more such equations exist between circles i and k, and between circles j and k, giving a system of linear equations which may be solved for x and y, which are the real and imaginary components of the complex reflection coefficient.

The analytical results have been obtained with MATLAB using the formulation and solutions discussed elsewhere and illustrated in Figures 7.18 and 7.19. Figure 7.18 demonstrates an imperfect power circle intersection, which could be the result of measurement noise and/or calibration inaccuracies. The three large circles represent the power measurement circles, which separated by approximately 120° in phase and at an equal distance from the origin. The radical center is located at $(0.27 - j0.21)$ inside the unit circle. The error bound of the measurement can be considered the unique triangle, which has the three power circles as its excircles. Once again, the radical center is the unique point with equal power to all three measurement circles.

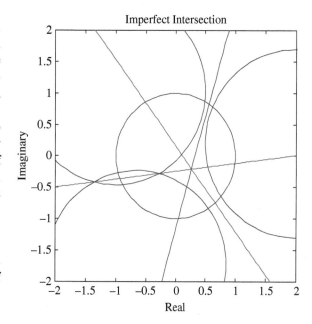

Figure 7.18 Illustration of imperfect power circle for complex reflection coefficient determination.

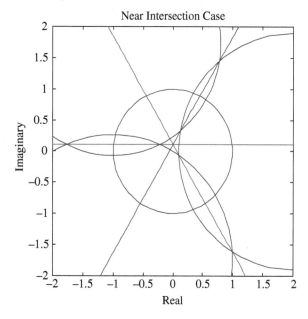

Figure 7.19 Illustration of near intersection power circle for complex reflection coefficient determination.

Figure 7.19 illustrates the case of a near intersection of the three power circles for complex reflection coefficient determination. The radical axes are clearly shown as the lines passing through the common chords of each circle intersection. The error triangle in this case is much smaller, and represents the most likely measurement scenario when using accurately calibrated detectors.

One additional area of concern is in the calibration of multiport and multistate reflectometers. In calibration, there is no simple linearization, and it is often the case that numerical methods are required to calculate the calibration constants. Since the inception of the first six-port reflectometers, there have been several breakthroughs in reducing the complexity of the calibration process. Most of these improvements come in the form of realizing hidden relationships between the complex constant parameters of the six-port or multiport network.

7.4 Combiners and Dividers

Power combiners, dividers (splitters), and phase inverters have widespread use in RF applications. Power combiners and splitters are passive devices where signals need to be combined or split with required insertion loss, good amplitude, and phase balance. These devices are used in applications including RF/microwave amplifiers, transmitters and receivers, and antenna-array feed networks.

The concept of combiner is well explained by Wilkinson [28]. The same concept can be applied in the design of splitters too. Wilkinson describes a circularly symmetric N-way hybrid power divider having an excellent isolation between individual ports. Wilkinson's power splitter/combiner concept can be realized using lumped elements or distributed elements.

7.4.1 Analysis of Combiners and Dividers

$n + 1$ power combiner consists of 1 input port and n output ports, as shown in Figure 7.20 whereas $n + 1$ power divider has n input ports and 1 output port, as shown in Figure 7.21.

Analysis of power dividers using the circuit shown in Figure 7.22 is given by Wilkinson [28]. The input port is defined as node a, where input resistor R_a is connected. The voltage at node a is defined as V_a. Ports 1 through n are defined as the output ports, and voltage at these nodes are designated as V_n. Each transmission line with a characteristic impedance of Z_o and length $l = \lambda/4$ is connected to a common node designated by n via resistor R_x. This is illustrated in Figure 7.23.

Currents out of output ports are defined as I_{an}, and currents going into input ports are defined as I'_{an}. The transmission lines are basically a quarter of a wavelength away from the input port, as illustrated in Figures 7.22 and 7.23. The voltage and current at any point on the transmission lines can be calculated from

$$V(z) = V_0^+ e^{-\gamma z} + V_0^- e^{+\gamma z} \tag{7.120}$$

$$I(z) = \frac{V_0^+}{Z_0} e^{-\gamma z} - \frac{V_0^-}{Z_0} e^{+\gamma z} \tag{7.121}$$

"The forward and reflected wave are related through reflection coefficient with the following equations

$$\Gamma = \frac{V_0^-}{V_0^+} = \frac{Z_L - Z_0}{Z_L + Z_0} \tag{7.122}$$

Assuming the transmission line is lossless, (7.120) and (7.121) can be rewritten in terms of reflection coefficient as

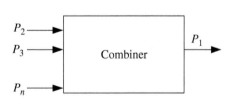

Figure 7.20 $n + 1$ port power combiner.

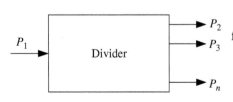

Figure 7.21 $n + 1$ power divider.

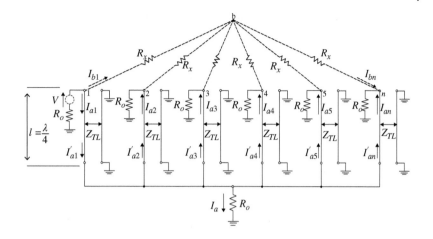

Figure 7.22 Equivalent Wilkinson power divider circuit [28].

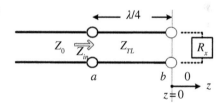

Figure 7.23 Quarter wave transmission line connection in the combiner circuit.

$$V(z) = V_0^+ \left(e^{-\beta z} + \Gamma e^{+\beta z}\right) \tag{7.123}$$

$$I(z) = \frac{V_0^+}{Z_0} \left(e^{-\beta z} - \Gamma e^{+\beta z}\right) \text{ or } I(z) = I_0^+ \left(e^{-\beta z} - \Gamma e^{+\beta z}\right) \tag{7.124}$$

At node a, the input voltage can be then expressed using (7.123) as

$$V(z = -l) = V_1 = V_a \left(e^{\frac{\pi}{2}} + \Gamma e^{-\frac{\pi}{2}}\right) \tag{7.125}$$

Since

$$Z_{TL} = Z_L = Z_o = R_o, \text{ then } \Gamma = 0 \tag{7.126}$$

When (7.126) is substituted into (7.125), it reduces to

$$V_1 = jV_a \tag{7.127}$$

The current flowing out of node n can be expressed from (7.124) as

$$I(z = -l) = I'_{an} = I_{an} \left(e^{\frac{\pi}{2}} - \Gamma e^{-\frac{\pi}{2}}\right) \tag{7.128}$$

When (7.126) is substituted into (7.128), it reduces to

$$I'_{an} = jI_{an} \tag{7.129}$$

The node voltages at node n and a are also equal to

$$V_n = I_{an} Z_{TL} \tag{7.130}$$

$$V_a = I'_{an} Z_{TL} \tag{7.131}$$

When (7.130) is substituted into (7.127), at port 1

$$V_1 = jV_a = jI'_{a1}Z_{TL} \tag{7.132}$$

In addition, voltage at port a can be written in terms of output current as

$$V_a = jI_{an}Z_{TL} \tag{7.133}$$

Current at node 1 can be expressed applying KCL as

$$I'_{a1} = I_a + I'_{an}(n-1) = \frac{V_a}{R_o} + I'_{an}(n-1) \tag{7.134a}$$

Similarly

$$I_{an} = \frac{V_n}{R_o} - I_{bn} = \frac{V_n}{R_o} - \frac{I_{b1}}{(n-1)} \tag{7.134b}$$

When Eq. (7.134b) is substituted into Eq. (7.131), we obtain

$$V_a = jV_n = j\left(\frac{V_n}{R_o} - \frac{I_{b1}}{(n-1)}\right)Z_{TL}$$

The voltage at node b can be written as

$$V_b = V_1 - I_{b1}R_x \tag{7.135}$$

or

$$V_b = V_n + I_{bn}R_x \tag{7.136}$$

(7.135) and (7.136) lead to

$$V_1 - I_{b1}R_x = V_n + I_{bn}R_x \quad \text{or} \quad V_1 - V_n = I_{b1}R_x + I_{bn}R_x \tag{7.137}$$

Equation (7.137) can be modified as

$$V_1 - V_n = R_x(I_{b1} + I_{bn}) \tag{7.138a}$$

or

$$V_1 - V_n = R_x\left(I_{b1} + \frac{I_{b1}}{n-1}\right) = R_xI_{b1}\left(\frac{n}{n-1}\right) \tag{7.138b}$$

Then

$$I_{b1} = \frac{(V_1 - V_n)}{R_x}\left(\frac{n-1}{n}\right) \tag{7.139}$$

From Eq. (7.134a), I'_{a1} can be written as

$$I'_{a1} = -\frac{jV_1}{Z_{TL}} \tag{7.140}$$

Substitution of Eq. (7.129) into (7.130) gives

$$I'_{an} = j\frac{V_n}{Z_{TL}} \tag{7.141}$$

When Eqs. (7.140) and (7.141) are substituted into (7.134a), we obtain

$$-\frac{jV_1}{Z_{TL}} = \frac{V_a}{R_o} + j\frac{V_n}{Z_{TL}}(n-1) \tag{7.142}$$

Rewriting (7.142) gives

$$V_a + jV_1\frac{R_o}{Z_{TL}} + j(n-1)V_n\frac{R_o}{Z_{TL}} = 0 \tag{7.143}$$

For perfect isolation $V_n = 0$, hence Eq. (7.143) is simplified to

$$V_a = -jV_1\frac{R_o}{Z_{TL}} \tag{7.144}$$

Substitution of Eq. (7.134b) into (7.133) gives

$$V_a = j\left(\frac{V_n}{R_o} - \frac{I_{b1}}{(n-1)}\right)Z_{TL} \tag{7.145}$$

When Eq. (7.139) is substituted into (7.145), we obtain

$$V_a = j\left(\frac{V_n}{R_o} - \frac{(V_1 - V_n)}{nR_x}\right)Z_{TL} \tag{7.146}$$

or

$$V_a - jV_n\frac{Z_{TL}}{R_o} + jV_1\frac{Z_{TL}}{nR_x} - jV_n\frac{Z_{TL}}{nR_x} = 0 \tag{7.147}$$

which can be further simplified to

$$V_a + jV_1\frac{Z_{TL}}{nR_x} - jV_n\left(\frac{nR_xZ_{TL} + R_oZ_{TL}}{nR_oR_x}\right) = 0 \tag{7.148}$$

For perfect isolation $V_n = 0$, so Eq. (7.148) is simplified to

$$V_a = -jV_1\frac{Z_{TL}}{nR_x} \tag{7.149}$$

When Eqs. (7.144) and (7.149) are compared, we obtain the following relation

$$\frac{R_o}{Z_{TL}} = \frac{Z_{TL}}{nR_x} \tag{7.150}$$

or

$$Z_{TL} = \sqrt{nR_xR_o} \tag{7.151}$$

Since for matched condition, $R_x = R_o$ and then Eq. (7.151) can be expressed as

$$Z_{TL} = \sqrt{n}R_o \tag{7.152}$$

Equation (7.152) defines the characteristic impedance of an N-way power combiner/divider when source impedance is R_o. We can now determine the input impedance of the system using Figure 7.22 when $z = -l$ from

$$Z_i = Z_{TL}\frac{R_0 + jZ_{TL}\tan\beta l}{Z_{TL} + jR_0\tan\beta l} \tag{7.153}$$

When the length of the transmission line is $l = \lambda/4$, Eq. (7.153) is simplified to

$$Z_i = \frac{1}{n} Z_{TL} \frac{R_0 + j Z_{TL} \tan\left(\frac{2\pi}{\lambda}\frac{\lambda}{4}\right)}{Z_{TL} + j R_0 \tan\left(\frac{2\pi}{\lambda}\frac{\lambda}{4}\right)} = \frac{Z_{TL}^2}{nR_0} \tag{7.154}$$

Since $Z_{TL} = \sqrt{n}R_0$ from Eq. (7.152), (7.154) can be written as

$$Z_i = \frac{Z_{TL}^2}{nR_0} = \frac{nR_0^2}{nR_0} = R_0 \tag{7.155}$$

The isolation between one of the output ports with respect to any other port and VSWR at the input of an N-way hybrid power combiner/divider is calculated with the application of superposition using even- and odd-mode analysis. Figure 7.21 is modified based on the results using the formulations (7.120)–(7.155) and redrawn and shown in Figure 7.24.

The characteristic impedances of the transmission lines in Figure 7.24 are equal to $Z_{TL} = Z_o = \sqrt{N}R_0$ as given by Eq. (7.152) and are a quarter wavelength long at the center frequency.

The equivalent four-port network for an N-way divider circuit is shown in Figure 7.25. The even mode corresponds to an open circuit at symmetry plane when voltage source $(+V)$ is placed in series at port 4 whereas odd

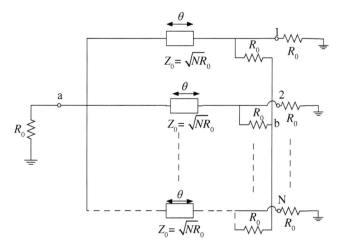

Figure 7.24 *N*-way Wilkinson power divider circuit.

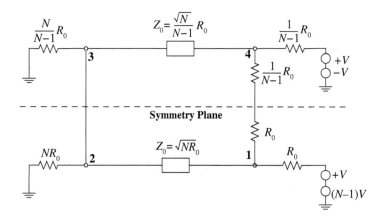

Figure 7.25 Four-port network for an *N*-way power divider.

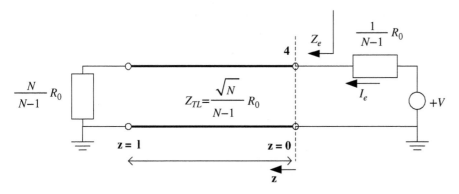

Figure 7.26 Even-mode network for an *N*-way divider.

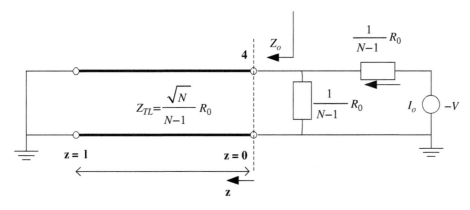

Figure 7.27 Odd-mode network for an *N*-way divider.

mode corresponds to a short circuit when $(-V)$ is placed in series at port 4. The networks for even and odd modes are illustrated in Figures 7.26 and 7.27, respectively.

The analysis of an even-mode network in Figure 7.26 gives the even-mode impedance as

$$Z_e = Z_{TL} \frac{\left(\frac{N}{N-1}R_0\right) + jZ_{TL}\tan(\theta)}{Z_{TL} + j\left(\frac{N}{N-1}R_0\right)\tan(\theta)} \tag{7.156}$$

Since

$$Z_{TL} = \frac{\sqrt{N}}{N-1}R_o \tag{7.157}$$

Equation (7.156) can be simplified as

$$Z_e = \left(\frac{\sqrt{N}}{N-1}R_o\right)\frac{\left(\frac{N}{N-1}R_0\right) + j\left(\frac{\sqrt{N}}{N-1}R_o\right)\tan(\theta)}{\left(\frac{\sqrt{N}}{N-1}R_o\right) + j\left(\frac{N}{N-1}R_0\right)\tan(\theta)} = \left(\frac{\sqrt{N}}{N-1}R_o\right)\frac{1 + j\left(\frac{1}{\sqrt{N}}\right)\tan(\theta)}{\left(\frac{1}{\sqrt{N}}\right) + j\tan(\theta)}$$

or

$$Z_e = \left(\frac{\sqrt{N}}{N-1}R_o\right)\frac{\sqrt{N}\cot(\theta) + j}{\cot(\theta) + j\sqrt{N}} \tag{7.158}$$

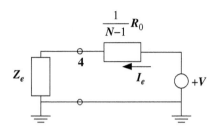

Figure 7.28 Simplified even-mode network for an N-way divider.

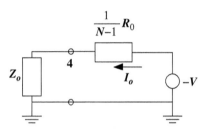

Figure 7.29 Simplified odd-mode network for an N-way divider.

The even-mode network can be then further reduced to the circuit given in Figure 7.28.

The even-mode current is now found from

$$V = I_e \left(\left(\frac{\sqrt{N}}{N-1} R_o \right) \frac{\sqrt{N} \cot(\theta) + j}{\cot(\theta) + j\sqrt{N}} + \frac{R_o}{N-1} \right) \tag{7.159}$$

as

$$I_e = \frac{V}{R_o} \frac{1}{\left[\left(\frac{\sqrt{N}}{N-1} \right) \frac{\sqrt{N} \cot(\theta) + j}{\cot(\theta) + j\sqrt{N}} + \frac{1}{N-1} \right]} \tag{7.160}$$

We perform the odd-mode analysis similarly and find the odd-mode impedance as

$$Z_o = \frac{\left(\frac{R_o}{N-1} \right) (jZ_{TL} \tan(\theta))}{\left(\frac{R_o}{N-1} \right) + (jZ_{TL} \tan(\theta))} \tag{7.161}$$

where Z_{TL} is given by Eq. (7.158). The odd-mode network can be then further reduced to the circuit given in Figure 7.29.

The odd-mode current is found from

$$V = -I_o \left(\frac{R_o}{N-1} \right) \left(1 + \frac{jZ_{TL} \tan(\theta)}{\frac{R_o}{N-1} + jZ_{TL} \tan(\theta)} \right) \tag{7.162}$$

as

$$I_o = -V \left(\frac{N-1}{R_o} \right) \left[\frac{\cot(\theta) + j\sqrt{N}}{\cot(\theta) + j2\sqrt{N}} \right] \tag{7.163}$$

The total current now is found by adding the odd- and even-mode currents as given by Eq. (7.164).

$$I_t = I_e + I_o = \frac{V}{R_o} \left[\frac{1}{\left[\left(\frac{\sqrt{N}}{N-1} \right) \frac{\sqrt{N} \cot(\theta) + j}{\cot(\theta) + j\sqrt{N}} + \frac{1}{N-1} \right]} - (N-1) \left[\frac{\cot(\theta) + j\sqrt{N}}{\cot(\theta) + j2\sqrt{N}} \right] \right] \tag{7.164}$$

The power delivered to each of $(N-1)$ ports with load resistance R_o is then equal to

$$P = |I_t|^2 \left(\frac{R_o}{N-1} \right) \left(\frac{1}{N-1} \right) = |I_t|^2 \left(\frac{R_o}{(N-1)^2} \right) \tag{7.165}$$

The power that is available from the excitation port is defined as P_a and given by

$$P_a = \frac{(NV)^2}{4R_o} \tag{7.166}$$

The isolation of one port from others is defined as

$$\text{Isolation (dB)} = 10 \log \left(\frac{P_a}{P} \right) \tag{7.167}$$

Substitution of (7.165) and (7.166) into (7.167) gives

$$\text{Isolation (dB)} = 10\log\left(\frac{N^2}{4\left|\left[\frac{\cot(\theta)+j\sqrt{N}}{(N+1)\cot(\theta)+j2\sqrt{N}}\right]-\left[\frac{\cot(\theta)+j\sqrt{N}}{\cot(\theta)+j2\sqrt{N}}\right]\right|^2}\right) \tag{7.168}$$

The input VSWR of the system which is calculated at node a for an N-way divider is found from

$$\text{VSWR} = \frac{1+|\Gamma|}{1-|\Gamma|} \tag{7.169}$$

where

$$\Gamma = \frac{Z_i - R_o}{Z_i + R_o} \tag{7.170}$$

$$Z_i = \frac{1}{N}\left(Z_{TL}\frac{R_0+jZ_{TL}\tan\theta}{Z_{TL}+jR_0\tan\theta}\right) = \left(\frac{R_0}{\sqrt{N}}\right)\frac{1+j\sqrt{N}\tan\theta}{\sqrt{N}+j\tan\theta} \tag{7.171}$$

The insertion loss at each port is defined as

$$\text{Insertion loss (dB)} = 10\log\left(\frac{N}{1-\Gamma^2}\right) \tag{7.172}$$

The response of the calculated isolation versus electrical length is given in Figure 7.30. The isolation goes toward infinite when the electrical length is $\theta = 90°$ or $l = \lambda/4$ as expected.

The response of the calculated input VSWR and insertion loss versus electrical length are given in Figures 7.31 and 7.32, respectively. The insertion loss at a quarter wavelength transmission line is tabulated and shown in Table 7.4. The important note on the divider design that is given is that when there is no phase difference between ports it can be also used as a power combiner by simply applying signals at the output ports that will be added at the single input port.

Figure 7.30 Isolation response versus electrical length for an N-way divider.

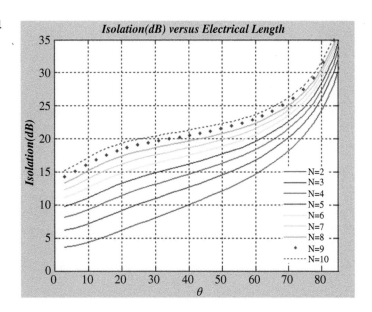

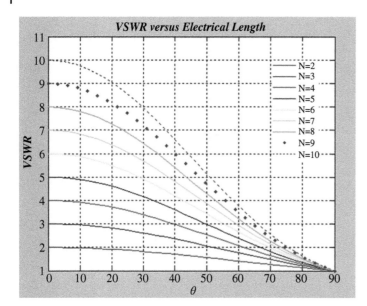

Figure 7.31 VSWR response versus electrical length for an *N*-way divider.

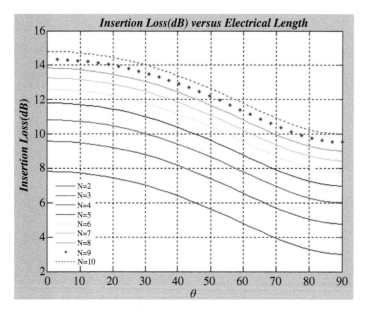

Figure 7.32 Insertion loss response versus electrical length for an *N*-way divider.

7.4.2 Analysis of Dividers with Different Source Impedance

The analysis that was given in Section 7.4.2 was based on the assumption that the source and load impedances were equal. However, there might be cases where the source impedance is different from the load impedance, as shown in Figure 7.33. When this is the case, the characteristic impedance of the transmission line is then defined from (7.151) as

$$Z_{TL} = \sqrt{nR_xR_o} = \sqrt{nR_gR_o} \tag{7.173}$$

where $R_x = R_g$ is the source impedance.

Table 7.4 Calculated insertion loss (dB) for $\theta = 90°$.

Number of ports	Calculated insertion loss (dB)
2	3.0103
3	4.7712
4	6.0206
5	6.9897
6	7.7815
7	8.4510
8	9.0309
9	9.5424
10	10

Figure 7.33 *N*-way Wilkinson power divider circuit with different source impedance.

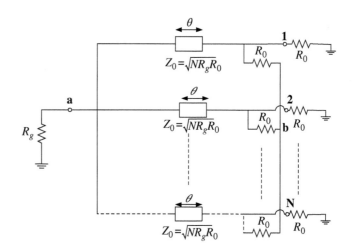

Figure 7.34 Four-port network for an *N*-way power divider with different source impedance.

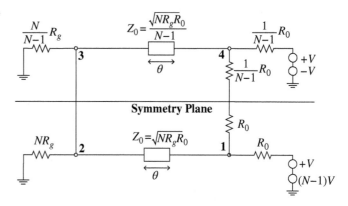

The equivalent four-port network can be established similarly, as shown in Figure 7.34, for even- and odd-mode analysis based on the application of superposition.

The even- and odd-mode networks obtained from the circuit given in Figure 7.34 are illustrated in Figures 7.35 and 7.36, respectively.

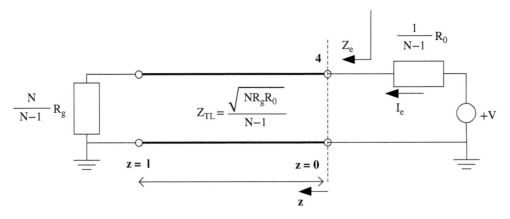

Figure 7.35 Even-mode network for an *N*-way divider with different source impedance.

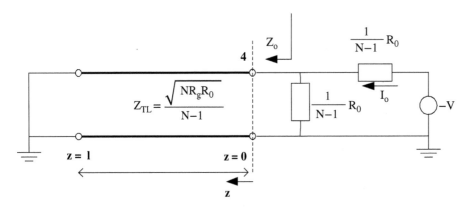

Figure 7.36 Odd-mode network for an *N*-way divider with different source impedance.

When similar analysis performed in Section 7.4.2 is performed, the even- and odd-mode currents are obtained as

$$I_e = \frac{V}{R_0} \left[\frac{1}{\left(\frac{\sqrt{N\frac{R_g}{R_0}}}{N-1} \right) \left[\frac{\sqrt{N\frac{R_g}{R_0}}\cot(\theta) + j}{\cot(\theta) + j\sqrt{N\frac{R_g}{R_0}}} \right] + \frac{1}{N-1}} \right] \tag{7.174}$$

and

$$I_o = -V \left(\frac{N-1}{R_0} \right) \left[\frac{\cot(\theta) + j\sqrt{N\frac{R_g}{R_0}}}{\cot(\theta) + j2\sqrt{N\frac{R_g}{R_0}}} \right] \tag{7.175}$$

The total current from superposition is then equal to

$$I_t = I_e + I_o = \frac{V}{R_0} \left[\frac{1}{\left[\left(\frac{\sqrt{N\frac{R_g}{R_0}}}{N-1} \right) \frac{\sqrt{N\frac{R_g}{R_0}}\cot(\theta) + j}{\cot(\theta) + j\sqrt{N\frac{R_g}{R_0}}} + \frac{1}{N-1} \right]} - (N-1) \left[\frac{\cot(\theta) + j\sqrt{N\frac{R_g}{R_0}}}{\cot(\theta) + j2\sqrt{N\frac{R_g}{R_0}}} \right] \right] \tag{7.176}$$

Isolation and insertion loss are found from Eqs. (7.168) and (7.172), respectively. The input impedance with different source impedance is given by

$$Z_i = \frac{1}{N}\left(Z_{TL}\frac{R_0 + jZ_{TL}\tan\theta}{Z_{TL} + jR_0\tan\theta}\right) = \left(\sqrt{\frac{R_g R_0}{N}}\right)\frac{1 + j\sqrt{N\frac{R_g}{R_o}}\tan\theta}{\sqrt{N\frac{R_g}{R_o}} + j\tan\theta} \tag{7.177}$$

The reflection coefficient at the input is calculated using

$$\Gamma_{in} = \frac{Z_i - R_g}{Z_i + R_g} \tag{7.178}$$

The input VSWR is then equal to

$$VSWR_{in} = \frac{1 + |\Gamma_{in}|}{1 - |\Gamma_{in}|} \tag{7.179}$$

Available power, P_a, is found from (7.166). The isolation is obtained with the substitution of (7.176) into (7.167) as

$$\text{Isolation (dB)} = 10\log\left(\frac{N^2}{4\left|\left[\frac{\cot(\theta) + j\sqrt{N\frac{R_g}{R_o}}}{\left(N\frac{R_g}{R_o} + 1\right)\cot(\theta) + j2\sqrt{N\frac{R_g}{R_o}}}\right] - \left[\frac{\cot(\theta) + j\sqrt{N\frac{R_g}{R_o}}}{\cot(\theta) + j2\sqrt{N\frac{R_g}{R_o}}}\right]\right|^2}\right) \tag{7.180}$$

The VSWR at the output ports is calculated using the output reflection coefficient from

$$\Gamma_o = \left[\frac{2\sqrt{N\frac{R_g}{R_o}}\left(\frac{R_g}{R_o} - 1\right) + j\cot(\theta)\left(N\frac{R_g}{R_o} + 1 - 2\frac{R_g}{R_o}\right)}{2\sqrt{N\frac{R_g}{R_o}}\left(2 + N\frac{R_g}{R_o}\right) + j\cot(\theta)\left(N\frac{R_g}{R_o}(4\tan^2(\theta) - 1) - 1\right)}\right] \tag{7.181}$$

Then, output VSWR is found by

$$VSWR_{out} = \frac{1 + |\Gamma_{out}|}{1 - |\Gamma_{out}|} \tag{7.182}$$

The response of isolation for several values of (R_g/R_o) are given by Figures 7.37 and 7.38. Figure 7.37 illustrates the isolation versus electrical length when $R_g > R_o$, whereas Figure 7.38 gives the isolation versus electrical length when $R_g < R_o$. It has been observed the isolation between ports for an N-way power divider with different source impedance gets better when $R_g < R_o$. The input and output VSWR versus electrical length when $R_g > R_o$ and $R_g < R_o$ are given in Figures 7.39–7.42. The insertion loss response is given in Figures 7.43 and 7.44 for $R_g > R_o$ and $R_g < R_o$.

It should be noted that the variation in insertion loss versus electrical length is much less when $R_g < R_o$. Hence, it is possible to obtain broadband response when $R_g < R_o$.

Example 7.2   Combiner Design

Design an eight-way combiner at 100 MHz by assuming (a) $R_g = R_o = 50\,\Omega$ and (b)$R_g = 50\,\Omega$, $R_o = 25\,\Omega$ and find (i) isolation, (ii) insertion loss, (iii) input VSWR, and (iv) output VSWR when $\theta = 90°$ and $\theta = 70°$ using analytical formulation and compare values with simulation results.

Solution

a) In part (a) of the example, a balanced N-way divider is analyzed when $R_g = R_o = 50\,\Omega$.
   i) Isolation for a balanced N-way divider, $R_g = R_o = 50\,\Omega$, is found from (7.168) as

(a)

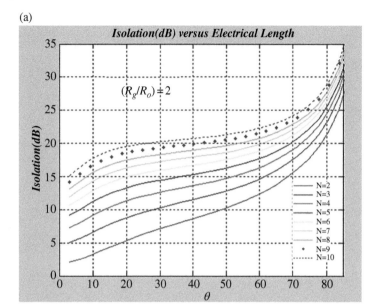

Figure 7.37 Isolation response versus electrical

Figure 7.37 Isolation response versus electrical length for an *N*-way divider with different source impedance: (a) $(R_g/R_o) = 2$; (b) $(R_g/R_o) = 4$.

(b)

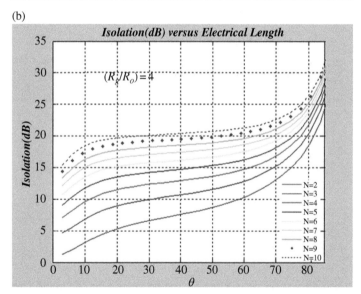

$$\text{Isolation(dB)} = 10\log\left(\frac{N^2}{4\left|\left[\frac{\cot(\theta) + j\sqrt{N}}{(N+1)\cot(\theta) + j2\sqrt{N}}\right] - \left[\frac{\cot(\theta) + j\sqrt{N}}{\cot(\theta) + j2\sqrt{N}}\right]\right|^2}\right)$$

Isolation (dB) = 339.3119 @ $\theta = 90°$ and isolation (dB) = 25.03 @ $\theta = 70°$

ii) Insertion loss is calculated from (7.172) as

$$IL(dB) = 10\log\left(\frac{N}{1 - \Gamma^2}\right)$$

Insertion loss (dB) = 9.0309 @ $\theta = 90°$ and insertion Loss (dB) = 9.75 @ $\theta = 70°$

Figure 7.38 Isolation response versus electrical length for an *N*-way divider with different source impedance: (a) $(R_g/R_o) = 0.5$; (b) $(R_g/R_o) = 0.25$.

(a)
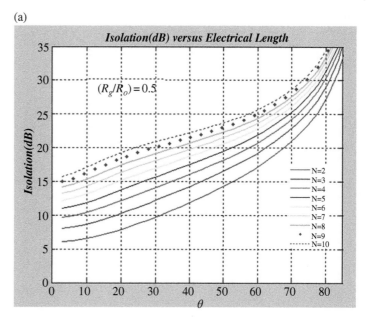

(b)
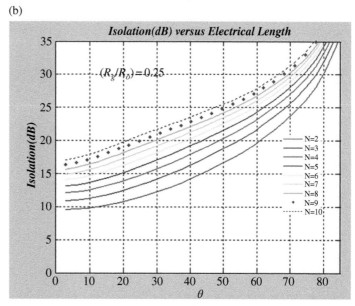

iii) Input VSWR = 1 @ $\theta = 90°$ and input VSWR = 2.27 @ $\theta = 70°$

The simulation when $R_g = R_o = 50\,\Omega$ has been performed by Ansoft Designer and the results are shown below for each section. The circuit that is simulated is shown in Figure 7.45. In Figure 7.45, port 1 represents the input port and port 2–9 represent the output ports.

iv) The isolation versus frequency when electrical length of each transmission line is $\theta = 90°$ is given in Figure 7.46. The isolation is very high as expected when $f = 100\,\text{MHz}$.

The isolation versus frequency when the electrical length of each transmission line is $\theta = 70°$ is given in Figure 7.47. The isolation is 25.96 dB at $f = 100\,\text{MHz}$.

(a)

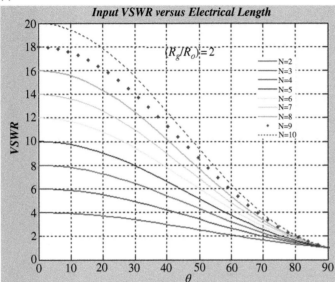

(b)

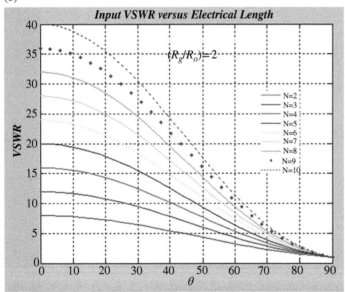

Figure 7.39 Input VSWR response versus electrical length for an *N*-way divider with different source impedance: (a) $(R_g/R_o) = 2$; (b) $(R_g/R_o) = 4$.

v) The insertion loss versus frequency when the electrical length of each transmission line is $\theta = 90°$ is given in Figure 7.48. The insertion loss is 9.03 dB when $f = 100$ MHz

The insertion loss is 9.57 dB when $f = 100$ MHz for $\theta = 70°$, as shown in Figure 7.49.

vi) The input and output VSWR versus frequency when $\theta = 90°$ is given in Figure 7.50.

The VSWR versus frequency response when $\theta = 70°$ is given in Figure 7.51. As illustrated, all the simulated values are very close to the calculated values.

b) In part (b) of the example, an unbalanced *N*-way divider is analyzed when $R_g = 50\ \Omega$ and $R_o = 25\ \Omega$

Figure 7.40 Input VSWR response versus electrical length for an *N*-way divider with different source impedance: (a) $(R_g/R_o) = 0.5$; (b) $(R_g/R_o) = 0.25$.

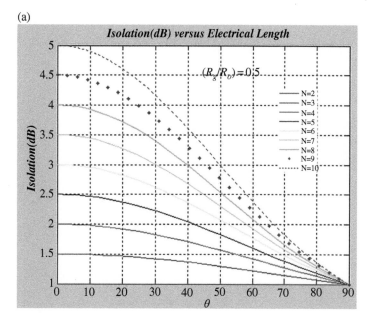

(a)

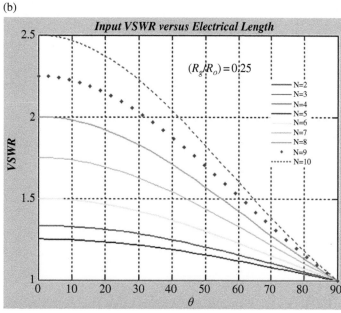

(b)

i) Isolation for a balanced eight-way divider when $R_g = 50\ \Omega$ and $R_o = 25\ \Omega$ is found from (7.180) as

$$
\text{Isolation (dB)} = 10\log\left(\cfrac{N^2}{4\left|\left[\cfrac{\cot(\theta) + j\sqrt{N\dfrac{R_g}{R_o}}}{\left(N\dfrac{R_g}{R_o} + 1\right)\cot(\theta) + j2\sqrt{N\dfrac{R_g}{R_o}}}\right] - \left[\cfrac{\cot(\theta) + j\sqrt{N\dfrac{R_g}{R_o}}}{\cot(\theta) + j2\sqrt{N\dfrac{R_g}{R_o}}}\right]\right|^2}\right)
$$

(a)

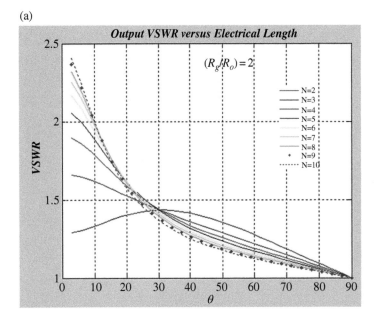

Figure 7.41 Output VSWR response versus electrical length for an *N*-way divider with different source impedance: (a); (R_g/R_o) = 2 (b) (R_g/R_o) = 4.

(b)

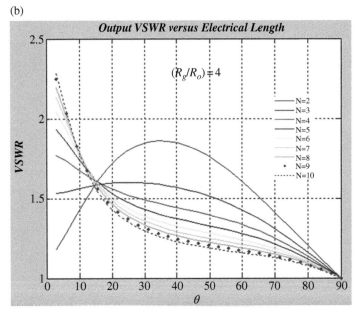

$$\text{Isolation (dB)} = 10\log\left(\frac{8^2}{4\left|\left[\dfrac{\cot(\theta) + j\sqrt{8\dfrac{50}{25}}}{\left(8\dfrac{50}{25} + 1\right)\cot(\theta) + j2\sqrt{8\dfrac{50}{25}}}\right] - \left[\dfrac{\cot(\theta) + j\sqrt{8\dfrac{50}{25}}}{\cot(\theta) + j2\sqrt{8\dfrac{50}{25}}}\right]\right|^2}\right)$$

Isolation (dB) = 336.30 @ $\theta = 90°$ and isolation (dB) = 22.8294 @ $\theta = 70°$

Figure 7.42 Output VSWR response versus electrical length for an *N*-way divider with different source impedance: (a) (R_g/R_o) = 0.5; (b) (R_g/R_o) = 0.25.

(a)

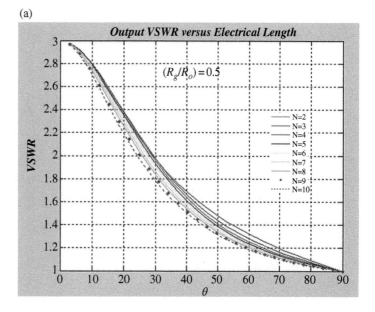

(b)

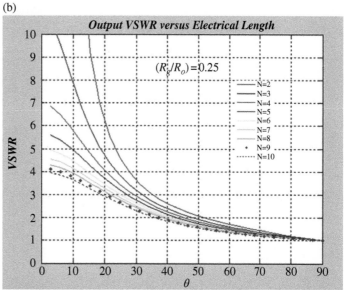

ii) Insertion loss is calculated from (7.172) using (7.181) as

$$\text{Insertion loss (dB)} = 10\log\left(\frac{N}{1-\Gamma^2}\right)$$

Insertion loss (dB) = 9.0309 @ θ = 90° and insertion loss (dB) = 10.5269 @ θ = 70°

(a)

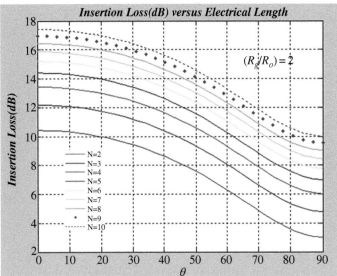

(b)

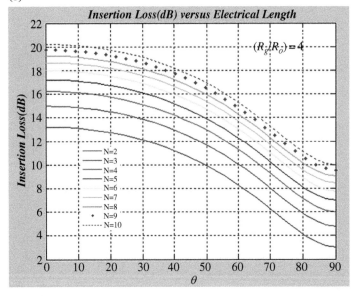

Figure 7.43 Insertion loss response versus electrical length for an N-way divider with different source impedance: (a) $(R_g/R_o) = 2$; (b) $(R_g/R_o) = 4$.

iii) Input VSWR = 1 @/ θ = 90° and input VSWR = 3.3461 @ θ = 70°. The simulation when R_g = 50 Ω and R_o = 25 Ω has been performed by Ansoft Designer and the results are presented. The circuit that is simulated is shown in Figure 7.52.

iv) The simulated isolation versus frequency when the electrical length of each transmission line is θ = 90° and θ = 70° is given in Figures 7.53 and 7.54, respectively. The simulated isolation values are given on the figures.

Figure 7.44 Insertion loss response versus electrical length for an *N*-way divider with different source impedance: (a) $(R_g/R_o) = 0.5$; (b) $(R_g/R_o) = 0.25$.

(a)

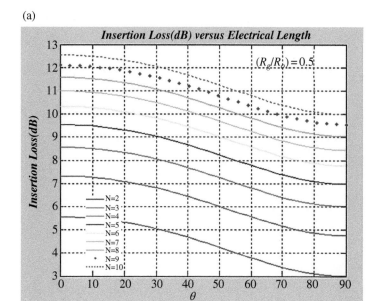

(b)

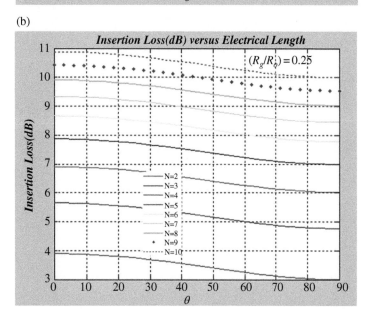

v) The simulated insertion loss versus frequency when the electrical length of each transmission line is $\theta = 90°$ and $\theta = 70°$ is given in Figures 7.55 and 7.56, respectively. The simulated insertion loss values are given on the figures.

vi) The input and output VSWR versus frequency when $\theta = 90°$ and $\theta = 70°$ is given in Figures 7.57 and 7.58, respectively.

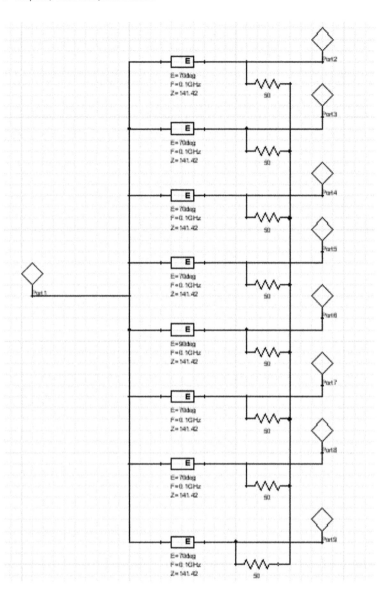

Figure 7.45   Simulated eight-way balanced divider.

The simulated values are in agreement with the calculated values, as shown in part (a) and part (b) of the example. This confirms the accuracy of the analytical formulation. As a result, MATLAB GUI has been developed to design balanced and unbalanced *N*-way dividers using the method presented here. The program requests source impedance, load impedance, center frequency, number of ports, electrical length, and type of divider, and it outputs isolation, insertion loss, and input and output VSWRs at the desired electrical length. The program also gives divider characteristic curves including isolation, insertion loss, and input and output VSWRs versus electrical length. The program outputs for balanced and unbalanced eight-way dividers are given in Figures 7.59 and 7.60, respectively. Results agree with the previously calculated and simulated results.

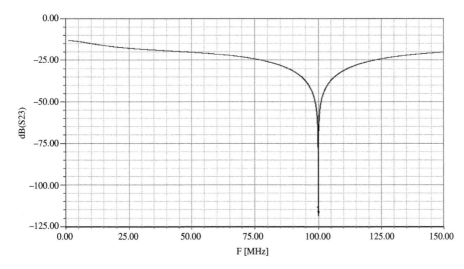

Figure 7.46 Isolation versus frequency for an eight-way divider when θ = 90°.

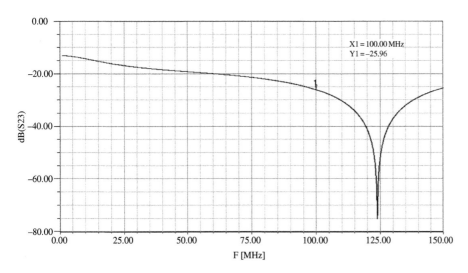

Figure 7.47 Isolation versus frequency for an eight-way divider when θ = 70°.

7.4.3 Microstrip Implementation of Combiners/Dividers

Consider the N-way combiner circuit given in Figure 7.61.

θ in Figure 7.61 is the physical length of a transmission line and equal to a quarter of a wavelength, $\lambda/4$. Because of the increased physical length of the distributed components in HF range, we transform distributed components to lumped-element components. The network showing transformation from distributed elements to lumped elements for a quarter-wavelength-long transmission line is given in Figure 7.62.

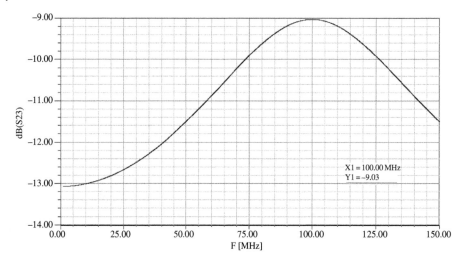

Figure 7.48 Insertion Loss versus frequency for an eight-way divider when $\theta = 90°$.

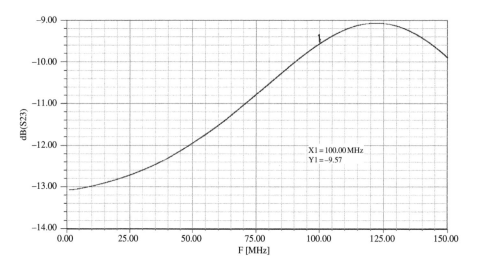

Figure 7.49 Insertion Loss versus frequency for an eight-way divider when $\theta = 70°$.

In Figure 7.62, the element values for the lumped components can be found using the following formulas.

$$L = \frac{Z_0}{2\pi f}, C = \frac{1}{2\pi f Z_0} \tag{7.183}$$

The network in Figure 7.62 also performs impedance transformation from R_1 to R_2 at each distribution port on the combiner. The lumped-element transformation network shown in Figure 7.63 is a π network and it consists of three reactive elements. The quality factor for this network can be found from

Figure 7.50 The input and output VSWR versus frequency when for $\theta = 90°$ (a) Input VSWR (b) Output VSWR.

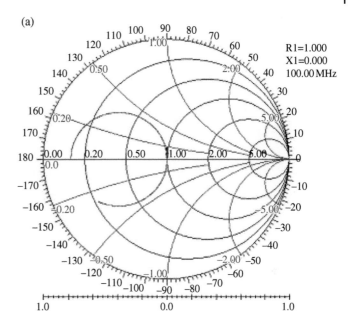

(a)

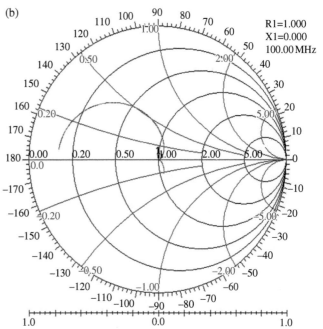

(b)

$$Q = R_1/X_c \tag{7.184}$$

The number of reactive elements can be reduced by transforming the π network to the L network. Q of the π network can be used to obtain the corresponding element values for the L network when it is transformed. The equivalent L network is given in Figure 7.63 when $R_1 \geq R_2$.

(a)

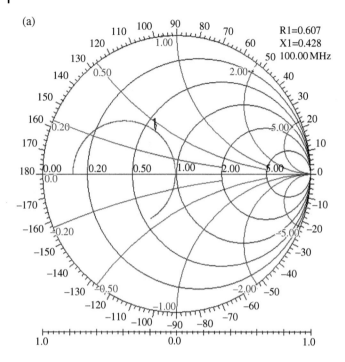

R1=0.607
X1=0.428
100.00 MHz

Figure 7.51 The input and output VSWR versus frequency when for $\theta = 70°$ (a) Input VSWR (b) Output VSWR.

(b)

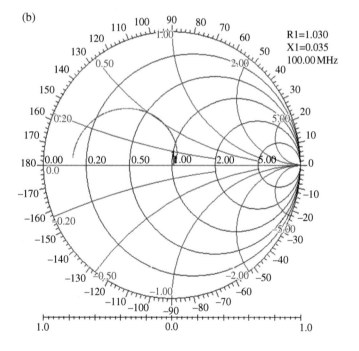

R1=1.030
X1=0.035
100.00 MHz

Figure 7.52 Simulated eight-way unbalanced divider.

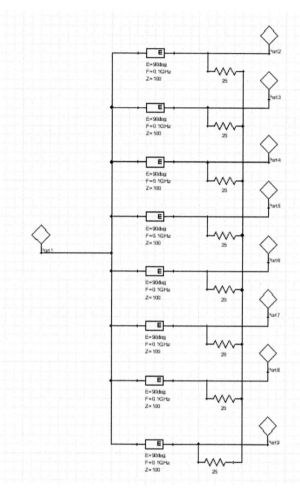

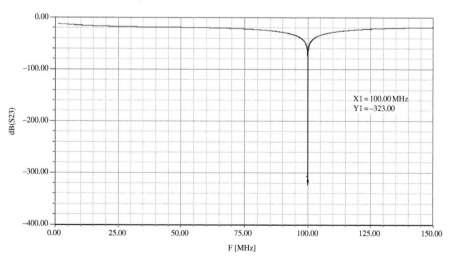

Figure 7.53 Isolation versus frequency for eight-way unbalanced divider when $\theta = 90°$.

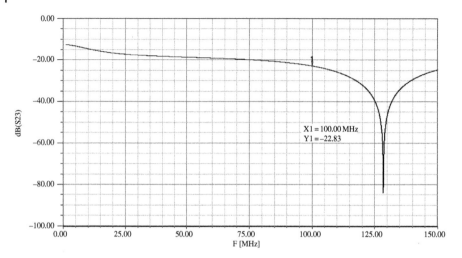

Figure 7.54 Isolation versus frequency for eight-way unbalanced divider when $\theta = 70°$.

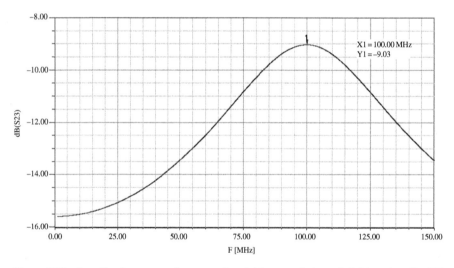

Figure 7.55 Insertion loss versus frequency for eight-way unbalanced divider when $\theta = 90°$.

As a result, each distributed element in Figure 7.61 can be replaced with its equivalent lumped-element L network, as shown in Figure 7.63b.

7.5 Engineering Application Examples

This section discusses real world engineering design examples for microstrip and transformer directional couplers and combiners. Their design, simulation, and implementation methods are detailed.

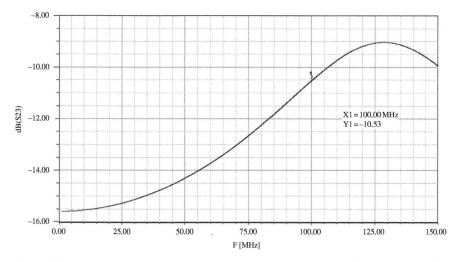

Figure 7.56 Insertion loss versus frequency for eight-way unbalanced divider when $\theta = 70°$.

Design Example 7.1   Microstrip Directional Coupler Design
Design, simulate, and implement 15 dB two- and three-line directional couplers operating at 300 MHz. Use AlO_2 with dielectric constant 9.8 as the substrate.

Solution
The design of three-line couplers can be implemented by following the steps outlined in Section 7.2. The design process begins with the creation of two-line couplers at 300 MHz for a -15 dB coupling level. The substrate material is chosen to be TMM10, which has a relative dielectric constant of 9.8. The MATLAB GUI shown in Figure 7.64 is used to generate the physical dimensions to simulate a two-line coupler. Several other parameters, including spacing and shape ratios versus relative dielectric constant, odd- and even-mode capacitance variation, length, etc., are calculated and illustrated in this GUI.

Once all the physical dimensions are obtained, simulation can be performed. The simulation is done by choosing the thickness of the material to be 100 mil due to the availability of material thickness in practice. This gives the calculated spacing to be 63 mil and the width to be 95.9 mil. The tolerance of the milling machine allows us to have 65 mil and 100 mil spacing and width instead of the calculated values. This small deviation is expected to produce minimal error. The layout of the simulated two-line microstrip coupler with Ansys Designer is illustrated in Figure 7.64b.

The simulation results for the two-line coupler illustrated in Figure 7.64 are shown in Figure 7.65. The results show that the coupling level is -17.266 dB and the isolation level is -27.458 dB. Now, the design steps described in Section 7.2.1.2 should be followed, and a three-line coupler is formed using the formulation with MATLAB GUI with results shown in Table 7.5 including mode impedances illustrated.

As expected, the spacing and shape ratios remain the same to produce the same level of coupling level. The layout of the simulated three-line microstrip coupler with Ansys Designer is illustrated in Figure 7.66.

The simulation results showing coupling and isolation levels at 300 MHz are given in Figure 7.67. The simulation results give almost identical coupling and isolation levels as -17.31 dB and -27.52 dB, respectively.

Since the analytical results and simulation results are in agreement, one can proceed to build prototype three-line directional couplers using the physical dimensions in the simulation based on the calculated values where the width of the main and coupling traces are 100 mil, the thickness of the material is 100 mil, and the spacing of the coupled line from the main line is 65 mil. The substrate used in the prototype is Rogers TMM10 material, as

(a)

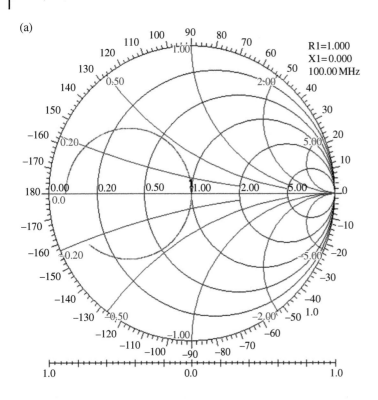

(b)

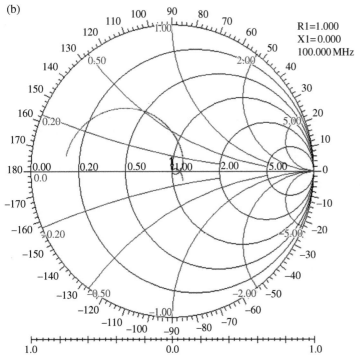

Figure 7.57 The input and output VSWR versus frequency when for θ = 90°: (a) input VSWR; (b) output VSWR.

Figure 7.58 The input and output VSWR versus frequency when for $\theta = 70°$: (a) input VSWR; (b) output VSWR.

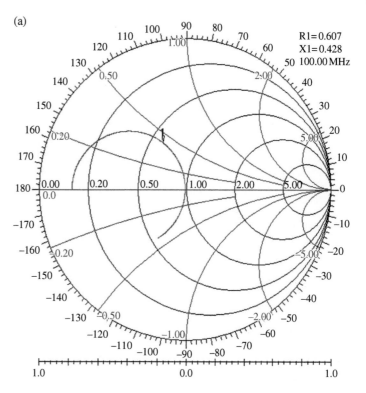

(a)

R1 = 0.607
X1 = 0.428
100.00 MHz

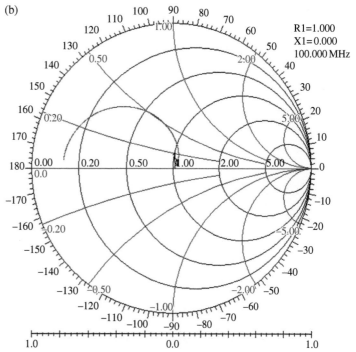

(b)

R1 = 1.000
X1 = 0.000
100.000 MHz

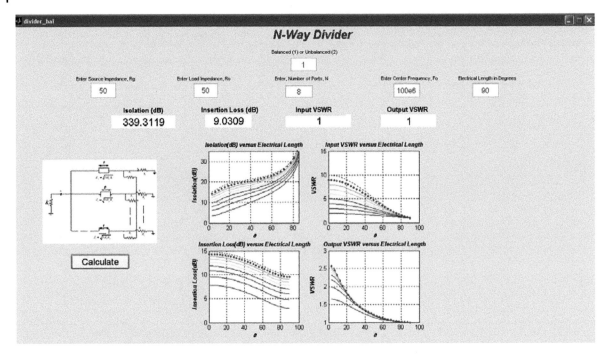

Figure 7.59   MATLAB GUI output for an eight-way balanced divider when θ = 90°.

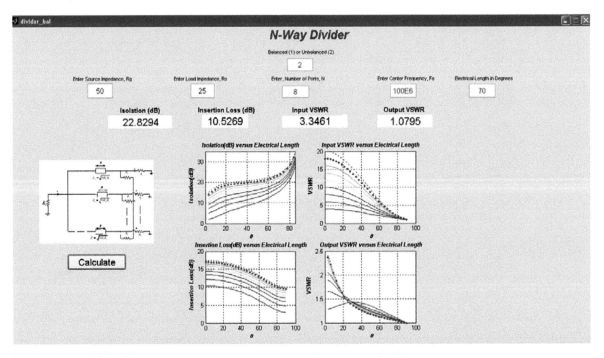

Figure 7.60   MATLAB GUI output for an eight-way unbalanced divider when θ = 70°.

Figure 7.61 *N*-way combiner circuit.

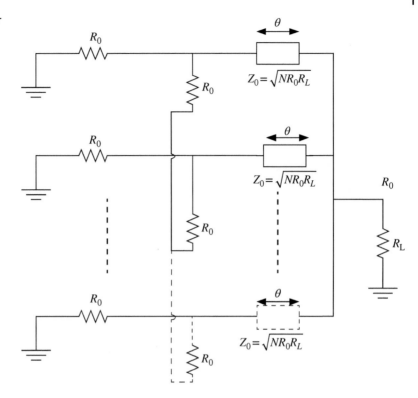

illustrated in Figure 7.67. The measurement results for coupling and isolation are shown in Figure 7.68. Measurement has been done using an E5063A Keysight network analyzer. The coupling and isolation levels are measured at 300 MHz and shown in Figure 7.69. The measured values of coupling and isolation levels are found to be −18.37 dB and −28.85 dB, respectively. These are close to the analytical and simulated values.

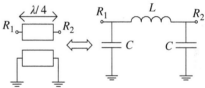

Figure 7.62 Distributed to lumped conversion.

Design Example 7.2 Transformer Directional Coupler Design

Design, simulate, build, and measure transformer directional coupler with 20 dB coupling and directivity better than 30 dB for high power RF application at 27.12 MHz when input voltage is $V_{in,peak} = 100\,\text{V}$. The port impedances are matched and given to be equal to $R = 50\,\Omega$. Compare your analytical results with simulation results in time domain and frequency domain. Use also scattering parameters and compare your results with the circuit analysis results.

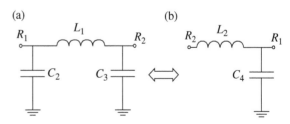

Figure 7.63 Transformation from (a) a π network to (b) an L network.

(a)

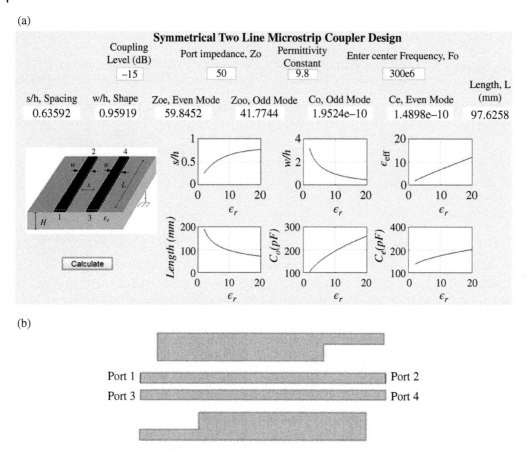

(b)

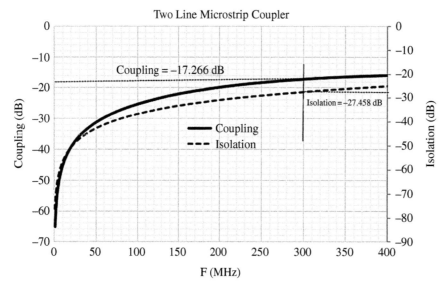

Figure 7.64 (a) Two-line microstrip coupler design for 15 dB coupling using alumina as a substrate. (b) Two-line microstrip coupler with two-dimensional (2D) view.

Figure 7.65 Simulation results of a two-line microstrip coupler at 300 MHz for coupling level = S_{13} = −17.266 dB and isolation level = S_{14} = −27.458 dB.

Table 7.5 Three-line microstrip coupler design parameters for TMM10.

Coupling level (dB)	F (MHz)	Material	ε_r	s/h	w/h
−15	300	TMM10	9.8	0.63592	0.95919
K12 (dB)	**K13(dB)**	**Zoe (Ω)**	**Zoo (Ω)**	**Zee (Ω)**	**L (mil)**
−15	−33.3976	50	43.6002	62.4608	3843.5433

Figure 7.66 Three-line microstrip coupler with 2D view.

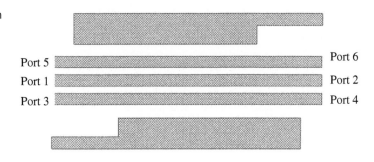

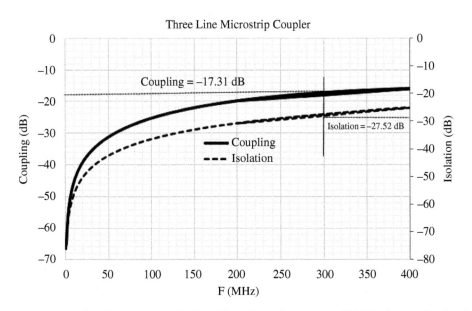

Figure 7.67 Simulation results of a three-line microstrip coupler at 300 MHz for coupling level = −17.31 dB and isolation level = −27.52 dB.

Figure 7.68 The prototype of a three-line directional coupler using TMM10 material.

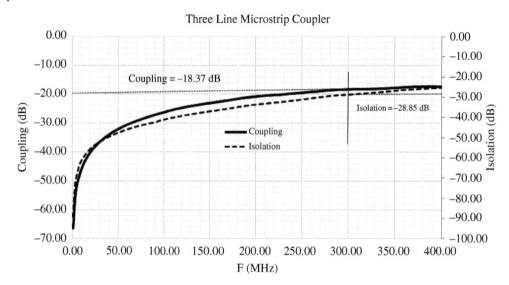

Figure 7.69 Measurement results for three-line coupler at 300 MHz for coupling level = −18.37 dB and isolation level = −28.85 dB.

Solution

A coupling of 20 dB can be obtained when the number of turns is equal to $N_1 = N_2 = 10$ using scattering parameters derived using the equations given in Section 7.2.1.2. MATLAB GUI has been developed to calculate several design critical parameters, including scattering parameters. The parameters obtained using scattering parameters are valid only when all the ports are matched, as is the case for the considered design problem. If ports are not matched, the design parameters obtained by circuit analysis in the MATLAB GUI are the values that should be used. The program performs the directional analysis coupler in the forward mode, and accepts excitation voltage at port 1, port impedances, and number of turns as inputs.

The program then calculates the coupling, directivity, isolation levels, and several other design critical parameters, such as program output. The program also produces design curves for coupling, directivity, and isolation levels versus number of turns. The screenshot of the program showing the calculated design values for the required couple design is given in Figure 7.70. Output, coupled and isolated voltages, power at the input and output, currents at the excitation and output port, coupling, isolation, directivity, and return loss are calculated and displayed. The transformer directional coupler is then simulated by Ansoft Designer in frequency domain using ideal transformers. It is good practice to compare the analytical values using scattering parameters with the simulated values with the application of scattering parameters. The circuit that is simulated is shown in Figure 7.71.

The simulation results for coupling, isolation, and directivity are given in Figures 7.72 and 7.73. The coupling level and isolation are found to be −20 dB and −66.02 dB at the frequency of interest, respectively. Directivity is found to be 46.02 dB, as shown in Figure 7.74. These values are in agreement with the values obtained by the program using the formulation in this section.

Since the calculated values and simulated values are in agreement, the transformer coupler can be built. However, the details about the transformer have to be determined. These include the type of material that will be used as a magnetic core, winding information, inductance information, etc. Hence, it is beneficial for the designer to use the calculated values of the transformer design in the time domain circuit simulator which will take into account coupling as it is the case in real world applications. The core material is chosen to be −7, which is Carbonyl TH with a permeability of 9 and has good performance for applications when the frequency of operation is between 3 and 35 MHz. The core dimensions are found to be OD = 1.75, ID = 0.94, and h = 0.48 cm. This core is T-68-7 with a white color code. Three cores are stacked and 16AWG wire is used for winding. Ten turns results in a transformer

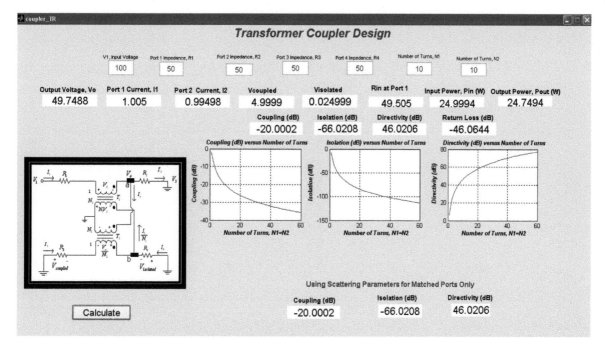

Figure 7.70 MATLAB GUI for transformer coupler design.

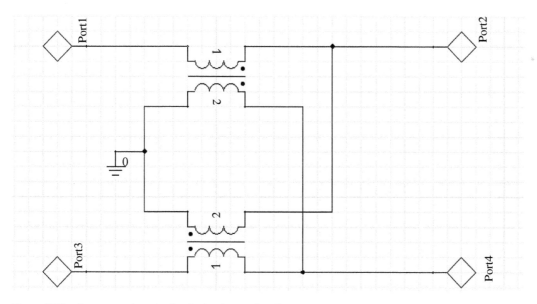

Figure 7.71 Frequency domain circuit simulator using S parameters.

inductance value of 1.61 μH, and this gives an impedance value that is more than five times higher than the higher impedance termination, which is 50 Ω. Hence, the transformer design should produce correct operating conditions at the frequency of operation. Since the required inductance value is determined, this value can now be used in a time domain circuit simulator to obtain all the parameters versus time and take into account actual approximate coupling. The circuit that is simulated by PSpice for the desired transformer coupler, as shown in Figure 7.75.

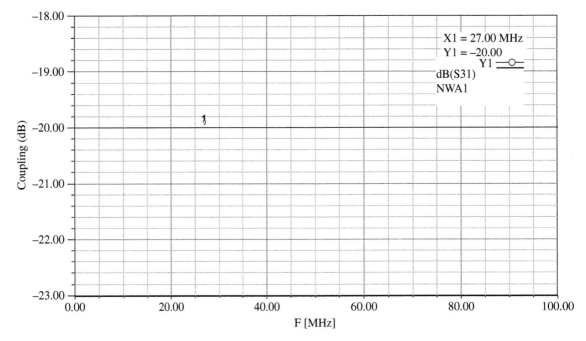

Figure 7.72 Simulated coupling level for a transformer coupler in frequency domain.

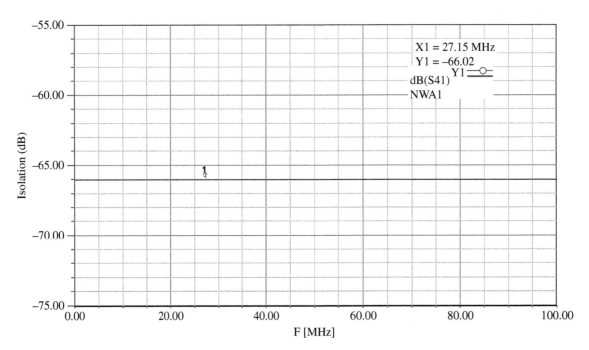

Figure 7.73 Simulated isolation level for a transformer coupler in frequency domain.

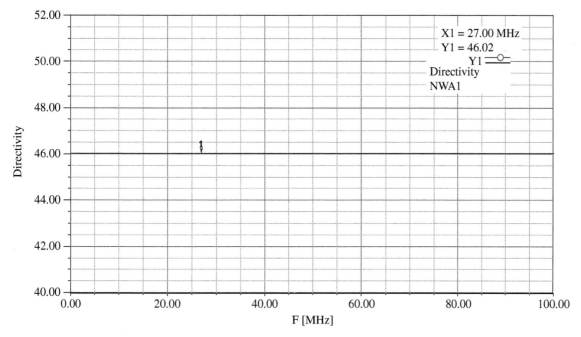

Figure 7.74 Simulated directivity level for a transformer coupler in frequency domain.

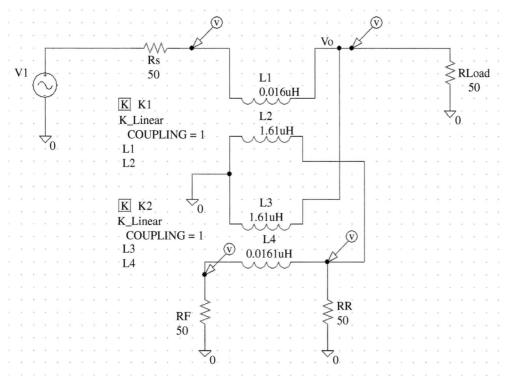

Figure 7.75 Time domain simulation for a transformer coupler.

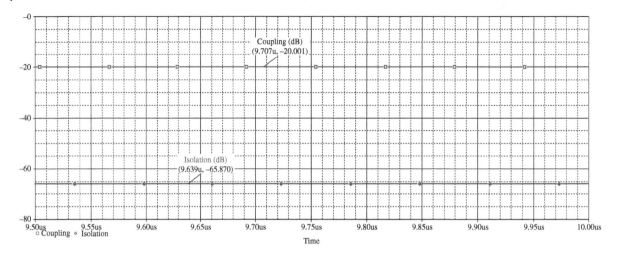

Figure 7.76 Simulated coupling and isolation levels for a transformer coupler in time domain.

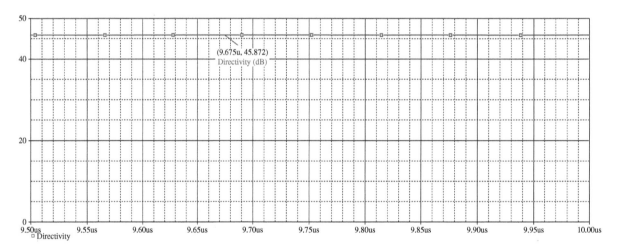

Figure 7.77 Simulated directivity level for a transformer coupler in time domain.

Simulation results showing the coupling, isolation, and directivity are given by Figures 7.76 and 7.77. Coupling, isolation, and directivity in PSpice are found using macros in the Probe section with the application of formulation given in Figure 7.78. The coupling level and isolation are found to be −20 dB and −65.87 dB, respectively. The directivity is found to be 45.87 dB based on the time domain simulation. Simulated values are in agreement with the frequency domain simulator and MATLAB GUI program. As a result, we can go ahead and build the coupler using the calculated values confirmed by the time and frequency domain simulators.

The constructed transformer coupler is shown in Figure 7.79. The single turn is accomplished by using semirigid coax cable with 50 Ω characteristic impedance, 0.25 in. outer conductor diameter, and 0.076 in. inner conductor diameter. PTFE is used as a dielectric material with 0.214 in. diameter. The length of the semirigid coax cable used in the construction of the coupler is 3.75 cm. The outline of a semirigid coax cable is illustrated in Figure 7.80. As it is planned, T-68-7 toroidal core is used with 10 turns and winding is done with a 16AWG enameled wire. Three-mil thick Teflon tape is used to insulate rigid cable from core and core windings. *N*-type connectors are used due to high

Figure 7.78 Macros used in PSpice for coupling, isolation, and directivity simulation.

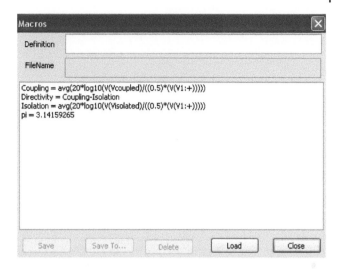

Figure 7.79 Constructed transformer coupler for 27.12 MHz operation.

Figure 7.80 Semirigid coax cable used in transformer coupler.

power application requirement. The measured frequency response of the coupler for coupling and isolation are given in Figures 7.81 and 7.82.

As seen, the measured coupling and isolation levels at 27.12 MHz are −20.2 dB and −55.147 dB. This gives a measured value of 34.947 dB directivity. The measured input impedance versus frequency is shown in Figure 7.83. It can be seen that the input impedance, which is calculated as R_{in}, is equal to $(49.691 + j3.2676)\Omega$. The real part is in agreement with the calculated

The reactance always exists because the used components are lossy. The measured inductive reactance can be eliminated by using compensation capacitor with ease. The measured input impedance corresponds to a VSWR of 1.07, which is an indication of a well-matched system. This can be further improved with the implementation of the compensation capacitor. The data for the measured values are extracted for coupling and isolation from the network analyzer, and Excel is used to see the results for comparison better, as illustrated in Figure 7.84.

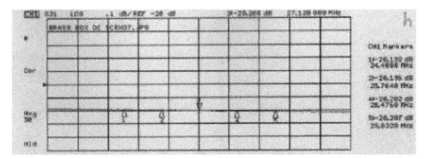

Figure 7.81 Measured coupling of a constructed transformer coupler.

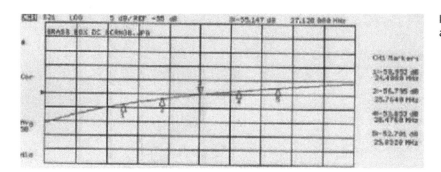

Figure 7.82 Measured isolation of a constructed transformer coupler.

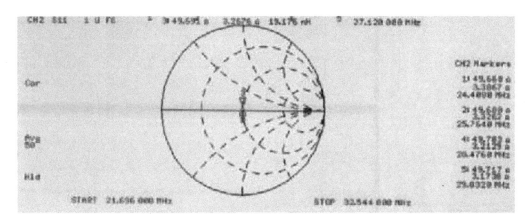

Figure 7.83 Measured input impedance of a constructed transformer coupler.

In addition, the directivity response of the coupler versus frequency is obtained in this way and shown in Figure 7.85. The measured values show that directivity and hence the isolation of the coupler get worse as frequency increases. The coupling level is maintained across the frequency bandwidth.

Overall, the designed system meets with requirements outlined in the problem statement. It can be seen that hybrid simulation of the coupler is crucial to meeting the design requirements in a cost-effective way.

Design Example 7.3 Combiner Design

Design, simulate, build, and measure a high power combiner using microstrip technology at 13.56 MHz to combine output of three PA modules with 25 Ω. The output of the combiner is to be 30 Ω.

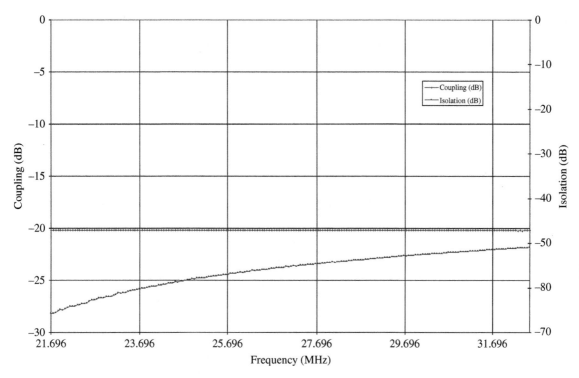

Figure 7.84 Measured isolation and coupling of a constructed transformer coupler.

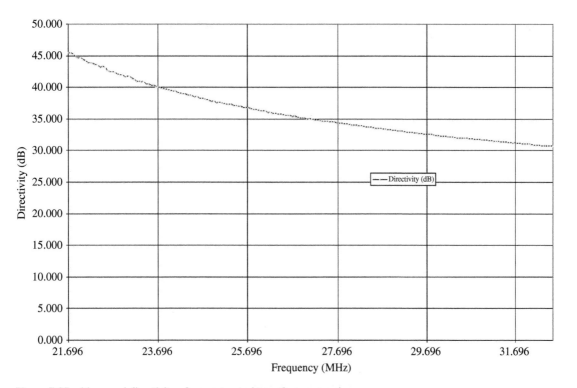

Figure 7.85 Measured directivity of a constructed transformer coupler.

Solution

The operational frequency is 13.56 MHz, which is a common frequency for ISM (industrial, scientific, and medical) applications, and the combiner should be capable of providing 12 000 W output power. The combiner is intended to combine the outputs of three PA modules. Each PA module presents 25 Ω impedance to the input of each distribution port on the combiner. The output of the combiner should be 30 Ω. The analytical formulation developed for dividers can be adapted to combiner analysis. The input parameters are entered into the MATLAB GUI developed for combiners. Insertion loss, isolation, VSWRs, and characteristic curves are obtained and shown in Figure 7.86. The program does not take into account any imperfections that might exist in the real system and hence theoretically gives perfect isolation when $\theta = 90°$. The design parameters for the three-way combiner are

$$R_0 = 25\Omega, R_L = 30\Omega, \text{ and } Z_0 = 47.43\Omega$$

Each PA module is required to provide 4000 W output power under a matched condition. The component values calculated for the π network in Figure 7.62 are

$$L_1 = 556.69 \text{ nH}, C_2 = C_3 = 247.46 \text{ pF}, Q = 1.898$$

The corresponding component values of the lumped elements for the L network given in Figure 7.63 are

$$L_3 = 472.2 \text{ nH}, C_4 = 210.5 \text{ pF}, \text{ and } Q = 1.612$$

In both circuits

$$R_1 = 90\Omega \text{ and } R_2 = 25\Omega$$

The lumped-element inductor in the L network is implemented as a spiral inductor on alumina substrate having a planar form. The form of the spiral inductor that will be used in our application is shown in Figure 7.87. It is a rectangular spiral inductor with rounded edges versus sharp edges. This type of implementation on the edges

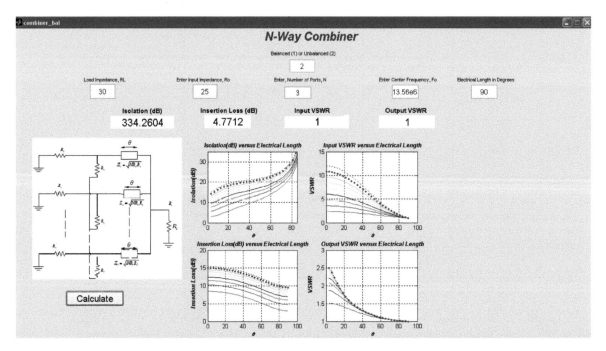

Figure 7.86 MATLAB GUI output for a three-way unbalanced combiner when $\theta = 90°$.

increases the effective arcing distance between traces. The physical dimensions for the spiral inductor are the width of the trace, w, the length of the outside edges, l_1 and l_2, and the spacing between the traces, s.

The simplified two-port lumped-element equivalent circuit for the spiral inductor shown in Figure 7.87 is illustrated in Figure 7.88. In Figure 7.88, L is the series inductance of the spiral and C is the substrate capacitance. This model ignores the losses in the substrate and the conductor. When the two-port equivalent model is inserted into Figure 7.63 to represent the lumped-element model for the spiral inductor, the overall network that represents each distribution circuit can be shown in Figure 7.89. In the proposed design method, first, the lumped-element values of the spiral inductor have to be obtained and then the physical dimensions of the spiral inductor, which give the desired series inductance, L, are calculated. The effective value of the inductance, L_{eff}, of the spiral inductor is measured as a one-port network, as shown in Figure 7.90. The lumped-element values for the spiral inductor to perform the required impedance transformation from $R_2 = 25\,\Omega$ to $R_1 = 90\,\Omega$ for the network in Figure 7.88 are calculated to be

$$L = 497\,\text{nH and } C = 43.6\,\text{pF}$$

Figure 7.90b depicts the impedance plot of the spiral inductor on a Smith chart. The impedance at point 1, P_1, on the Smith chart is purely inductive and equal to $Z = j50.2\,\Omega$. So, the effective value of the inductance for the spiral at the operational frequency using the one-port measurement network in Figure 7.90a is found using $Z = j\omega L_{eff}$ as

$$L_{eff} = 589.2\,\text{nH}$$

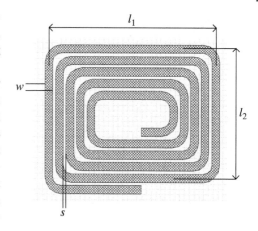

Figure 7.87 Spiral inductor layout.

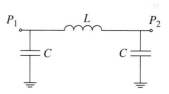

Figure 7.88 Simplified equivalent circuit for spiral inductor without loss factor.

The shunt capacitance C_4 in Figure 7.89 is referred to as C_4' in the existence of the spiral inductor to distinguish it from the value for the model in Figure 7.63b. When the spiral inductor model is used, as shown in Figure 7.63, C_4' is calculated accordingly as

$$C_4' = (210.5 - 43.6) = 166.9\,\text{pF}$$

The values of the lumped-element components for the distribution circuit in Figure 7.89 using the spiral inductor model are illustrated in Figure 7.91. This circuit represents the final form of the lumped-element L network distribution circuit that is used as a reference for electromagnetic simulation.

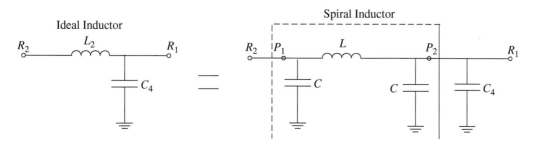

Figure 7.89 Spiral inductor model is inserted into an L network.

(a)

(b)

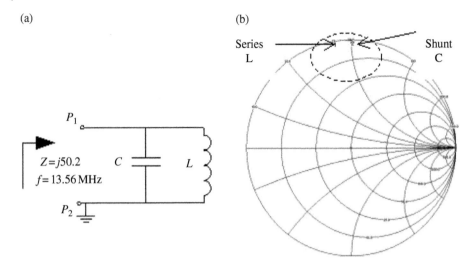

Series L

Shunt C

P_1

$Z=j50.2$

C

L

$f=13.56\,\text{MHz}$

P_2

Figure 7.90 (a) One-port measurement network and (b) its impedance plot for spiral inductor.

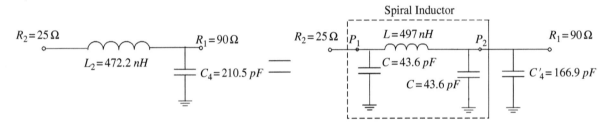

Spiral Inductor

$R_2=25\,\Omega$

$L_2=472.2\,nH$

$R_1=90\,\Omega$

$C_4=210.5\,pF$

$R_2=25\,\Omega$ P_1

$L=497\,nH$

P_2

$R_1=90\,\Omega$

$C=43.6\,pF$

$C=43.6\,pF$

$C'_4=166.9\,pF$

Figure 7.91 Final form of the lumped-element distribution circuit with spiral inductor.

The simulation results shown in Figures 7.92 and 7.93 compare the responses of π-network and L-network using ideal lumped elements with circuit simulator. In this figure, insertion loss of the combiner and the isolation between each distribution port with lumped elements are illustrated using Ansoft Designer. The lumped-element values obtained are used in the simulation. Based on the results, the insertion loss is not affected by the transformation of networks from π to L. Although the insertion loss for each network is found to be the same using both topologies, the isolation for each distribution port is much better using the π network at the operational frequency. So, for applications where good isolation between the distribution ports is required, a π network should be employed.

A three-way planar high power combiner is simulated with the method of moment field solver, Ansoft Designer using the initial lumped-element values. The spiral inductors are implemented in planar form on a 100-mil-thick Al_2O_3 substrate. The planar three-way combiner has three spiral inductors, lumped capacitors, and planar circuitry to combine the structure. The spiral inductor has a bridge-type connection from the excitation point to carry the input signal to the spiral. The simulated structure is shown in Figure 7.94. The simulation results for the whole combiner are shown in Figure 7.95. The current and near-field distribution for this structure at the center operational frequency, 13.56 MHz, are given in Figure 7.96. The simulation has been performed by importing the electromagnetic model into the circuit simulator and co-simulating both designs, as shown in Figure 7.97. The use of an electromagnetic simulator enables the designer to observe the electromagnetic coupling and high current distribution regions that need to be paid attention to during the implementation stage of the physical design.

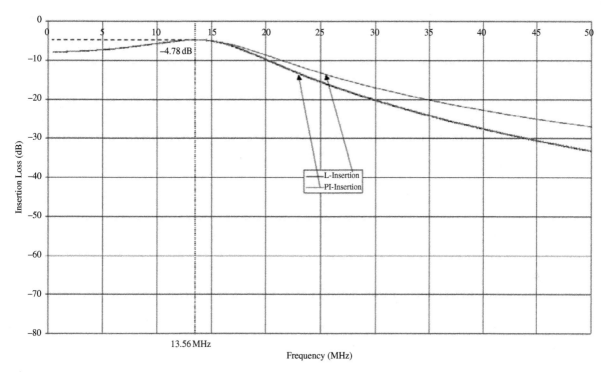

Figure 7.92   Simulation results for the insertion loss between each distribution port using π and L networks.

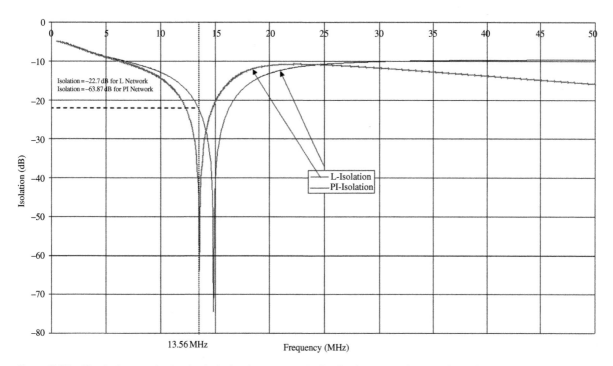

Figure 7.93   Simulation results for the isolation between each distribution port using transformation networks.

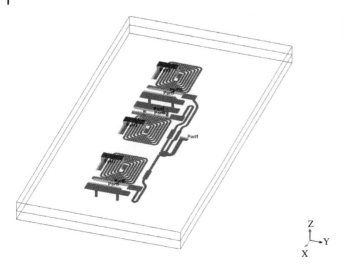

Figure 7.94 Simulated three-way combiner in planar form using L network topology.

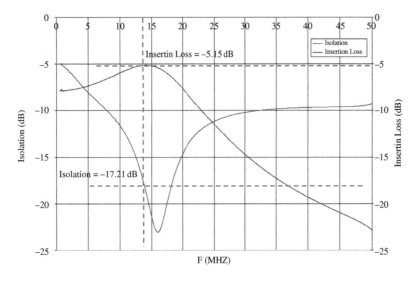

Figure 7.95 Simulation results for a three-way combiner in planar form using L network topology.

Based on the simulation results, the insertion loss and the isolation between each port are found to be −5.15 dB and −17.21 dB, respectively, at $f = 13.56$ MHz. When the impedance at the output port is measured for the three-way combiner using an electromagnetic (EM) simulator, it is found to be $(29.49 − j0.06)\Omega$ at the operational frequency. There is an additional planar circuitry to combine the signals at the output, as shown in Figure 7.94. Each leg of this circuit is connected to the output of the $L − C$ network. The impedance at the interface of each leg after the transformation by the $L − C$ network is $Z_c = 90\Omega$. The phase difference between each leg and the insertion loss introduced by it are adjusted to be minimum. This is required for a properly balanced combiner. Figure 7.98 shows the simulated planar circuitry to combine the output of each $L − C$ network. The simulation results for the planar circuit depicted in Figure 7.98 show that the insertion loss caused by each of the legs is −0.033 dB at 13.56 MHz. The phase difference between each of the legs at the operational frequency is given in Table 7.6.

Based on the electromagnetic simulation results, the inductance of the spiral inductor and the shunt capacitance on the L network are found to be

Figure 7.96 Current and near-field distribution for a three-way combiner.

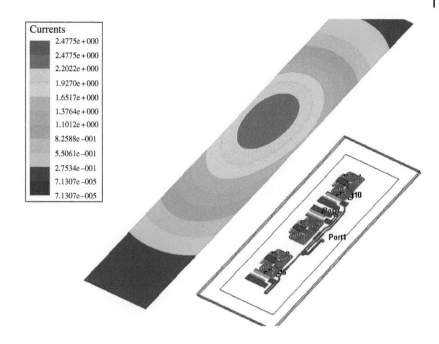

Figure 7.97 Co-simulation of a three-way combiner.

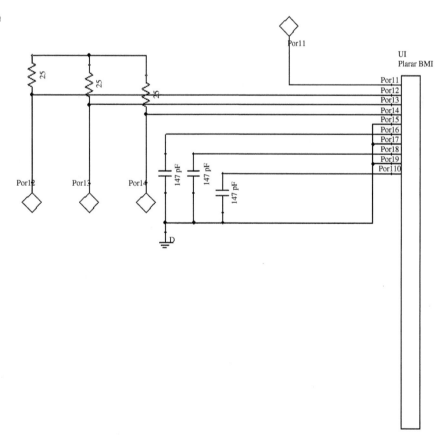

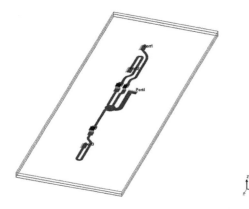

Figure 7.98 Simulated planar combining circuitry for a three-way combiner.

Table 7.6 Phase information.

Ports	Phase difference (deg)
1–2	0.13
1–3	−0.11
2–3	−0.24

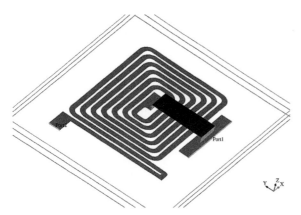

Figure 7.99 Spiral inductor that is simulated with method-of-moment-based electromagnetic solver has an inductance of $L = 588.6nH$.

$$L = 588.6\,\text{nH}, C'_4 = 147\,\text{pF}, \text{at}\, f = 13.56\,\text{MHz}$$

The spiral inductor that is simulated in the three-way combiner is shown in Figure 7.99. The simulated results for inductance value versus frequency of spiral inductor given in Figure 7.100 show that the resonance occurs at 37.55 MHz. The quality factor and the insertion loss of the spiral inductor are found to be 64.1 and −0.325 dB at the operational frequency, respectively. The simulated quality factor of the spiral inductor versus frequency is given in Figure 7.101. The dimensions of the spiral inductor and the substrate properties are given in Table 7.7. All dimensions are in thousandths of an inch (mil).

The simulated VSWRs at the combined output versus frequency are shown in Figure 7.102. The given impedances on the figures are normalized with respect to reference impedance. Based on the results

$$\text{Simulated combiner VSWR} = 1.695$$

The final constructed combiner is shown in Figure 7.103. The measurements are done using an HP- 8504A network analyzer. The measured isolation loss and insertion loss for the combiner are illustrated in Figures 7.104 and 7.105 and compared with the simulated results. The isolation loss and insertion loss at the operational frequency are measured to be −17.47 dB and −5.24 dB, respectively. The impedance at the output port of the combiner is measured when each distribution port is terminated with 25 Ω impedance. This reflects the real operating condition when each PA module is connected to the combiner. The impedance at the output port under this condition is measured to be $(30.23 - j0.06)\Omega$ at the operational frequency. This corresponds to

$$\text{Measured combiner VSWR} = 1.654$$

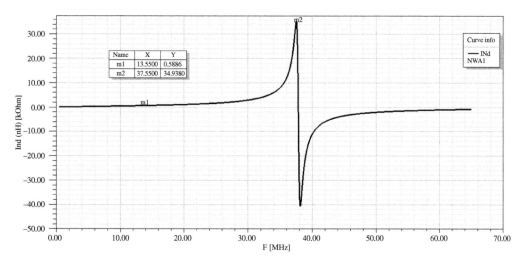

Figure 7.100 Simulated spiral inductor inductance versus frequency.

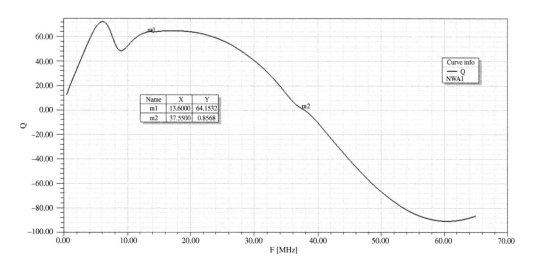

Figure 7.101 Simulated spiral inductor quality factor versus frequency.

Table 7.7 Physical dimensions of spiral inductor.

Trace width (w)	Spacing (S)	Horizontal trace length (l_1)	Vertical trace length (l_2)	Copper thickness (t)	
80	30	1870	1450	4.2	
Dielectric material	**Dielectric permittivity (ε_r)**	**Dielectric thickness (h)**	**Number of turns (n)**	**Bridge height (h_b)**	**Bridge width (w_b)**
Al_2O_3	9.8	100	6.375	100	350

Name	F	Ang	Mag	RX
m1	0.0136	179.7847	0.2580	0.5898 + 0.0012i

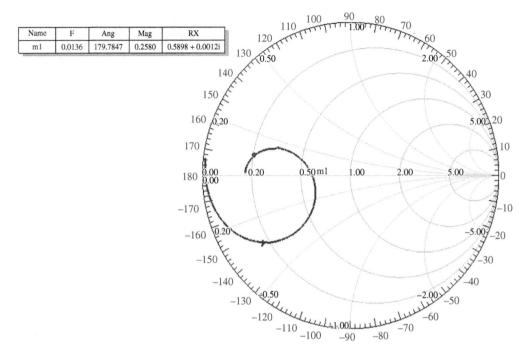

Figure 7.102 Input VSWR versus frequency for three-way microstrip combiner.

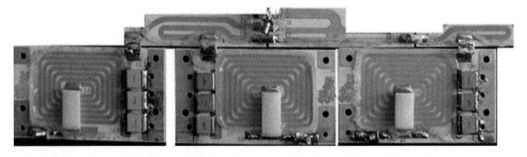

Figure 7.103 Three-way combiner implemented in planar form using *L* network topology.

The measured combiner port impedance is given in Figure 7.106. This value is very close to the targeted impedance value of 30 Ω at the output port.

At the operational frequency, the measured inductance value of the spiral inductor shown in Figure 7.107 is 594.91 nH. The measured inductance value of the spiral inductor versus frequency is given in Figure 7.108. The quality factor and the insertion loss for the spiral inductor are measured to be 58.6 and −0.295 dB, respectively. The measured insertion loss characteristics versus frequency is given in Figure 7.109. Table 7.8 tabulates the simulated and the measurement results for the three-way combiner.

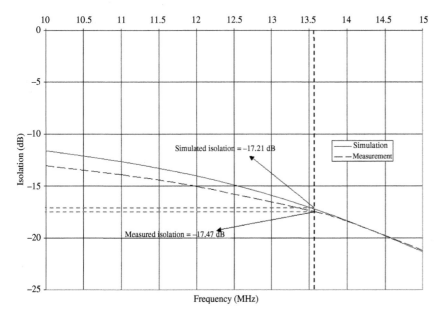

Figure 7.104 Measurement results for insertion loss of a three-way combiner in planar form using *L* network topology.

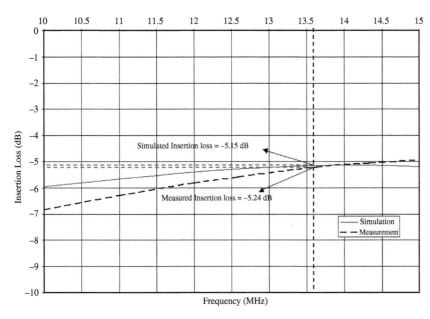

Figure 7.105 Measurement results for insertion loss of three-way combiner in planar form using *L* network topology.

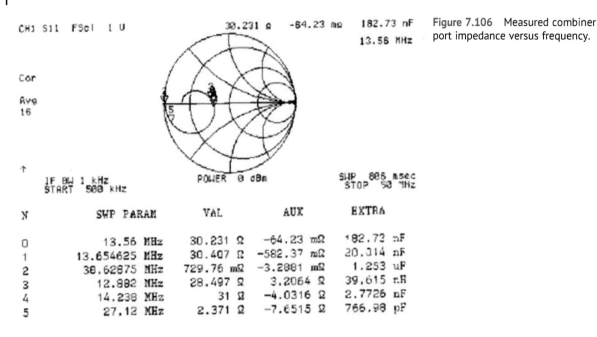

Figure 7.106 Measured combiner port impedance versus frequency.

N	SWP PARAM	VAL	AUX	EXTRA
0	13.56 MHz	30.231 Ω	−64.23 mΩ	182.72 nF
1	13.654625 MHz	30.407 Ω	−582.37 mΩ	20.314 nF
2	38.62875 MHz	729.76 mΩ	−3.2881 mΩ	1.253 uF
3	12.882 MHz	28.497 Ω	3.2064 Ω	39.615 nH
4	14.238 MHz	31 Ω	−4.0316 Ω	2.7726 nF
5	27.12 MHz	2.371 Ω	−7.6515 Ω	766.98 pF

Alumina Substrate   Figure 7.107 Spiral inductor on alumina substrate.

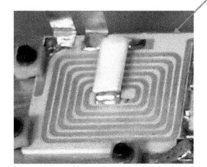

It is important to note the analytical model and GUI give similar results with the simulated and measured results when the electrical length is changed from $\theta = 90°$ to $\theta = 65°$, as shown in Figure 7.110. When the adjusted electrical length is inserted in the analytical design, the insertion loss, isolation, and VSWR become in agreement. This deterioration in the response is due to imperfections, loss, and the use of an *L*-type network instead of a *PI*-type network in the implementation. However, use of an *L* network simplifies the implementation significantly and reduces the parts counts and associated cost. As a result, the use of *L* or *PI* network topology is based on design requirements.

Figure 7.108 Measured inductance value of spiral inductor versus frequency.

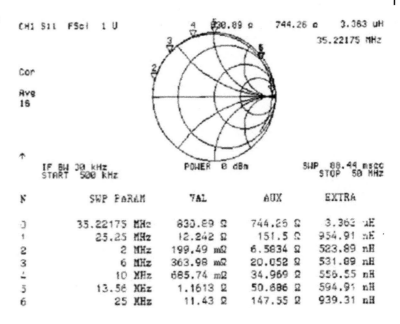

N	SWP PARAM	VAL	AUX	EXTRA
0	35.22175 MHz	830.89 Ω	744.26 Ω	3.363 uH
1	25.25 MHz	12.242 Ω	151.5 Ω	954.91 nH
2	2 MHz	199.49 mΩ	6.5834 Ω	523.89 nH
3	6 MHz	363.98 mΩ	20.052 Ω	531.89 nH
4	10 XHz	685.74 mΩ	34.969 Ω	556.55 nH
5	13.56 XHz	1.1613 Ω	50.686 Ω	594.91 nH
6	25 XHz	11.43 Ω	147.55 Ω	939.31 nH

Figure 7.109 Measured insertion loss of spiral inductor versus frequency.

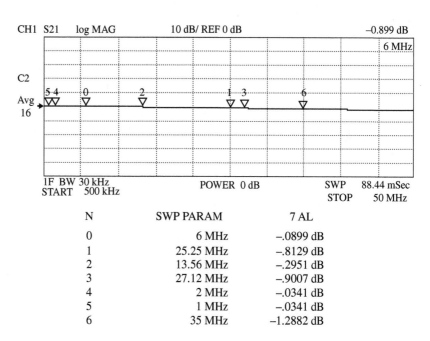

N	SWP PARAM	7 AL
0	6 MHz	−.0899 dB
1	25.25 MHz	−.8129 dB
2	13.56 MHz	−.2951 dB
3	27.12 MHz	−.9007 dB
4	2 MHz	−.0341 dB
5	1 MHz	−.0341 dB
6	35 MHz	−1.2882 dB

Table 7.8 Simulation and measurement results.

	Measured	Simulated	Measured	Simulated	Measured	Simulated
	Inductance (nH)	**Inductance (nH)**	**Insertion Loss (dB)**	**Insertion Loss (dB)**	**Quality Factor**	**Quality Factor**
Spiral inductor	594.91	588.6	−0.295	−0.325	58.6	64
	Capacitance (pF)	**Capacitance (pF)**				
Capacitor	133	147				
	Impedance (Ω)	**Impedance (Ω)**				
Output impedance of the combiner	$30.26 - j0.06$	$29.55 - j0.55$				
	Insertion Loss (dB)	**Insertion Loss (dB)**	**Isolation (dB)**	**Isolation (dB)**		
Combiner	−5.24	−5.25	−17.47	−17.21		

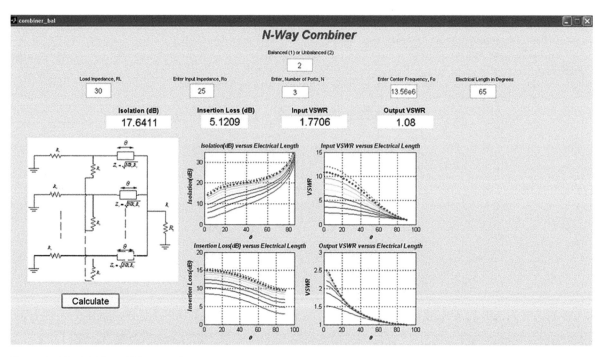

Figure 7.110 MATLAB GUI output for a three-way unbalanced combiner when $\theta = 65°$.

References

1 Eroglu, A. (2013). *RF Circuit Design Techniques for MF-UHF Applications*. CRC Press.

2 Eroglu, A. and Lee, J.K. (2008). The complete design of microstrip directional couplers using the synthesis technique. *IEEE Transactions on Instrumentation and Measurement* **57** (12): 2756–2761.

3 Monteath, G.D. (1955). Coupled transmission lines as symmetrical directional couplers. *Proceedings of the IEE* **102** (part B): 383–392.

4 Eroglu, A. and Ronnow, D. (2018). Design of multilayer and multiline microstrip directional coupler with closed form relations. *Progress In Electromagnetics Research C* **83**: 1–14.

5 Huang, J., Hao, Q, She, J. and Feng, Z (2005). A six-port coupler with high directivity for VSWR measurement, *2005 Asia-Pacific Microwave Conference Proceedings*, https://doi.org/10.1109/APMC.2005.1606811.

6 Hoer, C.A. (1972). The six-port coupler: a new approach to measuring voltage, current, power, impedance, and phase. *IEEE Transactions on Instrumentation and Measurement* **21** (4): 466–470.

7 Cullen, A.L., Judah, S.K., and Nikravesh, F. (1980). Impedance measurement using a 6-port directional coupler. *Proceedings of the IEE* **127** (2): 92–98.

8 Cooper, T.S., Baldwin, G., and Farrell, R. (2006). Six-port precision directional coupler. *Electronics Letters* **42** (21): 1232–1233.

9 Engen, G.F. and Hoer, C.A. (1972). Application of an arbitrary 6-port junction to power measurement problems. *IEEE Transactions on Instrumentation and Measurement* **21** (4): 470–474.

10 Engen, G.F. (1977). The six-port reflectometer; an alternative network analyser. *IEEE Transactions on Microwave Theory and Techniques* **MTT-25**: 1075–1080.

11 Ji, J.Y. and Yeo, S.P. (2008). Six-port reflectometer based on modified hybrid couplers. *IEEE Transactions on Microwave Theory and Techniques* **56** (2): 493–498.

12 Yao, J.J., Yeo, S.P., and Bialkowski, M.E. (2009). Modifying branch-line coupler design to enhance six-port reflectometer performance. In: *IEEE MTT S International Microwave Symposium Digest*, 1669–1672. IEEE.

13 Griffin, E.J. (1982). Six-port reflectometer circuit comprising three directional couplers. *Electronics Letters* **18** (12): 491–493.

14 Stumper, U. (1990). *Simple millimeter-wave six-port reflectometers, Conference on Precision Electromagnetic Measurements*, 51–52. IEEE https://doi.org/10.1109/CPEM.1990.109918.

15 Akyel, C. and Ghannouchi, F.M. (1994). A new design for high-power six-port reflectometers using hybrid stripline/waveguide technology. *IEEE Transactions on Instrumentation and Measurement* **43** (2): 316–321.

16 Collier, R.J. and El-Deeb, N.A. (1979). On the use of a microstrip three-line system as a six-port reflectometer. *IEEE Transactions on Microwave Theory and Techniques* **MTT-27**: 847–853.

17 Collier, R.J. and El-Deeb, N.A. (1980). Microstrip coupler suitable for use as a 6-port reflectometer. *Proceedings of the IEE* **127**, Pt. H (2) https://doi.org/10.1049/ip-h-1:19800021.

18 El-Deeb, N.A. (1983). The calibration and performance of a microstrip six-port Rdlectometer. *IEEE Transactions on Microwave Theory and Techniques* **MTT- 31** (7): 509–514.

19 Mohra, A.S.S. (2001). Six-port reflectometer realization using two microstrip three-section couplers, *18th National Radio Science Conference*, Egypt (March 2001).

20 Hansson, E.R.B. and Riblet, G.P. (1983). An ideal six-port network consisting of a matched reciprocal lossless five-port and a perfect directional coupler. *IEEE Transactions on Microwave Theory and Techniques* **31**: 284–288.

21 Dobrowolski, J.A. (1982). Improved six-port circuit for complex reflection coefficient measurements. *Electronics Letters* **18** (17): 748–750.

22 Alessandri, F. (2003). A new multiple-tuned six-port Riblet-type directional coupler in rectangular waveguide. *IEEE Transactions on Microwave Theory and Techniques* **51** (5): 1441–1448.

23 Ghannouchi, F.M., Bosisio, R.G., and Demers, Y. (1989). Load-pull characterization method using six-port techniques. In: *6th IEEE Instrumentation and Measurement Technology Conference*, 536–539. IEEE.

24 Lê, D., Poiré, P., and Ghannouchi, F.M. (1998). Six-port-based active source-pull measurement technique. *Measurement Science and Technology* **9** (8): 1336–1342.

25 Carr, J. (1999). Directional couplers. *Proceedings of RF Design* **August**: 676–677.

26 Kajfez (1999). Scattering matrix of a directional coupler with ideal transformers. *IEE Microwaves, Antennas and Propagation* **146** (4): 295–297.

27 Sui, Q., Wang, K., and Li, L. (2010). The measurement of complex reflection coefficient by means of an arbitrary four-port network and a variable attenuator. In: *2010 International Symposium on Signals Systems and Electronics (ISSSE)*, vol. **2**, 1–3.

28 Wilkinson, E.J. (1960). An *N*-way hybrid power divider. *IEEE Transactions on Microwave Theory and Techniques* **8** (1): 116–118.

29 Smith, J. (1999). *The Big One Thousand Oaks LA*. Sage Publishing.

Problems

Problem 7.1

Calculate even- and odd-mode impedances of a 10 dB directional coupler using coupled microstrip lines when the substrate permittivity constant is 4.4 and thickness is 100 mil. Assume the operational frequency is 1 GHz and the system impedance is 50 Ω.

Problem 7.2

For the 10 dB directional microstrip coupler which has the substrate permittivity constant is 4.4 and thickness is 100 mil and operational frequency of 1 GHz, calculate the shape, spacing ratios, and length.

Problem 7.3

For the 10 dB directional microstrip coupler which has the substrate permittivity constant is 4.4 and thickness is 100 mil and operational frequency of 1 GHz, calculate the even- and odd-mode capacitances of the coupler at the operating frequency and obtain the characteristics of the coupler versus dielectric constant of the substrate.

Problem 7.4

Derive the coupling level, isolation level, and directivity given by Eqs. (7.66)–(7.68) for the four-port transformer coupled directional coupler shown in Figure 7.13 using nodal analysis. The equivalent circuit is given as a reference in Figure 7.111.

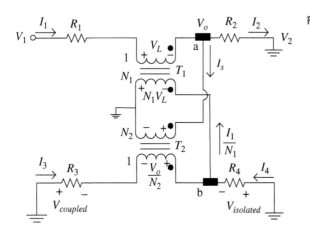

Figure 7.111 Problem 7.4.

Problem 7.5

Design a transformer coupled directional coupler with 20 dB coupling and directivity levels better than 30 dB for high power RF application at 27.12 MHz when input voltage is $V_{in,\ peak} = 100$ V. The port impedances are matched and given to be $R = 50\ \Omega$.

Problem 7.6

Design a four-way combiner at 1 GHz by assuming (a) $R_g = R_o = 75\ \Omega$; (b) $R_g = 100\ \Omega$, $R_o = 50\ \Omega$ and find (i) isolation, (ii) insertion loss, (iii) input VSWR, and (iv) output VSWR, when $\theta = 90°$ and $\theta = 45°$; (c) compare your results with simulation results.

Problem 7.7

In the RF system shown in Figure 7.112, the RF signal source can provide power output from 0 to 30 dBm. The RF signal is fed through 1 dB T-pad attenuator and 20 dB directional coupler where the sample of the RF signal is further attenuated by a 3 dB π-pad attenuator before power meter reading in dB. The "through" port of the directional coupler has 0.1 dB of loss before it is sent to the PA. The RF PA output is then connected to a 6 dB π-pad attenuator. If the power meter is reading 10 dBm, what is the power delivered to the load shown in Figure 7.112 in mW?

Figure 7.112 Problem 7.7.

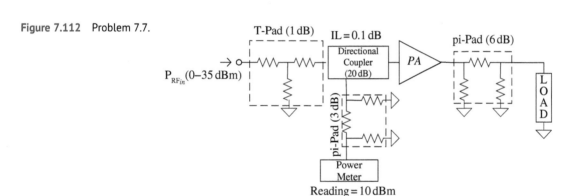

Problem 7.8

A directional coupler is characterized by the scattering matrix given below. Based on the characteristics given by the scattering matrix, calculate the coupling level, directivity, isolation, insertion loss, and return loss for this directional coupler when all the ports are matched.

$$S = \begin{bmatrix} 0.005 & j0.15 & 0.2 & 0.001 \\ j0.15 & 0.005 & 0.02 & 0.01 \\ 0.02 & 0.02 & 0.005 & -j0.02 \\ 0.001 & 0.01 & -j0.02 & 0.005 \end{bmatrix}$$

Problem 7.9

Consider the directional coupler illustrated in Figure 7.113. It has a coupling level of 10 dB, directivity of 40 dB, and an insertion loss of 0.1 dB. What are the power levels at P_2, P_3, and P_4, if the input power at $P_1 = 15$ dBm? Please note that P_2 is through port P_3 is a coupled port and P_4 is the isolation port.

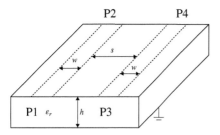

Figure 7.113 Problem 7.9.

Problem 7.10
Design a broadband four-way divider to be used as a test fixture to measure the performance of a four-way high power planar combiner that is used to combine the outputs of power amplifiers. The combiner has 25 Ω input port impedances and operates at 13.56 MHz.

Design Challenge 7.1 Microstrip Coupler Design

Design, simulate, and build a 15 dB microstrip symmetrical coupler using Teflon, ε_r = 2.08, and FR4, ε_r = 4.4, at 300 MHz. The thicknesses of substrates are 90 and 120 mil for Teflon and FR4, respectively. Compare, measure, simulate, and analyze results.

Design Challenge 7.2 Multilayer Microstrip Coupler Design

Design a 10 dB symmetrical microstrip two-line directional coupler at 300 MHz using the analytical method and simulate it with the values obtained for coupling and directivity levels. Use Teflon material with ε_r = 2.08 with a 120 mil thickness as a substrate. Then, modify your design to a two-layer configuration and simulate again. Compare directivity over frequency for both cases.

Design Challenge 7.3 Power Divider for Array Antennas

Use a Sonnet or Ansoft Designer to design a power divider in microstrip configuration starting from a 50 Ω source and feeding four antennas, each with an input impedance of 100 Ω, as shown in Figure 7.114. The inputs to each antenna are equal in magnitude and in phase. Use Duroid RT5880 with dielectric constant, ε_r = 2.2, substrate thickness, and h = 1/16″.

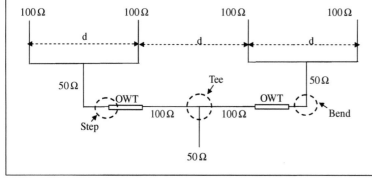

Figure 7.114 Power divider antenna feeder system.

8

Filters

8.1 Introduction

Filters are one of the indispensable components in radio frequency applications. An ideal filter provides perfect transmission of signal for frequencies in the required passband region, and infinite attenuation for frequencies in the stopband region. The critical RF filter parameters include high return loss, minimum attenuation distortion, a flat group delay in the passband, and high attenuation in the stopband. Basic filters that are used in RF applications are categorized as low pass, high pass, bandpass, and bandstop filters (BSFs) with their ideal filter characteristics, as shown in Figure 8.1.

8.2 Filter Design Procedure

Filters can be analyzed as lossless linear two-port networks using network parameters. The conventional filter design procedure for low pass, high pass, bandpass, or BSFs begins with a low pass filter prototype and then involves impedance and frequency scaling, and filter transformation to high pass, bandpass, or BSFs to obtain final component values at the frequency of operation. The design is then simulated and compared with specifications. The final step in the design of filters involves implementation and measurement of the filter response. This design procedure is shown as a block diagram in Figure 8.2. The low pass filter prototype is the basic building block in the filter design. The attenuation profile of the low pass filter is then a critical parameter in the design procedure. The attenuation profiles of the low pass filter can be binomial (Butterworth), Chebyshev, or elliptic (Cauer), as shown in Figure 8.3. Binomial filters provide a monotonic attenuation profile and need more components to achieve a steep attenuation transition from pass to stop band, whereas Chebyshev filters have a steeper slope and equal amplitude ripples in the pass band.

Elliptic filters have a steeper transition from pass to stop band similar than Chebyshev filters and exhibit equal amplitude ripples in the pass band and stop band. RF/microwave filters and filter components can be represented using a two-port network, as shown in Figure 8.4. The network analysis can be conducted using $ABCD$ parameters for each filter. The filter elements can be considered cascaded components and hence the overall $ABCD$ parameter of the network is just a simple matrix multiplication of $ABCD$ parameters for each element. The characteristics of the filter in practice is determined via insertion loss, S_{21}, and return loss, S_{11}. $ABCD$ parameters can be converted to scattering parameters, and the insertion loss and the return loss for the filter can be determined. The insertion loss of filter networks shown in Figure 8.5a,b can be analyzed using $ABCD$ parameters and the cutoff frequency for each network determined using transfer functions. Low pass filter (LPF) and high pass filter (HPF) circuits for transfer function derivation to obtain cutoff frequency are given by Figures 8.5a,b, respectively.

RF/Microwave Engineering and Applications in Energy Systems, First Edition. Abdullah Eroglu.
© 2022 John Wiley & Sons Ltd. Published 2022 by John Wiley & Sons Ltd.
Companion website: www.wiley.com/go/eroglu/rfmicrowave

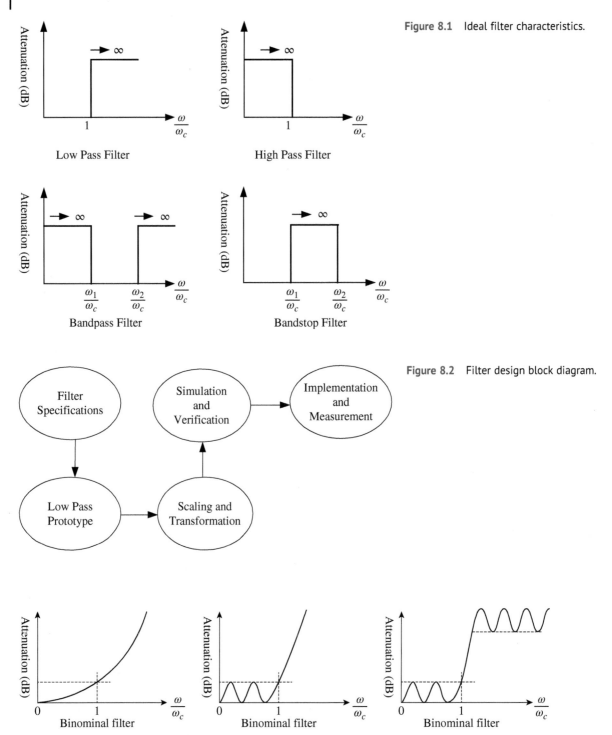

Figure 8.1   Ideal filter characteristics.

Figure 8.2   Filter design block diagram.

Figure 8.3   Attenuation profiles of a low pass filter.

Figure 8.4 Two-port network representation.

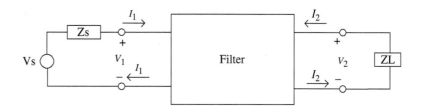

Figure 8.5 Transfer function analysis circuits: (a) low pass filter; (b) high pass filter.

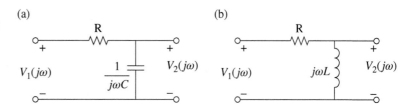

The cutoff frequency of the circuit for the LPF in Figure 8.5a is determined using the transfer function given by

$$H(j\omega) = \frac{V_2}{V_1} = \frac{1/RC}{j\omega + 1/RC} \qquad (8.1)$$

Cutoff frequency occurs when

$$|H(j\omega_c)| = \frac{1}{\sqrt{2}} H_{max} \qquad (8.2)$$

From Eq. (8.1)

$$|H(j\omega)| = \frac{\dfrac{1}{RC}}{\sqrt{\omega^2 + \left(\dfrac{1}{RC}\right)^2}} \qquad (8.3)$$

The maximum value of the transfer function is found as

$$H_{max} = |H(j0)| = 1 \qquad (8.4)$$

Using (8.3) and (8.4), we have

$$|H(j\omega_c)| = \frac{1}{\sqrt{2}}(1) = \frac{\dfrac{1}{RC}}{\sqrt{\omega_c^2 + \left(\dfrac{1}{RC}\right)^2}} \qquad (8.5)$$

Equation (8.5) is satisfied when

$$\omega_c = \frac{1}{RC} \qquad (8.6)$$

The cut-off frequency is found from (8.6) as

$$f_c = \frac{1}{2\pi RC} \qquad (8.7)$$

The same analysis can be conducted for the HPF circuit shown in Figure 8.5b. The transfer function of the circuit is obtained as

$$H(j\omega) = \frac{j\omega}{j\omega + \frac{R}{L}} \tag{8.8}$$

From Eq. (8.8)

$$|H(j\omega)| = \frac{\omega}{\sqrt{\omega^2 + \left(\frac{R}{L}\right)^2}} \tag{8.9}$$

and

$$H_{\max} = |H(j\infty)| = 1 \tag{8.10}$$

Then, the cutoff frequency for the HPF is found from

$$\frac{1}{\sqrt{2}} = |H(j\omega_c)| = \frac{\omega_c}{\sqrt{\omega_c^2 + \left(\frac{R}{L}\right)^2}} \tag{8.11}$$

which gives the cutoff frequency as

$$\omega_c = \frac{R}{L} \text{ or } f_c = \frac{1}{2\pi}\left(\frac{R}{L}\right) \tag{8.12}$$

The resonant frequency and cutoff frequencies of the bandpass filter shown in Figure 8.6a can be analyzed using the transfer function

$$H(j\omega) = \frac{\frac{R_L}{L}j\omega}{(j\omega)^2 + \left(\frac{R_L + R}{L}\right)j\omega + \frac{1}{LC}} \tag{8.13}$$

The magnitude of the transfer function is

$$|H(j\omega)| = \frac{\frac{R_L}{L}\omega}{\sqrt{\left(\frac{1}{LC} - \omega^2\right)^2 + \left(\omega\frac{R_L + R}{L}\right)^2}} \tag{8.14}$$

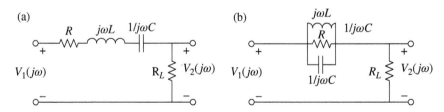

Figure 8.6 Transfer function analysis circuits: (a) bandpass filter; (b) bandstop filter.

The maximum value of the transfer function occurs at the resonant frequency f_o and is equal to

$$H_{max} = |H(j\omega_o)| = \frac{R}{R + R_i} \tag{8.15}$$

where

$$\omega_o = \frac{1}{\sqrt{LC}} \quad \text{or} \quad f_o = \frac{1}{2\pi\sqrt{LC}} \tag{8.16}$$

The cutoff frequencies for the bandpass filter are found as

$$\omega_{c1} = -\frac{R_L + R}{2L} + \sqrt{\left(\frac{R_L + R}{2L}\right)^2 + \frac{1}{LC}} \tag{8.17a}$$

$$\omega_{c2} = \frac{R_L + R}{2L} + \sqrt{\left(\frac{R_L + R}{2L}\right)^2 + \frac{1}{LC}} \tag{8.17b}$$

and it can be shown that

$$\omega_o = \sqrt{\omega_{c1} \cdot \omega_{c2}} \tag{8.18}$$

BSF can be analyzed similarly by replacing a series RLC circuit with a parallel RLC circuit, as shown in Figure 8.6b.

LPFs and HPFs have four sections and are cascaded, as shown in Figure 8.7. The insertion loss for the LPF in Figure 8.7a can be calculated using $ABCD$ parameters as

$$\begin{bmatrix} A & B \\ C & D \end{bmatrix} = \begin{bmatrix} 1 & Z_S \\ 0 & 1 \end{bmatrix} \begin{bmatrix} 1 & R \\ 0 & 1 \end{bmatrix} \begin{bmatrix} 1 & 0 \\ j\omega C & 1 \end{bmatrix} \begin{bmatrix} 1 & 0 \\ 1/Z_L & 1 \end{bmatrix} \tag{8.19a}$$

or

$$\begin{bmatrix} A & B \\ C & D \end{bmatrix} = \begin{bmatrix} 1 + (R + Z_S)\left(j\omega C + \dfrac{1}{Z_L}\right) & R + Z_S \\ j\omega C + \dfrac{1}{Z_L} & 1 \end{bmatrix} \tag{8.19b}$$

The insertion loss is then found from

$$\text{Insertion loss (dB)} = 20\log\left(|S_{21}|\right) \tag{8.20}$$

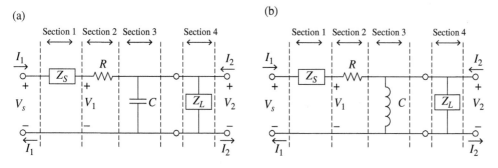

Figure 8.7 Network analysis of (a) low pass filter and (b) high pass filter.

S_{21} for LPF shown in Figure 8.7a when $Z_S = Z_L = Z_o$ is

$$S_{21} = \frac{2V_2}{V_S} = \frac{2}{A} = \frac{2}{1 + (R + Z_o)\left(j\omega C + \frac{1}{Z_o}\right)} \qquad (8.21)$$

The response is obtained by MATLAB using the formulation given by (8.20) and (8.21) and shown in Figure 8.6 for $C = 8$ pF, $R = 100\,\Omega$, and $Z_o = 50\,\Omega$ as illustrated in Figure 8.8.

A circuit with an LPF is simulated with a frequency domain circuit simulator and results are compared with the results obtained with MATLAB. Figure 8.9 is the LPF that is simulated and Figure 8.10 is the response versus frequency for insertion loss.

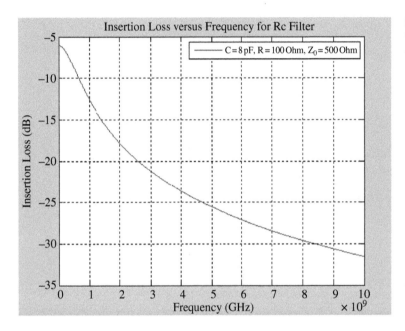

Figure 8.8 Insertion loss for low pass filter when $C = 8$ pF, $R = 100\,\Omega$, and $Z_o = 50\,\Omega$.

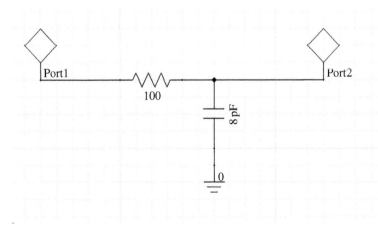

Figure 8.9 Low pass filter simulation when $C = 8$ pF, $R = 100\,\Omega$, and $Z_o = 50\,\Omega$.

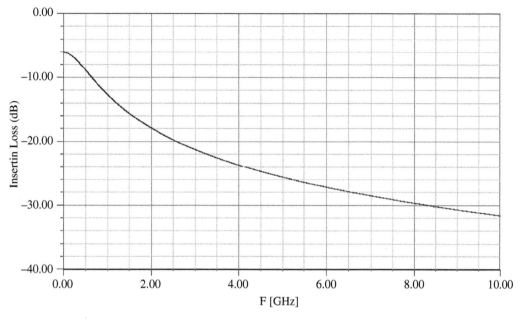

Figure 8.10 Simulated insertion loss for low pass filter when $C = 8$ pF, $R = 100\,\Omega$, and $Z_o = 50\,\Omega$.

ABCD parameters for HPF network shown in Figure 8.7b can be written as

$$\begin{bmatrix} A & B \\ C & D \end{bmatrix} = \begin{bmatrix} 1 & Z_S \\ 0 & 1 \end{bmatrix}\begin{bmatrix} 1 & R \\ 0 & 1 \end{bmatrix}\begin{bmatrix} 1 & 0 \\ \dfrac{1}{j\omega L} & 1 \end{bmatrix}\begin{bmatrix} 1 & 0 \\ \dfrac{1}{Z_L} & 1 \end{bmatrix} \tag{8.22}$$

or

$$\begin{bmatrix} A & B \\ C & D \end{bmatrix} = \begin{bmatrix} 1 + (R + Z_S)\left(\dfrac{1}{j\omega L} + \dfrac{1}{Z_L}\right) & R + Z_S \\[2mm] \dfrac{1}{j\omega L} + \dfrac{1}{Z_L} & 1 \end{bmatrix} \tag{8.23}$$

S_{21} for HPF shown in Figure 8.7b when $Z_S = Z_L = Z_o$ is

$$S_{21} = \frac{2V_2}{V_S} = \frac{2}{A} = \frac{2}{1 + (R + Z_o)\left(\dfrac{1}{j\omega L} + \dfrac{1}{Z_o}\right)} \tag{8.24}$$

Insertion loss is found from (8.20). The response is obtained by MATLAB using the formulation given in (8.20) and (8.21) and shown in Figure 8.11 for $L = 5$ nH, $R = 5\,\Omega$, and $Z_o = 50\,\Omega$. The results can be confirmed with a circuit simulator. Bandpass and BSF networks can be analyzed using the networks shown in Figure 8.12a,b, respectively. Both filters have three sections where Section 8.2 is a series RLC circuit for bandpass filter and parallel RLC circuit for the BSF.

The *ABCD* parameter of the bandpass filter network circuit in Figure 8.12a is

$$\begin{bmatrix} A & B \\ C & D \end{bmatrix} = \begin{bmatrix} 1 & Z_S \\ 0 & 1 \end{bmatrix}\begin{bmatrix} 1 & R + j\omega L + (1/j\omega C) \\ 0 & 1 \end{bmatrix}\begin{bmatrix} 1 & 0 \\ \dfrac{1}{Z_L} & 1 \end{bmatrix} \tag{8.25}$$

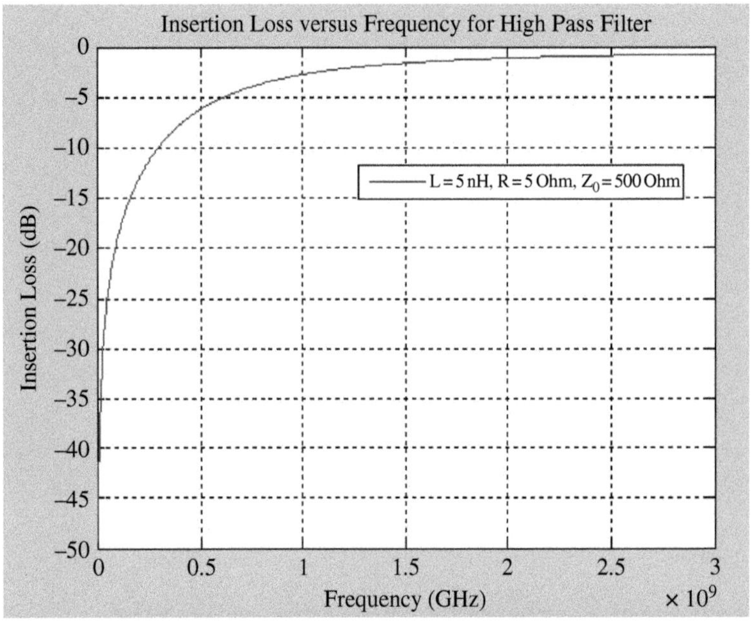

Insertion Loss versus Frequency for High Pass Filter

L = 5 nH, R = 5 Ohm, Z_0 = 500 Ohm

Insertion Loss (dB)

Frequency (GHz)

× 10⁹

Figure 8.11 Insertion loss for high pass filter when L = 5 nH, R = 5 Ω, and Z_o = 50 Ω.

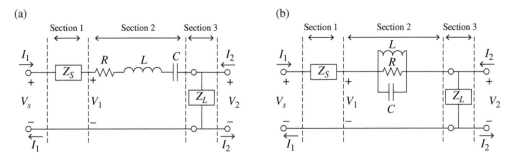

Figure 8.12 Network analysis of (a) bandpass filter and (b) bandstop filter.

or

$$\begin{bmatrix} A & B \\ C & D \end{bmatrix} = \begin{bmatrix} 1 + \dfrac{R + j\omega L + (1/j\omega C) + Z_s}{Z_L} & R + j\omega L + (1/j\omega C) + Z_S \\ \dfrac{1}{Z_L} & 1 \end{bmatrix} \tag{8.26}$$

S_{21} for bandpass filter shown in Figure 8.12a when $Z_S = Z_L = Z_o$ is

$$S_{21} = \frac{2V_2}{V_S} = \frac{2}{A} = \frac{2}{1 + \dfrac{R + j\omega L + (1/j\omega C) + Z_s}{Z_L}} \tag{8.27}$$

Insertion loss is found from (8.20). The response for insertion loss is obtained by MATLAB using the formulation given by (8.20) and (8.27) and is shown in Figure 8.13.

Figure 8.13 Insertion loss for bandpass filter when L = 6 nH, C = 1 pF, R = 5 Ω, and Z_0 = 50 Ω.

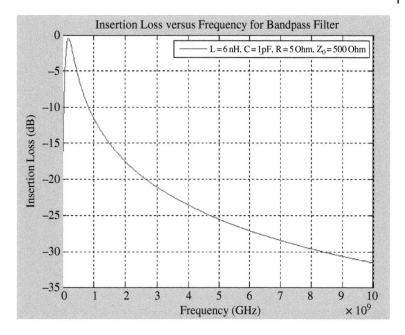

The resonant frequency for the bandpass filter from (8.16) is

$$f_o = \frac{1}{2\pi\sqrt{LC}} = \frac{1}{2\pi\sqrt{(6\times 10^{-9})(1\times 10^{-12})}} = 2.054\,\text{GHz} \qquad (8.28)$$

as confirmed with MATLAB results shown in Figure 8.13. The *ABCD* parameter of the BSF network circuit in Figure 8.12b can be derived similarly as

$$\begin{bmatrix} A & B \\ C & D \end{bmatrix} = \begin{bmatrix} 1 & Z_S \\ 0 & 1 \end{bmatrix} \begin{bmatrix} 1 & \dfrac{1}{G + j\omega C + (1/j\omega L)} \\ 0 & 1 \end{bmatrix} \begin{bmatrix} 1 & 0 \\ \dfrac{1}{Z_L} & 1 \end{bmatrix} \qquad (8.29)$$

or

$$\begin{bmatrix} A & B \\ C & D \end{bmatrix} = \begin{bmatrix} 1 + \dfrac{\dfrac{1}{G + j\omega C + (1/j\omega L)} + Z_s}{Z_L} & \dfrac{1}{G + j\omega C + (1/j\omega L)} + Z_S \\ \dfrac{1}{Z_L} & 1 \end{bmatrix} \qquad (8.30)$$

S_{21} for the BSF shown in Figure 8.12b when $Z_S = Z_L = Z_o$ is

$$S_{21} = \frac{2V_2}{V_S} = \frac{2}{A} = \frac{2}{1 + \dfrac{\dfrac{1}{G + j\omega C + (1/j\omega L)} + Z_s}{Z_L}} \qquad (8.31)$$

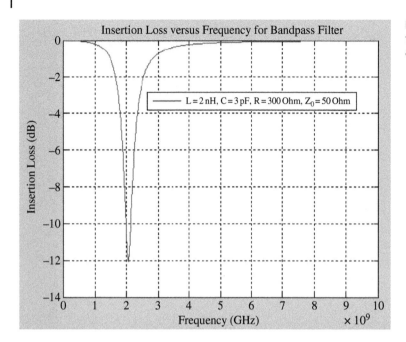

Insertion loss is found from (8.20). The response for the insertion loss is obtained by MATLAB using the formulation given by (8.20) and (8.31) and is shown in Figure 8.14. The resonant frequency for the BSF is again

$$f_o = \frac{1}{2\pi\sqrt{LC}} = \frac{1}{2\pi\sqrt{(2 \times 10^{-9})(3 \times 10^{-12})}} = 2.054\,\text{GHz} \tag{8.32}$$

Return loss for any filter discussed is obtained using the relation

$$S_{11} = \frac{Z_{in} - Z_o}{Z_{in} + Z_o} \tag{8.33}$$

For the BSF

$$Z_{in} = \frac{1}{G + j\omega C + (1/j\omega L)} + Z_s \tag{8.34}$$

The return loss is calculated from

$$\text{Return loss (dB)} = 20\log\left(|S_{11}|\right) \tag{8.35}$$

The return loss of the BSF shown in Figure 8.12 with $L = 2$ nH, $C = 3$ pF, $R = 300\,\Omega$, and $Z_0 = 50\,\Omega$ is obtained versus frequency and given in Figure 8.15.

8.3 Filter Design by the Insertion Loss Method

There are mainly two filter synthesis methods in the design of RF filters: the image parameter method and the insertion loss method. Although the design procedure with the image parameter method is straightforward and easy, it is not possible to realize an arbitrary frequency response with the use of that method. We will be

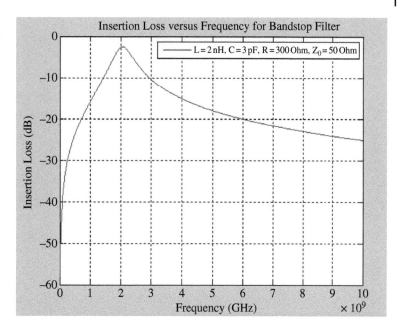

Figure 8.15 Return loss for bandstop filter when L = 2 nH, C = 3 pF, R = 300 Ω, and Z_0 = 50 Ω.

applying the insertion loss method to design and implement the filters in this section. Filter design with the insertion loss method begins with the complete filter specifications shown in Figure 8.2. Filter specifications are used to identify the prototype filter values, and the prototype filter circuit is synthesized. Scaling and transformation of the prototype values are performed to have the final filter component values. The prototype element values of the LPF circuit are obtained using a power loss ratio.

8.3.1 Low Pass Filters

Consider the two-element LPF prototype shown in Figure 8.16.

In the insertion loss method the filter response is defined by the power loss ratio, P_{LR}, which is defined by

$$P_{LR} = \frac{P_{incident}}{P_{load}} = \frac{1}{1 - |S_{11}|^2} \tag{8.36}$$

where

$$S_{11} = \frac{Z_{in} - 1}{Z_{in} + 1} \tag{8.37a}$$

$P_{incident}$ refers to the available power from the source and P_{load} represents the power delivered to the load. As explained elsewhere, attenuation characteristics of the filter fall into one of three categories: binomial (Butterworth), Chebyshev, and elliptic. The binomial or Butterworth response provides the flattest passband response for a given filter and is defined by

$$P_{LR} = 1 + k^2 \left(\frac{\omega}{\omega_c}\right)^{2N} \tag{8.37b}$$

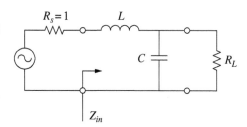

Figure 8.16 Two element low pass prototype circuit.

where $k = 1$ and N is the order of filter. Chebyshev filters provide steeper transition from passband to stopband, while they have equal ripples in the passband and their attenuation characteristics are defined by

$$P_{LR} \doteq 1 + k^2 T_N^2 \left(\frac{\omega}{\omega_c} \right) \tag{8.38}$$

where T_N is the Chebyshev polynomial. The prototype values of the filter circuit, L and C, shown in Figure 8.16, are found by solving the equations given in (8.36)–(8.38).

8.3.1.1 Binomial Filter Response

Binomial filter response can be explained using Figure 8.16 for a two-element LPF network. Source impedance and cutoff frequency for the circuit in Figure 8.16 are assumed to be 1 Ω and 1 rad/s, respectively. In the LPF network, $N = 2$, and power loss becomes

$$P_{LR} = 1 + \omega^4 \tag{8.39}$$

The input impedance is found as

$$Z_{in} = \frac{j\omega L \left(1 + \omega^2 R^2 C^2 \right) + R(1 - j\omega RC)}{\left(1 + \omega^2 R^2 C^2 \right)} \tag{8.40}$$

which leads to

$$S_{11} = \frac{\left[\dfrac{j\omega L \left(1 + \omega^2 R^2 C^2 \right) + R(1 - j\omega RC)}{\left(1 + \omega^2 R^2 C^2 \right)} \right] - 1}{\left[\dfrac{j\omega L \left(1 + \omega^2 R^2 C^2 \right) + R(1 - j\omega RC)}{\left(1 + \omega^2 R^2 C^2 \right)} \right] + 1} \tag{8.41}$$

When (8.41) is substituted into (8.36) with (8.39), the only L and C values satisfying the equation are found to be

$$L = C = 1.4142 \tag{8.42}$$

The same procedure is applied to LPF circuits with any number. The values obtained in this way are tabulated in Table 8.1. The two-element low pass proto circuit we analyzed is called a ladder network. Although, the LPF circuit in Figure 8.16 begins with a series inductor, our analysis applies when it is switched with a shunt capacitor. In

Table 8.1 Component values for binomial low pass filter response with $g_0 = 1$, $\omega_c = 1$.

N	g1	g2	g3	g4	g5	g6	g7	g8	g9
1	2.0000	1.0000							
2	1.4142	1.4142	1.0000						
3	1.0000	2.0000	1.0000	1.0000					
4	0.7654	1.8478	1.8478	0.7654	1.0000				
5	0.6180	1.6180	2.0000	1.6180	0.6180	1.0000			
6	0.5176	1.4142	1.9318	1.9318	1.4142	0.5176	1.0000		
7	0.4450	1.2470	1.8019	2.0000	1.8019	1.2470	0.4450	1.0000	
8	0.3902	1.1111	1.6629	1.9615	1.9615	1.6629	1.1111	0.3902	1.0000

(a)

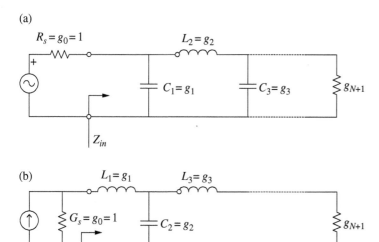

(b)

Figure 8.17 Low pass prototype ladder networks (a) 1st element shunt C and (b) 1st element series L.

addition, the number of elements can be increased to N and a ladder network can be generalized, as shown in Figure 8.17 where the table values shown in Table 8.1 can be used for binomial response.

In the low pass prototype circuits in Figure 8.17, g_0 represents the source resistance or conductance, whereas g_{N+1} represents the load resistance or conductance. g_N is an inductor for a series connected component and capacitor for a parallel connected component. Attenuation curves for the low pass prototype filters can be found from Eqs. (8.37) and (8.38) as

$$\text{Attenuation} = \log\left(P_{LR}\right) \tag{8.43}$$

$$\text{Attenuation (dB)} = 10\log\left(P_{LR}\right) \tag{8.44}$$

Attenuation curves for binomial response using (8.44) are obtained by MATLAB and given in (8.18). Once design filter specifications are given, the number of required elements to have the required attenuation are determined from the attenuation curves. In the second step, Table 8.1 is used to determine the normalized component values for the required number of elements found in the previous stage. Then, the scaling and transformation step is performed and the final filter component values obtained.

Since the original normalized component values of the (LPF) filter are designated as L, C, and R_L, the final scaled components' values of the filter with source impedance R_0 is found from

$$R_s' = R_o \tag{8.45}$$

$$R_L' = R_o R_L \tag{8.46}$$

$$L' = R_o \frac{L_n}{\omega_c} \tag{8.47}$$

$$C' = \frac{C_n}{R_o \omega_c} \tag{8.48}$$

Example 8.1 Maximally Flat Low Pass Filter Response
Design binomial LPF with cutoff frequency $f_c = 1.5\,\text{GHz}$ to have a minimum 35 dB attenuation at 4.5 GHz when source and load impedances are 50 Ω.

Solution

Step 1 – Use Figure 8.18 to determine the required number of elements to achieve minimum 35 dB attenuation at $f = 4.5\,\text{GHz}$.

$$\left|\frac{\omega}{\omega_c}\right| - 1 = \frac{4.5}{1.5} - 1 = 2 \rightarrow N = 4$$

Step 2 – Use Table 8.1 to determine the normalized LPF component values. We choose to begin with the series inductor, as shown in Figure 8.19. The normalized component values for the filter are obtained from the table as

$$L_1 = g_1 = 0.7654, L_2 = g_3 = 1.8478, C_1 = g_2 = 1.8478$$

Step 3 – Apply impedance and frequency scaling

$$R'_s = R_o = 50\,\Omega$$

$$R'_L = R_oR_L = 50(1) = 50\,\Omega$$

$$L'_1 = R_o\frac{L_n}{\omega_c} = 50\frac{0.7654}{(2\pi \times 1.5 \times 10^9)} = 4.06\,\text{nH}\quad C'_2 = \frac{C_n}{R_o\omega_c} = \frac{1.8478}{50(2\pi \times 1.5 \times 10^9)} = 3.92\,\text{pF}$$

$$L'_3 = R_o\frac{L_n}{\omega_c} = 50\frac{1.8478}{(2\pi \times 1.5 \times 10^9)} = 9.8\,\text{nH}$$

$$C'_4 = \frac{C_n}{R_o\omega_c} = \frac{0.7654}{50(2\pi \times 1.5 \times 10^9)} = 1.62\,\text{pF}$$

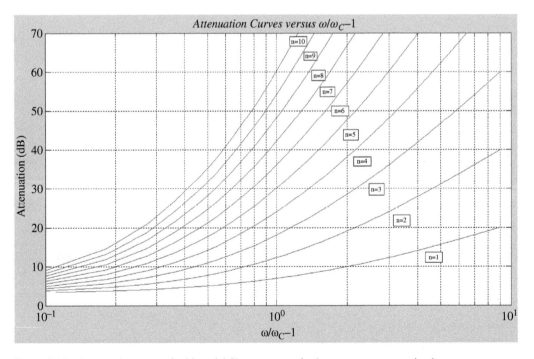

Figure 8.18 Attenuation curves for binomial filter response for low pass prototype circuits.

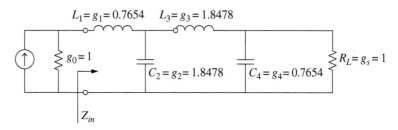

Figure 8.19 Fourth-order normalized LPF for binomial response.

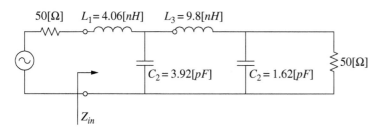

Figure 8.20 Final LPF with binomial response.

The final LPF circuit having binomial filter response is shown in Figure 8.20. The circuit shown in Figure 8.20 is analyzed using network parameters, and the insertion loss is obtained using *ABCD* parameters for the cascaded components, as previously discussed

$$
\begin{bmatrix} A & B \\ C & D \end{bmatrix} = \begin{bmatrix} 1 & Z_S \\ 0 & 1 \end{bmatrix} \begin{bmatrix} 1 & j\omega L'_1 \\ 0 & 1 \end{bmatrix} \begin{bmatrix} 1 & 0 \\ j\omega C'_2 & 1 \end{bmatrix} \begin{bmatrix} 1 & j\omega L'_3 \\ 0 & 1 \end{bmatrix} \begin{bmatrix} 1 & 0 \\ j\omega C'_4 & 1 \end{bmatrix} \begin{bmatrix} 1 & 0 \\ \dfrac{1}{Z_L} & 1 \end{bmatrix}
\tag{8.49}
$$

The insertion loss is obtained from (8.20) and (8.21). MATLAB is used to plot the insertion loss from (8.49), as shown in Figure 8.21. The cutoff frequency with MATLAB is 1.5 GHz and it is confirmed with the frequency domain circuit simulator. The circuits simulated with Ansoft Designer for fourth-order LPF and simulation results are shown in Figures 8.22 and 8.23.

Based on the filter responses in Figures 8.21–8.23, the filter has a cutoff frequency at 1.5 GHz with 3 dB attenuation and meets the attenuation requirement at 4.5 GHz by having a −38.15 dB attenuation at that level.

8.3.1.2 Chebyshev Filter Response

A Chebyshev filter response can be obtained similarly using the two-element low pass prototype circuit given in Figure 8.16. The power loss from (8.38) takes the following form when $N = 2$

$$
P_{LR} = 1 + k^2 T_2^2 \left(\frac{\omega}{\omega_c} \right)
\tag{8.50}
$$

where

$$
T_2 \left(\frac{\omega}{\omega_c} \right) = 2 \left(\frac{\omega}{\omega_c} \right)^2 - 1
\tag{8.51}
$$

Substituting (8.51) into (8.50) gives

$$
P_{LR} = 1 + k^2 \left(4\omega^4 - 4\omega^2 + 1 \right)
\tag{8.52}
$$

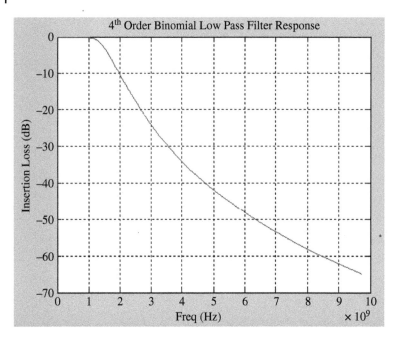

Figure 8.21 MATLAB results for fourth-order LPF with binomial response.

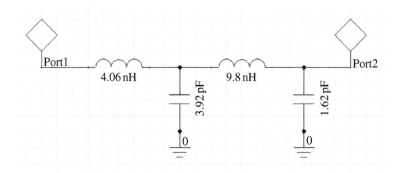

Figure 8.22 Simulated fourth-order LPF.

The input impedance and S_{11} is given by Eqs. (8.40) and (8.41). When (8.41) is substituted into (8.36) with (8.52), the L and C values satisfying the equation are found. Chebyshev polynomials up to seventh order are given in Table 8.2. The polynomials given in Table 8.2 can be used to obtain design tables giving component values for various ripple values, as shown in Table 8.3.

Chebyshev polynomials defined by a three-term recursion where

$$T_0(x) = 1, \qquad T_1(x) = x, \qquad T_{n+1}(x) = 2xT_n(x) - T_{n-1}(x), \qquad n = 1, 2 \ldots \tag{8.53}$$

where $x = \omega/\omega_c$. The attenuation curves for a Chebyshev LPF response is obtained from (8.38) to (8.44) as

$$\text{Attenuation (dB)} = 10 \log \left(1 + \varepsilon^2 T_N^2 \left(\frac{\omega}{\omega_c} \right)' \right) \tag{8.54}$$

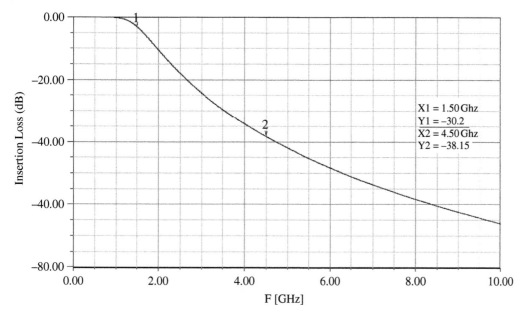

Figure 8.23 Simulation results for fourth-order LPF.

Table 8.2 Chebyshev polynomials up to seventh order.

Order of polynomial, N	$T_N\left(\dfrac{\omega}{\omega_c}\right)$
1	$\dfrac{\omega}{\omega_c}$
2	$2\left(\dfrac{\omega}{\omega_c}\right)^2 - 1$
3	$4\left(\dfrac{\omega}{\omega_c}\right)^3 - 3\left(\dfrac{\omega}{\omega_c}\right)$
4	$8\left(\dfrac{\omega}{\omega_c}\right)^4 - 8\left(\dfrac{\omega}{\omega_c}\right)^2 + 1$
5	$16\left(\dfrac{\omega}{\omega_c}\right)^5 - 20\left(\dfrac{\omega}{\omega_c}\right)^3 + 5\left(\dfrac{\omega}{\omega_c}\right)$
6	$32\left(\dfrac{\omega}{\omega_c}\right)^6 - 48\left(\dfrac{\omega}{\omega_c}\right)^4 + 18\left(\dfrac{\omega}{\omega_c}\right)^2 - 1$
7	$64\left(\dfrac{\omega}{\omega_c}\right)^7 - 112\left(\dfrac{\omega}{\omega_c}\right)^5 + 58\left(\dfrac{\omega}{\omega_c}\right)^3 - 7\left(\dfrac{\omega}{\omega_c}\right)$

Table 8.3 Component values for Chebyshev low pass filter response with $g_0 = 1$, $\omega_c = 1$, $N = 1$ to 7.

Ripple = 0.01 dB

N	g1	g2	g3	g4	g5	g6	g7	g8
1	0.096	1						
2	0.4488	0.4077	1.1007					
3	0.6291	0.9702	0.6291	1				
4	0.7128	1.2003	1.3212	0.6476	1.1007			
5	0.7563	1.3049	1.5773	1.3049	0.7563	1		
6	0.7813	1.36	1.6896	1.535	1.497	0.7098	1.1007	
7	0.7969	1.3924	1.7481	1.6331	1.7481	1.3924	0.7969	1

Ripple = 0.1 dB

N	g1	g2	g3	g4	g5	g6	g7	g8
1	0.3052	1						
2	0.843	0.622	1.3554					
3	1.0315	1.1474	1.0315	1				
4	1.1088	1.3061	1.7703	0.818	1.3554			
5	1.1468	1.3712	1.975	1.3712	1.1468	1		
6	1.1681	1.4039	2.0562	1.517	1.9029	0.8818	1.3554	
7	1.1811	1.4228	2.0966	1.5733	2.0966	1.4228	1.1811	1

Ripple = 0.5 dB

N	g1	g2	g3	g4	g5	g6	g7	g8
1	0.6986	1						
2	1.4029	0.7071	1.9841					
3	1.5963	1.0967	1.5963	1				
4	1.6703	1.1926	2.3661	0.8419	1.9841			
5	1.7058	1.2296	2.5408	1.2296	1.7058	1		
6	1.7254	1.2479	2.6064	1.3137	2.4758	0.8696	1.9841	
7	1.7372	1.2683	2.6381	1.3444	2.6381	1.2583	1.773 72	1

Ripple = 1 dB

N	g1	g2	g3	g4	g5	g6	g7	g8
1	1.0177	1						
2	1.8219	0.685	2.6599					
3	2.0236	0.9941	2.0236	1				
4	2.0991	1.0644	2.8311	0.7892	2.6599			
5	2.1349	1.0911	3.0009	1.0911	2.1349	1		
6	2.1546	1.1041	3.0634	1.1518	2.9387	0.8101	2.6599	
7	2.1664	1.1116	3.0934	1.1736	3.0934	1.1116	2.1664	1

Table 8.3 (Continued)

Ripple = 3 dB

N	g1	g2	g3	g4	g5	g6	g7	g8
1	1.9953	1						
2	3.1013	0.5339	5.8095					
3	3.3487	0.7117	3.3487	1				
4	3.4389	0.7483	4.3471	0.592	5.8095			
5	3.4817	0.7618	4.5381	0.7618	3.4817	1		
6	3.5045	0.7685	4.6061	0.7929	4.4641	0.6033	5.8095	
7	3.5182	0.7723	4.6386	0.8039	4.8388	0.7723	3.5182	1

where

$$\left(\frac{\omega}{\omega_c}\right)' = \left(\frac{\omega}{\omega_c}\right)\cosh(B) \tag{8.55}$$

$$B = \frac{1}{N}\cosh^{-1}\left(\frac{1}{\varepsilon}\right) \tag{8.56}$$

$$\varepsilon = \sqrt{10^{\frac{ripple(dB)}{10}} - 1} \tag{8.57}$$

The attenuation curves are obtained and given for several ripple values using MATLAB, as shown in Figures 8.24–8.28.

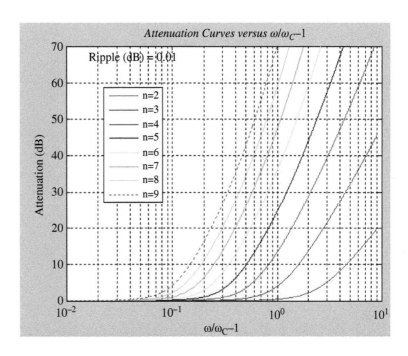

Figure 8.24 Attenuation curves for Chebyshev filter response for 0.01 dB ripple.

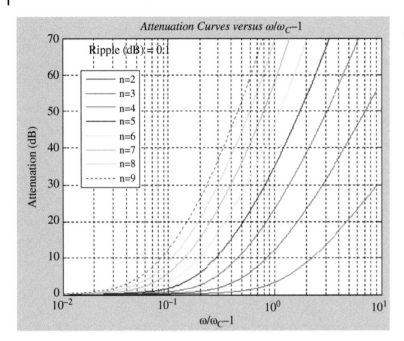

Figure 8.25 Attenuation curves for Chebyshev filter response for 0.1 dB ripple.

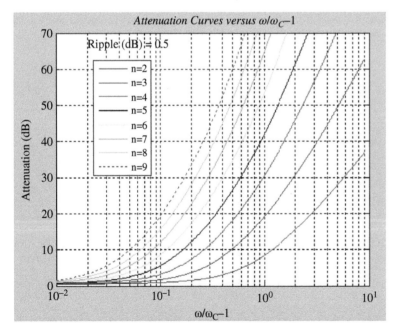

Figure 8.26 Attenuation curves for Chebyshev filter response for 0.5 dB ripple.

Figure 8.27 Attenuation curves for Chebyshev filter response for 1 dB ripple.

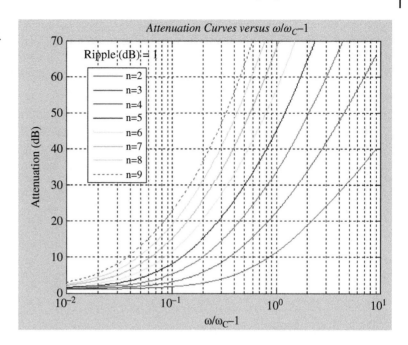

Figure 8.28 Attenuation curves for Chebyshev filter response for 3 dB ripple.

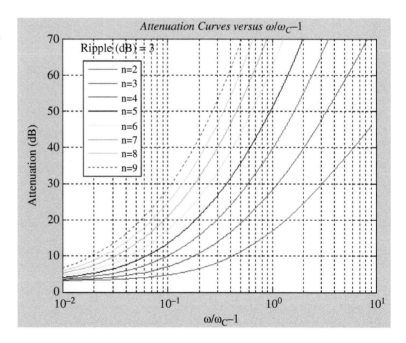

Example 8.2 F/2 Low Pass Filter Design for RF Power Amplifiers

Design an F/2 filter for RF power amplifier that is operating at 13.56 MHz. The filter should have no impact during the normal operation of the amplifier. It should have at least a 20 dB attenuation at F/2 frequency. The passband ripple should not exceed 0.1 dB ripple. It is given that amplifier is presenting a 30 Ω impedance to the load line.

Solution

In RF power amplifier applications, signals having a frequency of F/2 may become an important problem that affects the signal purity and the amount of power delivered to the load. This problem can be resolved by eliminating signals using LPFs commonly called F/2 filters. An F/2 filter is connected off-line to the load line of the amplifier and presents high impedance at the center frequency but matched impedance at F/2. The analysis begins with identifying the F/2 frequency as

$$\frac{F}{2} = 6.78 \text{ MHz}$$

The cutoff frequency of the filter is selected to be 25–35% higher than F/2, as a rule of thumb. The attenuation at the cutoff frequency is expected to be 3 dB, as shown below.

$$\text{Attenuation} = 3 \text{ dB@} f_c = 9 \text{ MHz}$$

Now, we can apply the steps that we used before to design the filter. Since ripple requirement in the passband is mentioned, a Chebyshev filter is used for design and implementation.

Step 1 – Use Figure 8.25 to determine the required number of elements to achieve minimum 20 dB attenuation at $f = 13.56$ MHz.

$$\left|\frac{\omega}{\omega_c}\right| - 1 = \frac{13.56}{9} - 1 = 0.5 \rightarrow N = 5$$

Step 2 – Use Table 8.3 to determine the normalized LPF component values as

Ripple = 0.1 dB

N	g1	g2	g3	g4	g5	g6
5	1.1468	1.3712	1.975	1.3712	1.1468	1

The normalized component values for the filter shown in Figure 8.29 are obtained from the table as

$$L_1 = L_5 = 1.1468, L_3 = 1.975, C_2 = C_4 = 1.3712$$

Step 3 – Apply impedance and frequency scaling

$$R_s' = R_0 = 30 \, \Omega$$

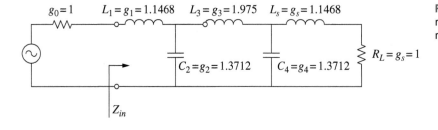

Figure 8.29 Fifth-order normalized LPF for Chebyshev response.

Figure 8.30 Final LPF with Chebyshev response.

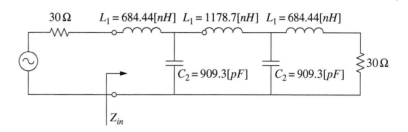

$$R'_L = R_o R_L = 30(1) = 30 \ \Omega$$

$$L'_1 = L'_5 = R_o \frac{L_n}{\omega_c} = 30 \frac{1.1468}{(2\pi \times 8 \times 10^6)} = 684.44 \ \text{nH} \quad C'_2 = C'_4 = \frac{C_n}{R_o \omega_c} = \frac{1.3712}{30(2\pi \times 8 \times 10^6)} = 909.3 \ \text{pF}$$

$$L'_3 = R_o \frac{L_n}{\omega_c} = 30 \frac{1.975}{(2\pi \times 8 \times 10^6)} = 1178.7 \ \text{nH}$$

The final LPF circuit having a Chebyshev filter response is shown in Figure 8.30. The final circuit shown in Figure 8.30 is analyzed using network parameters, and insertion loss is obtained using *ABCD* parameters for the cascaded components as previously discussed.

$$\begin{bmatrix} A & B \\ C & D \end{bmatrix} = \begin{bmatrix} 1 & Z_S \\ 0 & 1 \end{bmatrix} \begin{bmatrix} 1 & j\omega L'_1 \\ 0 & 1 \end{bmatrix} \begin{bmatrix} 1 & 0 \\ j\omega C'_2 & 1 \end{bmatrix} \begin{bmatrix} 1 & j\omega L'_3 \\ 0 & 1 \end{bmatrix} \begin{bmatrix} 1 & 0 \\ j\omega C'_4 & 1 \end{bmatrix} \begin{bmatrix} 1 & j\omega L'_5 \\ 0 & 1 \end{bmatrix} \begin{bmatrix} 1 & 0 \\ \frac{1}{Z_L} & 1 \end{bmatrix} \tag{8.58}$$

The insertion loss in the passband and stopband are obtained using MATLAB from (8.58) and shown in Figures 8.31 and 8.32.

Figure 8.31 Passband ripple response for fifth-order LPF with Chebyshev filter response.

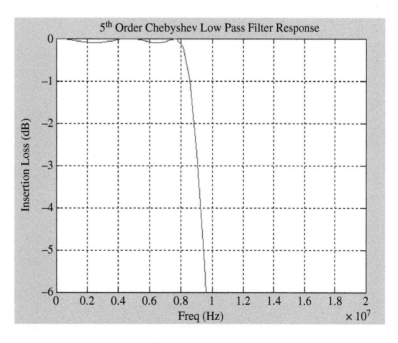

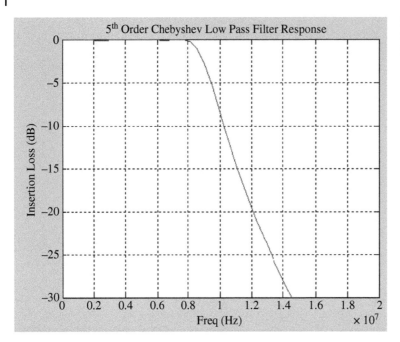

Figure 8.32 Attenuation response for fifth-order LPF with Chebyshev filter response.

The passband ripple is less than 0.1 dB and the cutoff frequency is around 9 MHz, as shown in Figure 8.31. In addition, we have more than 25 dB attenuation at 13.56 MHz, as illustrated in Figure 8.32. The circuit is simulated with Ansoft Designer for accuracy using the circuit shown in Figure 8.33. The passband ripple, attenuation at cutoff frequency, and operational frequency are given in Figure 8.34 and in agreement with MATLAB results obtained.

The input impedance for the filter designed is given in Figure 8.35. Based on the results on the Smith chart, filter input impedance is $(29.58-j8.48)\Omega$ at F/2 and $(0.06 + j43.02)\,\Omega$. Hence the filter presents a very close matched load to amplifier at F/2 and terminates the F/2 frequency content and presents very high inductance and acts like an open load at the operational frequency and does not have any impact on amplifier performance.

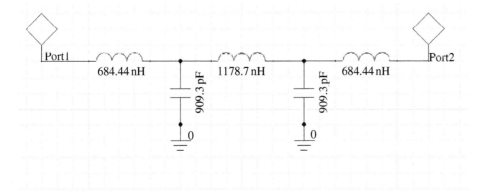

Figure 8.33 Simulated fifth-order LPF.

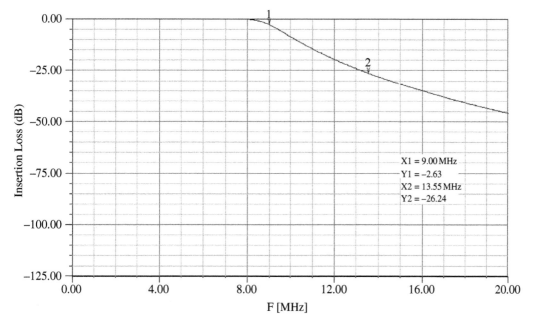

Figure 8.34 Simulation results for fifth-order LPF.

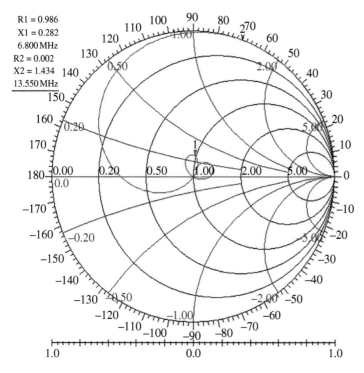

Figure 8.35 Input impedance of fifth-order LPF.

8.3.2 High Pass Filters

HPFs are designed from LPF prototypes using the frequency transformation given by

$$-\frac{\omega_c}{\omega} \rightarrow \omega \tag{8.59}$$

This transformation converts an LPF to an HPF with the following frequency and impedance scaling relations for L and C.

$$L'_n = \frac{R_0}{\omega_c C_n} \tag{8.60}$$

$$C'_n = \frac{1}{\omega_c R_0 L_n} \tag{8.61}$$

The design begins with the low pass prototype by finding L_n and C_n, and then applying (8.60) and (8.61). The transformation of the components from LPF to HPF is illustrated in Figure 8.36.

Example 8.3   High Pass Filter Design

Design an HPF with a 3 dB equal ripple response and a cutoff frequency of 1 GHz. Source and load impedances are given to be 50 Ω and attenuation at 0.6 GHz is required to be a minimum of 40 dB.

Solution

Step 1 – Use Figure 8.28 to determine the required number of elements to achieve minimum 40 dB attenuation at $f = 0.6$ MHz. Apply (8.59) and obtain

$$\left| \left(\frac{\left(-\frac{\omega_c}{\omega} \right)}{\omega_c} \right) \right| - 1 = \frac{1}{0.6} - 1 = 0.667 \rightarrow N = 5$$

Step 2 – Use Table 8.3 to determine the normalized LPF component values as

Ripple = 3 dB						
N	g1	g2	g3	g4	g5	g6
5	3.4817	0.7618	4.5381	0.7618	3.4817	1

We would like to begin with shunt capacitor as the first element in LPF prototype. However, the LPF has to be transformed to an HPF, as shown in Figure 8.37.

Now, the HPF component values can be calculated using (8.60)–(8.61) as

$$L'_1 = L'_5 = \frac{50}{(2\pi \times 1 \times 10^9)(3.4817)} 2.28 \text{ nH}$$

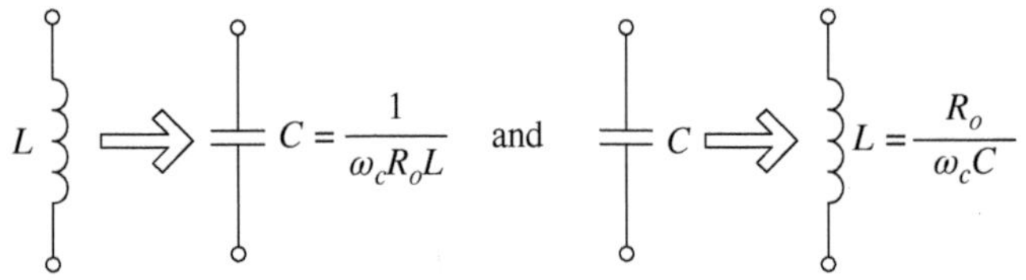

$$L \Rightarrow C = \frac{1}{\omega_c R_o L} \quad \text{and} \quad C \Rightarrow L = \frac{R_o}{\omega_c C}$$

Figure 8.36   LPF component to HPF component transformation.

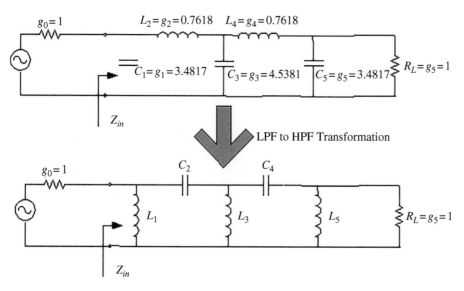

Figure 8.37 LPF Prototype circuit to HPF transformation.

$$C_2' = C_4' = \frac{1}{(2\pi \times 1 \times 10^9)(50)(0.7618)} = 4.18 \text{ pF}$$

$$L_3' = \frac{50}{(2\pi \times 1 \times 10^9)(4.5381)} 1.75 \text{ nH}$$

The final circuit is shown in Figure 8.38 and its response is obtained using MATLAB as it was done before using network parameters as

$$\begin{bmatrix} A & B \\ C & D \end{bmatrix} = \begin{bmatrix} 1 & Z_S \\ 0 & 1 \end{bmatrix} \begin{bmatrix} 1 & 0 \\ 1/j\omega L_1' & 1 \end{bmatrix} \begin{bmatrix} 1 & 1/j\omega C_2' \\ 0 & 1 \end{bmatrix} \begin{bmatrix} 1 & 0 \\ 1/j\omega L_3' & 1 \end{bmatrix} \begin{bmatrix} 1 & 1/j\omega C_4' \\ 0 & 1 \end{bmatrix} \begin{bmatrix} 1 & 0 \\ 1/j\omega L_5' & 1 \end{bmatrix} \begin{bmatrix} 1 & 0 \\ \frac{1}{Z_L} & 1 \end{bmatrix} \quad (8.62)$$

The attenuation response obtained by MATLAB is shown in Figure 8.39. The attenuation at cutoff frequency 1 and 0.6 GHz are around 3 and 40 dB as expected. The circuit is simulated with Ansoft Designer for accuracy using the circuit shown in Figure 8.40. The passband ripple, attenuation at cutoff frequency and operational frequency is given on Figure 8.41. It has been shown that 3 dB attenuation at 1 GHz and 42.14 dB attenuation at 0.6 GHz are obtained with the HPF designed.

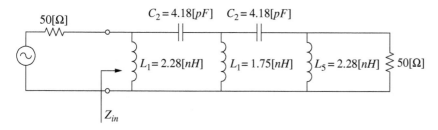

Figure 8.38 Final HPF filter.

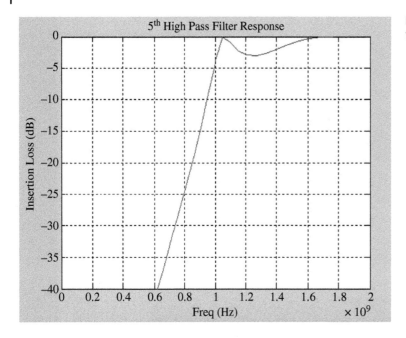

Figure 8.39 Attenuation response for fifth-order HPF.

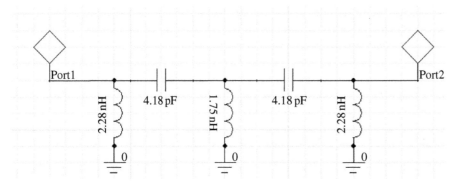

Figure 8.40   Simulated fifth-order HPF.

8.3.3   Bandpass Filters

Bandpass filters are designed from LPF prototypes using the frequency transformation given by

$$\frac{\omega_o}{\omega_{c2} - \omega_{c1}} \left(\frac{\omega}{\omega_o} - \frac{\omega_o}{\omega} \right) \rightarrow \omega \tag{8.63}$$

The term $(\omega_{c2} - \omega_{c1})/\omega_o$ is called a fractional bandwidth and ω_o is called a resonant or center frequency and defined by Eq. (8.18). ω_{c2} and ω_{c1} are the upper and lower cutoff frequencies and are defined by Eqs. (8.17a) and (8.17b). The transformation given by Eq. (8.63) maps the series component of an LPF prototype circuit to

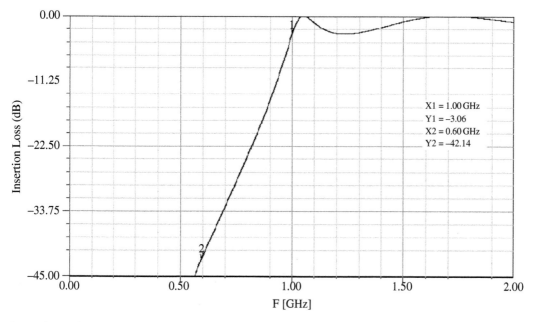

Figure 8.41 Simulation results for fifth-order LPF.

the series LC circuit and shunt component of an LPF prototype circuit to a shunt LC circuit in a bandpass filter. The component values of the series LC circuit are calculated as

$$L'_n = \frac{R_o L_n}{\left(\dfrac{\omega_o}{\omega_{c2} - \omega_{c1}}\right)^{-1} \omega_o} \tag{8.64}$$

$$C'_n = \frac{\left(\dfrac{\omega_o}{\omega_{c2} - \omega_{c1}}\right)^{-1}}{\omega_o R_o L_n} \tag{8.65}$$

The component values of the shunt LC circuit are calculated as

$$L'_n = \frac{\left(\dfrac{\omega_o}{\omega_{c2} - \omega_{c1}}\right)^{-1} R_o}{\omega_o C_n} \tag{8.66}$$

$$C'_n = \frac{C_n}{\left(\dfrac{\omega_o}{\omega_{c2} - \omega_{c1}}\right)^{-1} R_o \omega_o} \tag{8.67}$$

The transformation of the components from LPF to BPF is illustrated in Figure 8.42.

Example 8.4 Bandpass Filter Design

Design a bandpass filter with 5% fractional bandwidth and a center frequency of 2 GHz. The filter is to have a maximally flat response in the passband and four sections. The source impedance and load impedances are given to be 50 Ω.

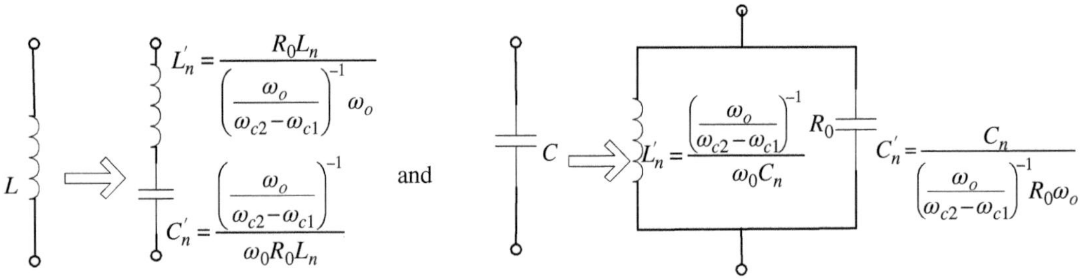

$$L_n' = \dfrac{R_0 L_n}{\left(\dfrac{\omega_o}{\omega_{c2}-\omega_{c1}}\right)^{-1}\omega_o}$$

$$C_n' = \dfrac{\left(\dfrac{\omega_o}{\omega_{c2}-\omega_{c1}}\right)^{-1}}{\omega_o R_0 L_n}$$

and

$$L_n' = \dfrac{\left(\dfrac{\omega_o}{\omega_{c2}-\omega_{c1}}\right)^{-1}}{\omega_o C_n}$$

$$C_n' = \dfrac{C_n}{\left(\dfrac{\omega_o}{\omega_{c2}-\omega_{c1}}\right)^{-1} R_0 \omega_o}$$

Figure 8.42 LPF component to BPF component transformation.

Solution

The filter specifications mention that the BPF filter has four sections and maximally flat bandpass response. As a result, we will be using binomial LPF filter prototype circuit to design BPF circuit to meet these specifications. The fractional bandwidth is given to be $\dfrac{\omega_o}{\omega_{c2}-\omega_{c1}} = 0.05$

Step 1 – The required number of sections is defined as 4. This requires the LPF prototype to have four components. So, $N = 4$.

Step 2 – Use Table 8.1 to determine the normalized LPF component.

N	g1	g2	g3	g4	g5
4	0.7654	1.8478	1.8478	0.7654	1.0000

We would like to begin with a series inductor as the first element in an LPF prototype. However, the LPF has to be transformed to a BPF, as shown in Figure 8.43.

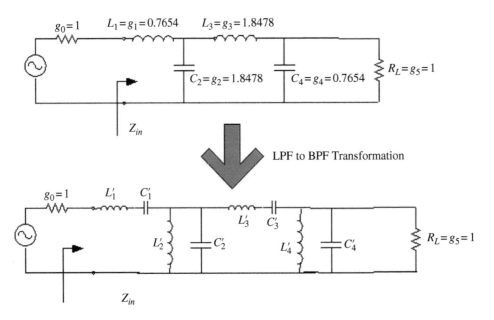

Figure 8.43 LPF prototype circuit to BPF transformation.

BPF component values can be calculated using (8.64)–(8.67) as

$$L_1' = \frac{50(0.7654)}{0.05(2\pi \times 2 \times 10^9)} = 60.908\,\text{nH}$$

$$C_1' = \frac{0.05}{(2\pi \times 2 \times 10^9)50(0.7654)} = 0.10396\,\text{pF}$$

$$L_2' = \frac{(0.05)50}{(2\pi \times 2 \times 10^9)(1.8478)} = 0.107\,\text{nH}$$

$$C_2' = \frac{1.8478}{(0.05)(50)(2\pi \times 2 \times 10^9)} = 58.817\,\text{pF}$$

$$L_3' = \frac{50(1.8478)}{0.05(2\pi \times 2 \times 10^9)} = 147.04\,\text{nH}$$

$$C_3' = \frac{0.05}{(2\pi \times 2 \times 10^9)50(1.8478)} = 0.043\,\text{pF}$$

$$L_4' = \frac{(0.05)50}{(2\pi \times 2 \times 10^9)(0.7654)} = 0.258\,\text{nH}$$

$$C_4' = \frac{0.7654}{(0.05)(50)(2\pi \times 2 \times 10^9)} = 24.363\,\text{pF}$$

The final circuit is analyzed using MATLAB as it was done before using network parameters, and the response is obtained, as shown in Figure 8.44. The BPF is simulated with Ansoft Designer using the circuit in Figure 8.45. The Ansoft Designer attenuation response confirming the results of MATLAB for BPF is given in Figure 8.46.

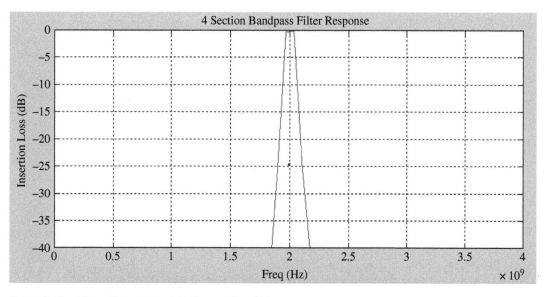

Figure 8.44 Attenuation response for four-section BPF.

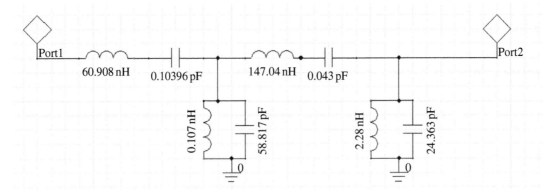

Figure 8.45 Simulated four-section BPF.

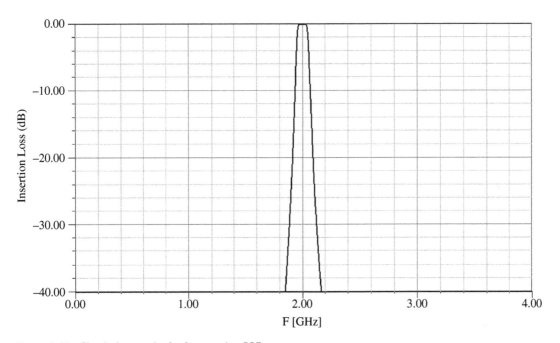

Figure 8.46 Simulation results for four-section BPF.

8.3.4 Bandstop Filters

BSFs are designed from LPF prototypes using the frequency transformation given by

$$\frac{\omega_{c2} - \omega_{c1}}{\omega_o} \left(\frac{\omega}{\omega_o} - \frac{\omega_o}{\omega} \right)^{-1} \rightarrow \omega \tag{8.68}$$

This transformation maps the series component of the LPF prototype circuit to the shunt LC circuit and the shunt component of the LPF prototype circuit to the series LC circuit in the BSF. The component values of the shunt LC circuit are calculated as

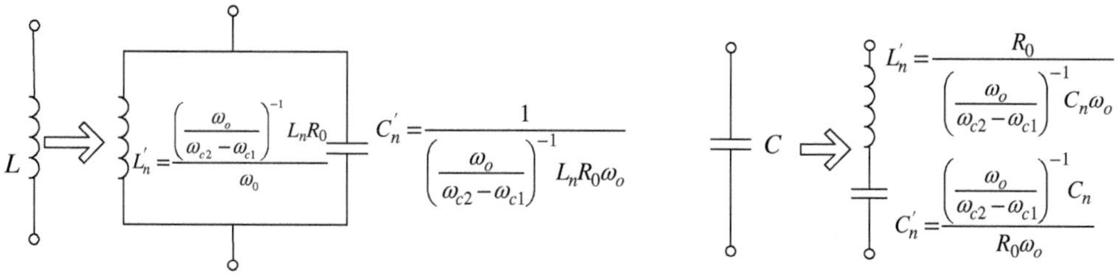

Figure 8.47 LPF component to BSF component transformation.

$$L'_n = \frac{\left(\dfrac{\omega_o}{\omega_{c2} - \omega_{c1}}\right)^{-1} L_n R_o}{\omega_o} \tag{8.69}$$

$$C'_n = \frac{1}{\left(\dfrac{\omega_o}{\omega_{c2} - \omega_{c1}}\right)^{-1} L_n R_o \omega_o} \tag{8.70}$$

The component values of the series LC circuit are calculated as

$$L'_n = \frac{R_o}{\left(\dfrac{\omega_o}{\omega_{c2} - \omega_{c1}}\right)^{-1} C_n \omega_o} \tag{8.71}$$

$$C'_n = \frac{\left(\dfrac{\omega_o}{\omega_{c2} - \omega_{c1}}\right)^{-1} C_n}{R_o \omega_o} \tag{8.72}$$

The transformation of the components from LPF to BSF is illustrated in Figure 8.47.

8.4 Stepped Impedance Low Pass Filters

A stepped impedance filter is made up of high and low impedance sections of a transmission line, as shown in Figure 8.48. Using transmission line theory, the high impedance sections and the low impedance sections are implemented to realize LPFs.

The two-port Z parameter matrix for a transmission line in Figure 8.48 is

$$Z = \begin{bmatrix} -jZ_O \cot(\beta\ell) & -jZ_O \csc(\beta\ell) \\ -jZ_O \csc(\beta\ell) & -jZ_O \cot(\beta\ell) \end{bmatrix} \tag{8.73}$$

where

$$Z_{11} = Z_{22} = -jZ_O \cot(\beta\ell) \text{ and } Z_{12} = Z_{21} - jZ_O \csc(\beta\ell) \quad (8.74)$$

An equivalent T connected network can be used to represent the two-port transmission line network in Figure 8.48. The equivalent T connected network representing the transmission line is shown in Figure 8.49.

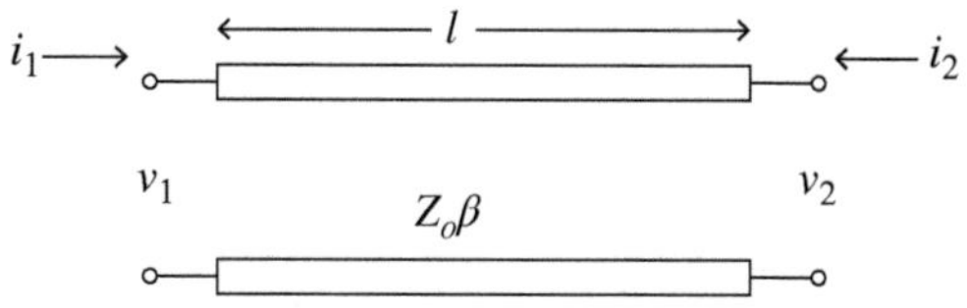

Figure 8.48 Transmission line model.

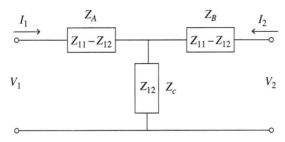

Figure 8.49 *T* network equivalent circuit.

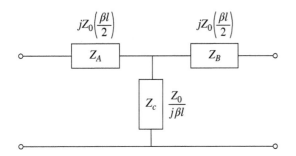

Figure 8.50 *T* network representation with transmission lines.

Where the components of the *T* network are defined as

$$Z_A = Z_B = Z_{11} - Z_{12} = jZ_o \tan\left(\frac{\beta l}{2}\right) \quad (8.75)$$

$$Z_C = Z_{12} = -jZ_o \csc(\beta \ell) \quad (8.76)$$

When the electrical length, $\beta\ell$, is small, the following approximation can be done.

$$\sin(\beta l) \approx \beta l, \cos(\beta l) \approx 1, \text{and } \tan(\beta l) \approx \beta l \quad (8.77)$$

Approximations given by (8.77) lead to the following element values illustrated in Figure 8.50 for the *T* network

$$Z_A = Z_B = Z_{11} - Z_{12} \approx jZo\left(\frac{\beta l}{2}\right) \quad (8.78)$$

$$Z_C = Z_{12} \approx \frac{Z_0}{j\beta\ell} \quad (8.79)$$

Consider the case when the characteristic impedance Z_o is very high. We denote this impedance as Z_{High}. For the shunt component, since $\beta\ell$ is very small, the impedance will be very large. In fact it can be considered an open circuit. This results in an approximate circuit impedance of series component, $jZ_{HIGH}\beta\ell$, as

$$\frac{Z_{High}}{j\beta\ell} \rightarrow \infty \text{ when } \beta\ell << 1 \text{ and } Z_{High} >> Z_o \quad (8.80)$$

High impedance condition transforms the T – network to equivalent series connected L network, as shown in Figure 8.51. Now consider the case when the characteristic impedance is low (Z_{Low}). This time, the series components have a very low impedance and can be considered shorted. The resulting approximate circuit impedance is that of the shunt component alone or $\frac{Z_{Low}}{j\beta\ell}$ as

$$jZ_{Low}\left(\frac{\beta l}{2}\right) \rightarrow 0 \text{ when } \beta\ell << 1 \text{ and } Z_{Low} << Z_o \quad (8.81)$$

As a result, low impedance condition transforms the *T* network to an equivalent shunt connected *C* network, as shown in Figure 8.52. The physical length of component values for series and shunt connected elements are found from (8.80) and (8.81). The length for inductive elements can be obtained from

$$X_L = j\omega L = jZ_{HIGH}\beta\ell \quad (8.82)$$

So

$$\ell_{High} = \frac{\omega L}{Z_{HIGH}\beta} \quad (8.83)$$

The reactance of capacitive elements is found from

$$X_C = \frac{1}{j\omega C} = \frac{Z_{LOW}}{j\beta\ell} \quad (8.84)$$

(a)

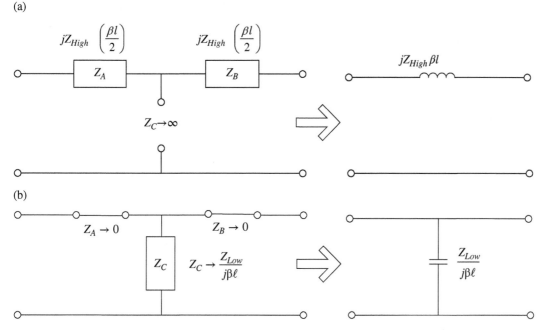

(b)

Figure 8.51 (a) High impedance transformation of T network; (b) low impedance transformation of T-network.

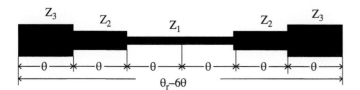

Figure 8.52 Three-section SIR bandpass filter.

and the length is

$$\ell_{Low} = \frac{Z_{LOW}\omega C}{\beta} \tag{8.85}$$

L and C values are the values obtained using an LPF prototype circuit based on the filter specifications. In Eqs. (8.82) and (8.85), the phase constant is defined as

$$\beta = \frac{\omega}{v_p} \tag{8.86}$$

where v_p is a phase velocity as defined by

$$v_p = \frac{c}{\sqrt{\varepsilon_e}} \tag{8.87}$$

ε_e is the effective permittivity constant of the microstrip line. The high and low impedance values should be

$$Z_{Low} < Z_o < Z_{High} \tag{8.88}$$

The selection of Z_{Low} and Z_{High} values carries importance for the response of the filter. The ratio of Z_{High} to Z_{Low} should be kept as large as possible to get more accurate results. We can define approximate limits for Z_{Low} and Z_{High} based on the assumption that the electrical length is small if

$$\beta \ell < \frac{\pi}{4} \tag{8.89}$$

Then, the impedance limit for Z_{Low} is

$$Z_{Low} < \frac{\pi}{4\omega_c C} \tag{8.90}$$

and the impedance limit for Z_{High}

$$Z_{High} > \frac{4\omega_c L}{\pi} \tag{8.91}$$

Once Z_{Low} and Z_{High} are defined, the width of each line can be obtained using a microstrip line equation defined by

$$\frac{W}{d} = \begin{cases} \dfrac{8e^A}{e^{2A} - 2} & \text{for} \quad W/d < 2 \\[2ex] \dfrac{2}{\pi}\left[B - 1 - \ln(2B-1) + \dfrac{\varepsilon_r - 1}{2\varepsilon_r}\left\{ \ln(B-1) + 0.39 - \dfrac{0.61}{\varepsilon_r} \right\} \right] & \text{for} \quad W/d > 2 \end{cases} \tag{8.92}$$

where

$$A = \frac{Z_o}{60}\sqrt{\frac{\varepsilon_r + 1}{2}} + \frac{\varepsilon_r - 1}{\varepsilon_r + 1}\left(0.23 + \frac{0.11}{\varepsilon_r} \right) \tag{8.93}$$

$$B = \frac{377\pi}{2Z_0\sqrt{\varepsilon_r}} \tag{8.94}$$

8.5 Stepped Impedance Resonator Bandpass Filters

Conventional parallel coupled bandpass filters suffer drastically from the spurious harmonics. The stepped impedance resonator (SIR) bandpass filters can be used to realize high performance bandpass filters by suppressing the spurious harmonics to overcome this problem. One of the key features of an SIR bandpass filter is that its resonant frequencies can be tuned by adjusting the impedance ratios of the high Z and low Z sections. The symmetrical tri-section SIR bandpass filter used in bandpass filter design is shown in Figure 8.52.

In the symmetrical SIR bandpass filter structure, each section should have the same electrical length. Then, it can be shown that the resonance occurs when it is equal to

$$\theta = \tan^{-1}\left(\sqrt{\frac{K_1 K_2}{K_1 + K_2 + 1}} \right) \tag{8.95}$$

where

$$K_1 = \frac{-(\cos\alpha)(\cos\beta) + \sqrt{(\cos\alpha)^2(\cos\beta)^2 + 4(\sin b)^2(\cos(ab))^2}}{2(\cos(ab))^2} \tag{8.96}$$

$$K_2 = \frac{1 + K_1}{\tan^2(ab) - K_1} \tag{8.97}$$

The design parameters in (8.96) and (8.97) are found from

$$a = \frac{f_{s1}}{f_o} \tag{8.98}$$

$$b = \frac{\pi}{2} \frac{f_o}{f_{s2}} \tag{8.99}$$

$$\alpha = \frac{\pi}{2} \frac{f_{s1} + f_o}{f_{s2}} \tag{8.100}$$

$$\beta = \frac{\pi}{2} \frac{f_{s1} - f_o}{f_{s2}} \tag{8.101}$$

The terminating impedance of the SIR bandpass filter at the input and output should be $Z_3 = 50\ \Omega$. Once the operating frequencies, f_o, f_{s1}, and f_{s2}, of the bandpass filter and terminating impedance, Z_3, of the SIR bandpass filter are identified, the line impedances, Z_1 and Z_2, are found from

$$Z_2 = \frac{Z_3}{K_1} \tag{8.102}$$

$$Z_1 = \frac{Z_2}{K_2} \tag{8.103}$$

The physical length and the width of the transmission lines in tri-section SIR bandpass filter are found using microstrip line equations. The symmetrical SIR bandpass filter illustrated in Figure 8.52 has

$$\theta_1 = 2\theta, \theta_2 = \theta, \theta_3 = \theta \tag{8.104}$$

The physical length for each section in the SIR bandpass filter can be found from

$$l_n = \frac{\lambda_n \theta_n}{2\pi}, n = 1, 2, 3 \tag{8.105}$$

The width of the sections in the SIR bandpass filter is obtained from (8.92)–(8.94). The performance of bandpass filters with SIR bandpass filters can be improved by using the configuration given in Figure 8.53. The bandpass filter in Figure 8.53 provides triple band filter characteristics with the coupling scheme shown in Figure 8.54. In Figure 8.53, the coupled lines' equivalent circuit is represented by two single transmission lines of electrical length θ, and characteristic impedance Z_o and admittance inverter parameter J, as shown in Figure 8.55. Inverter parameter J is an important design parameter because it is directly proportional to the coupling strength of the coupled lines.

This parameter is found using network synthesis from the equivalent circuit and is given by

$$J_{01} = Y_0 \sqrt{\frac{2k\theta_0}{g_0 g_1}} \tag{8.106a}$$

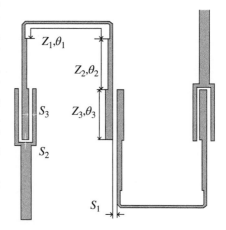

Figure 8.53 Triple band bandpass filter using SIR bandpass filters.

(a)

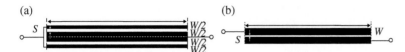

(b)

Figure 8.54 Coupling schemes: (a) improved coupling scheme; (b) conventional coupling scheme.

Figure 8.55 Equivalent circuit of parallel coupled lines.

$$J_{j,j+1} = Y_0 \frac{2k\theta_0}{\sqrt{g_j g_0 g_1}} \tag{8.106b}$$

$$J_{n,n+1} = Y_0 \sqrt{\frac{2k\theta_0}{g_n g_{n+1}}} \tag{8.106c}$$

As the ratio J/Y between the coupled lines increases, the coupling strength also increases.

8.6 Edge/Parallel-coupled, Half-wavelength Resonator Bandpass Filters

A parallel-coupled, half-wavelength bandpass filter is shown in Figure 8.56. For a filter of order n, there are $n+1$ couplings between half-wavelength resonators. The first step of the design begins with finding the low pass prototype circuit component values.

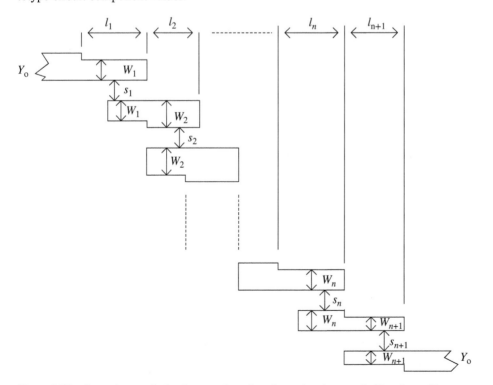

Figure 8.56 General setup for implementation of a microstrip edge-coupled bandpass filter.

Then, the low pass prototype filter undergoes transformations to achieve the desired bandpass filter characteristics, as described in Section 8.3.3. To configure the circuit in a parallel microstrip implementation, the filter must first be represented as a series of cascaded J inverters. The equations to perform this are given by

$$\frac{J_{01}}{Y_0} = \sqrt{\frac{\pi}{2} \frac{FBW}{g_0 g_1}} \tag{8.107}$$

$$\frac{J_{j,j+1}}{Y_0} = \frac{\pi FBW}{2} \frac{1}{\sqrt{g_j g_{j+1}}} \quad \text{for } j = 1 \text{ to } n-1 \tag{8.108}$$

$$\frac{J_{n,n+1}}{Y_0} = \sqrt{\frac{\pi}{2} \frac{FBW}{g_n g_{n+1}}} \tag{8.109}$$

Where $g_0 - g_n$ are the normalized impedance elements of the low pass prototype filter, the fractional bandwidth, *FBW* in (8.109), of the filter of 15%, and Y_0 is the characteristic admittance. These coefficients can be calculated in MATLAB. The next important step in the filter design is to calculate the even- and odd-mode impedance values for the coupled microstrip lines. These values are also calculated in the MATLAB script from the following equations.

$$(Z_{0e})_{j,j+1} = \frac{1}{Y_0} \left[1 + \frac{J_{j,j+1}}{Y_0} + \left(\frac{J_{j,j+1}}{Y_0}\right)^2 \right] \quad \text{for } j = 0 - n \tag{8.110}$$

$$(Z_{0o})_{j,j+1} = \frac{1}{Y_0} \left[1 - \frac{J_{j,j+1}}{Y_0} + \left(\frac{J_{j,j+1}}{Y_0}\right)^2 \right] \quad \text{for } j = 0 - n \tag{8.111}$$

These even- and odd-mode impedances are directly used to find the dimensions of the microstrips in the bandpass filter. After even- and odd-mode impedances are determined, we use the synthesis technique introduced in [1, 2] and in Chapter 6 to give us accurate physical dimensions. Based on the synthesis method, we find the spacing ratio between the coupled lines using

$$s/h = \frac{2}{\pi} \cosh^{-1} \left[\frac{\cosh\left[\frac{\pi}{2}\left(\frac{w}{h}\right)_{se}\right] + \cosh\left[\frac{\pi}{2}\left(\frac{w}{h}\right)'_{so}\right] - 2}{\cosh\left[\frac{\pi}{2}\left(\frac{w}{h}\right)'_{so}\right] - \cosh\left[\frac{\pi}{2}\left(\frac{w}{h}\right)_{se}\right]} \right] \tag{8.112}$$

$(w/h)_{se}$ and $(w/h)_{so}$ are the shape ratios for the equivalent single case corresponding to even- and odd-mode geometry, respectively. $(w/h)'_{so}$ is the second term for the shape ratio. (w/h) is the shape ratio for the single microstrip line and it is expressed as

$$\frac{w}{h} = \frac{8\sqrt{\left[\exp\left(\frac{R}{42.4}\sqrt{(\varepsilon_r + 1)}\right) - 1\right]\frac{7 + (4/\varepsilon_r)}{11} + \frac{1 + (1/\varepsilon_r)}{0.81}}}{\left[\exp\left(\frac{R}{42.4}\sqrt{\varepsilon_r + 1}\right) - 1\right]} \tag{8.113}$$

where

$$R = \frac{Z_{oe}}{2}$$

or

$$R = \frac{Z_{oo}}{2} \tag{8.114}$$

Z_{ose} and Z_{oso} are the characteristic impedances corresponding to the single microstrip shape ratios $(w/h)_{se}$ and $(w/h)_{so}$, respectively. They are given as

$$Z_{ose} = \frac{Z_{oe}}{2} \tag{8.115}$$

$$Z_{oso} = \frac{Z_{oo}}{2} \tag{8.116}$$

and

$$(w/h)_{se} = (w/h) \Big|_{R = Z_{ose}} \tag{8.117}$$

$$(w/h)_{so} = (w/h) \Big|_{R = Z_{oso}} \tag{8.118}$$

The term $(w/h)'_{so}$ in Eq. (8.112) is given as

$$\left(\frac{w}{h}\right)'_{so} = 0.78 \left(\frac{w}{h}\right)_{so} + 0.1 \left(\frac{w}{h}\right)_{se} \tag{8.119}$$

After the spacing ratio s/h for the coupled lines is found, we can proceed to find w/h for the coupled lines. The shape ratio for the coupled lines is

$$\left(\frac{w}{h}\right) = \frac{1}{\pi} \cosh^{-1}(d) - \frac{1}{2}\left(\frac{s}{h}\right) \tag{8.120}$$

where

$$d = \frac{\cosh\left[\frac{\pi}{2}\left(\frac{w}{h}\right)_{se}\right](g+1) + g - 1}{2} \tag{8.121}$$

$$g = \cosh\left[\frac{\pi}{2}\left(\frac{s}{h}\right)\right] \tag{8.122}$$

The physical length of the directional coupler is obtained using

$$l = \frac{\lambda}{4} = \frac{c}{4f\sqrt{\varepsilon_{eff}}} \tag{8.123}$$

The calculation of the effective permittivity constant using odd- and even mode capacitances is detailed in [1, 2].

Example 8.5 Edge-coupled Resonator Bandpass Filter

Design and simulate a fifth-order, edge-coupled half-wavelength resonator bandpass filter with 0.1 dB passband ripple, $f_C = 10$ GHz, $\varepsilon_r = 10.2$, FBW = 0.15, and dielectric thickness of 0.635 mm.

Solution

The first step is to design the low pass Chebyshev prototype filter coefficients. They are determined from Table 8.3. The filter coefficients are

$$g_0 = 1 = g_6, g_1 = 1.1468 = g_5, g_2 = 1.3712 = g_4, g_3 = 1.9750$$

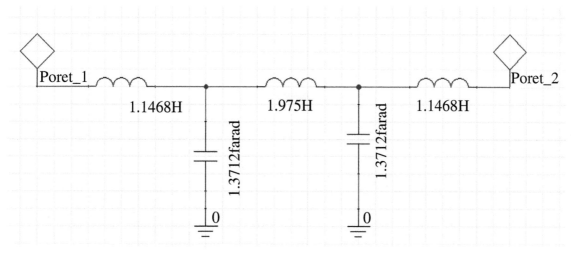

Figure 8.57   Low pass prototype circuit for bandpass filter.

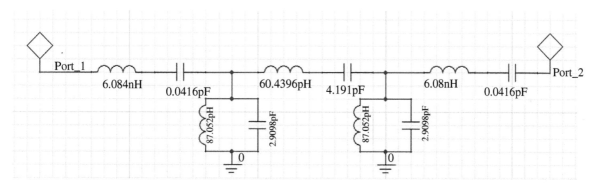

Figure 8.58   Bandpass filter with final lumped element component values.

The prototype circuit is then transformed into an equivalent bandpass filter using the transformation circuits given in Figure 8.57. The bandpass filter with the final component values is illustrated in Figure 8.58. The frequency response of the filter is simulated with Ansoft Designer and the insertion and return loss are given in Figure 8.59.

A MATLAB program has been written to obtain the physical dimensions for the edge-coupled bandpass filter using the formulation given. In addition, the MATLAB program is used to obtain the filter response using *ABCD* two-port parameters. The calculated inverter and corresponding even- and odd-mode impedance values are given in Table 8.4.

The frequency response of the lumped element bandpass filter is also obtained with the program and shown in Figure 8.60. Results obtained in Figure 8.60 match the results obtained with Ansoft Designer. The physical

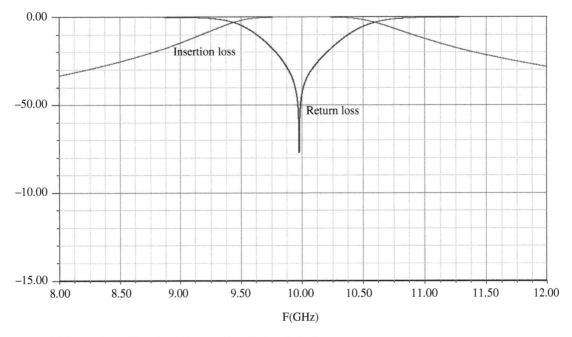

Figure 8.59 Bandpass filter simulation results with Ansoft Designer.

Table 8.4 Even- and odd-mode impedance values.

j	$J_{j, j+1}$	$(Z_{0e})_{j, j+1}$	$(Z_{0o})_{j, j+1}$
0	0.4533	82.9367	37.6092
1	0.1879	61.1600	42.3705
2	0.1432	58.1839	43.8661
3	0.1432	58.1839	43.8661
4	0.1879	61.1600	42.3705
5	0.4533	82.9367	37.6092

dimensions calculated with the MATLAB program are used to simulate the microstrip edge-coupled filter with a Sonnet planar electromagnetic simulator, as shown in Figure 8.61.

The Sonnet simulation results are illustrated in Figure 8.62. Since the material properties are entered, the simulation results are slightly different from the ideal results obtained using the lumped elements in Ansoft and MATLAB. Overall, the insertion loss requirement, center frequency, and attenuation profile for edge-coupled filters are achieved with the method used.

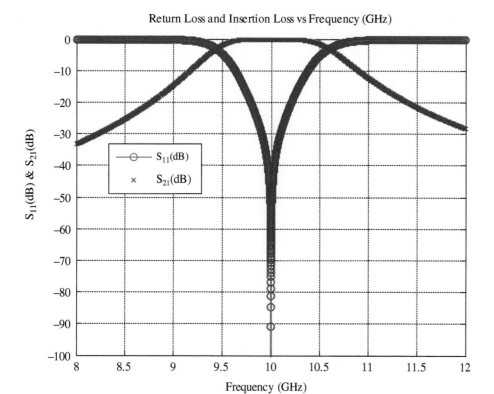

Figure 8.60   Bandpass filter simulation results with MATLAB.

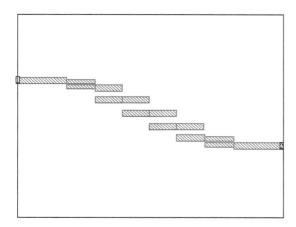

Figure 8.61   Simulated edge-coupled microstrip circuit with Sonnet.

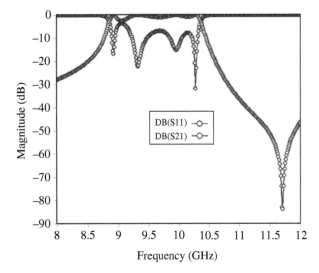

Figure 8.62 Edge-coupled bandpass filter simulation results with Sonnet.

8.7 End-Coupled, Capacitive Gap, Half-Wavelength Resonator Bandpass Filters

The general configuration of an end-coupled microstrip bandpass filter is shown in Figure 8.63.

The gap between two adjacent open ends is capacitive and can be represented by inverters. J inverters tend to reflect high impedance levels to the ends of each half-wavelength resonator, causing the resonator to act like a shunt resonator type of filter. The design equations for the inverters are given by Eqs. (8.107)–(8.109). The gap between each resonator can be represented by the equivalent circuit shown in Figure 8.64.

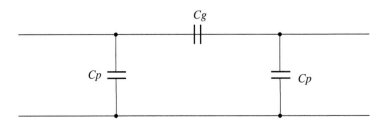

Figure 8.63 End-coupled microstrip bandpass filter.

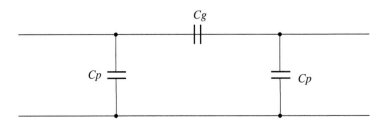

Figure 8.64 Capacitive-gap equivalent circuit.

Assuming the capacitive gap acts perfectly, the susceptance of the series capacitance discontinuities can be found from

$$\frac{B_{j,j+1}}{Y_0} = \frac{\frac{J_{j,j+1}}{Y_0}}{1 - \left(\frac{J_{j,j+1}}{Y_0}\right)^2} \tag{8.124}$$

and

$$\theta_j = \pi - \frac{1}{2}\left[\tan^{-1}\left(\frac{2B_{j-1,j}}{Y_0}\right) + \tan^{-1}\left(\frac{2B_{j,j+1}}{Y_0}\right)\right] \tag{8.125}$$

θ in (8.125) is given in radians. Thus, the final length of each resonator can be found from

$$\ell_j = \frac{\lambda_{g0}}{2\pi}\theta_j - \Delta\ell_j^{e1} - \Delta\ell_j^{e2} \tag{8.126}$$

where

$$\Delta\ell_j^{e1} = \frac{\omega_0 C_p^{j-1,j}}{Y_0}\frac{\lambda_{g0}}{2\pi} \tag{8.127}$$

$$\Delta\ell_j^{e2} = \frac{\omega_0 C_p^{j,j+1}}{Y_0}\frac{\lambda_{g0}}{2\pi} \tag{8.128}$$

The coupling gap between each resonator can be found such that the resultant series capacitance is equal to

$$C_g^{j,j+1} = \frac{B_{j,j+1}}{\omega_0} \tag{8.129}$$

The gap dimensions can be calculated using the closed-form expressions given. Planar electromagnetic simulators can also be utilized to obtain the capacitance values shown in Figure 8.64, with the simulation of a two-port microstrip gap shown in Figure 8.65. The two-port parameters can be obtained from the simulation and can be represented in Y parameters as

$$Y = \begin{bmatrix} Y_{11} & Y_{12} \\ Y_{21} & Y_{22} \end{bmatrix} \tag{8.130}$$

Using the simulated Y parameters, the following capacitance values are obtained.

$$C_g = -\frac{\text{Im}(Y_{21})}{\omega_0} \tag{8.131}$$

$$C_p = -\frac{\text{Im}(Y_{11} + Y_{21})}{\omega_0} \tag{8.132}$$

Example 8.6 End-coupled Resonator Bandpass Filter
Design and simulate end-coupled capacitive gap microstrip bandpass filter with order of $n = 30.1$ dB passband ripple. The center frequency of the filter is at 6 GHz, and the filter has to meet a bandwidth requirement of 2.8%. The filter has to be inserted into a 50 Ω characteristic line impedance. For the microstrip implementation it is given that the dielectric constant is $\varepsilon r = 10.8$, the thickness of the substrate is 1.27 mm, and the width is 1.1 mm.

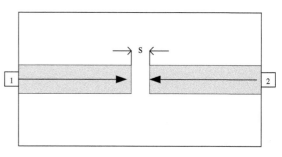

Figure 8.65 Layout of microstrip gap for Sonnet simulation.

Solution

The equivalent circuit of the bandpass filter is derived from the use of the LPF prototype illustrated in Figure 8.66. The *ABCD* parameters of the entire network are found from cascading the *ABCD* parameters of each circuit component. The frequency response of the filter is obtained from converting the network *ABCD* parameters to scattered parameters.

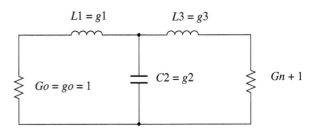

Figure 8.66 Low pass filter prototype.

Chebyshev filter prototype values are determined from Table 8.3 [1]. The normalized component values of the filter are

$g_0 = g_4 = 1.0$, $g_1 = g_3 = 1.0316$, and $g_2 = 1.1474$

The LPF prototype illustrated in Figure 8.66 is used to design the equivalent circuit for the bandpass filter. The low pass prototype filter is converted to the bandpass filter shown in Figure 8.67 with application of the transformation circuits given in Figure 8.42.

In order to determine the *ABCD* parameters of the overall network, the *ABCD* parameters of each component can be cascaded. To obtain the frequency response of the filter, the *ABCD* parameters are converted into scattered parameters as

$$\begin{bmatrix} A & B \\ C & D \end{bmatrix} = \begin{bmatrix} 1 & j\omega L'_s \\ 0 & 1 \end{bmatrix} \begin{bmatrix} 1 & \dfrac{1}{j\omega C_s} \\ 0 & 1 \end{bmatrix} \begin{bmatrix} 1 & 0 \\ \dfrac{1}{j\omega L'_p} & 1 \end{bmatrix} \begin{bmatrix} 1 & 0 \\ j\omega C'_p & 1 \end{bmatrix} \begin{bmatrix} 1 & j\omega L'_s \\ 0 & 1 \end{bmatrix} \begin{bmatrix} 1 & \dfrac{1}{j\omega C'_s} \\ 0 & 1 \end{bmatrix} \tag{8.133}$$

$$\begin{bmatrix} S_{11} & S_{12} \\ S_{21} & S_{22} \end{bmatrix} = \begin{bmatrix} \dfrac{A + \dfrac{B}{Z_0} - CZ_0 + D}{\psi_7} & \dfrac{2(AD - BC)}{\psi_7} \\ \dfrac{2}{\psi_7} & \dfrac{-A + \dfrac{B}{Z_0} - CZ_0 + D}{\psi_7} \end{bmatrix} \tag{8.134}$$

where

$$\psi_7 = A + \frac{B}{Z_0} + CZ_0 + D \tag{8.135}$$

Insertion and return loss can be obtained from S_{21}. The final component values of the filter are calculated and shown in Figure 8.68.

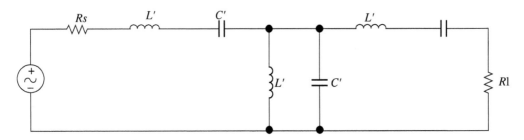

Figure 8.67 Equivalent circuit bandpass filter.

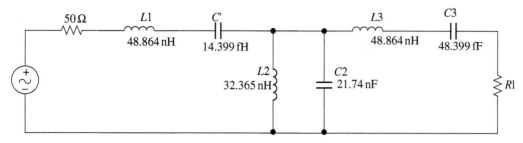

Figure 8.68 Equivalent bandpass filter schematic.

Figure 8.69 Insertion loss of the equivalent bandpass filter.

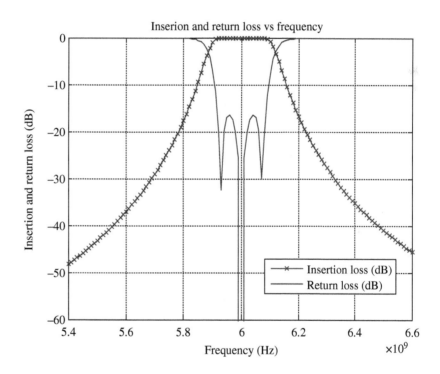

The insertion and return loss are obtained using MATLAB and shown in Figure 8.69.

In order to determine the length of each capacitive gap and the value of each parallel capacitor, several simulations of different gap lengths are performed using Sonnet with the configuration shown in Figure 8.63. In the simulation, the width of the microstrip is set to 1.1 mm, and its thickness is taken to be 1.27 mm. The dielectric constant of the material is given to be 10.8. The Y parameters of the microstrip, operating at 6 GHz, are extracted for each simulation. In addition, the series and parallel capacitor values are calculated using Eqs. (7.96) and (7.97). The results are shown in Table 8.5. The C_g and s values obtained from Table 8.5 are plotted, as shown in Figure 8.70. The equation showing relation between gap length and C_g is obtained via curve fitting and illustrated in Figure 8.70.

In the equation on the graph, x is used for capacitance, C_g, and y is the corresponding gap length. A similar approach is taken to determine the capacitance Cp terms for each equivalent gap. The curve fitting and the corresponding for Cp is shown in Figure 8.71.

Table 8.5 Simulation of microstrip gap.

s (mm)	$Y_{11} = Y_{22}$	$Y_{12} = Y_{21}$	Cg	Cp
0.05	0.004 578	4.412–3	1.1703–13	4.4033–15
0.1	0.003 912	3.594–3	9.5334–14	8.4352–15
0.2	0.003 286	2.695–3	7.1487 –14	1.5677–14
0.5	0.002 685	1.466–3	3.8887 –14	3.2335–14
0.8	0.002 524	8.8508–4	2.3477 –14	4.3474–14
1	0.002 481	6.4386–4	1.7079–14	4.8732–14

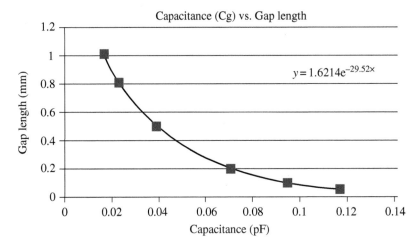

Figure 8.70 *Cg* vs. gap length from simulation.

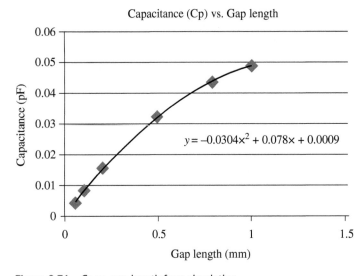

Figure 8.71 *Cp* vs. gap length from simulation.

The summary of the calculated gap lengths and corresponding capacitance values are given in Table 8.6. The end-coupled microstrip bandpass filter is simulated with the physical values calculated using generic MATLAB script with the results shown in Table 8.6. The Sonnet simulation circuit layout with physical dimensions is illustrated in Figure 8.72. The simulation results for insertion and return loss are given in Figure 8.73.

Table 8.6 Summary of the calculated lengths and capacitance values.

Port	C_p (pF)	C_g (pF)	S (mm)
01	0.0051	0.11442	0.055
12	0.0455	0.021482	0.86
23	0.0455	0.021482	0.86
34	0.0051	0.11442	0.055

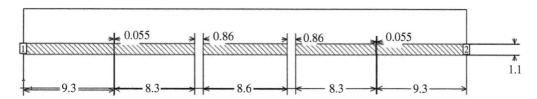

Figure 8.72 Simulation of end-coupled microstrip bandpass filter.

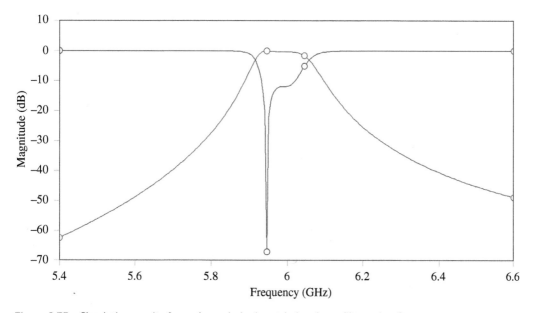

Figure 8.73 Simulation results for end-coupled microstrip bandpass filter using Sonnet.

8.8 Tunable Tapped Combline Bandpass Filters

Figure 8.74a shows a typical tapped combline structure. It has N number of sections according to the desired order of the filter and a lumped capacitive element used to achieve the desired frequency response for the filter. The spacing between the resonators determines the amount of coupling. Typically, all of the capacitor values, the overall length of the resonators, and the width of the resonators are equal. With this immense amount of symmetry, the

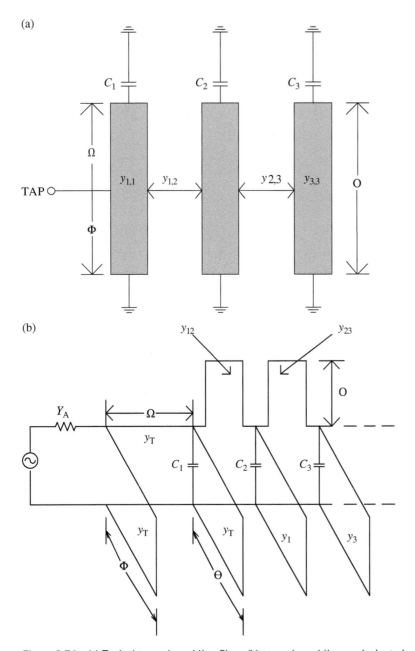

Figure 8.74 (a) Typical tapped combline filter; (b) tapped combline equivalent circuit used to develop design equations.

analysis and the design of the combline structure become less cumbersome. However, the typical combline design is not tunable, so often some alterations have to be made.

An equivalent circuit model is developed for that layout and given in Figure 8.74b. The design equations to find the dimension of the tapped filter in Figure 8.74a are given by

$$b = (Y_{a1}/2) \cdot [\theta_0/\sin^2(\theta_0) + \cot(\theta_0)] \tag{8.136}$$

$$J_{j,j+1}\big|_{j=1,N-1} = wb/\left(g_j g_{j+1}\right)^{1/2} \tag{8.137}$$

$$y_{j,j+1}\big|_{j=1,N-1} = J_{j,j+1} \cdot \tan(\theta_0) \tag{8.138}$$

$$\varphi_0 = \sin^{-1}\left\{[Y_{a1} \cdot w(\cos\theta_0 \sin\theta_0 + \theta_0)/2g_0 g_1 Y_A]^{1/2}\right\} \tag{8.139}$$

where the parameters in Eqs. (8.136)–(8.139) are defined in Table 8.7.

The design equations for the required lumped element filter component values are found from

$$C_1{}^S = Y_{a_1} \cdot \cot(\theta_0)/\omega_0 \tag{8.140}$$

$$y_T = Y_{a_1} - y_{12}^2/Y_{a_1} \tag{8.141}$$

$$C_c^s = Y_A^2 z_T \sin(\theta_0 - \varphi_0)/\left\{[(\sin\theta_0/\sin\varphi)^3 + (\sin\theta_0/\sin\varphi_0) \cdot Y_A^2 z_T^2 \sin^2(\theta_0 - \varphi_0)]\omega_0\right\} \tag{8.142}$$

$$C_{1TOTAL}^S = C_{NTOTAL}^S = C_1^s + C_c^s \text{ (combline) or } C_{1TOTAL}^S = C_{NTOTAL}^S = C_c^s \text{(interdigital)} \tag{8.143}$$

$$C_J^S\big|_{j=2,N-1} = C_1^s \tag{8.144}$$

Once the parameters in Table 8.7 are known, the physical dimensions of the microstrip transmission line parameters are calculated from

$$\frac{w}{d} = \frac{8e^A}{e^{2A} - 2} \text{ for } W/d < 2 \tag{8.145}$$

$$\frac{w}{d} = \frac{2}{\pi}\left[B - 1 - \ln(2B-1) + \frac{\varepsilon_r - 1}{2\varepsilon_r}\left\{\ln(B-1) + 0.39 - \frac{0.61}{\varepsilon_r}\right\}\right] \text{ for } W/d > 2 \tag{8.146}$$

$$\varepsilon_e = \frac{\varepsilon_r + 1}{2} + \frac{\varepsilon_r - 1}{2}\frac{1}{\sqrt{1 + 12(d/W)}} \tag{8.147}$$

$$v_p = \frac{c}{\sqrt{\varepsilon_e}} \tag{8.148}$$

$$\beta = \frac{\omega_0}{v_p} \tag{8.149}$$

$$\theta = \beta l_\theta \tag{8.150}$$

Table 8.7 Parameters in Eqs. (8.136)–(8.139).

N	Order of low pass prototype filter
g_i	Low pass prototype filter element values $i = 0$ to $N+1$
w	Fractional bandwidth of filter
Y_A	Source and load admittance
Y_{a1}	Resonator line admittance(same for all resonators)
Θ_0	Electrical length of resonator at center frequency, $\pi/2$ for interdigital

8.8.1 Network Parameter Representation of Tunable Tapped Filter

To mathematically represent the combline tapped filter with network parameters, consider the filter layout given in Figure 8.75. Note that the electrical lengths, θ, are functions of frequency. The varactors and biasing capacitors are represented as lumped capacitors; however, these are lossy capacitors. In later analysis steps, general impedances will be used in place of both the shorts and the capacitors.

Next, the transmission line equivalent circuit is drawn in Figure 8.76. Coupling between pairs of lines is denoted by dashed X's. The feedlines (Z_o) are of arbitrary length, as the source and load impedances are the same as the feedline characteristic impedance. The lower conductor of each transmission line is the ground plane.

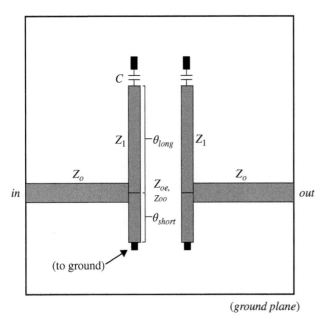

Figure 8.75 Microstrip layout of circuit.

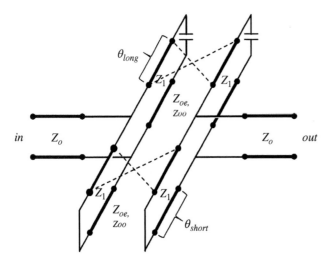

Figure 8.76 Transmission line equivalent circuit.

Figure 8.77 Network representation of circuit: (a) the two sets of coupled lines are each represented by two port networks; (b) the overall circuit is represented by the sum of the admittance matrices of the two coupled line networks.

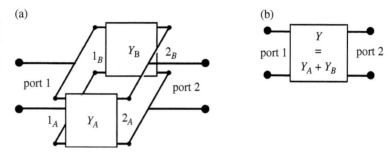

If each set of coupled lines is represented by a two-port network, the overall network is then the parallel connection of the two coupled line networks (Figure 8.77). This suggests that using admittance parameters is convenient.

As each set of coupled lines is terminated by some impedance, general impedance is used when computing $[Y_x]$. The general coupled line case is shown in Figure 8.78.

Even- and odd-mode analysis of the set of coupled lines in Figure 8.78 is now carried out. This is depicted in Figure 8.79. The driving concept of even- and odd-mode analysis is superposition. That is, the circuit is analyzed from two perspectives, the even mode and the odd mode, and the resulting voltages and currents are respectively added together, as given by Eq. (8.151).

$$\begin{cases} V_1 = V_1^e + V_1^o \\ V_2 = V_2^e + V_2^o \\ I_1 = I_1^e + I_1^o \\ I_2 = I_2^e + I_2^o \end{cases} \tag{8.151}$$

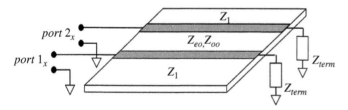

Figure 8.78 General coupled line case where the lines are excited from a common end and terminated by the same general impedance.

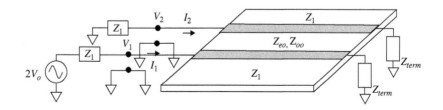

Figure 8.79 Overall excitation circuit for even- and odd-mode analysis.

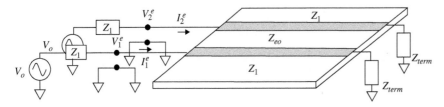

Figure 8.80 Even-mode excitation circuit.

The even-mode circuit is given in Figure 8.80.

Using transmission line theory, the input impedance at each port is given by

$$Z_{in_1}^e = Z_{in_2}^e = Z_{oe} \frac{Z_{term} + jZ_{oe} \tan(\theta)}{Z_{oe} + jZ_{term} \tan(\theta)} = Z_{in}^e \tag{8.152}$$

Therefore, the port voltages can be computed as the outputs of voltage dividers, and the currents from the input voltages and overall impedances.

$$\begin{cases} V_1^e = V_2^e = V_o \dfrac{Z_{in}^e}{Z_1 + Z_{in}^e} \\ I_1^e = I_2^e = \dfrac{V_o}{Z_1 + Z_{in}^e} \end{cases} \tag{8.153}$$

Similarly, the odd mode voltages and currents are computed from the odd mode excitation circuit (Figure 8.81). The resulting input impedances, voltages, and currents are

$$Z_{in_1}^o = Z_{in_2}^o = Z_{oe} \frac{Z_{term} + jZ_{oe} \tan(\theta)}{Z_{oe} + jZ_{term} \tan(\theta)} = Z_{in}^o \tag{8.154}$$

$$\begin{cases} V_1^o = V_2^o = V_o \dfrac{Z_{in}^o}{Z_1 + Z_{in}^o} \\ I_1^o = I_2^o = \dfrac{V_o}{Z_1 + Z_{in}^o} \end{cases} \tag{8.155}$$

Using the superposition equations, the total port voltages and currents are calculated. These are used to obtain the desired result: the admittance matrix.

$$\begin{bmatrix} I_1 \\ I_2 \end{bmatrix} = [Y^x] \begin{bmatrix} V_1 \\ V_2 \end{bmatrix} = \begin{bmatrix} Y_{11}^x & Y_{12}^x \\ Y_{21}^x & Y_{22}^x \end{bmatrix} \begin{bmatrix} V_1 \\ V_2 \end{bmatrix} \tag{8.156}$$

As the coupled lines are symmetric, the diagonal entries in the matrix will be equal.

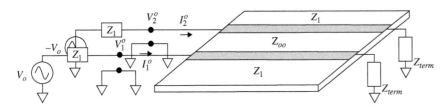

Figure 8.81 Odd-mode excitation circuit.

$$\begin{bmatrix} I_1 \\ I_2 \end{bmatrix} = \begin{bmatrix} Y^x_{11} & Y^x_{12} \\ Y^x_{12} & Y^x_{11} \end{bmatrix} \begin{bmatrix} V_1 \\ V_2 \end{bmatrix} \tag{8.157}$$

This gives two equations with two unknowns. Therefore, the admittance matrix describing a pair of coupled lines with general, but identical, terminations is obtained. Each of these is computed for the cases of the short terminated short lines and the varactor terminated long lines. The admittance matrices are added together to obtain the overall admittance matrix of the filter. This is converted to S parameters (specifically S_{21}) and the insertion loss is computed.

8.9 Dual Band Bandpass Filters using Composite Transmission Lines

In wireless front-end circuits employing multiple frequency bands, dual band components can reduce system complexity and cost. The second operating frequency of a conventional microwave filter employing entirely right-handed transmission lines (RHTL) is usually the first odd harmonic of the base frequency. Right-handed materials are not practical in dual band configurations, since current wireless standards do not employ operating frequencies separated by a factor of three. This limitation can be overcome by implementing components consisting of composite right/left-handed transmission lines (CRLH) as shown in Figure 8.82a,b. An equivalent circuit of the CRLH transmission line is shown in Figure 8.82. The left-hand section of the transmission line consists of two T type unit cells with series capacitors of value C_L and shunt inductors of value L_L, which are implemented with surface mount (SMT) components.

The synthesis procedure dual band filter for an arbitrary pair of frequencies f_1 and f_2 can be outlined as follows.

Step 1 – For a given f_1 and f_2, solve for P and Q using (8.158) and (8.159).
Step 2 – Choose a value of N.
Step 3 – Use Q and N to determine the $L_L C_L$ product in (8.160).
Step 4 – Solve for L_L and C_L with the product of $L_L C_L$ and the chosen characteristic impedance Z_{oL} using (8.161) and (8.162).
Step 5 – Use Pf_1 or Pf_2 to obtain the electrical length of the RHTL and, hence, its physical length l_R by using standard microstrip line formulas.
Step 6 – Calculate f_c^{LH} from (8.163). IF $f_c^{LH} < f_1$, the design is complete. Otherwise, choose a larger N in step 2 and repeat steps 3–6.

$$P \approx \frac{\pi}{2} * \frac{3*(f_2 - f_1)}{(f_2^2 - f_1^2)} \tag{8.158}$$

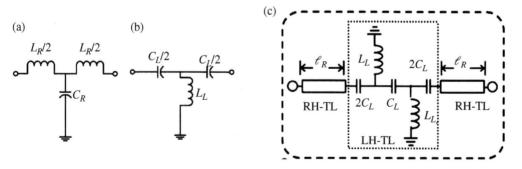

Figure 8.82   CRLH TLs: (a) unit cell RH TL; (b) unit cell left-handed transmission line (LH TL); (c) equivalent circuit.

$$Q \approx \frac{\pi}{2} * \frac{\dfrac{3}{f_2} - \dfrac{1}{f_1}}{\dfrac{1}{f_1^2} - \dfrac{1}{f_2^2}}$$

(8.159)

$$Q = \frac{N}{2\pi * \sqrt{L_L * C_L}}$$

(8.160)

$$L_L = Z_{OL}\sqrt{L_L C_L}$$

(8.161)

$$C_L = \frac{\sqrt{L_L C_L}}{Z_{OL}}$$

(8.162)

$$f_c^{LH} = \frac{1}{4\pi\sqrt{L_L C_L}}$$

(8.163)

Solving the preceding equations gives the user the capacitor and inductor values to be used in T type unit cells. A MATLAB script can be created to arbitrarily calculate the correct values and transmission line properties based on the desired center frequencies. To complete this calculation, the user only has to have knowledge of the desired center frequencies and characteristic impedance, as well as material thickness and relative dielectric constant.

8.10 Engineering Application Examples

In this section, several real-world filter design, simulation, and implementation examples are given.

Design Example 8.1   Design, Simulation, and Implementation of Stepped Impedance Filter
Design, simulate, and implement a stepped impedance LPF with cutoff frequency at 2 GHz. A filter is to provide a minimum 30 dB attenuation at 3 GHz. The source and load impedances of the filter are given to be 50 Ω. The ripple is defined to be not more than 0.5 dB in the passband. In addition, it is required to use FR4 as a substrate with dielectric constant of 3.7 and dielectric thickness of 60 mil.

Solution
It is mentioned in the filter specifications that there is a ripple requirement in the passband. This is an indication of a Chebyshev type filter.

Step 1 – Use Figure 8.26 to determine the required number of elements to achieve minimum 30 dB attenuation at $f = 3$ GHz.

$$\left|\frac{\omega}{\omega_c}\right| - 1 = \frac{3}{2} - 1 = 0.5 \rightarrow N = 7$$

Step 2 – Use Table 8.3 to determine the normalized LPF component values with 0.5 dB ripple as

Ripple = 0.5 dB							
N	g1	g2	g3	g4	g5	g6	g7	g8
7	1.7372	1.2583	2.6381	1.3444	2.6381	1.2583	1.77372	1

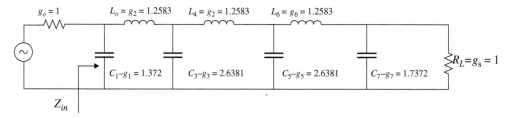

Figure 8.83 Seventh-order normalized LPF for Chebyshev response.

We begin a prototype LPF circuit with shunt C, as shown in Figure 8.83.
The normalized component values for the filter are obtained from the table as

$$C_1 = C_7 = 1.7372, C_3 = C_5 = 2.6381 \ L_2 = L_6 = 1.2583, L_4 = 1.3444$$

Step 3 – Apply scaling impedance and frequency.

$$R'_s = R_o = 50 \, \Omega$$

$$R'_L = R_o R_L = 50(1) = 50 \, \Omega$$

$$C'_1 = C'_7 = \frac{1.7372}{50(2\pi \times 2 \times 10^9)} = 2.764 \, \text{pF}$$

$$L'_2 = L'_6 = 50 \frac{1.2583}{(2\pi \times 2 \times 10^9)} = 5 \, \text{nH} \ C'_3 = C'_5 = \frac{2.6381}{50(2\pi \times 2 \times 10^9)} = 4.197 \, \text{pF}$$

$$L'_4 = 50 \frac{1.3444}{(2\pi \times 2 \times 10^9)} = 5.342 \, \text{nH}$$

The MATLAB response of this circuit is shown in Figure 8.84.

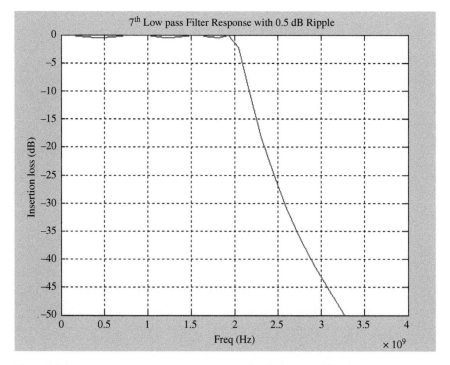

Figure 8.84 Attenuation response of seventh-order Chebyshev LPF.

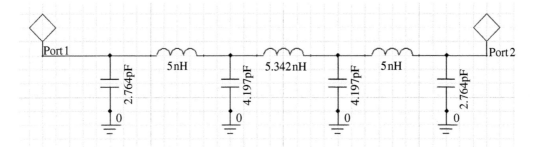

Figure 8.85 Simulated seventh-order Chebyshev LPF.

This circuit is simulated with Ansoft Designer for its frequency response and the simulated circuit. Its response versus frequency are illustrated in Figures 8.85 and 8.86, respectively. Simulation results confirm MATLAB results, and attenuation requirements at the cutoff frequency and 3 GHz are also met. At this point, we can go ahead and move to step 4 to transform our filter into a step impedance filter using the design method described.

Step 4 – Choose Z_{Low} and Z_{High}.

The low impedance limit is found to be

$$Z_{Low} < \frac{\pi}{4\omega_c C} \rightarrow Z_{Low} < 14.88\ \Omega$$

and the impedance limit is found to be

$$Z_{High} > \frac{4\omega_c L}{\pi} \rightarrow Z_{High} > 85.47\ \Omega$$

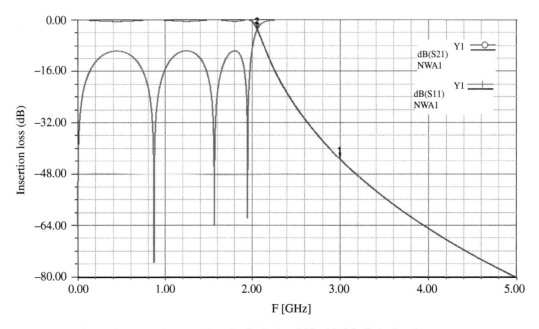

Figure 8.86 Simulation result for seventh-order Chebyshev LPF with 0.5 dB ripple.

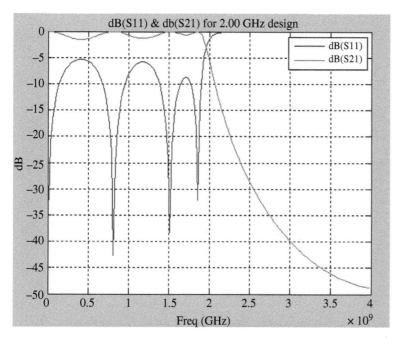

Figure 8.87 Attenuation profile for step impedance filter.

Based on the defined limits obtained, we set our impedances as

$$Z_{Low} = 14\,\Omega \text{ and } Z_{High} = 120\,\Omega$$

Step 5 – Determine the physical dimensions of the transmission lines.

A MATLAB program has been developed to obtain the physical dimensions of the transmission lines in a step impedance filter. The physical dimensions obtained by MATLAB are $l_1 = 9.5$ mm, $l_2 = 5.2$ mm, $l_3 = 14.4$ mm, $l_4 = 5.6$ mm, $l_5 = 14.4$ mm, $l_6 = 5.2$ mm, $l_7 = 9.5$ mm, $w_1 = 17.6$ mm,$w_2 = 0.5$ mm,$w_3 = 17.6$ mm,$w_4 = 0.5$ mm, $w_5 = 17.6$ mm,$w_6 = 0.5$ mm, and $w_7 = 17.6$ mm. The attenuation profile obtained by MATLAB is shown in Figure 8.87.

Comparison of the attenuation profiles obtained by the ideal Chebyshev LPF and step impedance filter response by MATLAB shows that they are in agreement. It is important to note that any slight difference between two profiles will be due to the approximation applied in the step impedance filter design.

Step 6 – Optimize and obtain final design with electromagnetic simulator.

We can now use an electromagnetic simulator to optimize this design before implementation. A Sonnet planar electromagnetic simulator is used to optimize for the final physical dimensions of the filter. The final dimensions of the filter are given in Table 8.8. The simulated structure is illustrated in Figure 8.88.

The simulation results of Sonnet are shown in Figure 8.89. It shows that the desired attenuation profile has been obtained with the optimized physical dimensions given in Table 8.4.

Step 7 – Implement the filter.

Since the final design now meets with the required attenuation profile, we can implement the design. The implemented structure is shown in Figure 8.90.

Table 8.8 Final dimensions of step impedance LPF.

Element	Width (mm)	Length (mm)
Rs	3.25	10
C1	11.5	7.8
L2	0.48	6.0
C3	11.5	11.8
L4	0.48	6.4
C5	11.5	11.8
L6	0.48	6.0
C7	11.5	7.8
RL	3.25	10

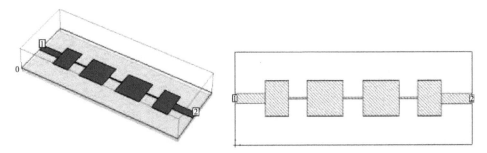

Figure 8.88 Simulated step impedance LPF structure.

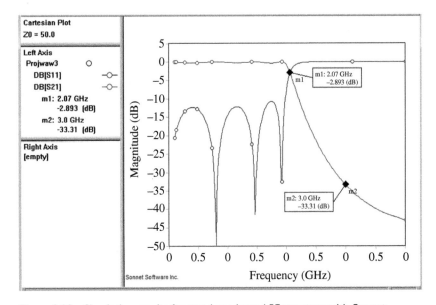

Figure 8.89 Simulation results for step impedance LPF structure with Sonnet.

Figure 8.90 Step impedance filter is implemented.

Figure 8.91 Measured results for step impedance filter.

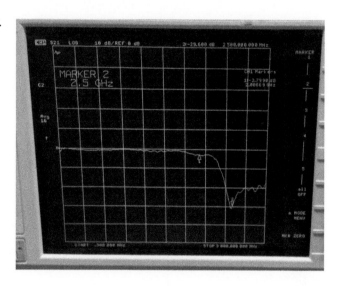

The measured values showing the insertion loss of the filter are given in Figure 8.91. It can be shown using Figure 8.60 that the attenuation at 2 GHz is 2.8 dB and more than 30 dB at 3 GHz.

Design Example 8.2 Stepped Impedance Resonator Bandpass Filters

Design a triple band bandpass filter with SIR bandpass filters using an improved coupling scheme. The center frequencies for each band are defined to be 1, 2.4, and 3.6 GHz. Use RO 4003 as a substrate with 32 mil thickness and 3.38 dielectric constant. The insertion loss in the passbands is required to be −3 dB or better. The return loss in the first and second bands should be −20 dB or lower. The third band stopband attenuation is specified to be −30 dB or lower. The ripple in the passband should not exceed 0.1 dB.

Table 8.9 The layout of the filter using the dimensions.

Given parameters	f_o (GHz)	f_{s1} (GHz)	f_{s2} (GHz)	d (mi)	ε_r	Z_3 (Ω)
	1	2.4	3.6	32	3.38	50
Calculated parameters	K_1	K_2	Z_1	Z_2		
	0.714	0.75	93.3	69.99		
Calculated physical dimensions	l_1 (in.)	l_2 (in.)	l_3 (in.)	w_1 (mil)	w_2 (mil)	w_3 (mil)
	1.044	0.512	0.502	22.57	41.48	74.1

Solution

The calculated design parameters for the given resonant frequency in each band using Eqs. (8.95)–(8.105) are given in Table 8.9.

The layout of the filter using the dimensions in Table 8.9 is shown in Figure 8.92.

The filter that is shown in Figure 8.92 is simulated with the planar electromagnetic simulator Sonnet, and then the prototype is built using the dimensions illustrated on the figure with Roger 4003. The final version of the filter that is constructed is illustrated in Figure 8.93. The 50 Ω SMA connectors are used for termination at the ports of the filter. The filter performance is measured with the Agilent 8753ES network analyzer. The simulation

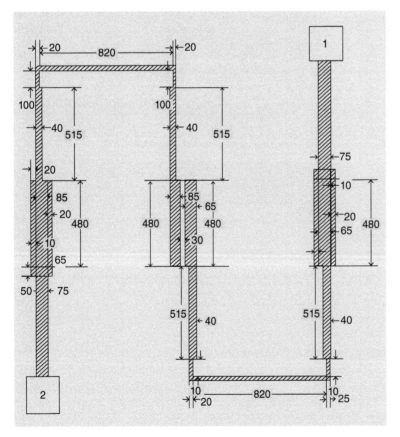

Figure 8.92 Layout of the triple band bandpass filter.

and measurement results showing overall performance of the filter up to 4 GHz for insertion loss, $|S_{21}|$, and return loss, $|S_{11}|$, are illustrated in Figure 8.94. The results for insertion loss are close in the passband and deviate slightly from each other in the stopband. Figures 8.95–8.97 take a closer look at the filter performance in the first, second, and third frequency bands. The most deviation between the simulated and measured results in the passband is observed in the second band at 2.4 GHz. The simulated and measured results for insertion loss are tabulated in Table 8.6. Based on the simulation and measurement results, the insertion loss specification is met, except in the second band, where it is slightly lower. The simulation and measured results for the return loss are found to be also very close. The return loss specification is met in all the frequency bands, as illustrated in Figures 8.67–8.70.

The effect of coupling between each resonator on the filter performance is studied using a planar electromagnetic simulator for three different cases in each frequency band. These cases represent different coupling distances between SIR bandpass filters and are designated by g. The coupling distance, g, is set to be 10, 30, and 60 mil. The simulation results up to 4 GHz are shown in Figure 8.98. Figures 8.99–8.101 take a closer look at the effect of coupling in each frequency band for the insertion loss. It has been observed that, as the coupling gap between SIR bandpass filters is decreased, a wider band-

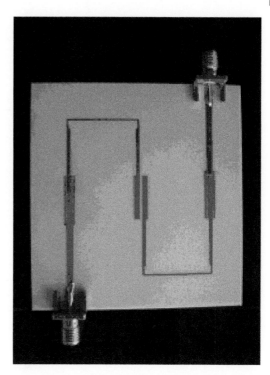

Figure 8.93 The constructed triple band tri-section bandpass filter using SIR bandpass filters.

width is obtained for each frequency band in the passband. However, although the wider bandwidth is obtained for the minimum coupling spacing, it results in ripples in the passband. As a consequence, the insertion loss is decreased more than 2 dB at the center frequency of the first frequency band. As the coupling spacing increases, the bandwidth gets narrower and insertion loss also decreases. The coupling spacing 30 mil between the coupled

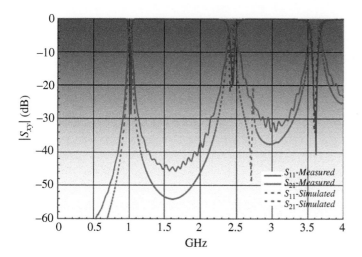

Figure 8.94 Measured and simulation results for insertion loss and return loss up to 4 GHz.

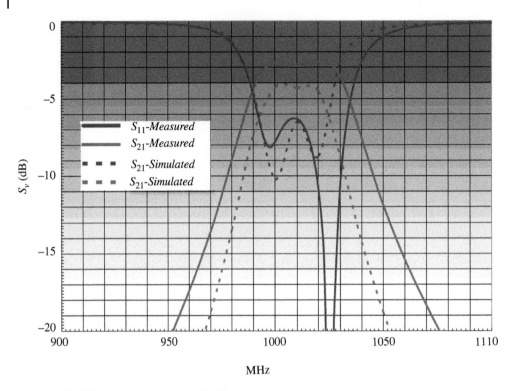

Figure 8.95 Filter performance in the first frequency band.

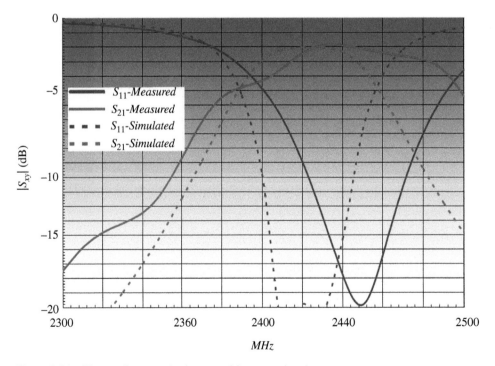

Figure 8.96 Filter performance in the second frequency band.

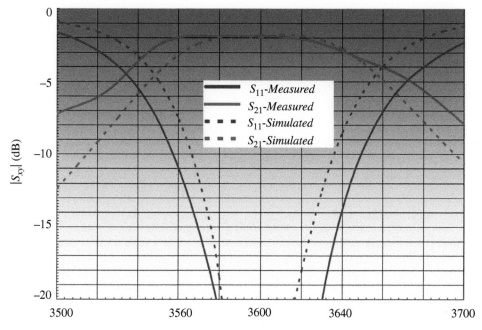

Figure 8.97   Filter performance in the third frequency band.

Figure 8.98   Coupling effect between SIR bandpass filters on insertion loss up to 4 GHz.

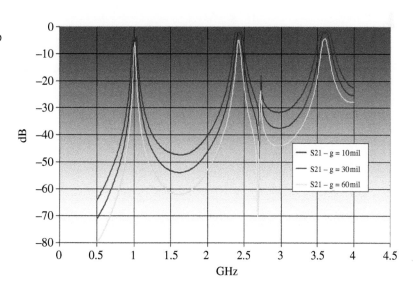

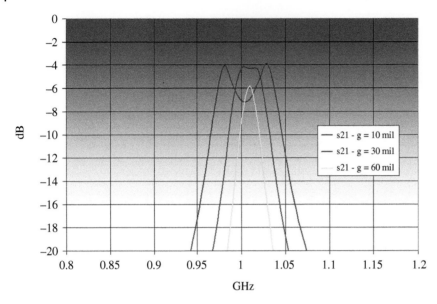

Figure 8.99   Effect of coupling in the first frequency band for tri-section triple band bandpass filter.

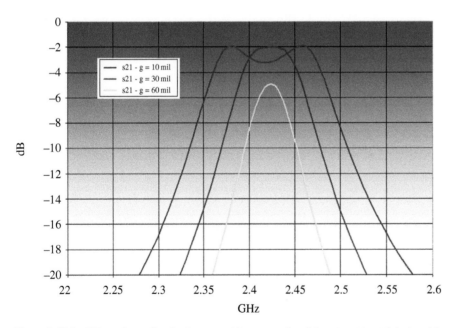

Figure 8.100   Effect of coupling in the second frequency band for tri-section triple band bandpass filter.

Figure 8.101 Effect of coupling in the third frequency band for tri-section triple band bandpass filter.

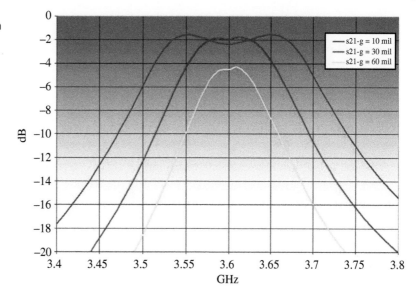

Figure 8.102 Coupling effect between SIR bandpass filters for return loss up to 4 GHz.

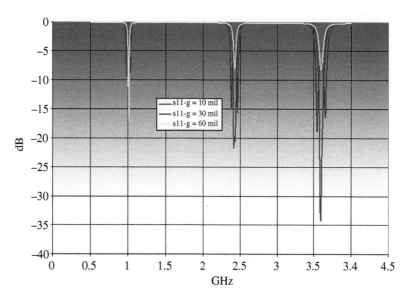

lines gives optimum filter performance in all frequency bands in terms of bandwidth and insertion loss level. This shows that, although the minimum coupling spacing between the coupling segments gives the highest coupling level, it does not necessarily give the best filter performance since the insertion loss characteristics become worse, unlike the improvements in the bandwidth.

The return loss of the filter in the stopband for each frequency band when the coupling gap is changed from 10 to 60 mil shows different characteristics. The minimum coupling spacing results in resonances in all frequency bands. When the coupling spacing is increased to 60 mil, the return loss improves in the first frequency band, but becomes worse in the second and third frequency band. So, in the second and third frequency bands, the return loss of the filter is improved as the gap distance decreases, whereas this phenomenon reverses in the first frequency band. The best overall filter performance is again obtained when the coupling spacing is 30 mil. This is illustrated in Figure 8.102. The improvement in the return loss with 30 mil spacing is over 15 dB in the third frequency band

Table 8.10 Tabulation of measurement and simulation results.

Frequency (GHz)	Simulation (dB)	Measurement (dB)
1	4	3
2.4	3	4.5
3.6	2	2

with respect to a return loss level with minimum spacing. As a result, the coupling distance between each SIR bandpass filter can be used as a tool to achieve the desired bandwidth characteristics in the passband and attenuation characteristics. The results are summarized in Table 8.10.

Design Example 8.3   Dual Band Filters Using CRLH TLs

Design dual band bandpass filter using CRHLs with center frequencies at 900 MHz and 2.4 GHz, which are the operational frequencies for many RFID and Wi-Fi systems. To implement this filter, use 1 mil DuPont Pyralux and 60 mil FR4.

Solution

The design equations (8.158)–(8.163) and MATLAB GUI can be used to calculate the component values, as shown in Figure 8.103.

Using the values obtained for C and L, the circuit was simulated using Ansoft Designer. The schematic used to simulate the bandpass is shown in Figure 8.104.

PCB layout was also created using the designer tools. The filter prototype is built using the milling machine with PCB layout and blueprints of components. Figure 8.105 shows the PCB layout for the desired filter. The upper and lower big green rectangles represent the transmission lines. The four small squares in blue represent the location of the inductors, and the four red rectangles represent the location of the capacitors.

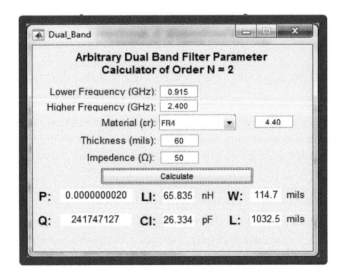

Figure 8.103   MATLAB GUI to calculate design parameters.

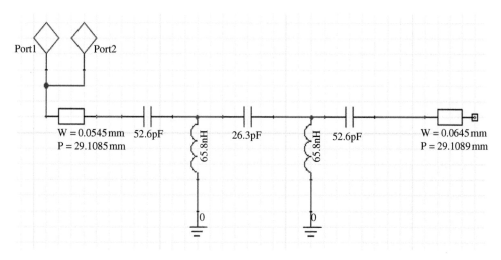

Figure 8.104 Dual band bandpass filter using CRLH TLs.

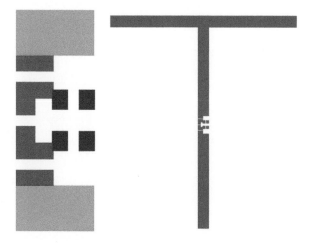

Figure 8.105 PCB layout.

Figure 8.106 shows the structure of the built filter, and the lamped element capacitors and inductors soldered to the board. This filter was tested using a network analyzer. The results are presented in Figures 8.107 and 8.108.

The summary of the simulation results for the bandpass filter are given in Table 8.11. The prototype is measured as show in Figure 8.109. The insertion loss, S_{21}, for FR4 is given in Figure 8.110. It will be observed that the response at a lower frequency was very good but the response was not as good at the higher frequency end.

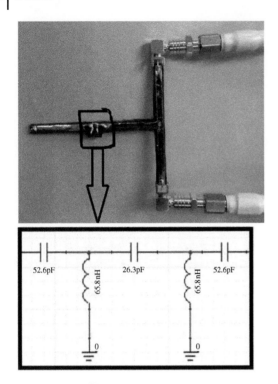

Figure 8.106 Filter prototype.

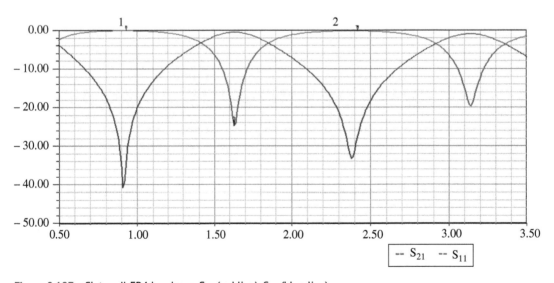

Figure 8.107 Sixty-mil FR4 bandpass, S_{21} (red line), S_{11} (blue line).

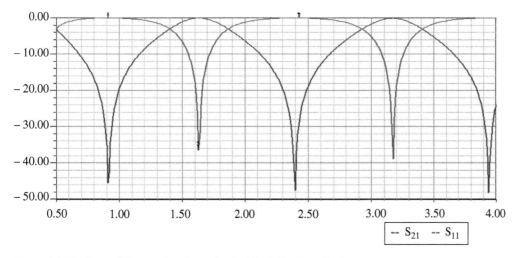

Figure 8.108   One-mil Pyralux bandpass, S_{21} (red line), S_{11} (blue line).

Table 8.11   Summary of bandpass responses.

	Bandpass filter	
	60 mil FR4	1 mil Pyralux AP
Lower BW (3 dB)	900 MHz	900 MHz
Upper BW (3 dB)	990 MHz	1.06 GHz
915 MHz IL	−0.07 dB	−0.01 dB
2.4 GHz IL	−0.02 dB	−0.03 dB
1.6 GHz RL	−24.74 dB	−36.36 dB

Figure 8.109   Measurement setup for filter response.

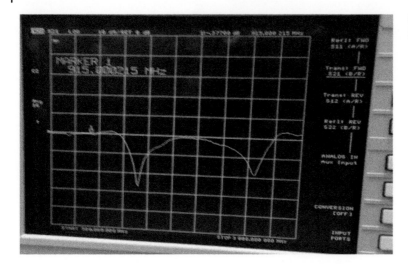

Figure 8.110 Sixty-mil FR4 bandpass insertion loss.

References

1 Eroglu, A. (2013). *RF Circuit Design Techniques for MF-UHF Applications*. CRC Press.
2 Eroglu, A. and Lee, J.K. (2008). The complete design of microstrip directional couplers using the synthesis technique. *IEEE Transactions on Instrumentation and Measurement* **57** (12): 2756–2761.

Problems

Problem 8.1
Design a binomial LPF with a cutoff frequency $f_c = 2$ GHz to have a minimum 30 dB attenuation at 5 GHz when source and load impedances are 50 Ω. Obtain the frequency characteristics of the filter via network parameters.

Problem 8.2
Compare the attenuation characteristics of fifth-order binomial and Chebyshev filters in the passband with a cutoff frequency $f_c = 2$ GHz when source and load impedances are 50 Ω.

Problem 8.3
Design a Chebyshev LPF with a cutoff frequency $f_c = 1.5$ GHz and a ripple 3 dB to have minimum 50 dB attenuation at 3 GHz when source and load impedances are 50 Ω. Obtain the frequency characteristics of the filter via network parameters.

Problem 8.4
Design a fifth-order HPF with a 3 dB equal ripple response, a cutoff frequency $f_c = 3$ GHz, and an impedance of 50 Ω. What is the resulting attenuation at 2 GHz? Obtain the frequency characteristics of the filter.

Problem 8.5
Design a five-section bandpass filter with a maximally flat group delay response at $f = 1$ GHz. Fractional bandwidth should be $\Delta = 5\%$. The system impedance is 50 Ω. Obtain the frequency characteristics of the filter.

Problem 8.6

Design a five-section bandpass filter with a Chebyshev response at $f = 1$ GHz with a 3 dB ripple. Fractional bandwidth should be $\Delta = 5\%$. The system impedance is 50 Ω. Obtain the frequency characteristics of the filter. Compare your frequency response results with the binomial response in Problem 8.5.

Problem 8.7

Design a third-order BSF with a 0.5 dB Chebyshev response operating at the center frequency of $f = 1.5$ GHz. The fractional bandwidth should be 5%. The system impedance is 50 Ω. Obtain the frequency characteristics of the filter and identify the attenuation at $f = 1.455$ GHz.

Problem 8.8

Design a fifth-order BSF with a 0.5 dB Chebyshev response operating at the center frequency of $f = 1.5$ GHz. The fractional bandwidth is to be 5%. The system impedance is 50 Ω. Obtain the frequency characteristics of the filter and identify the attenuation at $f = 1.455$ GHz. Compare your results with the results you obtained in Problem 8.7.

Problem 8.9

Design and simulate a fifth-order, edge-coupled half-wavelength resonator bandpass filter with 0.5 dB passband ripple, $f_c = 8$ GHz, $\varepsilon_r = 4.4$, FBW = 0.1, and dielectric thickness $h = 1$ mm. Obtain the frequency response of the filter, inverter parameters, even- and odd-mode impedances, and physical lengths.

Problem 8.10

Design and simulate a third-order end-coupled capacitive gap microstrip bandpass filter with order of $n = 3$ and 0.5 dB ripple. The center frequency of the filter is at 8 GHz, and the filter has to meet a bandwidth requirement of 5%. The filter has to be inserted into 50 Ω characteristic line impedance. Calculate the required gap capacitances.

Design Challenge 8.1 Wideband BPF Design

Design, simulate, and build a wideband bandpass filter with selectivity based on two parallel coupled lines resonators centered by a *T* inverted shape, as shown in Figure 8.111. The design goals for the case study are a 3.5 GHz center frequency, a 70% ultrawide fractional bandwidth and a minimum insertion loss of 0.3 dB. The filter design is composed of two parallel-coupled lines centered by a *T*-shaped transmission line with 50 Ω port impedances.

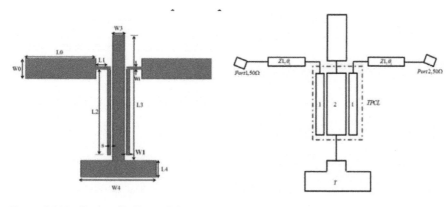

Figure 8.111 Design Challenge 8.1.

9

Waveguides

9.1 Introduction

Waveguides are used to transmit microwave power from one point to another. The electronic and magnetic fields are confined within waveguides, hence power loss due to radiation is prevented. There can be several modes of propagation inside waveguides, and these are found from Maxwell's equations. There is a certain frequency called *cutoff frequency* for each mode and wave can propagate when its frequency exceeds the cutoff frequency. The *dominant mode* in a waveguide is the mode with the lowest cutoff frequency. In this chapter, some of the most commonly used waveguides are discussed and their applications detailed.

9.2 Rectangular Waveguides

In this section, rectangular waveguide design procedure for isotropic, gyrotropic, and anisotropic media are given [1]. The wave propagation in waveguide can be analyzed by using two curl equations in the absence of electric and magnetic current density as

$$\nabla \times \overline{E} = i\omega\overline{\overline{\mu}} \cdot \overline{H} \tag{9.1}$$

$$\nabla \times \overline{H} = -i\omega\overline{\overline{\varepsilon}} \cdot \overline{E} \tag{9.2}$$

where $\nabla = \left(\hat{x}\dfrac{\partial}{\partial x} + \hat{y}\dfrac{\partial}{\partial y} + \hat{z}\dfrac{\partial}{\partial z} \right)$. Expansion of Eqs. (9.1) and (9.2) for electric and magnetic field components give two sets of equations.

$$\frac{\partial E_z}{\partial y} - \frac{\partial E_y}{\partial z} = i\omega \left[\mu_{11}H_x + \mu_{12}H_y + \mu_{13}H_z \right] \tag{9.3}$$

$$\frac{\partial E_x}{\partial z} - \frac{\partial E_z}{\partial x} = i\omega \left[\mu_{21}H_x + \mu_{22}H_y + \mu_{23}H_z \right] \tag{9.4}$$

$$\frac{\partial E_y}{\partial x} - \frac{\partial E_x}{\partial y} = i\omega \left[\mu_{31}H_x + \mu_{32}H_y + \mu_{33}H_z \right] \tag{9.5}$$

and

$$\frac{\partial H_z}{\partial y} - \frac{\partial H_y}{\partial z} = -i\omega \left[\varepsilon_{11}E_x + \varepsilon_{12}E_y + \varepsilon_{13}E_z \right] \tag{9.6}$$

RF/Microwave Engineering and Applications in Energy Systems, First Edition. Abdullah Eroglu.
© 2022 John Wiley & Sons Ltd. Published 2022 by John Wiley & Sons Ltd.
Companion website: www.wiley.com/go/eroglu/rfmicrowave

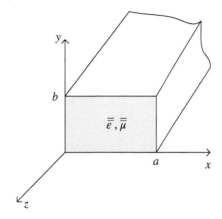

Figure 9.1 Geometry of the rectangular waveguide.

$$\frac{\partial H_x}{\partial z} - \frac{\partial H_z}{\partial x} = -i\omega\left[\varepsilon_{21}E_x + \varepsilon_{22}E_y + \varepsilon_{23}E_z\right] \tag{9.7}$$

$$\frac{\partial H_y}{\partial x} - \frac{\partial H_x}{\partial y} = -i\omega\left[\varepsilon_{31}E_x + \varepsilon_{32}E_y + \varepsilon_{33}E_z\right] \tag{9.8}$$

The geometry of the rectangular waveguide is given in Figure 9.1. The guide walls are assumed to be perfect conductor with $\sigma_c = \infty$. It is assumed that the wave propagation is in the z direction and the electric and magnetic field vector components are functions of x and y. Then, the electric and magnetic field vector amplitudes can be written as

$$F = F(x,y)e^{ikz} \tag{9.9}$$

where

$$F(x,y) = F_x(x,y)\hat{x} + F_y(x,y)\hat{y} + F_z(x,y)\hat{z} \tag{9.10}$$

F represent electric field intensity E or magnetic field intensity H. The time dependence $e^{-i\omega t}$ is assumed. The analysis will be based on the lossless dielectric filling.

9.2.1 Waveguide Design with Isotropic Media

Consider the rectangular waveguide illustrated in Figure 9.1. Assume that the dielectric filling inside the guide is isotropic with permittivity constant ε and permeability constant μ. So that

$$\bar{\bar{\varepsilon}} = \varepsilon\bar{\bar{I}} \text{ and } \bar{\bar{\mu}} = \mu\bar{\bar{I}} \tag{9.11}$$

When Eq. (9.11) is substituted into (9.3)–(9.8) with (9.9), they take the following form.

$$-ikE_y + \frac{\partial E_z}{\partial y} = i\omega\mu H_x \tag{9.12}$$

$$ikE_x - \frac{\partial E_z}{\partial x} = i\omega\mu H_y \tag{9.13}$$

$$\frac{\partial E_y}{\partial x} - \frac{\partial E_x}{\partial y} = i\omega\mu H_z \tag{9.14}$$

and

$$-ikH_y + \frac{\partial H_z}{\partial y} = -i\omega\varepsilon E_x \tag{9.15}$$

$$ikH_x - \frac{\partial H_z}{\partial x} = -i\omega\varepsilon E_y \tag{9.16}$$

$$\frac{\partial H_y}{\partial x} - \frac{\partial H_x}{\partial y} = -i\omega\varepsilon E_z \tag{9.17}$$

Now, we express transverse components of the field vectors, E_x, E_y, and H_x, H_y, in terms of the longitudinal components, E_z, H_z, which are the field vectors in the direction of propagation. By eliminating H_x from (9.12) and (9.16), and H_y from (9.13) and (9.15), we obtain

$$E_y = -\frac{1}{(\omega^2\mu\varepsilon - k^2)}\left[ik\frac{\partial E_z}{\partial y} - i\omega\mu\frac{\partial H_z}{\partial x}\right] \tag{9.18}$$

$$E_x = -\frac{1}{(\omega^2\mu\varepsilon - k^2)}\left[ik\frac{\partial E_z}{\partial x} + i\omega\mu\frac{\partial H_z}{\partial y}\right] \tag{9.19}$$

and

$$H_y = -\frac{1}{(\omega^2\mu\varepsilon - k^2)}\left[ik\frac{\partial H_z}{\partial y} + i\omega\varepsilon\frac{\partial E_z}{\partial x}\right] \tag{9.20}$$

$$H_x = -\frac{1}{(\omega^2\mu\varepsilon - k^2)}\left[ik\frac{\partial H_z}{\partial x} - i\omega\varepsilon\frac{\partial E_z}{\partial y}\right] \tag{9.21}$$

9.2.1.1 *TE*$_{mn}$ **Modes**

For *TE* modes, the longitudinal component of the electric filed should be zero, $E_z = 0$. Application of this constraint on the transverse components of the electric and magnetic field vectors given by (9.18)–(9.21) reduces them to

$$E_y = \frac{i\omega\mu}{(\omega^2\mu\varepsilon - k^2)}\frac{\partial H_z}{\partial x} \tag{9.22}$$

$$E_x = -\frac{i\omega\mu}{(\omega^2\mu\varepsilon - k^2)}\frac{\partial H_z}{\partial y} \tag{9.23}$$

and

$$H_y = -\frac{ik}{(\omega^2\mu\varepsilon - k^2)}\frac{\partial H_z}{\partial y} \tag{9.24}$$

$$H_x = -\frac{ik}{(\omega^2\mu\varepsilon - k^2)}\frac{\partial H_z}{\partial x} \tag{9.25}$$

When we take the derivative of (9.23) and (9.24) with respect to y and x, respectively

$$\frac{\partial E_y}{\partial x} = \frac{i\omega\mu}{(\omega^2\mu\varepsilon - k^2)}\frac{\partial^2 H_z}{\partial x^2} \tag{9.26}$$

$$\frac{\partial E_x}{\partial y} = -\frac{i\omega\mu}{(\omega^2\mu\varepsilon - k^2)}\frac{\partial^2 H_z}{\partial y^2} \tag{9.27}$$

and subtract (9.27) from (9.26), we obtain

$$\frac{\partial E_y}{\partial x} - \frac{\partial E_x}{\partial y} = \frac{i\omega\mu}{(\omega^2\mu\varepsilon - k^2)}\frac{\partial^2 H_z}{\partial x^2} + \frac{i\omega\mu}{(\omega^2\mu\varepsilon - k^2)}\frac{\partial^2 H_z}{\partial y^2}$$

or

$$\frac{\partial E_y}{\partial x} - \frac{\partial E_x}{\partial y} = \frac{i\omega\mu}{(\omega^2\mu\varepsilon - k^2)}\left[\frac{\partial^2 H_z}{\partial x^2} + \frac{\partial^2 H_z}{\partial y^2}\right] \tag{9.28}$$

Since from Eq. (9.14)

$$\frac{\partial E_y}{\partial x} - \frac{\partial E_x}{\partial y} = i\omega\mu H_z \tag{9.29}$$

then

$$\frac{i\omega\mu}{(\omega^2\mu\varepsilon - k^2)}\left[\frac{\partial^2 H_z}{\partial x^2} + \frac{\partial^2 H_z}{\partial y^2}\right] = i\omega\mu H_z$$

or

$$\frac{\partial^2 H_z}{\partial x^2} + \frac{\partial^2 H_z}{\partial y^2} + (\omega^2\mu\varepsilon - k^2)H_z = 0 \tag{9.30}$$

The same procedure can be applied to find the wave equation for TM waves. The solution of the second-order differential equation given in (9.30) takes the following form.

$$H_z(x,y) = A\cos(k_x x)\cos(k_y y) \tag{9.31}$$

Application of boundary conditions at $x = 0$, a and $y = 0$, b require tangential components of the magnetic field vectors to be zero, i.e. $H_z = 0$. When the boundary conditions are applied, we find from (9.31) that

$$k_x = \frac{m\pi}{a}, \qquad m = 0, 1, 2... \tag{9.32}$$

$$k_y = \frac{n\pi}{a}, \qquad n = 0, 1, 2... \tag{9.33}$$

So, Eq. (9.31) can be rewritten as

$$H_z(x,y) = A\cos\left(\frac{m\pi}{a}x\right)\cos\left(\frac{n\pi}{b}y\right) \tag{9.34}$$

The other components of the electric field vectors are found from (9.22)–(9.25).

$$E_x(x,y) = A\frac{-i\omega\mu}{(\omega^2\mu\varepsilon - k^2)}\frac{n\pi}{b}\cos\left(\frac{m\pi}{a}x\right)\sin\left(\frac{n\pi}{b}y\right) \tag{9.35a}$$

$$E_y(x,y) = A\frac{i\omega\mu}{(\omega^2\mu\varepsilon - k^2)}\frac{m\pi}{a}\sin\left(\frac{m\pi}{a}x\right)\cos\left(\frac{n\pi}{b}y\right) \tag{9.35b}$$

$$H_x(x,y) = A\frac{-ik}{(\omega^2\mu\varepsilon - k^2)}\frac{m\pi}{a}\sin\left(\frac{m\pi}{a}x\right)\cos\left(\frac{n\pi}{b}y\right) \tag{9.35c}$$

$$H_y(x,y) = A\frac{-ik}{(\omega^2\mu\varepsilon - k^2)}\frac{n\pi}{b}\cos\left(\frac{m\pi}{a}x\right)\sin\left(\frac{n\pi}{b}y\right) \tag{9.35d}$$

Substitution of (9.34) into (9.30) gives the dispersion relation for a rectangular waveguide filled with isotropic medium as

$$\omega^2\mu\varepsilon - k^2 = \left(\frac{m\pi}{a}\right)^2 + \left(\frac{n\pi}{b}\right)^2$$

or

$$k^2 = \omega^2\mu\varepsilon - \left[\left(\frac{m\pi}{a}\right)^2 + \left(\frac{n\pi}{b}\right)^2\right] \tag{9.36a}$$

Hence

$$k = \sqrt{\omega^2\mu\varepsilon - \left[\left(\frac{m\pi}{a}\right)^2 + \left(\frac{n\pi}{b}\right)^2\right]} \tag{9.36b}$$

Table 9.1 Field components, cutoff frequency, and dispersion relation for rectangular a waveguide filled with isotropic dielectric filling.

Mode	Dispersion relation	f_c cutoff frequency	E and H fields
TE_{mn}	$k = \sqrt{\omega^2 \mu \varepsilon - \left[\left(\frac{m\pi}{a}\right)^2 + \left(\frac{n\pi}{b}\right)^2\right]}$	$\frac{1}{2\pi\sqrt{\mu\varepsilon}}\sqrt{\left[\left(\frac{m\pi}{a}\right)^2 + \left(\frac{n\pi}{b}\right)^2\right]}$	$H_x(x,y) = A\frac{-ik}{(\omega^2\mu\varepsilon - k^2)}\frac{m\pi}{a}\sin\left(\frac{m\pi}{a}x\right)\cos\left(\frac{n\pi}{b}y\right)$ $H_y(x,y) = A\frac{-ik}{(\omega^2\mu\varepsilon - k^2)}\frac{n\pi}{b}\cos\left(\frac{m\pi}{a}x\right)\sin\left(\frac{n\pi}{b}y\right)$ $H_z(x,y) = A\cos\left(\frac{m\pi}{a}x\right)\cos\left(\frac{n\pi}{b}y\right)$ $E_x(x,y) = A\frac{-i\omega\mu}{(\omega^2\mu\varepsilon - k^2)}\frac{n\pi}{b}\cos\left(\frac{m\pi}{a}x\right)\sin\left(\frac{n\pi}{b}y\right)$ $E_y(x,y) = A\frac{i\omega\mu}{(\omega^2\mu\varepsilon - k^2)}\frac{m\pi}{a}\sin\left(\frac{m\pi}{a}x\right)\cos\left(\frac{n\pi}{b}y\right)$

Cutoff occurs when

$$\omega^2 \mu \varepsilon - \left[\left(\frac{m\pi}{a}\right)^2 + \left(\frac{n\pi}{b}\right)^2\right] = 0 \tag{9.37}$$

The cutoff frequency, f_c, is found from (8.37) as

$$f_c = \frac{1}{2\pi\sqrt{\mu\varepsilon}}\sqrt{\left[\left(\frac{m\pi}{a}\right)^2 + \left(\frac{n\pi}{b}\right)^2\right]} \tag{9.38}$$

A summary of the field components, cutoff frequency, and dispersion relation for rectangular a waveguide filled with isotropic dielectric medium in Figure 9.1 when the wave is propagating in y direction are given in Table 9.1.

9.2.2 Waveguide Design with Gyrotropic Media

Now, assume that the rectangular waveguide shown in Figure 9.1 is filled with the gyrotropic media in the existence of an external applied magnetic field, H_0, in y direction. This is illustrated in Figure 9.2.

The permittivity and permeability tensors of the gyrotropic medium are given in Chapter 3 as

$$\bar{\bar{\varepsilon}} = \varepsilon_1\left(\bar{\bar{I}} - \hat{b}_0\hat{b}_0\right) + i\varepsilon_2\left(\hat{b}_0 \times \bar{\bar{I}}\right) + \varepsilon_3\hat{b}_0\hat{b}_0 \tag{9.39}$$

$$\bar{\bar{\mu}} = \mu_1\left(\bar{\bar{I}} - \hat{b}_0\hat{b}_0\right) + i\mu_2\left(\hat{b}_0 \times \bar{\bar{I}}\right) + \mu_3\hat{b}_0\hat{b}_0 \tag{9.40}$$

where $\hat{b}_0$ shows the direction of the external applied magnetic field. When the external applied magnetic field is in y direction, $\hat{b}_0$ can be written as $\hat{b}_0 = [0 \ 1 \ 0]$. Then, (9.39) and (9.40) take the following matrix forms.

$$\bar{\bar{\varepsilon}} = \begin{bmatrix} \varepsilon_1 & 0 & -i\varepsilon_2 \\ 0 & \varepsilon_3 & 0 \\ i\varepsilon_2 & 0 & \varepsilon_1 \end{bmatrix} \tag{9.41}$$

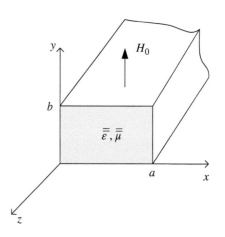

Figure 9.2 Geometry of the rectangular waveguide filled with transversely magnetized gyrotropic media.

and

$$\bar{\bar{\mu}} = \begin{bmatrix} \mu_1 & 0 & -i\mu_2 \\ 0 & \mu_3 & 0 \\ i\mu_2 & 0 & \mu_1 \end{bmatrix} \tag{9.42}$$

The waveguide analysis will be given for general gyrotropic medium using the permittivity and permeability tensors given by (9.41) and (9.42). The results can be simplified for gyromagnetic or gyroelectric medium by just replacing one of the tensors with a unit tensor. When the permeability and permittivity tensors given by (9.41)–(9.42) are substituted into the wave equations given by (9.3)–(9.8), we obtain

$$\frac{\partial E_z}{\partial y} - \frac{\partial E_y}{\partial z} = i\omega\mu_1 H_x - \omega\mu_2 H_z \tag{9.43}$$

$$\frac{\partial E_x}{\partial z} - \frac{\partial E_z}{\partial x} = i\omega\mu_3 H_y \tag{9.44}$$

$$\frac{\partial E_y}{\partial x} - \frac{\partial E_x}{\partial y} = \omega\mu_2 H_x + i\omega\mu_1 H_z \tag{9.45}$$

and

$$\frac{\partial H_z}{\partial y} - \frac{\partial H_y}{\partial z} = -i\omega\varepsilon_1 E_x + \omega\varepsilon_2 E_z \tag{9.46}$$

$$\frac{\partial H_x}{\partial z} - \frac{\partial H_z}{\partial x} = -i\omega\varepsilon_3 E_y \tag{9.47}$$

$$\frac{\partial H_y}{\partial x} - \frac{\partial H_x}{\partial y} = -\omega\varepsilon_2 E_x - i\omega\varepsilon_1 E_z \tag{9.48}$$

The wave equations can be put in the following forms assuming that the wave is propagating in z direction with the form that is given by Eq. (9.9).

$$\frac{\partial E_z}{\partial y} - ikE_y = i\omega\mu_1 H_x - \omega\mu_2 H_z \tag{9.49}$$

$$ikE_x - \frac{\partial E_z}{\partial x} = i\omega\mu_3 H_y \tag{9.50}$$

$$\frac{\partial E_y}{\partial x} - \frac{\partial E_x}{\partial y} = \omega\mu_2 H_x + i\omega\mu_1 H_z \tag{9.51}$$

and

$$\frac{\partial H_z}{\partial y} - ikH_y = -i\omega\varepsilon_1 E_x + \omega\varepsilon_2 E_z \tag{9.52}$$

$$ikH_x - \frac{\partial H_z}{\partial x} = -i\omega\varepsilon_3 E_y \tag{9.53}$$

$$\frac{\partial H_y}{\partial x} - \frac{\partial H_x}{\partial y} = -\omega\varepsilon_2 E_x - i\omega\varepsilon_1 E_z \tag{9.54}$$

The condition of no spatial dependence of the RF fields in the direction of the external applied magnetic fields, i.e. $\dfrac{\partial}{\partial y} = 0$, leads to separation of the waves into TE_{mo} and TM_{mo} modes for the rectangular waveguide when filled with the transversely magnetized gyrotropic medium.

9.2.2.1 TE_{m0} Modes

When the condition $\dfrac{\partial}{\partial y} = 0$ is applied for the wave equations given by (9.49)–(9.54), we obtain

$$-ikE_y = i\omega\mu_1 H_x - \omega\mu_2 H_z \tag{9.55}$$

$$\frac{\partial E_y}{\partial x} = \omega\mu_2 H_x + i\omega\mu_1 H_z \tag{9.56}$$

$$-i\omega\varepsilon_3 E_y = ikH_x - \frac{\partial H_z}{\partial x} \tag{9.57}$$

The field components H_x and H_z can be written in terms of E_y as

$$H_x = \frac{\mu_1}{\omega(\mu_1^2 - \mu_2^2)}\left(kE_y + \frac{\mu_2}{\mu_1}\frac{dE_y}{dx}\right) \tag{9.58}$$

$$H_z = -\frac{i\mu_1}{\omega(\mu_1^2 - \mu_2^2)}\left(k\frac{\mu_2}{\mu_1}E_y + \frac{dE_y}{dx}\right) \tag{9.59}$$

Substitution of (9.58) and (9.59) into (9.57) gives

$$\frac{d^2 E_y}{dx^2} + E_y\left(\omega^2\left(\frac{\mu_1^2 - \mu_2^2}{\mu_1}\right)\varepsilon_3 - k^2\right) = 0 \tag{9.60a}$$

or

$$\frac{d^2 E_y}{dx^2} + E_y k_\perp^2 = 0 \tag{9.60b}$$

where

$$k_\perp^2 = \omega^2\left(\frac{\mu_1^2 - \mu_2^2}{\mu_1}\right)\varepsilon_3 - k^2 \tag{9.61}$$

Solution of the wave equation in (9.60) is

$$E_y = A\sin(k_\perp x) \tag{9.62}$$

Applying the boundary condition at $x = 0$, $E_y = 0$ gives

$$k_\perp = \frac{m\pi}{a} \quad \text{for} \quad m = 1, 2, 3, \ldots \tag{9.63}$$

Hence, the dispersion relation for the rectangular waveguide filled with a transversely magnetized gyrotropic medium can be obtained from (9.61) and (9.63) as

$$k = \sqrt{\omega^2\left(\frac{\mu_1^2 - \mu_2^2}{\mu_1}\right)\varepsilon_3 - k_\perp^2} = \sqrt{\omega^2\left(\frac{\mu_1^2 - \mu_2^2}{\mu_1}\right)\varepsilon_3 - \left(\frac{m\pi}{a}\right)^2} \tag{9.64}$$

The cutoff frequency is calculated from

$$\omega^2\left(\frac{\mu_1^2 - \mu_2^2}{\mu_1}\right)\varepsilon_3 - \left(\frac{m\pi}{a}\right)^2 = 0 \tag{9.65}$$

Table 9.2 Field components, cutoff frequency, and dispersion relation for rectangular waveguide filled with transversely magnetized gyrotropic filling.

Mode	Dispersion relation	f_c cutoff frequency	E and H fields
TE_{m0}	$k = \sqrt{\omega^2 \left(\frac{\mu_1^2 - \mu_2^2}{\mu_1}\right)\varepsilon_3 - \left(\frac{m\pi}{a}\right)^2}$	$\dfrac{1}{2\pi\sqrt{\left(\frac{\mu_1^2 - \mu_2^2}{\mu_1}\right)\varepsilon_3}}\left(\frac{m\pi}{a}\right)$	$E_z = A \sin\left(\frac{m\pi}{a}x\right) H_x$ $= \frac{A\mu_1}{\omega(\mu_1^2 - \mu_2^2)}\left(\beta \sin\left(\frac{m\pi}{a}x\right) + \frac{\mu_2\left(\frac{m\pi}{a}\right)}{\mu_1}\cos\left(\frac{m\pi}{a}x\right)\right)$ $H_z = -\frac{iA\mu_1}{\omega(\mu_1^2 - \mu_2^2)}\left(k\frac{\mu_2}{\mu_1}\sin\left(\frac{m\pi}{a}x\right) + \left(\frac{m\pi}{a}\right)\cos\left(\frac{m\pi}{a}x\right)\right)$

So

$$f_c = \frac{1}{2\pi\sqrt{\left(\frac{\mu_1^2 - \mu_2^2}{\mu_1}\right)\varepsilon_3}}\left(\frac{m\pi}{a}\right) \tag{9.66}$$

The magnetic field components are found by substituting (9.62) into (9.58) and (9.59) as

$$H_x = \frac{A\mu_1}{\omega(\mu_1^2 - \mu_2^2)}\left(k \sin\left(\frac{m\pi}{a}x\right) + \frac{\mu_2\left(\frac{m\pi}{a}\right)}{\mu_1}\cos\left(\frac{m\pi}{a}x\right)\right) \tag{9.67}$$

$$H_z = -\frac{iA\mu_1}{\omega(\mu_1^2 - \mu_2^2)}\left(k\frac{\mu_2}{\mu_1}\sin\left(\frac{m\pi}{a}x\right) + \left(\frac{m\pi}{a}\right)\cos\left(\frac{m\pi}{a}x\right)\right) \tag{9.68}$$

Table 9.2. illustrates the results for the rectangular waveguide filled with a gyrotropic medium.

9.2.3 Waveguide Design with Anisotropic Media

In this section, we assume the dielectric filling for the rectangular waveguide illustrated in Figure 9.1 is a uniaxially anisotropic medium with the following permittivity and permeability tensors.

$$\bar{\bar{\varepsilon}} = \begin{bmatrix} \varepsilon_{11} & 0 & 0 \\ 0 & \varepsilon_{11} & 0 \\ 0 & 0 & \varepsilon_{33} \end{bmatrix}, \quad \bar{\bar{\mu}} = \mu_0 \bar{\bar{I}} \tag{9.69}$$

The optic axis of the anisotropic medium is assumed to be oriented in z direction. The dispersion relation for a vertically uniaxial anisotropic medium is obtained using

$$\bar{\bar{W}}_E \cdot \bar{E} = 0 \tag{9.70}$$

where

$$\bar{\bar{W}}_E = \left[\bar{k} \cdot \bar{\bar{\mu}}^{-1} \cdot \bar{k} + k_0^2 \bar{\bar{\varepsilon}}\right] \tag{9.71}$$

The characteristic values for ordinary and extraordinary waves are obtained when (9.69) is substituted into (9.70).

$$k_{\rho I}^2 = k_0^2 \varepsilon_{11} - k_z^2 \tag{9.72}$$

$$k_{\rho II}^2 = k_0^2 \varepsilon_{33} - k_z^2 \frac{\varepsilon_{33}}{\varepsilon_{11}} \tag{9.73}$$

where

$$k^2 = k_\rho^2 + k_z^2 = k_x^2 + k_y^2 + k_z^2$$

When (9.72) and (9.73) are substituted into (9.70), we obtain the corresponding characteristic field vectors for the characteristic values given by (9.72) and (9.73). The characteristic field vectors for ordinary and extraordinary waves are

$$\overline{E}_1 = \hat{x} A \cos\left(\tan^{-1}\left(\frac{k_y}{k_x}\right)\right) + \hat{y} B \sin\left(\tan^{-1}\left(\frac{k_y}{k_x}\right)\right) \tag{9.74}$$

$$\overline{E}_2 = \frac{\varepsilon_{33} k_z}{\varepsilon_{11}\sqrt{\varepsilon_{33} k_0^2 - k_z^2 \frac{\varepsilon_{33}}{\varepsilon_{11}}}} \left[\hat{x} \cos\left(\tan^{-1}\left(\frac{k_y}{k_x}\right)\right) E_{2z} + \hat{y} B \sin\left(\tan^{-1}\left(\frac{k_y}{k_x}\right)\right) E_{2z}\right] + \hat{z} E_{2z} \tag{9.75}$$

It is clear from (9.74) and (9.75) that the ordinary wave, $\overline{E}_1$, represents *TE* waves and the extraordinary wave, $\overline{E}_2$, represent *TM* waves. Application of boundary conditions gives

$$k_{x1} = \frac{m_1 \pi}{a} \quad \text{for} \quad m_1 = 1, 2, 3, \dots \tag{9.76}$$

$$k_{y1} = \frac{n_1 \pi}{b} \quad \text{for} \quad n_1 = 0, 1, 2, 3, \dots \tag{9.77}$$

for *TE* or ordinary waves and

$$k_{x2} = \frac{m_2 \pi}{a} \quad \text{for} \quad m_2 = 1, 2, 3, \dots \tag{9.78}$$

$$k_{y2} = \frac{n_2 \pi}{b} \quad \text{for} \quad n_2 = 0, 1, 2, 3, \dots \tag{9.79}$$

for *TM* or extraordinary waves. As a result, the substitution of (9.76) and (9.77) into (9.72) gives the dispersion relation for *TE* waves as

$$k_{\rho I}^2 = k_0^2 \varepsilon_{11} - k_z^2 = \left(\frac{m_1 \pi}{a}\right)^2 + \left(\frac{n_1 \pi}{b}\right)^2 \tag{9.80}$$

or

$$k_z = \sqrt{k_0^2 \varepsilon_{11} - \left[\left(\frac{m_1 \pi}{a}\right)^2 + \left(\frac{n_1 \pi}{b}\right)^2\right]} \tag{9.81}$$

The cutoff frequency for *TE* waves is

$$f_c = \frac{1}{2\pi\sqrt{\mu_0 \varepsilon_{11}}} \sqrt{\left[\left(\frac{m\pi}{a}\right)^2 + \left(\frac{n\pi}{b}\right)^2\right]} \tag{9.82}$$

Similarly, substitution of (9.78) and (9.79) into (9.73) gives the dispersion relation for *TM* waves as

$$k_{\rho II}^2 = k_0^2 \varepsilon_{33} - k_z^2 \frac{\varepsilon_{33}}{\varepsilon_{11}} = \left(\frac{m_2 \pi}{a}\right)^2 + \left(\frac{n_2 \pi}{b}\right)^2 \tag{9.83}$$

Table 9.3 Cutoff frequency and dispersion relation for rectangular waveguide filled with vertically uniaxial anisotropic filling.

Mode	Dispersion relation	f_c cutoff frequency
TE_{mn}	$k_z = \sqrt{k_0^2 \varepsilon_{11} - \left[\left(\dfrac{m_1 \pi}{a}\right)^2 + \left(\dfrac{n_1 \pi}{b}\right)^2\right]}$	$f_c = \dfrac{1}{2\pi\sqrt{\mu_0 \varepsilon_{11}}} \sqrt{\left[\left(\dfrac{m\pi}{a}\right)^2 + \left(\dfrac{n\pi}{b}\right)^2\right]}$
TM_{mn}	$k_z = \sqrt{k_0^2 \varepsilon_{11} - \dfrac{\varepsilon_{11}}{\varepsilon_{33}}\left[\left(\dfrac{m_2 \pi}{a}\right)^2 + \left(\dfrac{n_2 \pi}{b}\right)^2\right]}$	$f_c = \dfrac{1}{2\pi\sqrt{\mu_0 \varepsilon_{33}}} \sqrt{\left[\left(\dfrac{m\pi}{a}\right)^2 + \left(\dfrac{n\pi}{b}\right)^2\right]}$

Table 9.4 Standard rectangular waveguide dimensions.

Waveguide standard	Inside dimension, a (in.)	Inside dimension b (in.)
WR-2300	23	11.5
WR-2100	21	10.5
WR-1800	18	9
WR-1500	15	7.5
WR-1150	11.5	5.75
WR-1000	9.975	4.875
WR-770	7.7	3.385
WR-650	6.5	3.25
WR-430	4.3	2.15

or

$$k_z = \sqrt{k_0^2 \varepsilon_{11} - \frac{\varepsilon_{11}}{\varepsilon_{33}}\left[\left(\frac{m_2 \pi}{a}\right)^2 + \left(\frac{n_2 \pi}{b}\right)^2\right]} \tag{9.84}$$

The cutoff frequency for *TM* waves is

$$f_c = \frac{1}{2\pi\sqrt{\mu_0 \varepsilon_{33}}} \sqrt{\left[\left(\frac{m\pi}{a}\right)^2 + \left(\frac{n\pi}{b}\right)^2\right]} \tag{9.85}$$

Table 9.3 gives all the design parameters for a rectangular waveguide filled with a uniaxially anisotropic medium. Some of the standard rectangular waveguide dimensions used in RF applications are given in Table 9.4.

Example 9.1 Rectangular Waveguide
Calculate the cutoff frequencies for a WR-430 waveguide for isotropic, gyrotropic, and anisotropic cases.

Solution
i) Isotropic case

The calculated cutoff frequencies of the TE_{mn} modes for the rectangular waveguide WR-430 when filled with air are given in Table 9.5.

The frequency response of the propagation constant using (9.36b) is illustrated in Figure 9.3. It is obvious that the lowest cutoff frequency occurs for TE_{10} mode. The cutoff frequencies in Figure 9.2b match the calculated cutoff frequencies given in Table 9.5 for all the propagating modes.

Table 9.5 TE_{mn} modes for WR-430.

Waveguide standard	TE_{mn} mode	Calculated cutoff frequency f_c (GHz)
WR-430	TE_{10}	1.3727
	TE_{01}	2.7455
	TE_{11}	3.0696
	TE_{20}	2.7455
	TE_{02}	5.4910
	TE_{21}	3.8827
	TE_{12}	5.6600
	TE_{22}	6.1391

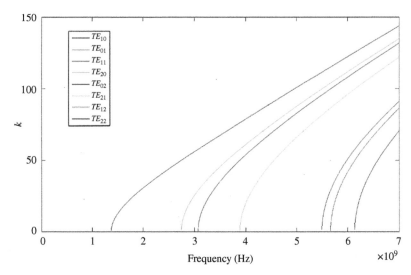

Figure 9.3 Frequency response of the propagation constant for TE_{mn} modes.

ii) Gyrotropic case

In this case, the standard rectangular waveguide, WR-430, is filled with a magnetically gyrotropic medium, i.e. ferrite. The dispersion relation given by Eq. (9.64) is a function of the term $\dfrac{\mu_1^2 - \mu_2^2}{\mu_1}$. This term is also referred to as the effective permeability constant of the ferrite material. The response of this term versus external magnetic field intensity is given in Figures 9.4–9.6 for various saturated magnetization levels at 6 GHz, 8 GHz, and 10 GHz, respectively. The values of $\dfrac{\mu_1^2 - \mu_2^2}{\mu_1}$ can be found from the graph provided for the given saturated magnetization level, magnetic field intensity, and operational frequency. It can then be used in (9.64) to calculate the frequency response of the propagation constant, k, and the corresponding cutoff frequency.

At this point, we can obtain the frequency response of the propagation constant with the design curves given in Figures 9.4–9.6. The ferrite material which has $\varepsilon_r = 12$ is used as a filling for the rectangular waveguide, WR-430, and biased with $H_0 = 1200$ Oe at $f = 8$ GHz with 1800 G saturated magnetization level. The frequency

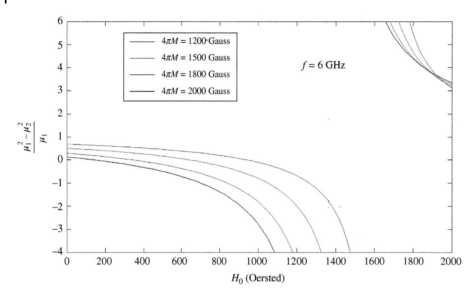

Figure 9.4 Permeability parameters versus magnetic field intensity for various saturated magnetization levels at f = 6 GHz.

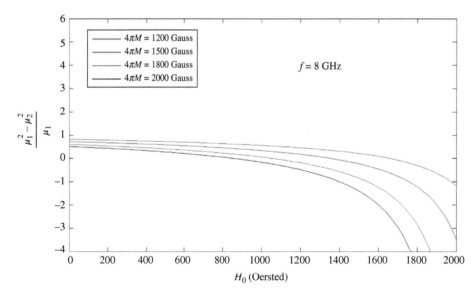

Figure 9.5 Permeability parameters versus magnetic field intensity for various saturated magnetization levels at f = 8 GHz.

response of k in the frequency range of 0–15 GHz for four *TE* modes is obtained and illustrated in Figure 9.7. The calculated cutoff frequencies for the *TE* modes for various values of the magnetic field intensities when saturated magnetization levels are 1800 G and 2200 G and the operational frequency is 10 GHz are shown in Tables 9.6 and 9.7.

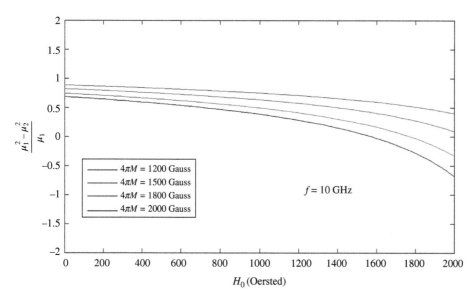

Figure 9.6 Permeability parameters versus magnetic field intensity for various saturated magnetization levels at f = 10 GHz.

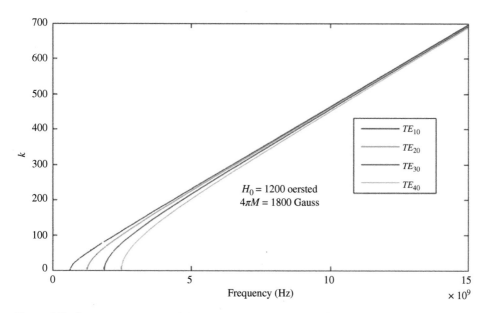

Figure 9.7 Frequency response of the propagation constant for TE_{mn} modes for a rectangular waveguide filled with magnetically gyrotropic medium.

Table 9.6 TE_{mn} modes for WR-430 waveguide filled with a magnetically gyrotropic medium for $4\pi M$ = 1800 Gauss.

Saturated magnetization	H_0 (Oe)	TE_{mn} mode	Calculated cutoff frequency f_c (GHz)
$4\pi M = 1800$ Gauss $f = 10$ GHz	600	TE_{10}	0.5040
		TE_{20}	1.0080
		TE_{30}	1.5120
		TE_{40}	2.0160
	800	TE_{10}	0.5288
		TE_{20}	1.0576
		TE_{30}	1.5864
		TE_{40}	2.1152
	1000	TE_{10}	0.5639
		TE_{20}	1.1279
		TE_{30}	1.6919
		TE_{40}	2.2559
	1200	TE_{10}	0.6187
		TE_{20}	1.2375
		TE_{30}	1.8563
		TE_{40}	2.4750

Table 9.7 TE_{mn} modes for WR-430 waveguide filled with a magnetically gyrotropic medium for $4\pi M$ = 2200 Gauss.

Saturated magnetization	H_0 (Oe)	TE_{mn} mode	Calculated cutoff frequency f_c (GHz)
$4\pi M = 2200$ Gauss $f = 10$ GHz	600	TE_{10}	0.594
		TE_{20}	1.189
		TE_{30}	1.784
		TE_{40}	2.379
	800	TE_{10}	0.658
		TE_{20}	1.316
		TE_{30}	1.974
		TE_{40}	2.632
	1000	TE_{10}	0.772
		TE_{20}	1.544
		TE_{30}	2.317
		TE_{40}	3.089
	1200	TE_{10}	1.067
		TE_{20}	2.135
		TE_{30}	3.203
		TE_{40}	4.270

Table 9.8 TE_{mn} and TM_{mn} modes for WR-430 waveguide filled with positively uniaxial anisotropic medium, sapphire.

Waveguide standard	TE_{mn} mode	Calculated cutoff freq. f_c (GHz)	TM_{mn} mode	Calculated cutoff frequency f_c (GHz)
WR-430	TE_{10}	0.447	TM_{10}	0.403
	TE_{01}	0.8954	TM_{01}	0.806
	TE_{11}	1.001	TM_{11}	0.901
	TE_{20}	0.895	TM_{20}	0.806
	TE_{02}	1.791	TM_{02}	1.612
	TE_{21}	1.266	TM_{21}	1.14
	TE_{12}	1.846	TM_{12}	1.662
	TE_{22}	2.002	TM_{22}	1.802

iii) Anisotropic case

The calculated cutoff frequencies for the ordinary waves, TE modes, and extraordinary waves, TM modes, for the standard rectangular waveguide, WR-430, when filled with positively uniaxial anisotropic medium, sapphire, are given in Table 9.8.

The permittivity tensor of the sapphire is given as

$$\overline{\overline{\varepsilon}} = \begin{bmatrix} 9.4 & 0 & 0 \\ 0 & 9.4 & 0 \\ 0 & 0 & 11.6 \end{bmatrix}$$

The frequency response of the propagation constant using Eqs. (9.81) and (9.84) are given in Figures 9.8 and 9.9 for TE and TM modes, respectively.

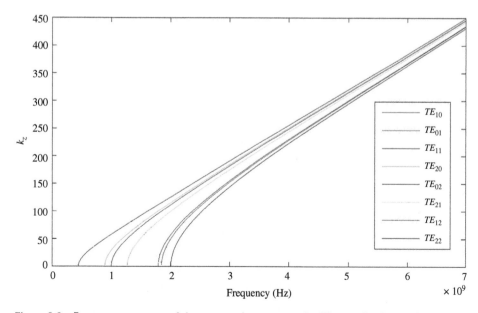

Figure 9.8 Frequency response of the propagation constant for TE_{mn} modes for sapphire.

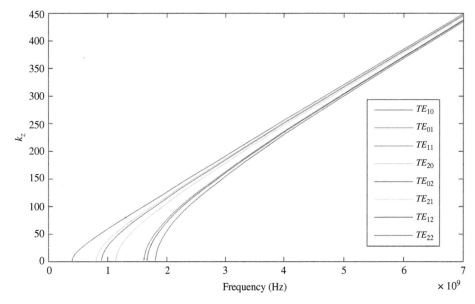

Figure 9.9 Frequency response of the propagation constant for TM_{mn} modes for sapphire.

Table 9.9 TE_{mn} and TM_{mn} modes for WR-430 waveguide filled with negatively uniaxial anisotropic medium, ceramic impregnated Teflon.

Waveguide standard	TE_{mn} mode	Calculated cut-off freq. f_c (GHz)	TM_{mn} mode	Calculated cut-off freq. f_c (GHz)
WR-430	TE_{10}	0.380	TM_{10}	0.429
	TE_{01}	0.761	TM_{01}	0.859
	TE_{11}	0.851	TM_{11}	0.961
	TE_{20}	0.761	TM_{20}	0.859
	TE_{02}	1.522	TM_{02}	1.719
	TE_{21}	1.076	TM_{21}	1.215
	TE_{12}	1.569	TM_{12}	1.772
	TE_{22}	1.702	TM_{22}	1.922

Now, we change the rectangular waveguide filling to negatively uniaxial anisotropic medium, ceramic impregnated Teflon, with the following permittivity tensor

$$\bar{\bar{\varepsilon}} = \begin{bmatrix} 13 & 0 & 0 \\ 0 & 13 & 0 \\ 0 & 0 & 10.2 \end{bmatrix}$$

The calculated cutoff frequencies for the ordinary waves, TE modes, and extraordinary waves, TM modes, for the standard rectangular waveguide, WR-430, when filled with a negatively uniaxial anisotropic medium, ceramic impregnated Teflon, are given in Table 9.9.

The frequency responses of the propagation constant using Eqs. (**9.81**) and (**9.84**) are given in Figures **9.10** and **9.11** for **TE** and **TM** modes, for WR-430 when filled with ceramic impregnated Teflon. The numerical and analytical values of the cutoff frequencies given in the tables and figures agree as **illustrated.**

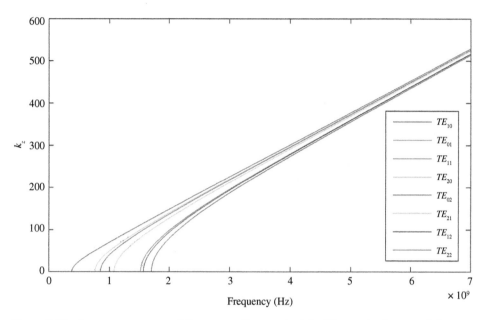

Figure 9.10 Frequency response of the propagation constant for TE_{mn} modes for ceramic impregnated Teflon.

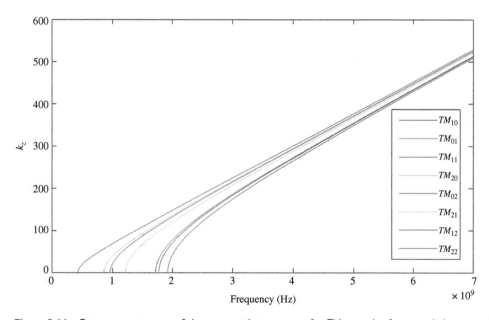

Figure 9.11 Frequency response of the propagation constant for TM_{mn} modes for ceramic impregnated Teflon.

9.3 Cylindrical Waveguides

Consider the cylindrical waveguide shown in Figure 9.12. In a cylindrical coordinate system, the transverse field components can be written in terms of longitudinal components as [1]

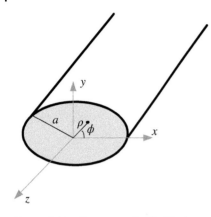

Figure 9.12 Geometry of cylindrical waveguide.

$$E_\rho = -\frac{j}{k_c^2}\left(\beta\frac{\partial E_z}{\partial\rho} + \frac{\omega\mu}{\rho}\frac{\partial H_z}{\partial\phi}\right)e^{-j\beta z} \tag{9.86}$$

$$E_\phi = -\frac{j}{k_c^2}\left(\frac{\beta}{\rho}\frac{\partial E_z}{\partial\phi} - \omega\mu\frac{\partial H_z}{\partial\rho}\right)e^{-j\beta z} \tag{9.87}$$

$$H_\rho = \frac{j}{k_c^2}\left(\frac{\omega\varepsilon}{\rho}\frac{\partial E_z}{\partial\phi} - \beta\frac{\partial H_z}{\partial\rho}\right)e^{-j\beta z} \tag{9.88}$$

$$H_\phi = -\frac{j}{k_c^2}\left(\omega\varepsilon\frac{\partial E_z}{\partial\rho} + \frac{\beta}{\rho}\frac{\partial H_z}{\partial\phi}\right)e^{-j\beta z} \tag{9.89}$$

$$k_c^2 = k^2 - \beta^2 \tag{9.90}$$

9.3.1 *TE* Modes

For *TE* mode propagation, $E_z = 0$, and we assume

$$H_z(\rho,\phi,z) = f_z(\rho,\phi)e^{-j\beta z} \tag{9.91}$$

We are looking for a solution of the equation given by

$$\nabla^2 H_z + k^2 H_z = 0 \tag{9.92}$$

Expanding (9.92) gives

$$\left(\frac{\partial^2}{\partial\rho^2} + \frac{1}{\rho}\frac{\partial}{\partial\rho} + \frac{1}{\rho^2}\frac{\partial^2}{\partial\phi^2} + k_c^2\right)f_z(\rho,\phi) = 0 \tag{9.93}$$

We can solve (9.93) using separation of variables where we define

$$f_z(\rho,\phi) = S(\rho)T(\phi) \tag{9.94}$$

Substitution of (9.94) into (9.93) gives

$$\begin{aligned}\frac{1}{S}\frac{d^2 S}{d\rho^2} + \frac{1}{\rho S}\frac{dS}{d\rho} + \frac{1}{\rho^2 T}\frac{d^2 T}{d\phi^2} + k_c^2 &= 0, \\ \frac{\rho^2}{S}\frac{d^2 S}{d\rho^2} + \frac{\rho}{S}\frac{dS}{d\rho} + \rho^2 k_c^2 &= -\frac{1}{T}\frac{d^2 T}{d\phi^2}\end{aligned} \tag{9.95}$$

which leads to

$$-\frac{1}{T}\frac{d^2 T}{d\phi^2} = k_\phi^2 \rightarrow \frac{d^2 T}{d\phi^2} + k_\phi^2 T = 0 \tag{9.96}$$

$$\rho^2\frac{d^2 S}{d\rho^2} + \rho\frac{dS}{d\rho} + \left(\rho^2 k_c^2 - k_\phi^2\right)S = 0 \tag{9.97}$$

The general solution for (9.96) is

$$T(\phi) = A \sin k_\phi \phi + B \cos k_\phi \phi \tag{9.98}$$

Since

$$f_z(\rho, \phi) = f_z(\rho, \phi \pm 2n\pi) \tag{9.99}$$

and

$$k_\phi = n \rightarrow \text{integer}$$

Hence

$$T(\phi) = A \sin n\phi + B \cos n\phi \tag{9.100}$$

Equation (9.97) is a Bessel differential equation and its solution is

$$S(\rho) = C J_n(k_c \rho) + D Y_n(k_c \rho) \tag{9.101}$$

where J_n is the Bessel function of first kind and Y_n is the Bessel function of second kind. Hence, the solution is

$$f_z^n(\rho, \phi) = (A \sin(n\phi) + B \cos(n\phi)) J_n(k_c \rho) \tag{9.102}$$

To determine k_c, we apply the boundary condition and obtain the cutoff frequency and phase constant as

$$\beta_{nm} = \sqrt{k^2 - \left(\frac{p'_{nm}}{a}\right)^2} \tag{9.103}$$

$$f_{c,nm} = \frac{k_{c,nm}}{2\pi} v_0 = \left(\frac{p'_{nm}}{2\pi a}\right) v_0 \tag{9.104}$$

or

$$\lambda_{c,nm} = \frac{2\pi a}{p'_{nm}} \tag{9.105}$$

Hence, the field vectors for TE_{11} dominant mode can be obtained as

$$
\begin{aligned}
E_\rho &= \frac{-j\omega\mu n}{k_c^2 \rho}(A \cos n\phi - B \sin n\phi) J_n(k_c \rho) e^{-j\beta z} \\
E_\phi &= \frac{j\omega\mu}{k_c}(A \sin n\phi + B \cos n\phi) J'_n(k_c \rho) e^{-j\beta z} \\
H_\rho &= \frac{-j\beta}{k_c}(A \sin n\phi + B \cos n\phi) J'_n(k_c \rho) e^{-j\beta z} \\
H_\phi &= \frac{-j\beta n}{k_c^2 \rho}(A \cos n\phi - B \sin n\phi) J_n(k_c \rho) e^{-j\beta z}
\end{aligned}
\tag{9.106}
$$

The attenuation due to conductor loss is obtained from

$$\alpha_c = \frac{P_l}{2P_o} \tag{9.107}$$

where power transmitted and loss are

$$P_0 = \frac{1}{2} \operatorname{Re} \int_{\rho=0}^{a} \int_{\phi=0}^{2\pi} \overline{E} \times \overline{H}^* \cdot \hat{z} \rho \, d\rho \, d\phi = \frac{\omega\mu |A|^2 \operatorname{Re}(\beta)}{4k_c^4}\left(p'^2_{11} - 1\right) J_1^2(k_c a) \tag{9.108}$$

and

$$P_l = \frac{R_s}{2} \int_{\phi=0}^{2\pi} |J_s|^2 a d\phi = \frac{|A|^2 R_s a}{2} \left(1 + \frac{\beta^2}{k_c^4 a^2}\right) J_1^2(k_c a) \tag{9.109}$$

9.3.2 TM Modes

Similar analysis can be done to find the cutoff frequency and phase constant by finding the solution of

$$\left(\frac{\partial^2}{\partial \rho^2} + \frac{1}{\rho} \frac{\partial}{\partial \rho} + \frac{1}{\rho^2} \frac{\partial^2}{\partial \phi^2} + k_c^2\right) f_z(\rho, \phi) = 0 \tag{9.110}$$

where

$$f_z(\rho, \phi) = (A \sin n\phi + B \cos n\phi) J_n(k_c \rho) \tag{9.111}$$

Application of boundary condition gives

$$E_z(\rho, \phi) = 0 \ \text{ at } \ \rho = a \tag{9.112}$$

$$J_n(k_c a) = 0 \rightarrow k_c = \frac{p_{nm}}{a} \tag{9.113}$$

Then, we obtain

$$\beta_{nm} = \sqrt{k^2 - k_c^2} = \sqrt{k^2 - \left(\frac{p_{nm}}{a}\right)^2}, f_{cnm} = \frac{k_c}{2\pi\sqrt{\mu\varepsilon}} = \frac{p_{nm}}{2\pi a \sqrt{\mu\varepsilon}} \tag{9.114}$$

The field components for TM modes are then found as

$$\begin{aligned}
E_\rho &= \frac{-j\beta}{k_c}(A \sin n\phi + B \cos n\phi) J'_n(k_c \rho) e^{-j\beta z} \\[4pt]
E_\phi &= \frac{-j\beta n}{k_c^2 \rho}(A \cos n\phi - B \sin n\phi) J_n(k_c \rho) e^{-j\beta z} \\[4pt]
H_\rho &= \frac{j\omega\varepsilon n}{k_c^2 \rho}(A \cos n\phi - B \sin n\phi) J_n(k_c \rho) e^{-j\beta z} \\[4pt]
H_\phi &= \frac{-j\omega\varepsilon}{k_c}(A \sin n\phi + B \cos n\phi) J'_n(k_c \rho) e^{-j\beta z}
\end{aligned} \tag{9.115}$$

9.4 Waveguide Phase Shifter Design

One of the application of the waveguides are their use as phase shifters. Nonreciprocal phase shifters such as ferrite phase shifters have excellent electrical performances and are commonly used as phasing elements in phased array antennas and in many other RF/microwave systems, owing to their advantages of high Q value, high power handling capability, etc. The geometry of a nonreciprocal phase shifter using ferrite as a thin slab in a rectangular waveguide is illustrated in Figure 9.13.

TE_{m0} modes are derived in Section 0 for a rectangular waveguide filled with gyrotropic medium. For this problem, we simplify the material to be magnetically gyrotropic, such as ferrite. This material is placed in region 2 of the rectangular waveguide shown in Figure 9.13. Regions 1 and 3 in the waveguide are assumed to be filled with air.

The field components for TE_{10} modes can be obtained. In the regions filled with air, region 1 and 3, the electric fields are

$$E_{y1} = C \sin(k_a x) \tag{9.116}$$

$$E_{y3} = D \sin(k_a(a-x)) \tag{9.117}$$

The dispersion relation in these regions is defined as

$$k_a^2 = \omega^2 \mu_0 \varepsilon_0 - k^2 \tag{9.118}$$

The magnetic field components are found following the procedure already outlined and given as

$$H_{z1} = \frac{-iC}{\omega \mu_3} \cos(k_a x) \tag{9.119}$$

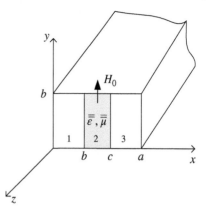

Figure 9.13 Nonreciprocal phase shifter.

$$H_{z2} = \frac{-\mu_1}{\omega(\mu_1^2 - \mu_2^2)} \left[\begin{array}{l} Ak_\perp[\cos(k_\perp x) - i\sin(k_\perp x)] - Bk_\perp[\cos(k_\perp x) + i\sin(k_\perp x)] \\ + - \frac{ik\mu_2}{\mu_1}[\cos(k_\perp x) - i\sin(k_\perp x)] - Bk_\perp[\cos(k_\perp x) + i\sin(k_\perp x)] \end{array} \right] \tag{9.120}$$

$$H_{z3} = \frac{iDk_a}{\omega \mu_0} \cos(k_a(a-x)) \tag{9.121}$$

$$H_{z1} = \frac{-Ck}{\omega \mu_0} \sin(k_a x) \tag{9.122}$$

$$H_{z2} = -\frac{\mu_1}{\omega(\mu_1^2 - \mu_2^2)} \left[\begin{array}{l} -k_\perp[A[\cos(k_\perp x) - i\sin(k_\perp x)] - B[\cos(k_\perp x) + i\sin(k_\perp x)]] \\ + - \frac{ik\mu_2}{\mu_1}[A[\cos(k_\perp x) - i\sin(k_\perp x)] + B[\cos(k_\perp x) + i\sin(k_\perp x)]] \end{array} \right] \tag{9.123}$$

$$H_{z3} = \frac{iDk_a}{\omega \mu_0} \cos(k_a(a-x)) \tag{9.124}$$

$$H_{x2} = \frac{\mu_1}{\omega(\mu_1^2 - \mu_2^2)} \left[\begin{array}{l} -k[A[\cos(k_\perp x) - i\sin(k_\perp x)] + B[\cos(k_\perp x) + i\sin(k_\perp x)]] \\ - \frac{ik_\perp \mu_2}{\mu_1}[A[\cos(k_\perp x) - i\sin(k_\perp x)] - B[\cos(k_\perp x) + i\sin(k_\perp x)]] \end{array} \right] \tag{9.125}$$

$$H_{x3} = -\frac{Dk}{\omega \mu_0} \sin(k_a(a-x)) \tag{9.126}$$

Using the boundary conditions on the walls of the dielectric in the waveguide, we can write

$$E_{y1} = E_{y2}, \qquad x = b \tag{9.127a}$$

$$E_{y3} = E_{y2}, \qquad x = c \tag{9.127b}$$

Application of the boundary conditions given by (9.127) on the fields given by (9.62), (9.107), and (9.108) lead to the following equation.

$$\frac{k_a k_\perp \mu_1}{\mu_0(\mu_1^2 - \mu_2^2)}[\tan(k_a b) + \tan k_a(a-c)]\cot(k_\perp(b-c))$$

$$- + \frac{k_a \mu_2 k}{\mu_0(\mu_1^2 - \mu_2^2)}[\tan(k_a b) - \tan k_a(a-c)]$$

$$- \left[\left(\frac{k_\perp \mu_1}{(\mu_1^2 - \mu_2^2)}\right)^2 + \left(\frac{k\mu_2}{(\mu_1^2 - \mu_2^2)}\right)^2\right]\tan(k_a b)\tan(k_a(a-c)) + \left(\frac{k_a}{\mu_0}\right)^2 = 0 \tag{9.128}$$

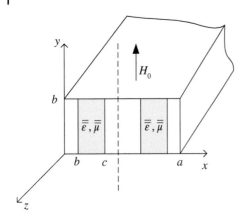

Figure 9.14 Two-slab nonreciprocal phase shifter.

The propagation constant or phase constant is found by solving the equation given in (9.128) with an algorithm developed. The phase shift effect can be extended to the symmetrical structure shown in Figure 9.14. The analysis is similar to what is presented for the single slab phase shifter.

9.5 Engineering Application Examples

In this section, rectangular waveguide and coaxial waveguide filter design are given and simulated with HFSS and implemented.

Design Example I   Rectangular Waveguide Design
Calculate cutoff frequencies for TE_{10}, TE_{20}, TE_{10}, and TE_{01} modes for the rectangular waveguide illustrated in Figure 9.15. Then, use HFSS to simulate the rectangular waveguide with physical dimensions illustrated and compare your analytical results.

Solution
Analytical Derivation and Plots
The equation for cutoff frequency was

$$F_c = \frac{c}{2\pi\sqrt{\varepsilon_r}}\sqrt{\left(\frac{\pi}{a}\right)^2 + \left(\frac{\pi}{b}\right)^2}$$

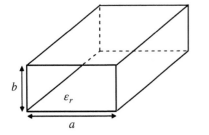

$a = 0.9"$ (2.286 cm)
$b = 0.4"$ (1.016 cm)
$\varepsilon_r = 1.0$

Figure 9.15   Dimensions of rectangular waveguide.

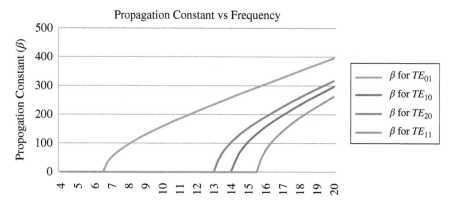

Figure 9.16 Propagation vs. frequency from theoretical equations.

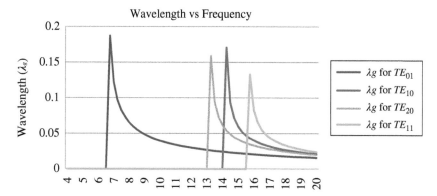

Figure 9.17 Wavelength vs frequency from theoretical equations.

So, the calculated values as shown in Figures 9.16 and 9.17 are

$$F_c \text{ for } TE_{10} = 6.56 \text{ GHz}$$
$$F_c \text{ for } TE_{01} = 14.14 \text{ GHz}$$
$$F_c \text{ for } TE_{02} = 13.11 \text{ GHz}$$
$$F_c \text{ for } TE_{11} = 15.59 \text{ GHz}$$

Simulation and Results from Ansoft HFSS

The simulated structure is illustrated in Figure 9.18. The simulation results are given in Figures 9.19 and 9.20.

Design Example 2   Coaxial High Pass Filter Design

Design high power coaxial high pass filter with cutoff frequency, $f_c = 450$ MHz. The amount of attenuation desired in the stopband at 60 MHz is < -60 dB.

Solution

The theory of electromagnetic wave propagation in hollow metal tubes was well described by Barrow [2] over 70 years ago. The concept of the proposed filter design is based on the operational principle of rectangular hollow

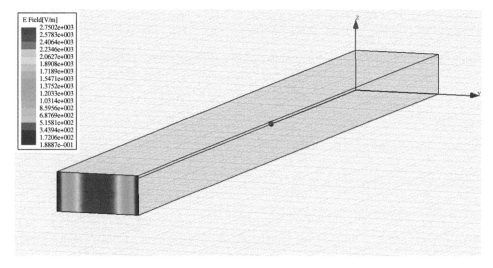

Figure 9.18 Simulated rectangular waveguide and *E* field.

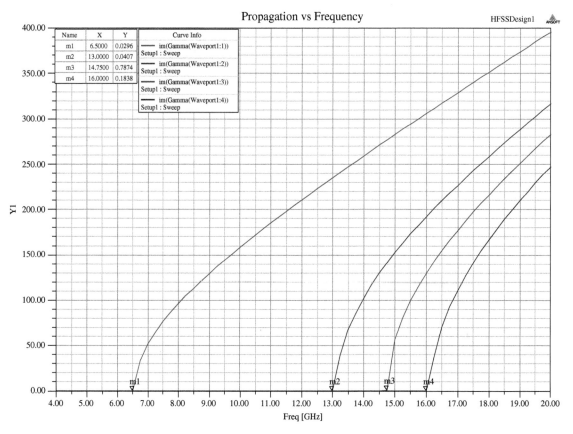

Figure 9.19 Propagation vs. frequency from Ansoft HFSS.

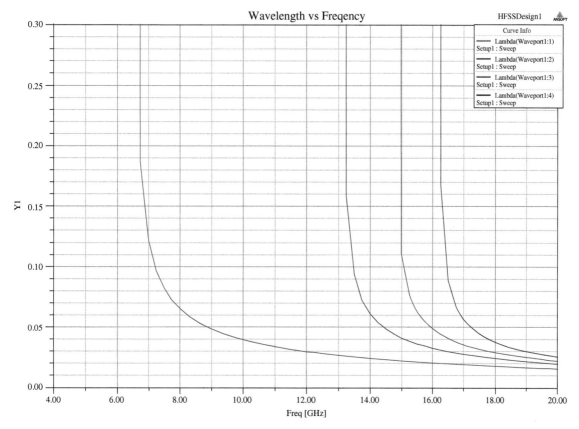

Figure 9.20 Wavelength vs. frequency from Ansoft HFSS.

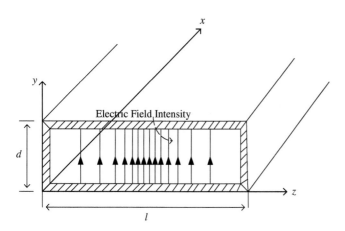

Figure 9.21 Hollow rectangular waveguide.

waveguides. TE_{01} waves in hollow rectangular waveguides have the simplest configuration. It has only one transverse component of electric field intensity, and it has the lowest critical frequency. In addition, its critical wavelength is only dependent on one side of the waveguide, as illustrated in Figure 9.21 and given by

$$\lambda_0 = 2l \tag{9.129}$$

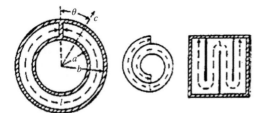

Figure 9.22 Cross sections of derivative hollow guides.

So, the rectangular hollow waveguide with TE_{01} waves might have a much smaller cross-sectional area by making one side of the waveguide arbitrarily small. If the waveguide illustrated in Figure 9.21 is bent or folded transversely by keeping the separation between the top and bottom sides unchanged, as shown in Figure 9.22, then a hollow conductor of smaller overall dimensions can be obtained with the same properties as a rectangular hollow waveguide [3].

Length l in the deformed shapes in Figure 9.22 is shown by the dotted lines. In each of the deformed shapes, there is a wave with a critical wavelength equal to $2l$ as given by Eq. (9.129). A novel high pass filter can be obtained by deforming the rectangular hollow waveguide into a coaxial structure involving two concentric hollow cylinders with a septum located between them at a single azimuthal angle [3]. It is a known fact that short circuiting both ends of the hollow tubes will give pronounced resonance characteristics resulting in a high Q structure. Based on this principle, we closed one end of the coaxial structure to obtain better attenuation characteristics. The excitation is done perpendicular to the direction of the propagation. The new filter structure is simulated with a 3D electromagnetic simulator, HFSS, and results are then compared with the measured results.

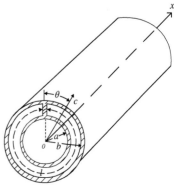

Figure 9.23 Base coaxial structure that will be used as a filter.

Analytical

The deformed coaxial structure that will be used as a high pass filter is illustrated in Figure 9.23. In the structure shown in Figure 9.23, each cylindrical pipe is hollow or can be considered filled with air whose permittivity and permeability constants are ε_0 and μ_0, respectively. The conductors are assumed to be perfect conductors. We are looking for the solutions of the longitudinal components for the field vectors $\overline{E}$ and $\overline{H}$ in the following form.

$$\left.\begin{array}{c} E_x \\ H_x \end{array}\right\} = \sin(k\theta + C_3)\left\{C_1 J_k\left(r\sqrt{\gamma^2 + \left(\frac{\omega}{c}\right)^2}\right) + C_2 Y_k\left(r\sqrt{\gamma^2 + \left(\frac{\omega}{c}\right)^2}\right)\right\} e^{i\omega t} e^{-\gamma x} \tag{9.130}$$

where C_1–C_3 are constants and found from boundary conditions.

TE and TM wave solutions are obtained when $E_x = 0$ and $H_x = 0$, respectively. The solutions for TE waves are found as

$$H_x = \cos\left(\frac{n\pi}{\theta_0}\theta\right)\left[C_1 J_k\left(r\sqrt{\lambda^2 + \left(\frac{\omega}{c}\right)^2}\right) + C_2 Y_k\left(r\sqrt{\lambda^2 + \left(\frac{\omega}{c}\right)^2}\right)\right] e^{-\gamma x + i\omega t} \tag{9.131}$$

$$H_\theta = \frac{\gamma\left(\frac{n\pi}{\theta_0}\right)}{r\left[\gamma^2 + \left(\frac{\omega}{c}\right)^2\right]} \sin\left(\frac{n\pi}{\theta_0}\theta\right)\left[C_1 J_k\left(r\sqrt{\lambda^2 + \left(\frac{\omega}{c}\right)^2}\right) + C_2 Y_k\left(r\sqrt{\lambda^2 + \left(\frac{\omega}{c}\right)^2}\right)\right] e^{-\gamma x + i\omega t} \tag{9.132}$$

Figure 9.24 *E* field plots for coaxial structure with septum.

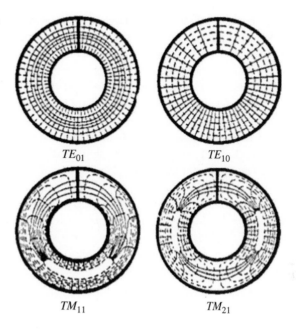

$$H_r = \frac{-\gamma}{\left[\gamma^2 + \left(\frac{\omega}{c}\right)^2\right]} \cos\left(\frac{n\pi}{\theta_0}\theta\right) \frac{d\left[C_1 J_k\left(r\sqrt{\lambda^2 + \left(\frac{\omega}{c}\right)^2}\right) + C_2 Y_k\left(r\sqrt{\lambda^2 + \left(\frac{\omega}{c}\right)^2}\right)\right]}{dr} e^{-\gamma x + i\omega t} \tag{9.133}$$

$$E_\theta = \frac{i\omega\gamma}{\left[\gamma^2 + \left(\frac{\omega}{c}\right)^2\right]} \cos\left(\frac{n\pi}{\theta_0}\theta\right) \frac{d\left[C_1 J_k\left(r\sqrt{\lambda^2 + \left(\frac{\omega}{c}\right)^2}\right) + C_2 Y_k\left(r\sqrt{\lambda^2 + \left(\frac{\omega}{c}\right)^2}\right)\right]}{dr} e^{-\gamma x + i\omega t} \tag{9.134}$$

$$E_r = \frac{i\omega\gamma}{r\left[\gamma^2 + \left(\frac{\omega}{c}\right)^2\right]} \sin\left(\frac{n\pi}{\theta_0}\theta\right) \left[C_1 J_k\left(r\sqrt{\lambda^2 + \left(\frac{\omega}{c}\right)^2}\right) + C_2 Y_k\left(r\sqrt{\lambda^2 + \left(\frac{\omega}{c}\right)^2}\right)\right] e^{-\gamma x + i\omega t} \tag{9.135}$$

The field plots are obtained in [12] and illustrated in Figure 9.24. The condition that makes the cross-section of the coaxial structure with septum similar to a rectangular hollow waveguide is the most important observation. That happens when $a \to b \to \infty$ as $\theta_0 \to 0$. Under this condition, waves in the structure shown in Figure 9.23 degenerate into TE_{n0} waves. The solution for these types of waves can be found for $n = 1$, $\theta_0 = 2\pi$, and $k = 1/2$. TE_{10} waves are important because they have the lowest possible cutoff frequency for this kind of structure.

The cutoff frequency for TE_{10} waves is found numerically from

$$\tan\left[(b-a)\sqrt{\gamma^2 + \left(\frac{\omega}{c}\right)^2}\right] = \frac{2(b-a)\sqrt{\gamma^2 + \left(\frac{\omega}{c}\right)^2}}{1 + 4ab\left(\gamma^2 + \left(\frac{\omega}{c}\right)^2\right)} \tag{9.136}$$

Simulation

In this section, a novel high pass filter is designed and simulated using the theory outlined in this example. The simulated results will then be compared with experimental results. The structure illustrated in Figure 9.23 has both ends open, and this figure will be used for initial simulation for the design verification before implementing the filter. We close one end of the coaxial structure to obtain better attenuation characteristics. The other end will be used as an output port. The structure will be excited perpendicular to the longitudinal axis. This excitation port can be considered an input port. The design parameters are the cutoff frequency and the amount of attenuation in the stopband. The cutoff frequency, f_c, is given as 450 MHz. The amount of attenuation desired in the stopband at 60 MHz is < -60 dB. When the desired cutoff frequency is substituted into Eq. (9.136), we obtain the dimensions of the coaxial high pass filter as $a = 152.4$ mm and $b = 66.23$ mm. This gives 50 Ω characteristic impedance for the coaxial structure. The dimensions obtained from Eq. (9.136) are used to simulate the filter with HFSS. The length of the filter is chosen to be 813 mm. The simulated structure and simulation result are shown in Figure 9.25a.

The simulation result in Figure 9.25b shows that the attenuation at 60 MHz is -65 dB. The attenuation at the cutoff frequency $f_c = 450$ MHz is -3.25 dB. The shape factor for this filter is around 0.4. So this is a very good high pass filter with a sharp stopband to bandpass transition. In addition, the analytical and simulated results are very close to each other for the given design parameters. Before implementing this filter, we close one end and use the other end of the filter as an output. The output should have 50 Ω impedance. The signal is excited from the side walls of the conductor via the 50 Ω connector. This final structure is again simulated with HFSS. The final simulated filter configuration and simulation result are given in Figure 9.26a.

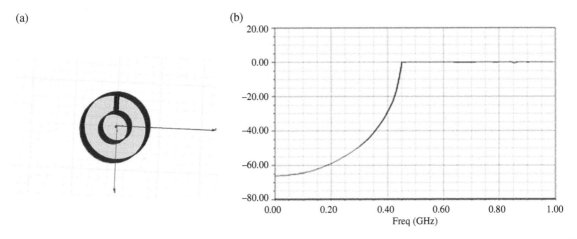

Figure 9.25 (a) Simulated coaxial filter with both ends open. (b) Simulation results for open ended coaxial high pass filter.

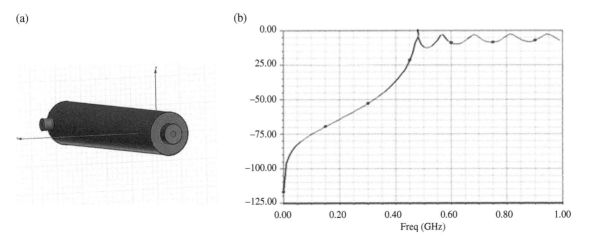

Figure 9.26 (a) Simulated final coaxial filter configuration. (b) Simulation results for the final coaxial filter configuration.

As seen from the simulation result in Figure 9.26b, the final configuration gave much better attenuation characteristics in comparison to the original coaxial filter shown in Figure 9.25a. The attenuation at 60 MHz is −80 dB. This is a −15 dB or 23% improvement in the attenuation characteristics.

The cutoff frequency is simulated to be 480 MHz in this new structure. The shape factor is improved and equal to 0.52. This shows a steeper transition from stopband to passband. The final configuration illustrated in Figure 9.26a is constructed and measured. The construction is done using N type 50 Ω connectors as simulated with HFSS for the excitation port and output port with the same physical dimensions.

Prototype

The constructed filter geometry is illustrated in Figure 9.27. Figure 9.28. illustrates the S parameter measurement and measurement results. A picture of the experimental setup is given in Figure 9.29.

The ripple which is ringing in the passband for the simulated result using a 3D electromagnetic simulator is found to be greater than the measured result. This ringing is a result of resonances due to imperfections in the modeled filter structure where it does not exist in the constructed one. The ripple goes up to −10 dB in the simulated structure, whereas it is less than −8 dB in the constructed filter. The attenuation is measured to be around −70 dB at 60 MHz in the constructed filter. This value is found to be −80 dB at the same frequency based on the simulation results.

Overall, the simulated results give results which have good proximity to the final measured results. The technique presented here can be used to design high performance high pass filters with steep slopes, which provide a fast transition from stopband to passband with low passband ripple.

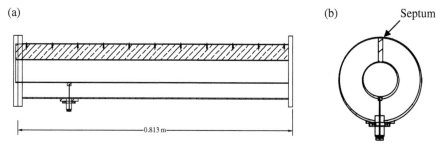

Figure 9.27 Constructed filter geometry: (a) side view; (b) end view.

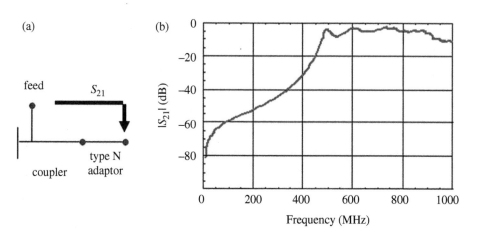

Figure 9.28 (a) S parameter measurement set-up. (b) Measured result for the final coaxial filter configuration.

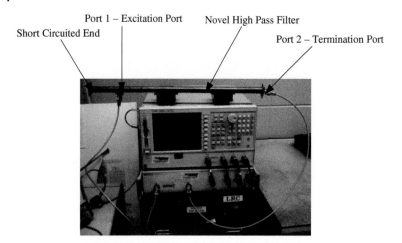

Figure 9.29 Experimental setup.

References

1 Eroglu, A. (2010). *Wave Propagation and Radiation in Gyrotropic and Anisotropic Media*, 1e. Springer.
2 Barrow, W.L. (1936). Transmission of electromagnetic waves in hollow tubes of metal. *Proceedings of Institute of Radio Engineers* **24** (10): 1298–1330.
3 Barrow, W.L. and Schaevitz, H. (1941). Hollow pies of relatively small dimension. *Transactions of the American Institute of Electrical Engineers* **60**: 119–122.

Problems

Problem 9.1

What are the field equations for the TM_1 mode in a parallel plate waveguide?

Problem 9.2

A parallel-plate waveguide made of PEC infinite planes shown in Figure 9.30 are spaced 5 cm apart in air operates at a frequency of 12 GHz. What is the maximum time-averaged power than can be transmitted per unit width of guide with no voltage breakdown for TEM mode and TM_1 mode?

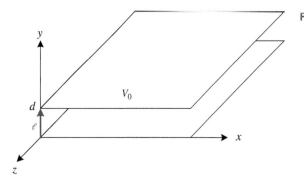

Figure 9.30 Problem 9.2.

Problem 9.3
If it is desired for a parallel-plate guide to propagate only TE_2 and TM_2 modes for frequencies greater than 12 GHz, what would be the maximum plate separation distance, d, if FR4, $\varepsilon_r = 4.4$, is placed between the plates?

Problem 9.4
Calculate the cutoff frequency for a rectangular waveguide with dimensions $a = 0.04$ m and $b = 0.02$ m for the first four TE propagating modes when it is air filled.

Problem 9.5
Rectangular waveguide with dimensions $a = 2.286$ cm and $b = 1.016$ cm is designed at 16 GHz. Calculate the guide wavelength, phase velocity, group velocity, and impedance for the propagating modes.

Problem 9.6
An air-filled waveguide with dimensions $a = 0.018$ m and $b = 0.006$ m is used to transmit a signal at $f = 10$ GHz, which has a narrow bandwidth. What is the length of the waveguide if the signal is delayed by 1 μs in comparison to its transmission in air?

Problem 9.7
Plot E field components for TE_{10} mode for an air-filled rectangular waveguide with dimensions $a = 0.04$ m and $b = 0.02$ m, when a signal is propagating with $f = 5$ GHz and $H = 0.25 \angle 0°$ A/m.

Problem 9.8
Find the maximum power that can be transmitted for a signal at $f = 10$ GHz in a rectangular waveguide with dimensions $a = 1.75$ cm and $b = 0.875$ cm, and when it is filled with air without any dielectric breakdown.

Problem 9.9
Assume a rectangular waveguide, WR90, operating at $f = 8$ GHz is open circuited. What should be the distance from the waveguide termination so that the input impedance of the waveguide becomes a short circuit?

Problem 9.10
Transmit a 12 GHz signal using a hollow circular metallic pipe. Find the inside radius of the pipe to set the cutoff frequency 40% below the desired frequency given.

Design Challenge 9.1

Identify the type of transmission lines, waveguide frequencies, applications, and lengths of the lines for (a) DC low voltage measurement, length 4 m; (b) DC, 1.5 MV buried power line of length 10 km; (c) 50 Hz of low voltage, 10 kW line of length 50 m; (d) 3 MHz, buried 50 Ω line to 50 kW transmitter antenna; (e) 55–889 MHz 300 Ω line from a shortwave antenna to a TV set through EM noise with a distance of 50 m; (f) 3.5 GHz, *S* band line or guide with 12 MHz BW from 100 kW radar transmitter to an antenna for a distance of 10 m.

10

Power Amplifiers

10.1 Introduction

This chapter discusses some of the common terminologies used in RF power amplifier (PA) design. For this discussion, consider the simplified RF PA block diagram given in Figure 10.1. In Figure 10.1, RF PA is simply considered a three-port network where the RF input signal port, DC input port, and RF signal output port constitute the ports of the network as illustrated. Power that is not converted to the RF output power, P_{out}, is dissipated as heat and designated by P_{diss}, as shown in Figure 10.1. The dissipated power, P_{diss}, is found from

$$P_{diss} = (P_{in} + P_{dc}) - P_{out} \qquad (10.1)$$

10.2 Amplifier Parameters

10.2.1 Gain

The RF PA gain is defined as the ratio of the output power to the input power, as given by

$$G = \frac{P_{out}}{P_{in}} \qquad (10.2)$$

It can be defined in terms of dB as

$$G(\text{dB}) = 10 \log \left(\frac{P_{out}}{P_{in}} \right) \text{dB} \qquad (10.3)$$

RF PA amplifier gain is higher at lower frequencies. This can be illustrated based on the measured data for a switched-mode RF amplifier operating at HF (high frequency) range in Figure 10.2 for several applied DC supply voltages.

It is possible to obtain a higher gain level when multiple amplifiers are cascaded to obtain a multistage amplifier configuration, as shown in Figure 10.3.

The overall gain of the multistage amplifier system for the one shown in Figure 10.3 can then be found from

$$G_{tot}(\text{dB}) = G_{PA_1}(\text{dB}) + G_{PA_2}(\text{dB}) + G_{PA_3}(\text{dB}) \qquad (10.4)$$

RF/Microwave Engineering and Applications in Energy Systems, First Edition. Abdullah Eroglu.
© 2022 John Wiley & Sons Ltd. Published 2022 by John Wiley & Sons Ltd.
Companion website: www.wiley.com/go/eroglu/rfmicrowave

$$P_{diss} = (P_{in} + P_{dc}) - P_{out}$$

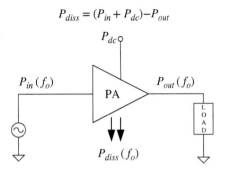

Figure 10.1 RF PA as a three-port network.

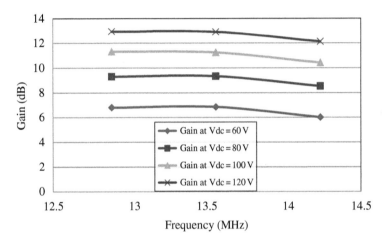

Figure 10.2 Measured gain variation versus frequency for a switched-mode RF amplifier.

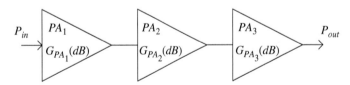

Figure 10.3 Multistage RF amplifiers.

The unit of the gain is given in terms of (dB) because it is ratio of the output power to input power. It is important to note that dB is not the unit to define the power. In amplifier terminology, dBm is used to define the power. dBm is found from

$$dBm = 10 \log \left(\frac{P}{1 \, mW} \right) \tag{10.5}$$

Example 10.1 RF System

In the RF system shown in Figure 10.4, the RF signal source can provide power output from 0 to 30 dBm. The RF signal is fed through a 1 dB T pad attenuator and 20 dB directional coupler where the sample of the RF signal is further attenuated by a 3 dB π pad attenuator before power meter reading in dB. The "through" port of the

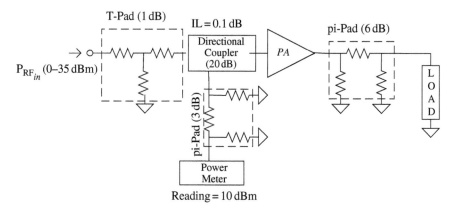

Figure 10.4 RF system with coupler and attenuation pads.

directional coupler has a 0.1 dB of loss before it is sent to the PA. The RF signal is fed through a 1 dB T pad attenuator and 20 dB directional coupler where the sample of the RF signal is further attenuated by a 3 dB π pad attenuator before power meter reading in dB. If the power meter is reading 10 dBm, what is the power delivered to the load shown in Figure 10.4 in mW?

Solution

We need to find the source power first. The loss from the power meter to the RF signal source is

$$\text{Loss from power meter to source} = 3\,\text{dB} + 20\,\text{dB} + 1\,\text{dB} = 23\,\text{dB}$$

Then, the power at the source is

$$\text{RF source signal} = 23\,\text{dBm} + 10\,\text{dBm} = 33\,\text{dBm}$$

The total loss toward PA is due to the T pad attenuator (1 dB) and directional coupler (0.1 dB) = 1.1 dB. So, the transmitted RF signal at the PA is

$$\text{RF signal at PA} = 33\,\text{dBm} - 1.1\,\text{dBm} = 31.9\,\text{dBm}$$

Hence, the power delivered to the load is found from
Power delivered to the load = 31.9 − 6 = 25.9 dBm
Power delivered in mW is found from

$$P(\text{mW}) = 10^{\frac{dBm}{10}} = 10^{\frac{25.9}{10}} = 389.04\,\text{mW}$$

10.2.2 Efficiency

In practical applications, RF PA is implemented as a subsystem and consumes most of the DC power from the supply. As a result, minimal DC power consumption for the amplifier becomes important and can be accomplished by having a high RF PA efficiency. RF PA efficiency is one of the critical and most important amplifier performance parameters. Amplifier efficiency can be used to define the drain efficiency for MOSFET (metal oxide semiconductor field-effect transistor) or collector efficiency for bipolar junction transistor (BJT). Amplifier efficiency is defined as the ratio of the RF output power to the power supplied by the DC source and can be expressed as

$$\eta(\%) = \frac{P_{out}}{P_{DC}} \times 100 \tag{10.6}$$

Efficiency in terms of gain can be put in the following form.

$$\eta(\%) = \frac{1}{1 + \left(\dfrac{P_{diss}}{P_{out}}\right) - \left(\dfrac{1}{G}\right)} \times 100 \tag{10.7}$$

The maximum efficiency is possible when there is no dissipation, i.e. $P_{diss} = 0$. The maximum efficiency from (10.7) is then equal to

$$\eta(\%) = \frac{1}{1 - \left(\dfrac{1}{G}\right)} \times 100 \tag{10.8}$$

When RF input power is included in the efficiency calculation, the efficiency is then called power added efficiency, η_{PAE}, and can be found from

$$\eta_{PAE}(\%) = \frac{P_{out} - P_{in}}{P_{DC}} \times 100 \tag{10.9a}$$

or

$$\eta_{PAE}(\%) = \eta\left(1 - \frac{1}{G}\right) \times 100 \tag{10.9b}$$

Example 10.2   RF Power Amplifier Efficiency

RF PA delivers 200 W to a given load. If the input supply power for this amplifier is given to be 240 W and the power gain of the amplifier 15 dB, find (a) drain efficiency and (b) power added efficiency.

Solution

a) The drain efficiency is found from (10.6) as

$$\eta(\%) = \frac{P_{out}}{P_{DC}} \times 100 = \frac{200}{240} \times 100 = 83.33\%$$

b) Power added efficiency is found from (10.9) as

$$\eta_{PAE}(\%) = \eta\left(1 - \frac{1}{G}\right) \times 100 = 83\left(1 - \frac{1}{15}\right) = 77.47\%$$

10.2.3   Power Output Capability

The power output capability of an amplifier is defined as the ratio of the output power of the amplifier to the maximum values of the voltage and current that the device experiences during the operation of the amplifier. When there is more than one transistor due to amplifier configuration, this is reflected in the denominator of the following equation

$$c_p = \frac{P_o}{NI_{max}V_{max}} \tag{10.10}$$

10.2.4   Linearity

Linearity is a measure of an RF amplifier's output as it follows the amplitude and phase of its input signal. In practice, the linearity of an amplifier is measured in a very different way. The linearity of an amplifier is measured by comparing the set power of an amplifier with the output power. The gain of the amplifier is then adjusted to

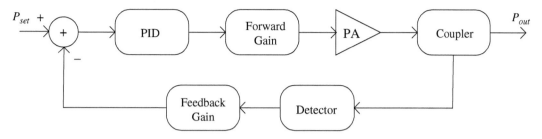

Figure 10.5 Typical closed loop control for an RF power amplifier for linearity control.

compensate one of the closed loop parameters, such as gain. The typical closed loop control system that is used to adjust the linearity of the amplifier through closed loop parameters is shown in Figure 10.5. When the linearity of the amplifier is accomplished, the linear curve shown in Figure 10.6 is obtained. The experimental setup that is used to calibrate RF PAs to have linear characteristics is given in Figure 10.7.

In Figure 10.7, RF amplifier output is measured by a thermocouple-based power meter via directional coupler. Directional coupler output is terminated by a 50 Ω load. The set power is adjusted by the user and output forward power, P_{fwr}, and reverse power is measured with the power meter. If the set power and output power are different, the control closed loop parameters are then modified.

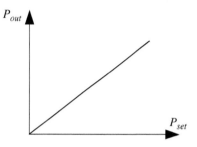

Figure 10.6 Linear curve for an RF amplifier.

10.2.5 1 dB Compression Point

The compression point for an amplifier is the point where amplifier gain becomes 1 dB below its ideal linear gain, as shown in Figure 10.8. Once the 1 dB compression point is identified for the corresponding input power range, the amplifier can be operated in linear or nonlinear mode. Hence, the 1 dB compression point can also be conveniently used to identify the linear characteristics of the amplifier.

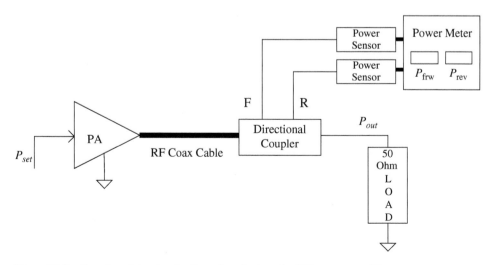

Figure 10.7 Experimental setup for linearity adjustment of RF power amplifiers.

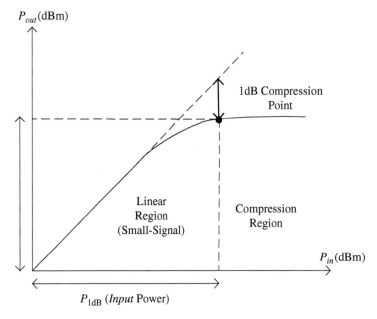

P_{out}(dBm)

1dB Compression
Point

Linear
Region
(Small-Signal)

Compression
Region

P_{in}(dBm)

P_{1dB} (*Input* Power)

Figure 10.8 1 dB compression point for amplifiers.

The gain at the 1 dB compression point can be found from

$$P_{1dB,out} - P_{1dB,in} = G_{1dB} = G_0 - 1 \tag{10.11}$$

where G_0 is the small signal or linear gain of the amplifier at fundamental frequency. The 1 dB compression point can also be expressed using the input and output voltages and their coefficients. The gain of the amplifier at the fundamental frequency when $v_i(t) = \beta \cos \omega t$ is found as

$$G_{1dB} = 20 \log \left| \alpha_1 + 3\alpha_3 \frac{\beta^2}{4} \right| \tag{10.12}$$

$$G_0(\text{linear/small signal gain}) = 20 \log |\alpha_1| \tag{10.13}$$

As a result, the 1 dB compression point in Figure 10.8 can be calculated from

$$20 \log \left| \alpha_1 + 3\alpha_3 \frac{\beta_{in,1dB}^2}{4} \right| = 20 \log |\alpha_1| - 1 \text{ dB} \tag{10.14}$$

where $\beta_{1 \text{ dB}}$ is the amplitude of the input voltage at the 1 dB compression point. The solution of (10.14) for $\beta_{1 \text{ dB}}$ leads to

$$\beta_{1dB} = \sqrt{0.145 \left| \frac{\alpha_1}{\alpha_3} \right|} \tag{10.15}$$

10.2.6 Harmonic Distortion

Harmonic distortion for an amplifier can be defined as the ratio of the amplitude of the $n\omega$ component to the amplitude of the fundamental component. The second- and third-order harmonic distortions can then be expressed as

$$HD_2 = \frac{1}{2}\frac{\alpha_2}{\alpha_1}\beta \tag{10.16}$$

$$HD_3 = \frac{1}{4}\frac{\alpha_3}{\alpha_1}\beta^2 \tag{10.17}$$

From (10.16) to (10.17), it is apparent that the second harmonic distortion is proportional to signal amplitude, whereas third-order amplitude is proportional to the square of the amplitude. Hence, when the input signal is increased by 1 dB, HD_2 increases by 1 dB and HD_3 increases by 2 dB. The total harmonic distortion (THD) in the amplifier can be found from

$$THD = \sqrt{(HD_2)^2 + (HD_3)^2 + \dots} \tag{10.18}$$

Example 10.3   RF Amplifier Harmonic Distortion

RF signal $v_i(t) = \beta\cos\omega t$ is applied to a linear amplifier and then to the nonlinear amplifier given in Figure 10.9 with output response $v_o(t) = \alpha_0 + \alpha_1\beta\cos\omega t + \alpha_2\beta^2\cos^2\omega t + \alpha_3\beta^3\cos^3\omega t$. Assume input and output impedances are equal to R. (a) Calculate and plot gain for linear amplifier. (b) Obtain second and third harmonic distortion for nonlinear amplifier when $\alpha_0 = 0$, $\alpha_1 = 1$, $\alpha_2 = 3$, $\alpha_3 = 1$ and $\beta = 1$, $\beta = 2$. Calculate THD for both cases.

Solution

a) For linear amplifier characteristics, the output voltage is expressed as

$$v_o(t) = \beta v_i(t) \tag{10.19}$$

which can also be written as

$$\frac{1}{2R}v_o^2(t) = \beta^2\frac{1}{2R}v_i^2(t) \tag{10.20}$$

or

$$P_o = \beta^2 P_i \tag{10.21}$$

When the power given Eq. (10.13) is given in dBm, (10.13) can be expressed as

$$10\log\left(\frac{P_o}{1\,mW}\right) = 10\log\left(\beta^2\frac{P_i}{1\,mW}\right) \tag{10.22}$$

or

$$P_o(dBm) = 10\log\left(\beta^2\right) + P_{in}(dBm) \tag{10.23}$$

Then, the power gain is obtained from (10.14) as

$$Gain(dBm) = G(dBm) = 10\log\left(\beta^2\right) = P_o(dBm) - P_{in}(dBm) \tag{10.24}$$

The relation between input and output power is plotted and illustrated in Figure 10.10.

b) The nonlinearity response of the amplifier using third-order polynomial can be expressed using (1.2) as

$$v_o(t) = \alpha_0 + \alpha_1 v_i(t) + \alpha_2 v_i^2(t) + \alpha_3 v_i^3(t) \tag{10.25a}$$

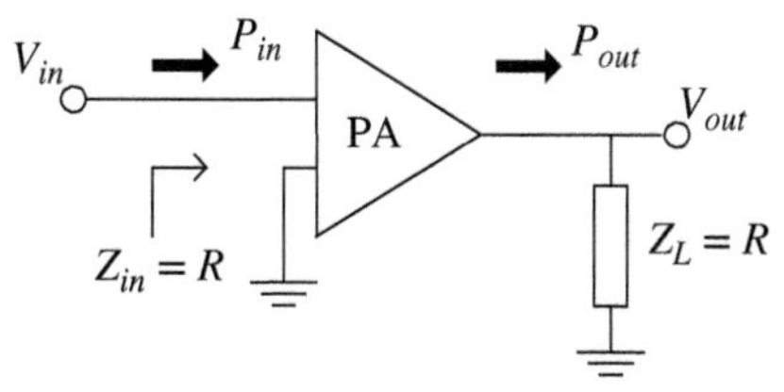

Figure 10.9   PA amplifier output response.

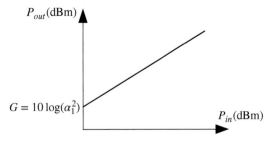

Figure 10.10   Power gain for linear operation.

or

$$v_o(t) = \alpha_0 + \alpha_1\beta\cos\omega t + \alpha_2\beta^2\cos^2\omega t + \alpha_3\beta^3\cos^3\omega t \tag{10.25b}$$

As can be seen from (10.25), we have fundamental, second-order harmonic, third-order harmonic, and a DC component in the output response of the amplifier. In (10.25b), it can also be seen that the DC component exists due to second harmonic content. Equation (10.25b) can be rearranged to give the following closed form relation

$$v_o(t) = \alpha_0 + \alpha_1\beta\cos\omega t + \frac{\alpha_2\beta^2}{2} + \frac{\alpha_2\beta^2}{2}\cos 2\omega t + 3\frac{\alpha_3\beta^3}{4}\cos\omega t + \frac{\alpha_3\beta^3}{4}\cos 3\omega t \tag{10.26a}$$

which can be simplified to

$$v_o(t) = \left(\alpha_0 + \frac{\alpha_2\beta^2}{2}\right) + \left(\alpha_1 + 3\frac{\alpha_3\beta^2}{4}\right)\beta\cos\omega t + \left(\frac{\alpha_2}{2}\right)\beta^2\cos 2\omega t + \frac{\alpha_3\beta^3}{4}\cos 3\omega t \tag{10.26b}$$

When $\alpha_1 = 1$, $\alpha_2 = 3$, $\alpha_3 = 1$ and $\beta = 1$, HD_2 and HD_3 are obtained from (10.16) to (10.17) as

$$HD_2 = \frac{1}{2}\frac{\alpha_2}{\alpha_1}\beta = \frac{1}{2}\frac{3}{1}(1) = 1.5 \tag{10.27}$$

$$HD_3 = \frac{1}{4}\frac{\alpha_3}{\alpha_1}\beta^2 = \frac{1}{4}\frac{1}{1}(1)^2 = 0.25 \tag{10.28}$$

When $\alpha_1 = 1$, $\alpha_2 = 3$, $\alpha_3 = 1$ and $\beta = 2$

$$HD_2 = \frac{1}{2}\frac{\alpha_2}{\alpha_1}\beta = \frac{1}{2}\frac{3}{1}(2) = 3 \tag{10.29}$$

$$HD_3 = \frac{1}{4}\frac{\alpha_3}{\alpha_1}\beta^2 = \frac{1}{4}\frac{1}{1}(2)^2 = 1 \tag{10.30}$$

The THD for this system is found from (10.18) as

$$\text{THD} = \sqrt{(1.5)^2 + (0.25)^2} = 1.5625 \tag{10.31a}$$

and

$$\text{THD} = \sqrt{(3)^2 + (1)^2} = 3.16 \tag{10.31b}$$

Example 10.4   1 dB Compression Point
The input of voltage for an RF circuit is given to be $v_{in}(t) = \beta\cos(\omega t)$. The RF circuit generates a signal at third harmonic as $V_3\cos(3\omega t)$. What is the 1 dB compression point?

Solution
Using (10.26b), the amplitude of the third harmonic component can be found from

$$\frac{\alpha_3\beta^3}{4} = V_3 \text{ or } \alpha_3 = \frac{4V_3}{\beta^3} \tag{10.32a}$$

Then, the 1 dB compression point, $\beta_{1\,dB}$, is found from (10.15) as

$$\beta_{1dB} = \sqrt{0.145 \left| \frac{\alpha_1}{\alpha_3} \right|} = \sqrt{\frac{0.145}{4} \left| \frac{\beta^3 \alpha_1}{V_3} \right|} = 0.19 \sqrt{\left| \frac{\beta^3 \alpha_1}{V_3} \right|} \tag{10.32b}$$

10.2.7 Intermodulation

When a signal comprising two cosine waveforms with different frequencies

$$v_i(t) = \beta_1 \cos \omega_1 t + \beta_2 \cos \omega_2 t \tag{10.33}$$

is applied to the input of an amplifier, the output signal consists of components of the self-frequencies and their products created by frequencies by ω_1 and ω_2 given in Eq. 10.34.

$$v_o(t) = \alpha_1 (\beta_1 \cos \omega_1 t + \beta_2 \cos \omega_2 t) + \alpha_2 (\beta_1 \cos \omega_1 t + \beta_2 \cos \omega_2 t)^2 + \alpha_3 (\beta_1 \cos \omega_1 t + \beta_2 \cos \omega_2 t)^3 \tag{10.34a}$$

or

$$v_o(t) = \alpha_1 (\beta_1 \cos \omega_1 t + \beta_2 \cos \omega_2 t) + \alpha_2 \left(\begin{array}{c} \frac{1}{2}\beta_1^2(1 + \cos 2\omega_1 t) + \frac{1}{2}\beta_2^2(1 + \cos 2\omega_2 t) \\ + \frac{1}{2}\beta_1\beta_2(\cos(\omega_1 + \omega_2)t + \cos(\omega_1 - \omega_2)t) \end{array} \right)$$

$$+ \alpha_3 \left(\begin{array}{c} \frac{3}{4}\beta_1^3(\cos \omega_1 t) + \frac{3}{2}\beta_1\beta_2^2(\cos \omega_1 t) + \frac{1}{4}\beta_1^3 \cos(3\omega_1 t) + \frac{3}{4}\beta_1\beta_2^2(\cos(\omega_1 - 2\omega_2)t) \\ + \frac{3}{4}\beta_1^2\beta_2(\cos(2\omega_1 - \omega_2)t) + \frac{3}{2}\beta_1^2\beta_2(\cos \omega_2 t) + \frac{3}{4}\beta_2^3(\cos \omega_2 t) + \frac{1}{4}\beta_2^3(\cos 3\omega_2 t) \\ + \frac{3}{4}\beta_1^2\beta_2(\cos(2\omega_1 + \omega_2)t) + \frac{3}{4}\beta_1\beta_2^2(\cos(\omega_1 + 2\omega_2)t) \end{array} \right) \tag{10.34b}$$

In (10.34), the DC component, α_0, is ignored. The components that will rise due to combinations of the frequencies ω_1 and ω_2 are given by Eq. (10.34a) and shown in Table 10.1. The corresponding frequency components in Table 10.1 are also illustrated in Figure 10.11.

In amplifier applications, intermodulation distortion products are undesirable components in the output signal. As a result, the amplifier needs to be tested using an input signal which is the sum of two cosines to eliminate these side products. This test is also known as a two-tone test. This specific test is important for an amplifier specifically when two frequencies, ω_1 and ω_2, are close to each other.

The second-order intermodulation distortion, IM_2, can be found from (10.18) and Table 10.1 when $\beta_1 = \beta_2 = \beta$. It is the ratio of the components at $\omega_1 \pm \omega_2$ to the fundamental components at ω_1 or ω_2.

$$IM_2 = \frac{\alpha_2}{\alpha_1} \beta \tag{10.35}$$

The third-order distortion, IM_3, can be found from the ratio of the component at $2\omega_2 \pm \omega_1$ (or $2\omega_1 \pm \omega_2$) to the fundamental components at ω_1 or ω_2.

$$IM_3 = \frac{3}{4} \frac{\alpha_3}{\alpha_1} \beta^2 \tag{10.36}$$

IM product frequencies are summarized in Table 10.2.

If Eqs. (10.16), (10.17) and (10.35), (10.36) are compared, *IM* products can be related to *HD* products as

$$IM_2 = 2HD_2 \tag{10.37}$$

Table 10.1 Intermodulation frequencies and corresponding amplitudes.

$\omega = \omega_1$	$\left(\alpha_1\beta_1 + \dfrac{3}{4}\alpha_3\beta_1^3 + \dfrac{3}{2}\alpha_3\beta_1\beta_2^2\right)\cos(\omega_1 t)$
$\omega = \omega_2$	$\left(\alpha_1\beta_2 + \dfrac{3}{4}\alpha_3\beta_2^3 + \dfrac{3}{2}\alpha_3\beta_2\beta_1^2\right)\cos(\omega_2 t)$
$\omega = \omega_1 + \omega_2$	$\dfrac{1}{2}(\alpha_2\beta_1\beta_2)\cos(\omega_1 + \omega_2)t$
$\omega = \omega_1 - \omega_2$	$\dfrac{1}{2}(\alpha_2\beta_1\beta_2)\cos(\omega_1 - \omega_2)t$
$\omega = 2\omega_1 + \omega_2$	$\left(\dfrac{3}{4}\alpha_3\beta_1^2\beta_2\right)\cos(2\omega_1 + \omega_2)t$
$\omega = 2\omega_1 - \omega_2$	$\left(\dfrac{3}{4}\alpha_3\beta_1^2\beta_2\right)\cos(2\omega_1 - \omega_2)t$
$\omega = 2\omega_2 + \omega_1$	$\left(\dfrac{3}{4}\alpha_3\beta_1^2\beta_2\right)\cos(2\omega_2 + \omega_1)t$
$\omega = 2\omega_2 - \omega_1$	$\left(\dfrac{3}{4}\alpha_3\beta_1^2\beta_2\right)\cos(2\omega_2 - \omega_1)t$

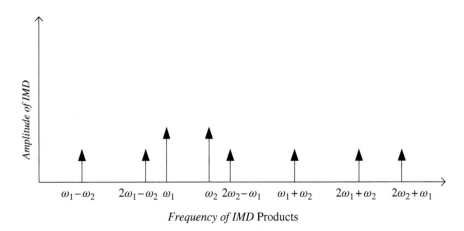

Figure 10.11 Illustration of IMD frequencies and products.

Table 10.2 Summary of *IM* product frequencies.

IM_2 frequencies	$\omega_1 \pm \omega_2$	
IM_3 frequencies	$2\omega_1 \pm \omega_2$	$2\omega_2 \pm \omega_1$
IM_5 frequencies	$3\omega_1 \pm 2\omega_1$	$3\omega_2 \pm 2\omega_1$

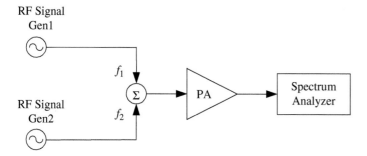

Figure 10.12 Simplified IMD measurement setup.

$$IM_3 = 3HD_3 \tag{10.38}$$

IM_3 distortion components at frequencies $2\omega_1 - \omega_2$ and $2\omega_2 - \omega_1$ are very close to the fundamental components. It is why IM_3 signal is measured most of time for intermodulation distortion (IMD) characterization of the amplifier. The simplified measurement set up for IMD testing is shown in Figure 10.12.

The point where the output components at the fundamental frequency and IM_3 intersect is called the intercept point, or IP_3. At this point, $IM_3 = 1$ and IP_3 are found from (10.36) as

$$IM_3 = 1 = \frac{3}{4}\frac{\alpha_3}{\alpha_1}(IP_3)^2 \tag{10.39}$$

or

$$IP_3 = \sqrt{\frac{4}{3}\frac{\alpha_1}{\alpha_3}} \tag{10.40}$$

which can also be written as

$$IP_3 = \frac{V_{in}}{\sqrt{IM_3}} \tag{10.41}$$

where V_{in} is the input voltage. Equation (10.41) can be expressed in terms of dB by taking log of both sides in (10.41) as

$$IP_3(dB) = V_{in}(dB) - \frac{1}{2}IM_3(dB) \tag{10.42}$$

The dynamic range (DR) is measured to understand the level of the output noise and is defined as

$$DR = \alpha_1 \frac{V_{in}}{V_{Nout}} = \frac{V_{in}}{V_{Nin}} \tag{10.43}$$

where the input noise is related to the output noise by

$$V_{Nin} = \frac{V_{Nout}}{\alpha_1} \tag{10.44}$$

So

$$DR(dB) = V_{in}(dB) - V_{Nin}(dB) \tag{10.45}$$

Intermodulation free dynamic range ($IMFDR_3$) is defined as the largest DR possible with no IM_3 product. For the third-order intermodulation distortion, V_{Nout} is defined by

$$V_{Nout} = \frac{3}{4} \alpha_3 V_{in}^3 \tag{10.46}$$

We can then obtain

$$V_{in} = \sqrt[3]{\frac{4}{3\alpha_3} V_{Nout}} \tag{10.47}$$

Substitution of (10.46) into (10.43) gives $IMFDR_3$ as

$$DR = IMFDR_3 = \alpha_1 \frac{V_{in}}{V_{Nout}} = V_{in} = \sqrt[3]{\frac{4}{3} \frac{\alpha_1^3}{\alpha_3} \frac{1}{V_{Nout}^2}} \tag{10.48}$$

Since, $V_{Nout} = \alpha_1 V_{Nin}$ from (10.41), Eq. (10.43) can be written in terms of input noise as

$$IMFDR_3 = \alpha_1 \frac{V_{in}}{V_{Nout}} = V_{in} = \sqrt[3]{\frac{4}{3} \frac{\alpha_1}{\alpha_3} \frac{1}{V_{Nin}^2}} \tag{10.49}$$

When Eqs. (10.40) and (10.49) are compared, $IMFDR_3$ can also be expressed using intercept point IP_3 as

$$IMFDR_3 = \left(\frac{IP_3}{V_{Nin}}\right)^{2/3} \tag{10.50}$$

or in terms of dB, (10.50) can be also given by

$$IMFDR_3(\text{dB}) = \frac{2}{3}(IP_3(\text{dB}) - V_{Nin}(\text{dB})) \tag{10.51}$$

The relationship between fundamental components and third-order distortion components via input and output voltages is illustrated in Figure 10.13. In Figure 10.13, −1 dB compression point is used to characterize the IM_3 product. A −1 dB compression point can be defined as the value of the input voltage, V_{in}, which is designated by $V_{in,1\,dBc}$, where fundamental component is reduced by 1 dB. $V_{in,1dBc}$ can be defined by

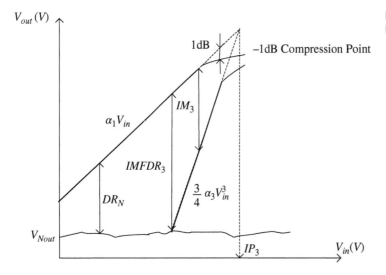

Figure 10.13 Illustration of the relation between fundamental components and IM_3.

$$V_{in,1dBc} = \sqrt{(0.122)\left(\frac{4}{3}\right)\left|\frac{\alpha_1}{\alpha_3}\right|} \tag{10.52}$$

which is also equal to

$$V_{in,1dBc} = \sqrt{(0.122)}IP_3 \tag{10.53}$$

Equation (10.53) can be expressed in dB as

$$V_{in,1dBc}(dB) = IP_3(dB) - 9.64dB \tag{10.54}$$

As a result, once IP_3 is determined, (10.53) can be used to calculate the 1 dB compression point for the amplifier.

Example 10.5   Time Domain and Frequency Domain Representation

Assume a sinusoid signal, $v_i(t) = \sin(\omega t)$, with a 5 Hz frequency is applied to a nonlinear amplifier which has output signal (a) $v_o(t) = 10\sin(\omega t) + 2\sin^2(\omega t)$, (b) $v_o(t) = 10\sin(\omega t) - 3\sin^3(\omega t)$, and (c) $v_o(t) = 10\sin(\omega t) + 2\sin^2(\omega t) - 3\sin^3(\omega t)$. Obtain the time domain representation of the input signal and frequency domain representation of the power spectra of the output signal of the amplifier.

Solution

a) The frequency spectrum for the input signal and power spectrum for the output signal of the amplifier are obtained using MATLAB script given in Figure 10.14. Based on the results shown in Figure 10.14, the amplifier output has components at DC, f, and $2f$.

b) As can be seen from the third-order response shown in Figure 10.15, the output signal does not have a DC component anymore. The third-order effect shows itself as clipping in the time domain signal and fundamental and third-order components at the output power spectra of the signal.

c) Using the modified MATLAB script in parts (a) and (b), the time domain and frequency domain signals are obtained and illustrated in Figure 10.16.

As illustrated, the output response has components at DC, f, $2f$, and $3f$. Overall, the level of the nonlinearity response of the amplifier strongly depends on the coefficients of the output signal.

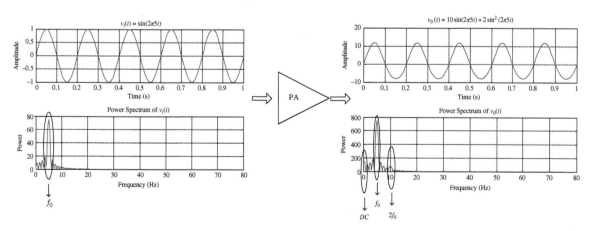

Figure 10.14   Second-order nonlinear amplifier output response which has components at DC, f, and $2f$.

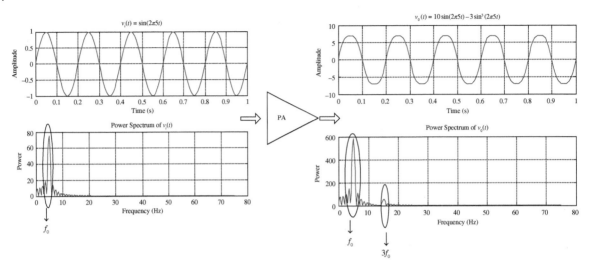

Figure 10.15 Third-order nonlinear amplifier output response which has components at *f* and 3*f*.

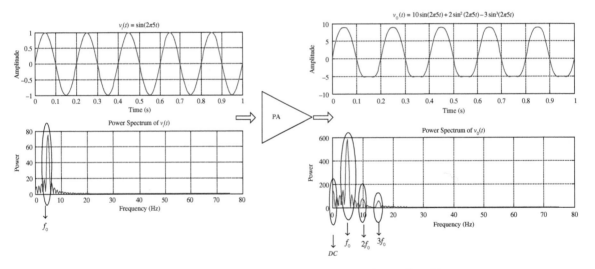

Figure 10.16 Nonlinear amplifier response which has second- and third-order nonlinearities.

10.3 Small Signal Amplifier Design

There are three critical design stages that need to be implemented in small signal amplifier designs. These are:

- Design of DC biasing circuit.
- Obtaining parameters of the transistor.
- Implementation of the small signal amplifier design.

The small signal amplifier design begins with designing a biasing circuit to choose the operating point, Q point, V_{ceq} and I_{cq} for BJT, and V_{dsq} and I_{dq} for field-effect transistor (FET). In general, V_{ceq} and V_{dsq} are usually taken at half the system supply voltage for a small signal amplifier design [1–3]. Hence, once the Q point is selected, the designer must determine I_{cq} and I_{dq}.

After operating point is determined via the DC biasing circuit, S parameters must be obtained at the operating Q point and the desired operating frequency because S parameters vary significantly on the operating conditions. S parameters are provided by the manufacturer of the transistor most of the time. Once the DC biasing circuit is designed and the S parameters obtained, the small signal amplifier is then designed following the design stages that are outlined in Section 10.3.1.

10.3.1 DC Biasing Circuits

Let's consider the BJT circuit given in Figure 10.17. When there is only DC source in the circuit, the i_c and v_{CE} curve can be obtained, as shown in Figure 10.18.

If the circuit shown in Figure 10.17 is revised and the AC sinusoidal source with amplitude ΔV_{BB} is introduced, as shown in Figure 10.19, the base current and collector current will change to $i_B + \Delta\, i_B$ Cos (ωt) and $i_C + \Delta\, i_C$ Cos (ωt), respectively. Similarly, the collector to emitter voltage will change as $v_{CE} + \Delta v_{CE}$ Cos (ωt).

Consider the operational conditions for the BJT circuit under DC excitation, as shown in Figure 10.18. Based on the given information, $I_B = 150\,\mu A$, $I_C = 23\,mA$, and collector to emitter voltage $V_{CE} = 7.5\,V$ at Q point. When the AC source is applied with a DC excitation, base current, collector, and collector to emitter voltage change to

$$i_B + \Delta\, i_B = 150 + 40\,\text{Cos}\,(\omega t)\,\mu A$$

$$i_c + \Delta\, i_c = 22 + 7\,\text{Cos}\,(\omega t)\,mA$$

$$v_{CE} + \Delta\, v_{CE} = 7.5 - 2.5\,\text{Cos}\,(\omega t)\,V$$

This indicates that the base current will swing between 110 and 190 μA. One important criterion is to keep the operating points on the load line during the swing of the current or voltage. When the base current reaches its peak value of 190 μA, the collector to emitter voltage, V_{CE}, reaches around 4.5 V and the collector current is equal to 29 mA. These changes can be illustrated in Figure 10.20.

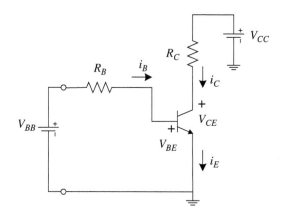

Figure 10.17 BJT circuit with only DC source.

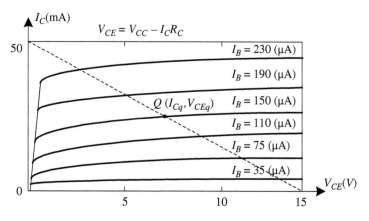

Figure 10.18 i_c and v_{CE} curve for DC bias.

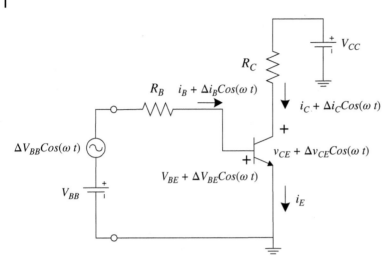

Figure 10.19   BJT circuit with DC and AC sources.

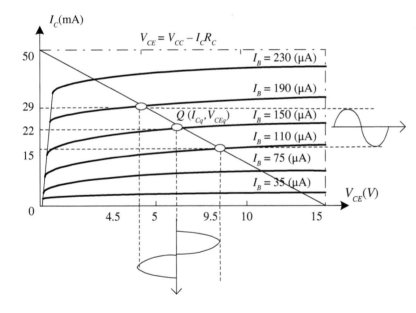

Figure 10.20 i_c and v_{CE} curve with DC and AC sources.

As seen from Figure 10.20, the signal is amplified with the implementation of AC source. This is valid when the transistor, BJT, operates in the active region. If the base current is increased further, the swing for the current will reach to the value, which will take the transistor into the cutoff region.

10.3.2   BJT Biasing Circuits

There are typical biasing circuits that are used for amplifiers with BJTs. These biasing circuits can be referred as (a) fixed bias, (b) stable bias, (c) emitter bias, and (d) self-bias.

10.3.2.1 Fixed Bias

Consider the common emitter configuration for the amplifier circuit given in Figure 10.21. Coupling capacitors are used to isolate the circuit from other circuits that are connected at the output and input of the amplifier. The common practice for this type of biasing circuit is to have V_{BB} and V_C equal to each other so that it can be supplied from the single source. The idea of using the simple biasing circuit is to set a constant base current. However, this makes it very sensitive to gain variations which then can be improved to some degree with R_1 and R_2. If BJT is assumed to be operating in active linear region and $V_{BB} = V_C$, the circuit can be analyzed as follows. In our analysis let's assume $R_1 \to 0$. So

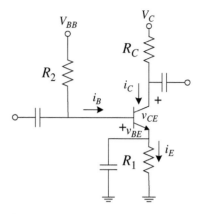

Figure 10.21 Fixed bias BJT circuit.

$$V_C = I_B R_2 + V_{BE} \tag{10.55}$$

which leads to

$$I_B = \frac{V_C - V_{BE}}{R_2} \tag{10.56}$$

So

$$I_C = \beta I_B = \beta \frac{V_C - V_{BE}}{R_2} \tag{10.57}$$

Collector to emitter voltage can be from

$$V_C = I_C R_C + V_{CE} \tag{10.58}$$

which leads to

$$V_{CE} = V_C - I_C R_C \tag{10.59}$$

When (10.57) is substituted into (10.59), we obtain

$$V_{CE} = V_C - R_C \left(\beta \frac{V_C - V_{BE}}{R_2} \right) \tag{10.60}$$

Example 10.6   RF Fixed Bias Circuit

Calculate with R_1 and R_C for the fixed biasing circuit given in Figure 10.21 when $\beta = 100$ and $V_C = 15$ V to have the Q point at $V_{CE} = 7.5$ V and $I_C = 25$ mA.

Solution

From (10.55)

$$R_2 = \frac{V_C - V_{BE}}{I_B} = \frac{15 - 0.7}{0.25 \times 10^{-3}} = 57.2 \text{ k}\Omega$$

R_C can be found from (10.59) as

$$R_C = \frac{V_C - V_{CE}}{I_C} = \frac{15 - 7.5}{25 \times 10^{-3}} = 300 \ \Omega$$

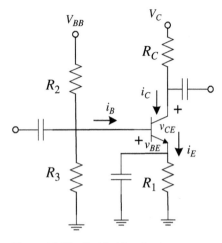

Figure 10.22 Stable bias circuit.

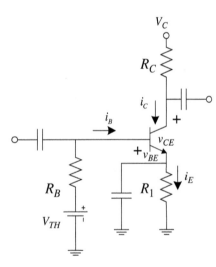

Figure 10.23 Simplified stable bias circuit.

10.3.2.2 Stable Bias

The stable bias circuit is different from the fixed bias circuit with the addition of the resistor between base and ground. R_3 can be adjusted with R_2 so that large improvement over stability can be achieved. Thévenin's theorem can be used to simplify the circuit shown in Figure 10.22, as illustrated in Figure 10.23. Typically, R_2 is 2 to 10 times greater than R_3.

In Figure 10.23

$$V_{TH} = \frac{R_3}{R_2 + R_3} V_{CC} \tag{10.61a}$$

and

$$R_B = \frac{R_2 R_3}{R_2 + R_3} \tag{10.61b}$$

The stability in the circuit shown in Figure 10.23 can be described as follows. If the temperature increases, this will cause β to increase and results in an increase in collector current, I_C, and emitter current, I_E beyond the desired operational values. However, since V_{TH} and R_B are fixed and don't vary, base current, I_B, will reduce and lower collector current, I_C, back to its original desired value. Hence, the stability of the biasing network is satisfied with this method. We can assume in the circuit that

$$I_E \approx I_C = \beta I_B \tag{10.62}$$

In addition

$$V_{TH} = I_B R_B + V_{BE} + I_E R_1 \tag{10.63}$$

So

$$I_B = \frac{V_{TH} - V_{BE}}{R_B + \beta R_1} \tag{10.64}$$

and

$$V_C = I_C R_C + V_{CE} + I_E R_1 \tag{10.65}$$

which leads to

$$V_{CE} = V_C - I_C (R_C + R_1) \tag{10.66}$$

For feedback to be effective to have the stability

$$R_B \ll \beta R_1 \tag{10.67}$$

Hence

$$I_B \approx \frac{V_{TH} - V_{BE}}{\beta R_1} \tag{10.68}$$

So

$$I_C \approx \frac{V_{TH} - V_{BE}}{R_1} \tag{10.69}$$

and

$$V_{CE} = V_C - I_C(R_C + R_1) \approx V_C - \frac{R_C + R_1}{R_1}(V_{TH} - V_{BE}) \tag{10.70}$$

It is clear from (10.70) that I_C and V_{CE} do not depend on β anymore which makes this circuit immune to the changes in β.

Example 10.7   RF Stable Bias Circuit

Design a stable bias circuit shown in Figure 10.23 with a Q point of $I_c = 1.5$ mA and $V_{CE} = 7.5$ V when the DC gain of the transistor, β, ranges from 50 to 200.

Solution

We first find the collector supply voltage, V_C. Since we desire Q point located in the middle of the load line

$$V_C = 2V_{CE} = 2(7.5) = 15 \text{ V}$$

R_C and R_1 can be found from (10.66) as

$$R_C + R_1 = \frac{V_C - V_{CE}}{I_C} = \frac{7.5}{1.5 \times 10^{-3}} = 5 \text{ k}\Omega$$

Designer can choose R_C and R_1 for the circuit. Let

$$R_C = 4 \text{ k}\Omega \text{ and } R_1 = 1 \text{ k}\Omega$$

We now need to find the base resistor, R_B, in Figure 10.23. It needs to satisfy the condition given in (10.67). Then, we can get the lowest value of $\beta = 50$, and

$$R_B \ll \beta R_1 \rightarrow R_B = 0.1(50)(1000) = 5 \text{ k}\Omega$$

From (10.69)

$$V_{TH} \approx I_C R_1 + V_{BE} = 1.5 \times 10^{-3}(1 \times 10^{-3}) + 0.7 = 2.2 \text{ V}$$

At this point, we can calculate R_3 and R_2 from (10.61) as

$$\frac{V_{TH}}{V_C} = \frac{R_3}{R_2 + R_3} = 0.147 \text{ and } R_B = \frac{R_2 R_3}{R_2 + R_3} = 5000$$

Solution of above equation for R_3 and R_2 gives

$$R_2 = 5.8R_3 \rightarrow R_2 = 24.734 \text{ k}\Omega \text{ and } R_3 = 4.264 \text{ k}\Omega$$

10.3.2.3   Self-bias

In self-bias, the Rc connector is connected through another resistor to the base of the circuit as a feedback resistor, as shown in Figure 10.24. Since the voltage at the collector is lower than the fixed bias where it is directly supplied from V_c, the value of R_2 needs to be smaller. Let's analyze this circuit by assuming that BJT is operating in the linear active region, and $Rc \rightarrow 0$. Since $I_B \ll I_C$, then

$$I_1 = I_C + I_B \approx I_C \tag{10.71}$$

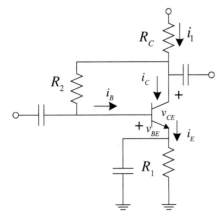

Figure 10.24 Self-bias circuit.

Since

$$V_C = I_C R_C + V_{BE} + I_B R_2 \qquad (10.72a)$$

or

$$V_C = I_C(R_C + R_2/\beta) + V_{BE} \qquad (10.72b)$$

Hence

$$I_C = \frac{V_C - V_{BE}}{R_C + R_2/\beta} \qquad (10.73)$$

If we assume, $R_B/\beta \ll R_C$, and $V_{BE} = V_\gamma$, then

$$I_C = \frac{V_C - V_\gamma}{R_C} \qquad (10.74)$$

Since collector current does not depend on β, the bias point remains stable. The operational point in the active region can be checked to see if $V_{CE} > V_\gamma$ using

$$V_{CE} = R_2 I_B + V_{BE} = R_2 I_B + V_\gamma > V_\gamma \qquad (10.75)$$

Equation (10.75) confirms that BJT is operating in the active region. The impact of the self-bias circuit can be understood better if we rewrite (10.72a) as

$$I_B = \frac{V_C - V_{BE} - I_C R_C}{R_2} = \frac{V_C - V_\gamma - I_C R_C}{R_2} \qquad (10.76)$$

If β increases due to temperature change, the collector current will increase and, eventually, the base current, I_B, will reduce. When I_B reduces, this will cause the collector current, I_C, to drop. Hence, this will self-stable the change due to variation in β.

10.3.2.4 Emitter Bias

The most stable operating point can be achieved if the base of BJT circuit with common emitter configuration is grounded via R_2 and emitter is connected to another supply voltage, as shown in Figure 10.25, as long as $R_2 << \beta R_1$. This can be verified by analyzing the circuit as

$$R_2 I_B + V_{BE} + R_1 I_E - V_E = 0 \qquad (10.77)$$

Since

$$I_E \approx I_C = \beta I_B \qquad (10.78)$$

Then

$$R_2 \frac{I_E}{\beta} + R_1 I_E = V_E - V_{BE} \text{ or } I_E = \frac{V_E - V_{BE}}{R_1 + R_2/\beta} \qquad (10.79)$$

With the condition as $R_2 << \beta R_1$, we also obtain

$$I_C \approx I_E \approx \frac{V_E - V_{BE}}{R_1} = k_1 = \text{constant} \qquad (10.80)$$

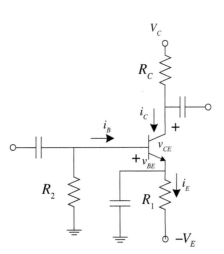

Figure 10.25 Emitter bias circuit.

So

$$V_C = I_C R_C + V_{CE} + I_E R_1 - V_E \qquad (10.81)$$

which leads to

$$V_{CE} = V_C + V_E - I_C(R_1 + R_C) = k_2 = \text{constant} \qquad (10.82)$$

Hence, it is obvious from (10.79), and (10.80)–(10.82) that $I_E, I_C,$ and V_{CE} do not depend on β, which shows a stable operation.

10.3.2.5 Active Bias Circuit

There are several other different biasing circuits that can be applied for BJTs and FETs. One method to compensate for temperature variation in the biasing circuit is to use diodes.

For better temperature compensation, one of the methods used is implementation of diodes for compensation. The compensation of the temperature is achieved by the reduction of the internal resistance of the diode against temperature increase. When the temperature increases, the diode's forward voltage is reduced. This causes the base to emit voltage to compensate for the effect of the increase in current due to the rise in temperature. The biasing circuit using two diodes to accomplish this task is illustrated in Figure 10.26. The circuit shown in Figure 10.26a can be enhanced and the number of components can be minimized by using active biasing controller IC, such as the Siemens BCR 400 shown in Figure 10.26b. This kind of biasing IC is able to supply stable bias current even at low supply voltage.

10.3.2.6 Bias Circuit using Linear Regulator

If the bias voltage should be independent of variations in the power supply, then the bias circuit with a linear regulator is a good option. In the sample bias circuit using linear regulator shown in Figure 10.27, temperature compensation is done using diode D_1.

10.3.3 FET Biasing Circuits

The biasing circuits for FETs can be constructed similar to circuits for BJTs. The bias circuit utilizing the linear regulator shown in Figure 10.27 can also be applied to FETs too. The important difference between FET and BJT circuits is the insignificantly small gate current flows with an application of input signal to gate, so the drain current can be assumed to always be equal to the source current. This enables FET circuits to be biased at their operating currents, which can be varied by the resistor connected to the source since it is always negative with respect to source voltage. The collection of similar biasing circuits for common source FETs are shown in Figure 10.28.

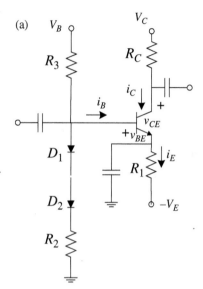

(a)

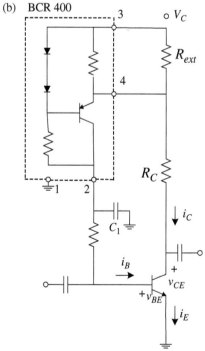

(b)

Figure 10.26 (a) Bias circuit with temperature compensating diodes. (b) Active bias circuit with BCR 400.

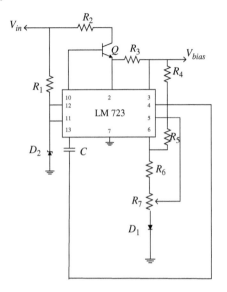

Figure 10.27 Bias circuit using linear regulator.

10.3.4 Small Signal Amplifier Design Method

The design process begins with the selection of the operating points for the transistor as it is detailed in Section 10.2.3. Since the operating points for voltage and current are identified for the transistor, S parameters that are supplied by the manufactured need to be obtained or measured. After S parameters are obtained or measured, it is now time to follow step-by-step design procedure to implement the small signal amplifier. There are specific quantities and terms that will be used in the design of small signal amplifiers. These terms and quantities are explained in Section 10.3.4.1.

10.3.4.1 Definitions Power Gains for Small Signal Amplifiers
Consider the generalized two port network shown in Figure 10.29. The illustration shown in Figure 10.30 in conjunction with Figure 10.29 can help to integrate the application of scattering parameters in amplifier design. In the design of small signal amplifiers, there are three power quantities that are used.

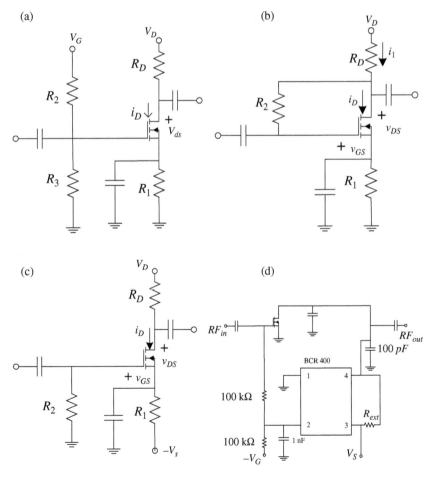

Figure 10.28 (a) Stable bias circuit. (b) Self-bias circuit. (c) Source bias circuit. (d) Active bias circuit.

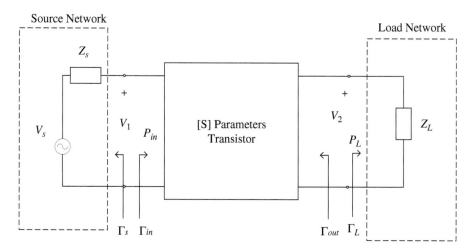

Figure 10.29 Generalized two-port network.

Figure 10.30 Integration of amplifier circuit.

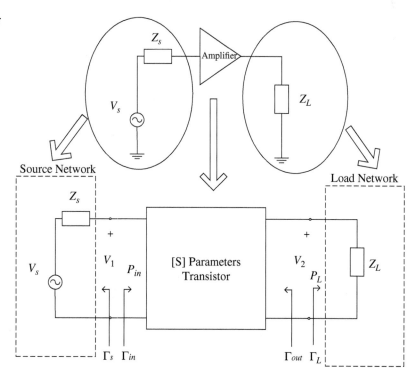

Let's obtain power relations for the circuit in Figure 10.29.

- At the input port

$$V_1 = V_s \frac{Z_{in}}{Z_{in} + Z_s} = V_1^+ + V_1^- = V_1^+ (1 + \Gamma_{in}) \tag{10.83}$$

In addition

$$\Gamma_{in} = \frac{Z_{in} - Z_0}{Z_{in} + Z_0} \tag{10.84}$$

$$\Gamma_s = \frac{Z_s - Z_0}{Z_s + Z_0} \tag{10.85}$$

$$\frac{Z_{in}}{Z_0} = \frac{1 + \Gamma_{in}}{1 - \Gamma_{in}} \tag{10.86}$$

So, from (10.83) to (10.86), we can obtain the following relation

$$V_1^+ = \frac{1}{1 + \Gamma_{in}} V_s \frac{Z_{in}}{Z_{in} + Z_s} \text{ or } V_1^+ = \frac{V_s(1 - \Gamma_s)}{2(1 - \Gamma_s \Gamma_{in})} \tag{10.87}$$

As a result, the input power can be written as

$$P_{in} = \frac{1}{2} \frac{|V_1^+|^2}{Z_0} \left(1 - |\Gamma_{in}|^2\right) \text{ or } P_{in} = \frac{1}{8} \frac{|V_s|^2}{Z_0} \frac{\left(1 - |\Gamma_{in}|^2\right)|1 - \Gamma_{in}|^2}{|1 - \Gamma_s \Gamma_{in}|^2} \tag{10.88}$$

Available power from source, P_{avs}, can be found when $\Gamma_{in} = \Gamma_s^*$ from

$$P_{avs}(\Gamma_s) = P_{in}|_{\Gamma_{in} = \Gamma_s^*} \tag{10.89a}$$

or

$$P_{avs}(\Gamma_s) = P_{in}|_{\Gamma_{in} = \Gamma_s^*} = \frac{1}{8} \frac{|V_s|^2}{Z_0} \frac{|1 - \Gamma_s|^2}{\left(1 - |\Gamma_s|^2\right)} \tag{10.89b}$$

- *At the output port*

$$V_2^- = S_{21} V_1^+ + S_{22} V_2^+ \tag{10.90a}$$

$$V_2^+ = \Gamma_L V_2^- \tag{10.90b}$$

Substitution of (10.90b) into (10.90a) gives

$$V_2^- = S_{21} V_1^+ + S_{22} \Gamma_L V_2^- \text{ or } V_2^- = \frac{S_{21} V_1^+}{1 - S_{22} \Gamma_L} = \frac{S_{21} V_s(1 - \Gamma_s)}{(1 - S_{22} \Gamma_L)(1 - \Gamma_s \Gamma_{in})} \tag{10.91}$$

Then

$$P_L = \frac{1}{2} \frac{|V_2^-|^2}{Z_0} \left(1 - |\Gamma_L|^2\right) \text{ or } P_L = \frac{1}{8} \frac{|V_s|^2}{Z_0} \frac{|S_{21}|^2 |1 - \Gamma_s|^2}{|1 - S_{22} \Gamma_L|^2 (1 - \Gamma_s \Gamma_{in})} \left(1 - |\Gamma_L|^2\right) \tag{10.92}$$

Available power from the network, P_{avn}, can be found when $\Gamma_L = \Gamma_{out}^*$ from

$$P_{avn}(\Gamma_{out}) = P_L|_{\Gamma_L = \Gamma_{out}^*} \tag{10.93a}$$

or

$$P_{avn}(\Gamma_{out}) = P_L|_{\Gamma_L = \Gamma_{out}^*} = \frac{1}{8} \frac{|V_s|^2}{Z_0} \frac{|S_{21}|^2 |1 - \Gamma_s|^2}{|1 - S_{22} \Gamma_{out}^*|^2 (1 - \Gamma_s \Gamma_{in})} \left(1 - |\Gamma_{out}|^2\right) \tag{10.93b}$$

Since

$$\Gamma_{in} = S_{11} + \frac{S_{12} S_{21} \Gamma_L}{1 - S_{22} \Gamma_L} = \frac{S_{11} - S_{11} S_{22} \Gamma_L + S_{12} S_{21} \Gamma_L}{1 - S_{22} \Gamma_L} \tag{10.94}$$

Substituting (10.94) into (10.93b) when $\Gamma_L = \Gamma_{out}^*$ gives

$$P_{avn}(\Gamma_{out}) = P_L|_{\Gamma_L = \Gamma_{out}^*} = \frac{1}{8}\frac{|V_s|^2}{Z_o}\frac{|S_{21}|^2|1-\Gamma_s|^2}{|1-S_{11}\Gamma_s|^2\left(1-|\Gamma_{out}|^2\right)^2} \tag{10.95}$$

There are three important power quantities that need to be known. They are: operating power gain, G_p, available power gain, G_A, and transducer power gain, G_T.

- *Transducer Gain*

  Transducer gain, G_T, is the ratio of the time-averaged power dissipated as the load to the maximally available time average power from the source

$$G_T = \frac{P_L}{P_{avs}} \tag{10.96}$$

Substitution of (10.92) and (10.89b) into (10.96) gives

$$G_T = \frac{P_L}{P_{avs}} = \underbrace{\frac{\left(1-|\Gamma_s|^2\right)}{|1-\Gamma_s\Gamma_{in}|^2}}_{source}|S_{21}|^2\underbrace{\frac{\left(1-|\Gamma_L|^2\right)}{|1-S_{22}\Gamma_L|^2}}_{load} \tag{10.97}$$

The transducer power gain, G_T, is the gain component that is used to allow us to understand the amplifying level of the amplifier. If there is a specific power gain requirement for the design of an amplifier, G_T is the quantity that is always referenced.

- *Operating Gain*

The operating gain, G_p, is the ratio of the time-averaged power dissipated as the load to the time-averaged power delivered to the network and is obtained from

$$G_p = \frac{P_L}{P_{in}} \tag{10.98}$$

Substituting (10.88) and (10.92) into (10.98) gives

$$G_p = \frac{P_L}{P_{in}} = \underbrace{\frac{1}{\left(1-|\Gamma_{in}|^2\right)}}_{Source}|S_{21}|^2\underbrace{\frac{\left(1-|\Gamma_L|^2\right)}{|1-S_{22}\Gamma_L|^2}}_{Load} \tag{10.99}$$

- *Available Gain*

  The available gain, G_A, is the ratio of the maximally available time-averaged power from the network to the maximally available time-averaged power from the source and is defined by

$$G_A = \frac{P_{avn}}{P_{avs}} \tag{10.100}$$

Substitution of (10.95) and (10.89b) into (10.100) gives

$$G_A = \frac{P_{avn}}{P_{avs}} = \underbrace{\frac{\left(1-|\Gamma_s|^2\right)}{|1-S_{11}\Gamma_s|^2}}_{source}|S_{21}|^2\underbrace{\frac{1}{\left(1-|\Gamma_{out}|^2\right)}}_{load} \tag{10.101}$$

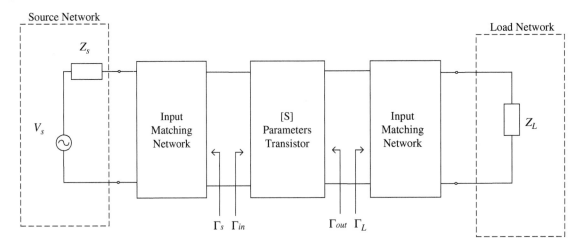

Figure 10.31 General two-port amplifier network.

Operating and available power gains, G_P and G_A, are used as knobs to meet the certain amplifier transducer gain, G_T.

Consider the more general and practical amplifier circuit shown in Figure 10.31. If the S parameters of the transistor, Γ_s, Γ_L of the circuit, are known, then the gain components of the amplifier: G_P, G_A, and G_T can be calculated using (10.97), (10.99), and (10.101).

The reflection coefficients that are used to calculate the gain parameters are

$$\Gamma_s = \frac{Z_s - Z_o}{Z_s + Z_o} \tag{10.102}$$

$$\Gamma_L = \frac{Z_L - Z_o}{Z_L + Z_o} \tag{10.103}$$

$$\Gamma_{in} = S_{11} + \frac{S_{12}S_{21}\Gamma_L}{1 - S_{22}\Gamma_L} \tag{10.104}$$

$$\Gamma_{out} = S_{22} + \frac{S_{12}S_{21}\Gamma_s}{1 - S_{11}\Gamma_s} \tag{10.105}$$

10.3.4.2 Design Steps for Small Signal Amplifier

The following steps will simplify the design of small signal amplifiers.

1) Design the biasing circuit based on a transistor manufacturer datasheet.
2) Obtain S parameters at bias conditions.
3) Investigate the stability of the transistor. Check the stability using Rollet's stability factor k. If $|k| > 1$ and $|\Delta| < 1$, this implies that transistor is unconditionally stable.
4) If the transistor is unconditionally stable, design the amplifier for gain.
5) If the condition mentioned in step 3 for unconditional stability is not met, the transistor is potentially stable. If the transistor is potentially stable, draw source and load stability circles by calculating the center points and radii for input and output stability circles on a Smith chart: r_{in}, C_{in}, and r_{out}, C_{out}.

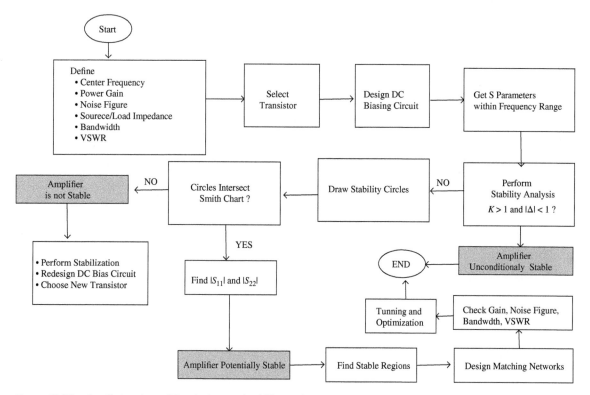

Figure 10.32   Small signal amplifier design method illustration.

6) Calculate if Γ_L and Γ_s lie in an unstable region. If they lie in a stable region, then nothing more needs to be done for stability.

7) If Γ_L and Γ_s lie in an unstable region, then shunt or series resistances need to be added at the input and/or output to move S_{22}^* and/or S_{11}^* into the stable region.

8) Adjust the gain of the amplifier if needed by Γ_L or Γ_s.

9) Plot the input and output gain circles for a unilateral case if $S_{12} \to 0$. Use the transducer power gain equation (10.97). The specified gain may be less than the maximum unilateral gain, $G_{TU\text{max}}$.

10) If $S_{12} \neq 0$, then design the amplifier using the unilateral method (to have control of selective mismatch at the input and output), which can be done by investigating the maximum error based on the unilateral figure of merit to determine if the method could be applied.

11) If the unilateral method cannot be applied, the designer should proceed with the bilateral design method.

These steps are simplified and illustrated in Figure 10.32.

10.3.4.3   Small Signal Amplifier Stability

Stability analysis is necessary to measure an amplifier's resistance against oscillations. The oscillations may occur if reflected signals at the input or output port increase their magnitudes while they are reflected between an active port and its termination continuously. Oscillations in an amplifier are not desired, because when they occur the characteristic of the amplifier changes drastically. Scattering parameters become no longer valid and hence the circuit does not perform as expected. This may result in a catastrophic failure and may damage active device and surrounding components.

10.3.4.3.1 Unconditional Stability

When an amplifier remains stable throughout the entire cycle under operating conditions and frequency, it is referred to as being unconditionally stable. Unconditional stability can be satisfied for any passive source and load when

$$|\Gamma_s| < 1 \tag{10.106a}$$

and

$$|\Gamma_L| < 1 \tag{10.106b}$$

Having negative resistance can be avoided if

$$|\Gamma_s \Gamma_{in}| < 1 \tag{10.107a}$$

$$|\Gamma_L \Gamma_{out}| < 1 \tag{10.107b}$$

So

$$|\Gamma_{out}| = \left| S_{22} + \frac{S_{12} S_{21} \Gamma_s}{1 - S_{11} \Gamma_s} \right| < 1 \tag{10.109}$$

A two-port network is said to be unconditionally stable at a given frequency if

$$\operatorname{Re}\{Z_{in}\} > 0 \quad \text{and} \quad \operatorname{Re}\{Z_{out}\} > 0 \tag{10.110}$$

In practice, Rollet's stability factor, k, and determinant of the scattering matrix of the active device, $|\Delta|$, can also be used to test the unconditional stability of two-port networks using the following equations.

$$k = \frac{1 + |\Delta|^2 - |S_{11}|^2 - |S_{22}|^2}{2|S_{21}||S_{12}|} > 1 \tag{10.111}$$

$$|\Delta| = |S_{22} S_{11} - S_{12} S_{21}| < 1 \tag{10.112}$$

10.3.4.3.2 Stability Circles

We can investigate the stability of the amplifier graphically by studying the stability circles on a Smith chart.

Output Stability Circle The Γ_L or load plane on a Smith chart is defined by output stability circles where the boundaries are defined as being between $|\Gamma_L| < 1$ (stable) and $|\Gamma_L| < 1$ (unstable). This can be found by solving between Γ_L in the following equation.

$$|\Gamma_{in}| = \left| S_{11} + \frac{S_{12} S_{21} \Gamma_L}{1 - S_{22} \Gamma_L} \right| = 1 \tag{10.113}$$

The solution of (10.113) lies on a circle with radius

$$r_L = \left| \frac{S_{12} S_{21}}{|S_{22}|^2 - |\Delta|^2} \right| \tag{10.114}$$

and center

$$c_L = \frac{\left(S_{22} - \Delta S_{11}^* \right)^*}{|S_{22}|^2 - |\Delta|^2} \tag{10.115}$$

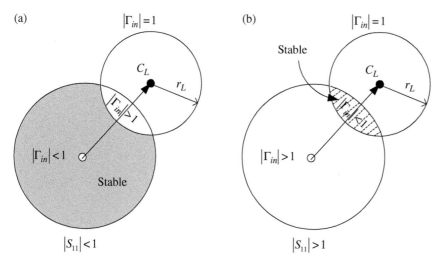

Figure 10.33 Smith chart illustrating output stability regions.

where

$$|\Delta| = |S_{22}S_{11} - S_{12}S_{21}| \tag{10.116}$$

The circles formed by Eqs. (10.113)–(10.115) establish the stability regions for the output. This can be done by plotting a Γ_L circle and investigating the intersection of this circle with a Smith chart. If the circle intersects the Smith chart, there is an instability region defined by the boundary. If there is no intersection, the device is unconditionally stable. The graphical illustration of the stability regions are given in Figure 10.33. The region inside the Smith chart where $|S_{11}| < 1$ represents the stable region. These stable regions are the shaded regions in the figure.

Input Stability Circle   The Γ_s or source plane on a Smith chart is defined by input stability circles where the boundaries are defined as being between $|\Gamma_s| < 1$ (stable) and $|\Gamma_s| < 1$ (unstable). This can be found by solving between Γ_L in the following equation.

$$|\Gamma_{out}| = \left| S_{22} + \frac{S_{12}S_{21}\Gamma_s}{1 - S_{11}\Gamma_s} \right| = 1 \tag{10.117}$$

The solution of (10.117) lies on a circle with radius and center are

$$r_s = \left| \frac{S_{12}S_{21}}{|S_{11}|^2 - |\Delta|^2} \right| \tag{10.118}$$

$$c_s = \frac{\left(S_{11} - \Delta S_{22}^*\right)^*}{|S_{11}|^2 - |\Delta|^2} \tag{10.119}$$

The circles formed by Eqs. (10.117)–(10.119) establish the stability regions for the input. This can be done by plotting a Γ_s circle and investigating the intersection of this circle with a Smith chart. The graphical illustration of the stability regions for input are given in Figure 10.34.

If the device is unconditionally stable, there is no intersection of the circle with the Smith chart. This is shown in Figure 10.35.

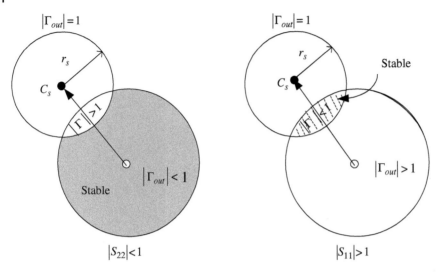

Figure 10.34 Smith chart illustrating input stability regions.

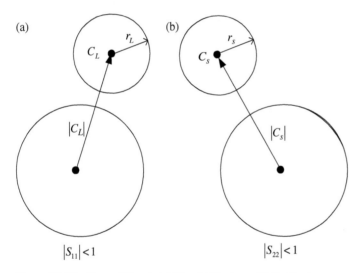

Figure 10.35 Unconditional stability: (a) Γ_L plane; (b) Γ_s plane.

Example 10.8 Stability Circles

S parameters of a transistor at 800 MHz are given to be $S_{11} = 0.68 < -72°$, $S_{12} = 0.18 < -14°$, $S_{21} = 4.5 < 82°$, and $S_{22} = 0.65 < -42°$. Determine the stability of the device and draw the stability circles if the device is potentially stable.

Solution Rollet's stability factor k and Δ are calculated from (10.111) and (10.112) and found to be $k = 1.0381 > 1$. However, since $\Delta > 1.2509$, the device is still potentially unstable. The stability circles will then be inside the Smith chart. In addition, the scattering parameters $|S_{11}| < 1$ and $|S_{22}| < 1$ show that inside these circles are stable and the point $\Gamma_L = 0$ remains in the stable region. The parameters of the stability circles from (10.113) to (10.119) have to be calculated to plot them. The calculation of the radius and center point for the input stability circle are

Figure 10.36 Stability circles for Example 10.8.

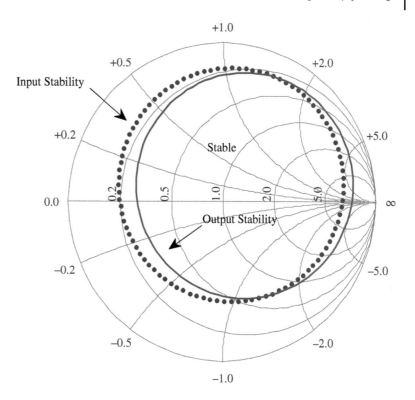

$$r_s = 0.7340 \text{ and } C_s = 0.5575 - \text{j}1.0801$$

The radius and center point for output stability circle are

$$r_L = 0.7083 \text{ and } C_L = -0.1927 - \text{j}1.0889$$

The results showing the stability circles and stable regions are shown in Figure 10.36

10.3.4.3.3 *Stabilization of an Amplifier*

Consider Figure 10.37. The stability in the circuit exits if

$$|\Gamma_{in}| = \left| S_{11} + \frac{S_{12}S_{21}\Gamma_L}{1 - S_{22}\Gamma_L} \right| < 1 \qquad (10.120a)$$

$$|\Gamma_{out}| = \left| S_{22} + \frac{S_{12}S_{21}\Gamma_s}{1 - S_{11}\Gamma_s} \right| < 1 \qquad (10.120b)$$

If $|\Gamma_{in}| > 1$ and $|\Gamma_{out}| > 1$, this indicates the condition for instability. Γ_{in} and Γ_{out} can also be represented in terms of impedances as

$$|\Gamma_{in}| = \left| \frac{Z_{in} - Z_o}{Z_{in} + Z_o} \right| > 1 \qquad (10.121a)$$

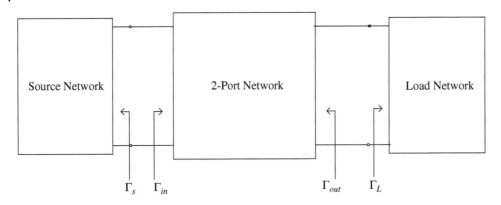

Figure 10.37 Two-port network for stabilization.

$$|\Gamma_{out}| = \left|\frac{Z_{out} - Z_o}{Z_{out} + Z_o}\right| > 1 \tag{10.121b}$$

The equations in (10.121a) imply that

$$\text{Re}\,\{Z_{in}\} < 0 \text{ or } \text{Re}\,\{Z_{out}\} < 0 \tag{10.122}$$

Re-expressing (10.121a) as

$$\Gamma_{in} = \frac{(R_{in} - Z_o) + jX_{in}}{(R_{in} + Z_o) + jX_{in}} \quad \rightarrow |\Gamma_{in}| = \sqrt{\frac{(R_{in} - Z_o)^2 + X_{in}^2}{(R_{in} + Z_o)^2 + X_{in}^2}} \tag{10.123}$$

Equation (10.123) shows that the amplifier can now be stabilized by adding a series resistance or shunt conductance to the source side to make the real part of the impedance positive. This can be similarly done for the load network, which gives

$$\Gamma_{in} = \frac{(R_{out} - Z_o) + jX_{out}}{(R_{out} + Z_o) + jX_{out}} \quad \rightarrow |\Gamma_{in}| = \sqrt{\frac{(R_{out} - Z_o)^2 + X_{out}^2}{(R_{out} + Z_o)^2 + X_{out}^2}} \tag{10.124}$$

The addition of the stability components at the source and load are illustrated in Figures 10.38 and 10.39. This can be better understood by a simple illustrative example. Assume we have an amplifier with a certain load impedance and load stability circle, as shown in Figure 10.40. If we insert a series resistance of 15 Ω, we can then limit the stability region as shown to be Z' where $R = 15$ circle is tangential to the output stability circle in Figure 10.40. The stabilization can also be done by using shunt conductance, as shown in Figure 10.41.

Assume we have an amplifier with a certain load impedance and load stability circle, as shown in Figure 10.41. If we insert a shunt resistance of 400 Ω, we can then limit the stability region as shown to be Y' in Figure 10.41, where the $G = 0.0025$ circle is tangential to the output stability circle.

10.3.4.4 Constant Gain Circles

Unilateral case   When $|S_{12}| \rightarrow 0$, the condition of the two-port network is called unilateral. It is advisable to obtain and plot the gain stability circles for unilateral devices. Under unilateral conditions, Γ_{in} from (10.104) becomes

$$\Gamma_{in} = S_{11} + \frac{S_{12}S_{21}\Gamma_L}{1 - S_{22}\Gamma_L} \rightarrow \Gamma_{in} = S_{11} \tag{10.125}$$

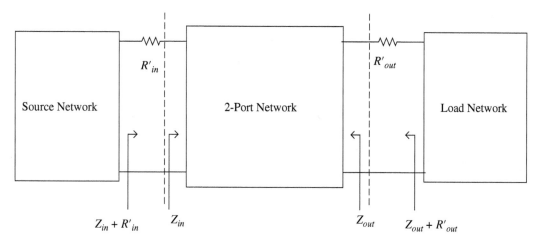

Figure 10.38 Stabilization network by adding series resistance.

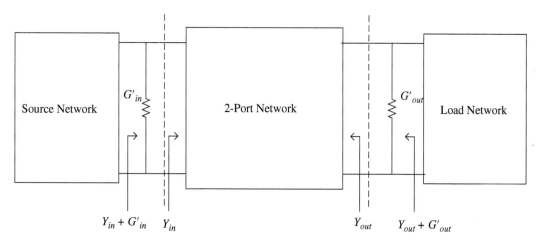

Figure 10.39 Stabilization network by adding shunt conductance.

Substituting (10.125) in (10.104) gives the transducer gain of the unilateral device as

$$G_T = \underbrace{\frac{\left(1 - |\Gamma_s|^2\right)}{|1 - \Gamma_s S_{11}|^2}}_{G_s} \underbrace{|S_{21}|^2}_{G_o} \underbrace{\frac{\left(1 - |\Gamma_L|^2\right)}{|1 - S_{22}\Gamma_L|^2}}_{G_L} \tag{10.126}$$

Equation (10.126) can be expressed as

$$G_T = G_s G_o G_L \tag{10.127a}$$

or

$$G_T(\text{dB}) = G_s(\text{dB}) + G_o(\text{dB}) + G_L(\text{dB}) \tag{10.127b}$$

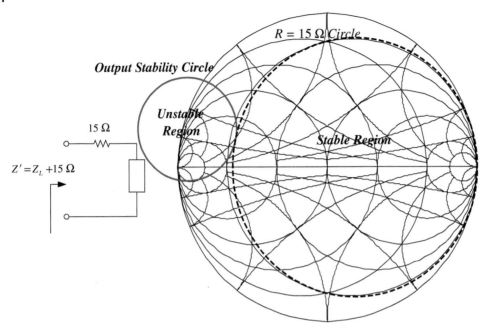

Figure 10.40 Stabilization with series resistor at the load.

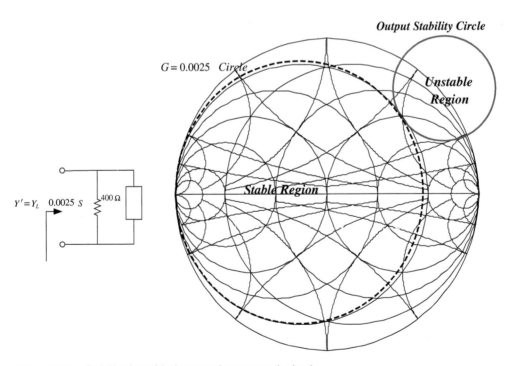

Figure 10.41 Stabilization with shunt conductance at the load.

where

$$G_s = \frac{(1-|\Gamma_s|^2)}{|1-\Gamma_s S_{11}|^2} \tag{10.128}$$

$$G_o = |S_{21}|^2 \tag{10.129}$$

$$G_L = \frac{(1-|\Gamma_L|^2)}{|1-S_{22}\Gamma_L|^2} \tag{10.130}$$

G_s and G_L are the gain contributions for input and output matching networks, respectively. G_L is the gain of the transistor, as shown in Figure 10.42. In terms of dB, Eq. (10.127) can be written as

$$G_T(\text{dB}) = G_s(\text{dB}) + G_o(\text{dB}) + G_L(\text{dB}) \tag{10.131}$$

The power transfer for the unilateral case can be maximized when both the input and the output ports of the amplifier are conjugately matched, as given below

$$\Gamma_s = S_{11}^* \tag{10.132}$$

$$\Gamma_L = S_{22}^* \tag{10.133}$$

When a simultaneous conjugate match is accomplished, the maximum value of the gain is obtained and the following relation holds.

$$G_{T\,\text{max}} = G_{p\,\text{max}} = G_{A\,\text{max}} \tag{10.134}$$

So

$$G_{T,\,\text{max}} = G_{s,\,\text{max}} G_o G_{L,\,\text{max}} = \frac{1}{|1-S_{11}|^2} |S_{21}|^2 \frac{1}{|1-S_{22}|^2} \tag{10.135}$$

where

$$G_{s,\,\text{max}} = \frac{1}{|1-S_{11}|^2} \tag{10.136}$$

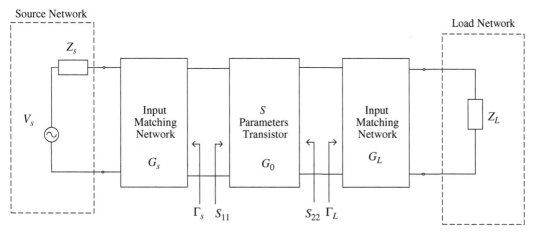

Figure 10.42 Unilateral amplifier design.

$$G_{L,\,\max} = \frac{1}{|1 - S_{22}|^2} \tag{10.137}$$

Expressions for G_s and G_L in (10.136) and (10.137) can be generalized as

$$G_i = \frac{1 - |\Gamma_i|^2}{|1 - S_{ii}\Gamma_i|^2} \tag{10.138a}$$

$$G_{i,\,\max} = \frac{1}{|1 - S_{ii}|^2} \tag{10.138b}$$

where $i = s$ or L and $i = 11$ or 22. Hence, the design for a specified amplifier gain can be done using (10.138). There are two cases to consider under the unilateral condition for constant gain amplifier design: $|S_{ii}| < 0 \rightarrow$ Unconditionally Stable Case and $|S_{ii}| > 0 \rightarrow$ Potentially Stable Case.

$|S_{ii}| < 1 \rightarrow$ Unilateral Unconditionally Stable Case The maximum gain for the source and load shown in (10.136) and (10.137) is obtained when simultaneous conjugate matching is done using

$$G_{i,\,\max} = \frac{1}{|1 - S_{ii}|^2} \tag{10.139}$$

$$\Gamma_i = S_{ii}^* \tag{10.140}$$

When $|\Gamma_i| = 1$, gain gets its lowest value from (10.138a). The range of the gain value for values different from $|\Gamma_i| \neq 1$ are

$$0 \leq G_i \leq G_{i,\,\max} \tag{10.141}$$

The constant gain circles lie inside a Smith chart and are obtained by solving

$$\left| \Gamma_i - C_{gi} \right| = r_{gi} \tag{10.142}$$

In (10.142), C_{gi} represents the center and r_{gi} is the radius of the constant gain circle, which are defined by

$$C_{gi} = \frac{g_i S_{ii}^*}{1 - |S_{ii}|^2 (1 - g_i)} \tag{10.143}$$

$$r_{gi} = \frac{\sqrt{1 - g_i}\,(1 - |S_{ii}|^2)}{1 - |S_{ii}|^2 (1 - g_i)} \tag{10.144}$$

where g_i is the normalized gain factor and expressed as

$$g_i = \frac{G_i}{G_{i,\,\max}} = \frac{1 - |\Gamma_i|^2}{|1 - S_{ii}\Gamma_i|^2} \left(1 - |S_{ii}|^2\right) \tag{10.145}$$

where

$$0 \leq g_i \leq 1 \tag{10.146}$$

The gain circles are obtained by

- Plot S_{ii}^* on the Smith chart. Make a line from the center of the Smith chart to S_{ii}^*. This is the point where gain is maximum.
- Calculate G_i and normalized g_i from (10.138a) to (10.145), respectively.

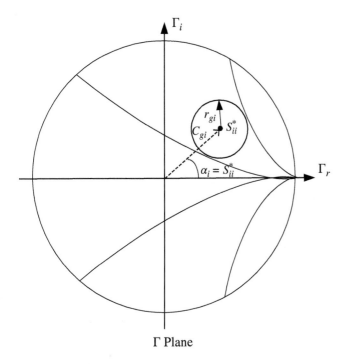

Γ Plane

Figure 10.43 Drawing constant gain circles.

- Calculate the center and radius points for corresponding G_i using (10.143) and (10.144) and draw the constant gain circles.

The process of drawing gain circles is illustrated in Figure 10.43.

$|S_{ii}| > 1 \rightarrow$ Unilateral Potentially Stable Case   When $|S_{ii}| > 1$, it is possible that constant gain value G_i can take infinite value for the critical value of Γ_i, which is defined by

$$\Gamma_{i,c} = \frac{1}{S_{ii}} \tag{10.147}$$

g_i in (10.145) can now take negative values due to the condition created by $|S_{ii}| > 1$. Hence, Γ_i should be chosen so that the real part of the termination impedance is greater than the negative resistance at the point defined by $1/S_{ii}^*$. The process for drawing the gain circles for unilateral potentially stable cases is similar to the process for drawing the gain circles for unilateral unconditionally stable cases.

10.3.4.5 Unilateral Figure of Merit
In practical cases, $S_{12} \neq 0$, so the criteria of unilateral design might not apply. However, it is possible to still use the unilateral design procedure by assuming that $S_{12} = 0$ with error. The error introduced by the approximation can be determined using the unilateral figure of merit. Consider the ratio of the transducer gain, G_T, to the unilateral transducer gain, G_{TU}. This ratio can be expressed by

$$\frac{G_T}{G_{TU}} = \frac{1}{|1 - X|^2} \tag{10.148}$$

where

$$X = \frac{S_{12}S_{21}\Gamma_s\Gamma_L}{(1 - S_{11}\Gamma_s)(1 - S_{22}\Gamma_L)} \tag{10.149}$$

The transducer ratio given in (10.148) is bounded by

$$\frac{1}{(1 + |X|)^2} < \frac{G_T}{G_{TU}} < \frac{1}{(1 - |X|)^2} \tag{10.150}$$

The maximum transducer gain occurs by simultaneous conjugate matching $\Gamma_s = S_{11}^*$ and $\Gamma_L = S_{22}^*$. So, the maximum error is introduced by

$$\frac{1}{(1 + U)^2} < \frac{G_T}{G_{T\,max}} < \frac{1}{(1 - U)^2} \tag{10.151}$$

where

$$U = \frac{|S_{12}||S_{21}||S_{11}||S_{22}|}{\left(1 - |S_{11}^*|\right)\left(1 - |S_{22}^*|\right)} \tag{10.152}$$

In (10.152), U is known as the figure of merit.

10.4 Engineering Application Examples

In this section, the design, simulation, and implementation of small signal amplifiers are discussed and presented.

Design Example 10.1   Low Noise Amplifier Design, Simulation and Implementation
Design, simulate, and build a low noise amplifier using the small signal amplifier design method operating at 915 MHz. The amplifier uses ATF-54143 HEMT (high electron mobility transistor) as a transistor, the S parameters of the device are given below. The amplifier needs to have a gain larger than 10 dB and a noise figure smaller than 2.5 dB with an operational bandwidth of 15%. ATF-5413 scattering parameters provided by the manufacturer are shown below. The complete S parameter dataset is available on the manufacture's website [4].

ATF-54143 Typical Scattering Parameters, $V_{DS} = 4$ V, $I_{DS} = 60$ mA.

Frequency (GHz)	S_{11} Mag.	S_{11} Ang.	S_{11} dB	S_{21} Mag.	S_{21} Ang.	S_{12} Mag.	S_{12} Ang.	S_{22} Mag.	S_{22} Ang.	MSG/MAG (dB)
0.1	0.99	−18.6	28.88	27.80	167.8	0.01	80.1	0.58	−12.6	34.44
0.5	0.81	−80.2	26.11	20.22	128.3	0.03	52.4	0.42	−52.3	28.29
0.9	0.71	−117.3	23.01	14.15	106.4	0.04	41.7	0.31	−73.3	25.49
1.0	0.69	−123.8	22.33	13.07	102.4	0.04	40.2	0.29	−76.9	25.14
1.5	0.64	−149.2	19.49	9.43	86.2	0.05	36.1	0.22	−89.4	22.76

Solution
We follow the small signal amplifier design process flow chart illustrated in Figure 10.32. S parameters are provided by the manufacturer, Avago, for several frequencies. Hence, a DC biasing circuit needs to be designed. This design problem is simulated by an Advanced Design System (ADS).

Step 1 – Select Transistor

The transistor selected to design the low noise amplifier is ATF-54143. This transistor was chosen as it has a wide frequency range, low noise figure, high gain, and is low cost. ATF-54143 is a high dynamic range, low noise, E-PHEMT (enhancement mode pseudomorphic high electron mobility transistor) type transistor that comes with various packages, such as an SC-70 (SOT-343, a small outline transistor) surface mount plastic package. It is ideal for cellular/PCS (personal communication service) base stations, a multichannel multipoint distribution service (MMDS), and other systems in the 450 MHz to 6 GHz frequency range. Before DC biasing can be done, we need to obtain the characteristics curve of the ATF-54143 transistor. The parameter values provided in ADS are compared with the parameters of the transistor parameters values provided in the datasheet by the manufacturer. The values obtained are in close agreement, as shown in Figure 10.44a.

Step 2 – Design DC Bias Circuit

The next stage is to design the DC biasing circuit for the transistor. The S parameters are provided at specific DC biasing voltages and currents, four biasing pairs with independent S parameter values. From the datasheet the S parameters, utilized to define the power flow, are defined over a wide range of frequencies. There is no specific S parameter set defined at 915 MHz but S parameters at 900 MHz are provided above. Since, the S parameter set provided is within 1.66% of the center frequency of 915 MHz, it is used in the design. The error observed is minimal with a maximum magnitude difference of 2 between the parameters provided at 1 GHz. There is a phase difference of less than 10°, which also presents a minimal difference in real and imaginary values. With the minimal error in magnitude and phase difference presented above, the usage of the 900 MHz S parameters is satisfactory.

The biasing circuit that is recommended to be used for this transistor is shown in Figure 10.44b. The equations given to derive component values in Figure 10.44b are given in Table 10.3.

Before biasing can be done, we need to obtain the characteristics curve of the ATF-54143 transistor. The transistor model parameters are given in Figure 10.45. The transistor I–V characteristics obtained in ADS are compared with the transistor I–V characteristics provided in the datasheet by the manufacturer, as shown in Figure 10.46. The I–V curves obtained are in close agreement, as shown in Figure 10.47, which are obtained using ADS.

The biasing circuit simulation results matched the I–V curves provided for the S parameter set in the datasheet when $V_{ds} = 4$V and $I_{ds} = 60$ mA. The component values of the final bias circuit and corresponding amplifier component values are illustrated in Table 10.4.

Step 3 – Get or Measure S Parameters within Frequency of Range

S parameters are given by the manufacturer as explained in Step 2 and given below as reference again to be used for the calculation.

$$S_{11} = 0.71 < -117.3°, S_{12} = 0.04 < 41.7°, S_{21} = 14.15 < 106.4°, \text{and } S_{22} = 0.31 < -73.8°.$$

Step 3 – Stability Analysis – K > 1 and | Δ | < 1?

Based on the calculation using (10.111) and (10.112)

$$k = \frac{1 + |\Delta|^2 - |S_{11}|^2 - |S_{22}|^2}{2|S_{21}||S_{12}|} = 0.4739 < 1 \text{ and } |\Delta| = |S_{22}S_{11} - S_{12}S_{21}| = 0.3697 < 1$$

Since, $k < 1$, the amplifier is potentially unstable.

Step 4 – Draw Stability Circles

Since the amplifier is potentially unstable, we need to draw the stability circles. We need to identify whether the amplifier design can be done using the unilateral design method. It is given that $S_{12} \neq 0$ so we need to determine the unilateral figure of merit and the maximum and minimum error associated with the assumption of unilateral design. From (10.151) and (10.152)

$$\frac{1}{(1 + U)^2} < \frac{G_T}{G_{T\max}} < \frac{1}{(1 - U)^2} \rightarrow -2.13 \text{ dB} < \frac{G_T}{G_{T\max}} < 2.82 \text{ dB}$$

(a)

ATF-54143 Die Model

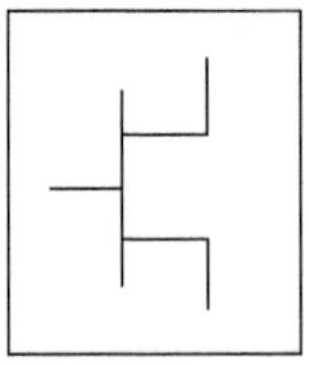

Advanced_Curtice2_Model
MESFETM1

NFET = yes	Rf =	Crf = 0.1 F	N =
PFET = no	Gscap = 2	Gsfwd =	Fnc = 1 MHz
Vto = 0.3	Cgs = 1.73 pF	Gsrev =	R = 0.08
Beta = 0.9	Cgd = 0.255 pF	Gdfwd =	P = 0.2
Lambda = 82e.3	Gdcap = 2	Gdrev =	C = 0.1
Alpha = 13	Fc = 0.65	R1 =	Taumdl = no
Tau =	Rgd = 0.25 Ohm	R2 =	wVgfwd =
Tnom = 16.85	Rd = 1.0125 Ohm	Vbi = 0.8	wBvgs =
Idstc =	Rg = 1.0 Ohm	Vbr =	wBvgd =
Ucnt = 0.72	Rs = 0.3375 Ohm	Vjr =	wBvds =
Vgexp = 1.91	Ld =	Is =	wIdsmax =
Gamds = 1e4	Lg = 0.18 nH	Ir =	wPmax =
Vtotc =	Ls =	Imax =	AllParams =
Betatce =	Cds = 0.27 pF	Xti =	
Rgs = 0.25 Ohm	Rc = 250 Ohm	Eg =	

(b)

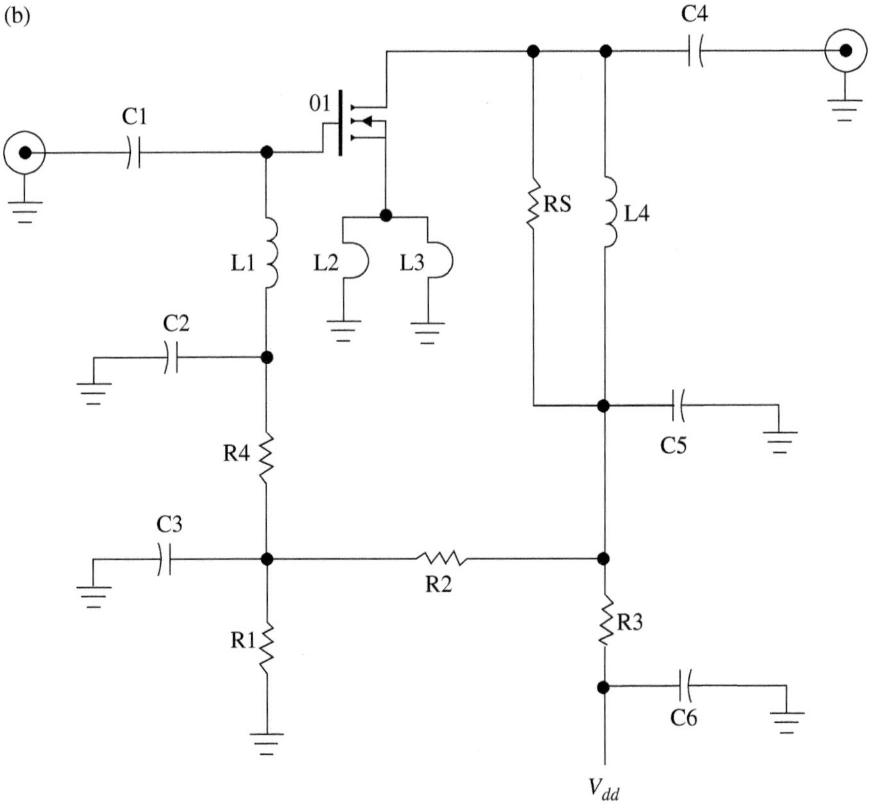

Figure 10.44 (a) ATF-54143 die model provided by Avago [4] (b) DC biasing circuit for ATF-54143 [4]

Table 10.3 DC bias circuit design equations.

Component	Equation
V_{gs} – Voltage gate to source	
V_{DD} – power supply voltage	
V_{DS} – voltage drain to source; target parameter	
I_{Ds} – current drain to source; target parameter	
$I_{BB} = 10*I_{gate\_leakage}$; $I_{gate\_leakage}$;	
$I_{gate\_leakage}$ – expected gate leakage current	
C1, L1	Input matching network
C4, L4	Output matching network
C2, C5	RF bypass
C3, C6	10 nF
R1	$R1 = \dfrac{V_{gs}}{I_{BB}}$
R2	$R2 = \dfrac{(V_{DS} - V_{GS})*R1}{V_{GS}}$
R3	$R3 = \dfrac{V_{DD} - V_{DS}}{I_{DS} + I_{BB}}$
R4	Undefined by datasheet

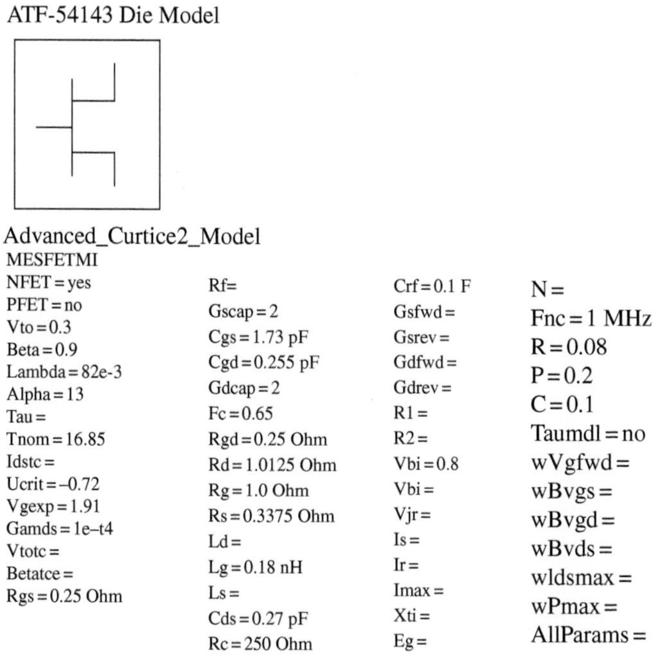

ATF-54143 Die Model

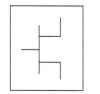

Advanced_Curtice2_Model

MESFETMI

NFET = yes	Rf =	Crf = 0.1 F	N =
PFET = no	Gscap = 2	Gsfwd =	Fnc = 1 MHz
Vto = 0.3	Cgs = 1.73 pF	Gsrev =	R = 0.08
Beta = 0.9	Cgd = 0.255 pF	Gdfwd =	P = 0.2
Lambda = 82e-3	Gdcap = 2	Gdrev =	C = 0.1
Alpha = 13	Fc = 0.65	R1 =	Taumdl = no
Tau =	Rgd = 0.25 Ohm	R2 =	wVgfwd =
Tnom = 16.85	Rd = 1.0125 Ohm	Vbi = 0.8	wBvgs =
Idstc =	Rg = 1.0 Ohm	Vbi =	wBvgd =
Ucrit = −0.72	Rs = 0.3375 Ohm	Vjr =	wBvds =
Vgexp = 1.91	Ld =	Is =	wldsmax =
Gamds = 1e−t4	Lg = 0.18 nH	Ir =	wPmax =
Vtotc =	Ls =	Imax =	AllParams =
Betatce =	Cds = 0.27 pF	Xti =	
Rgs = 0.25 Ohm	Rc = 250 Ohm	Eg =	

Figure 10.45 ATF-54143 die model provided by Avago.

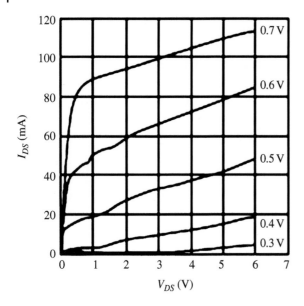

Figure 10.46 I_d vs V_{ds} provided by the manufacturer, Avago [4]

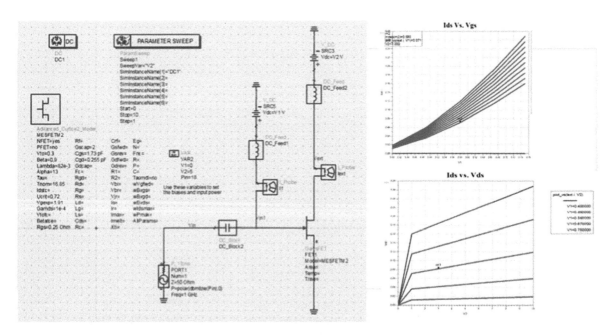

Figure 10.47 ATF-54143 I_d vs V_{gs} and I_d vs V_{ds} obtained from ADS.

Table 10.4 Final component values for the amplifier.

Part	Value	Quantity
C1, C4	5.6 pF (0805)	2
C2, C5	18 pF (0805)	2
C3, C6	10 nF (0805)	2
L1	6.8 nF (LL2012)	1
L2, L3	Shorting strip	2
L4	8.2 nH (LL2012)	1
R1	4.7 kΩ (0805)	1
R2	33 kΩ (0805)	1
R3	27 Ω (0805)	1
R4	56 Ω (0805)	1
R5	330 Ω (0805)	1
Q1	ATF-54143	1

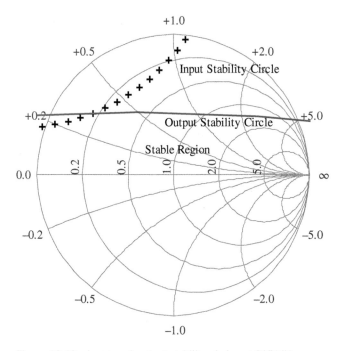

Figure 10.48 Input and output stability circles at 915 MHz.

where

$$U = \frac{|S_{12}||S_{21}||S_{11}||S_{22}|}{\left(1 - |S_{11}^*|\right)\left(1 - |S_{22}^*|\right)}$$

So, the error will be bounded between −2.13 and 2.82 dB if we assume and proceed with a unilateral design. For this design example, lets proceed with this error. However, it is important to note that this type of error is significant for applications where gain is critical.

The input and output stability circles are plotted using the equations from (10.114)–(10.119). The radius and center point calculations are done with MATLAB and plotted, as shown in Figure 10.48.

Step 5 – Design Matching Network

We need to match source impedance $Z_s = 50\,\Omega$, $Z_L = 50\,\Omega$ with $S_{11} = 0.71 < -117.3°$, and $S_{22} = 0.31 < 73.8°$. The unilateral design requires matching networks to be design to follow $\Gamma_i = S_{ii}^*$. This is used to determine matching network components. Hence

$$\Gamma_s = S_{11}^* = 0.71\angle117.3° \rightarrow Z_{sm} = Z_0\frac{1+\Gamma_s}{1-\Gamma_s} = 11.5038 + 29.2718i$$

Similarly, at the output

$$\Gamma_L = S_{22}^* = 0.311\angle73.8° \rightarrow Z_{Lm} = Z_0\frac{1+\Gamma_L}{1-\Gamma_L} = 48.9587 + 32.2482i$$

At the input, this requires a series C and shunt L as an input matching network, as shown in the Smith chart in Figure 10.49. After some tuning when interfaced in the final circuit simulated by ADS, the component values are obtained as $C = 5.6\,\text{pF}$ and $L = 6.8\,\text{nH}$.

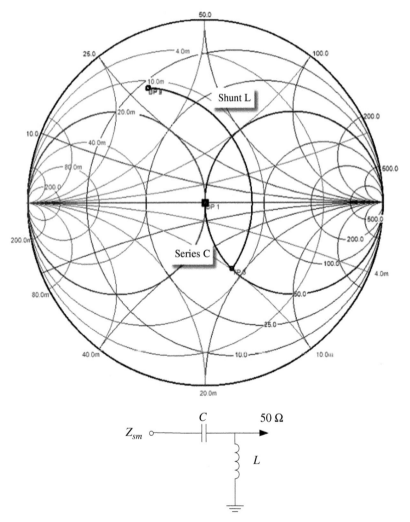

Figure 10.49 Input matching circuit.

Figure 10.50 Output matching circuit.

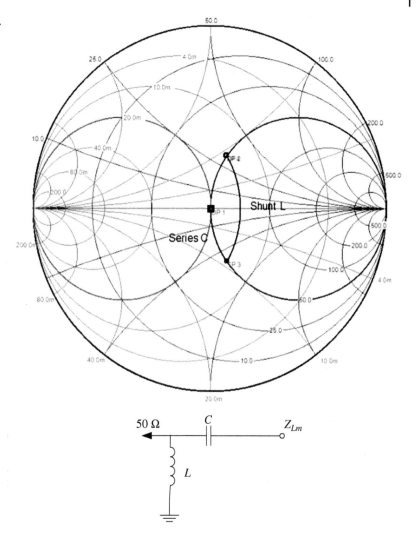

Similarly, output matching network has series C and shunt, as shown in the Smith chart shown in Figure 10.50. After some tuning when interfaced in the final circuit simulated by ADS in Figure 10.48, the component values are obtained as $C = 5.6\,\text{pF}$ and $L = 8.2\,\text{nH}$.

Step 6 – Gain, Bandwidth, VSWR

Gain, bandwidth, and VSWR of the amplifier is analyzed. The total theoretical transducer gain of the amplifier is found from Eq. (10.127b) as

$$G_T(\text{dB}) = G_s(\text{dB}) + G_o(\text{dB}) + G_L(\text{dB}) = 26.5\,\text{dB}$$

The constant gain circles for source and load are given in Figures 10.51 and 10.52, respectively.

Step 7 – Tuning and Optimization

That final circuit in Figure 10.53 is simulated with ADS and optimized, as shown below.

The gain obtained versus frequency also illustrating the bandwidth is given in Figure 10.54.

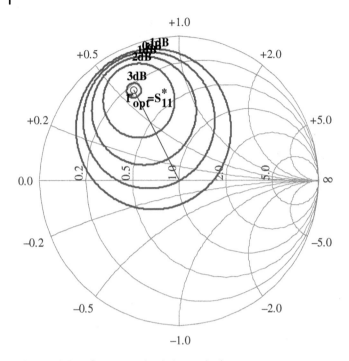

Figure 10.51 Constant gain circles at the input.

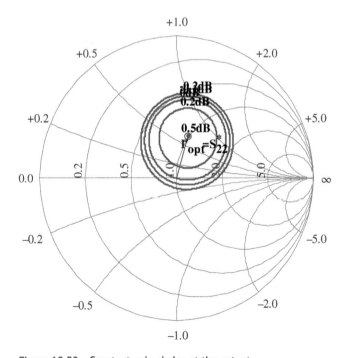

Figure 10.52 Constant gain circles at the output.

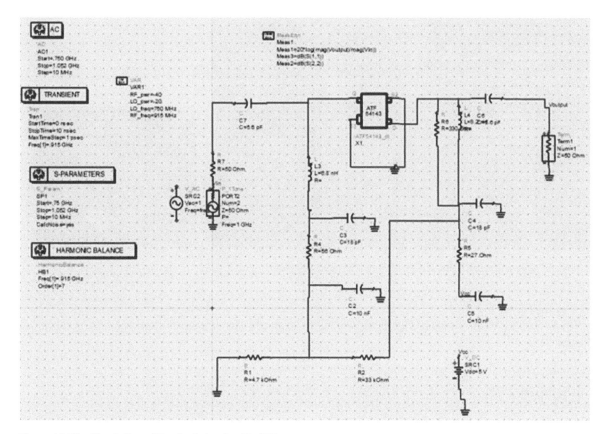

Figure 10.53 Simulation of the final circuit with ADS.

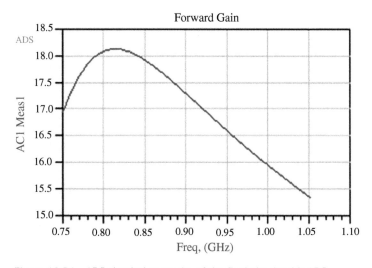

Figure 10.54 ADS simulation results of the final circuit with ADS.

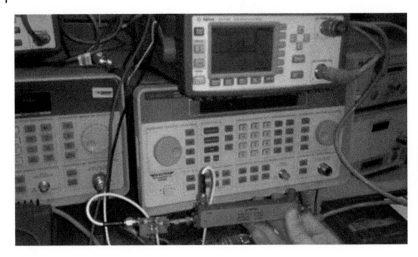

Figure 10.55 The prototype of the low noise amplifier build and tested.

Step 8 – Prototyping

The simulated circuit shown in Figure 10.54 is built and tested. The prototype of the amplifier is shown in Figure 10.55.

The measured results are obtained using an Agilent 4418B power meter with a 10 dB coupler, as illustrated in Table 10.4. A gain over 10 dB is obtained when the applied signal is greater than around 6 dBm.

A MATLAB program is used to perform the calculations and to plot the gain and stability circles.

MATLAB Script to Design Small Signal Amplifier

```
close all; % close all opened graphs .
clear all;

smith_chart;

prompt = {'Zo','S11Mag','S11Ang','S12Mag','S12Ang','S21Mag','S21Ang','S22Mag',
'S22Ang' };
dlg_title = 'Scattering Parameters ';
num_lines = 1;
def = {'50','0.71','-117.3','0.04','41.7','14.15','106.4','0.31','-73.8'};
answer = inputdlg(prompt,dlg_title,num_lines,def, 'on');

%convert the strings received from the GUI to numbers
valuearray=str2double(answer);

%Give variable names to the received numbers
Zo=valuearray(1);
S11Mag=valuearray(2);
S11Ang=valuearray(3);
S12Mag=valuearray(4);
```

```
S12Ang=valuearray(5);
S21Mag=valuearray(6);
S21Ang=valuearray(7);
S22Mag=valuearray(8);
S22Ang=valuearray(9);

% Enter angle in degrees, this converts to complex.

S11rad = S11Ang*pi/180;%radian conversion for complex notation
S12rad = S12Ang*pi/180;
S21rad = S21Ang*pi/180;
S22rad = S22Ang*pi/180;
S = [S11Mag*exp(1i*S11rad),S12Mag*exp(1i*S12rad); S21Mag*exp
(1i*S21rad),S22Mag*exp(1i*S22rad)];

input_stability(S, '+k');
output_stability(S, '-r');

delta=S(1,1)*S(2,2)-S(1,2)*S(2,1);
fprintf('Delta');
disp(abs(delta));
K=(1+(abs(delta))^2-(abs(S(1,1)))^2-(abs(S(2,2))^2))/(2*abs(S(1,2)*S(2,1)));
fprintf('\n Roulette K factor:');
disp(K);
if K > 1 && delta < 1
  fprintf('Unconditionally Stable');
else
  fprintf('Ptentially Unstable');
end

%Input Stabilty Circle Parameters

Cs=conj(S(1,1)-delta*conj(S(2,2)))/(abs(S(1,1))^2-abs(delta)^2);
Rs=abs(S(1,2)*S(2,1)/(abs(S(1,1))^2-abs(delta)^2));

%%Output Stabilty Circle Parameters

Cl=conj(S(2,2)-delta*conj(S(1,1)))/(abs(S(2,2))^2-abs(delta)^2);
Rl=abs(S(1,2)*S(2,1)/(abs(S(2,2))^2-abs(delta)^2));

%Optimum Termination Impedances

GammaS=conj(S(1,1));
GammaL=conj(S(2,2));
```

```
Zsm=Zo*((1+GammaS)/(1-GammaS)); %Source Impedance
Zlm=Zo*((1+GammaL)/(1-GammaL)); %Load Impedance

%Unilateral Figure of Merit, max and min errors

U=(S11Mag*S12Mag*S21Mag*S22Mag)/((1-S11Mag^2)*(1-S22Mag^2));
X=(S(1,2)*S(2,1)*GammaS*GammaL)/((1-S(1,1)*GammaS*(1-S(1,1)*GammaL)));
Max_Error=1/((1-U)^2);
Max_Error_dB=10*log10(Max_Error);
Min_Error=1/((1+U)^2);
Min_Error_dB=10*log10(Min_Error);
FoM=1/(abs(1-X))^2; % Ratio Gt/Gtu

% Calculation of GammaIn and GammaOut

GammaOut=S(2,2)+((S(1,2)*S(2,1)*GammaS)/(1-S(1,1)*GammaS));
GammaIn=S(1,1)+((S(1,2)*S(2,1)*GammaL)/(1-S(2,2)*GammaL));

% Source Side Gain Calculation

Gs_max=1/(1-abs(S(1,1))^2);
Gs_max_dB=10*log10(Gs_max);
Gs=(1-abs(GammaS)^2)/((abs(1-S(1,1)*GammaS))^2);
gs=Gs/Gs_max;

% Load Side Gain Calculation

Gl_max=1/(1-(abs(S(2,2)))^2);
Gl_max_dB=10*log10(Gl_max);
Gl=(1-(abs(GammaL))^2)/((abs(1-S(2,2)*GammaL))^2);
gl=Gl/Gl_max;

% Transistor Gain

G0=abs(S(2,1))^2;
G0_dB=10*log10(G0);

%Total Gain in dB

Gtu_dB=Gs_max_dB+Gl_max_dB+G0_dB;

%Source gain circles
smith_chart;
```

```
%Straight line connecting GammaS and the origin
hold on;
plot([0 real(GammaS)],[0 imag(GammaS)],'b');
plot(real(GammaS),imag(GammaS),'bo');

% specify the angle for the constant gain circles
a=(0:360)/180*pi;

gs_db=[-1 0 1 2 3]; % range of desired gains
gs=exp(gs_db/10*log(10))/Gs_max; % convert from dB to normal units

for n=1:length(gs)
  dg=gs(n)*conj(S(1,1))/(1-abs(S(1,1))^2*(1-gs(n)));
  rg=sqrt(1-gs(n))*(1-abs(S(1,1))^2)/(1-abs(S(1,1))^2*(1-gs(n)));
  plot(real(dg)+rg*cos(a),imag(dg)+rg*sin(a),'r','linewidth',2);
  text(real(dg)-0.05,imag(dg)+rg+0.05,strcat('\bf',sprintf('%gdB',gs_db(n))));

end;

text(real(GammaS)-0.05,imag(GammaS)-0.06,'\bf\Gamma_{opt}=S_{11}^*');

% Load Gain Circles

smith_chart;

%draw a straight line connecting Gs_opt and the origin
hold on;
plot([0 real(GammaL)],[0 imag(GammaL)],'b');
plot(real(GammaL),imag(GammaL),'bo');

% specify the angle for the constant gain circles
a=(0:360)/180*pi;

%plot source gain circles
gl_db=[-0.2 -0.1 0 0.2 0.5]; % range of desired gains
gl=exp(gl_db/10*log(10))/Gl_max; % convert from dB to normal units

for n=1:length(gl)
  dl=gl(n)*conj(S(2,2))/(1-abs(S(2,2))^2*(1-gl(n)));
  rl=sqrt(1-gl(n))*(1-abs(S(2,2))^2)/(1-abs(S(2,2))^2*(1-gl(n)));
  plot(real(dl)+rl*cos(a),imag(dl)+rl*sin(a),'r','linewidth',2);
  text(real(dl)-0.05,imag(dl)+rl+0.05,strcat('\bf',sprintf('%gdB',gl_db(n))));
end;

text(real(GammaL)-0.05,imag(GammaL)-0.06,'\bf\Gamma_{opt}=S_{22}^*');
```

```
 fprintf('\n Load Stability circle radius:');
disp(Rl);
fprintf('\n Load Stability circle center:');
disp(Cl);
fprintf('\n Source Stability circle radius:');
disp(Rs);
fprintf('\n Source Stability circle center:');
disp(Cs);
fprintf('\n GammaS:');
disp(GammaS);
fprintf('\n GammaL:');
disp(GammaL);
fprintf('Unilateral Figure of Merit:');
disp(U);
fprintf('\n Max Error: ');
disp(Max_Error_dB);
fprintf('\n Min Error: ');
disp(Min_Error_dB);
fprintf('\n GT/GTU:');
disp(FoM);
fprintf('\n Zs:');
disp(Zsm);
fprintf('\n ZL:');
disp(Zlm);
fprintf('\n Gtu in dB:');
disp(Gtu_dB);
fprintf('\n GammaIn:');
disp(GammaIn);
fprintf('\n GammaOut:');
disp(GammaOut);
```

References

1 Gozales, G. (1996). *Microwave Transistor Amplifiers: Analysis and Design*, 2e. Pearson.

2 Vendelin, G.D., Pavio, A.M., and Rohde, U.L. (2005). *Microwave Circuit Design Using Linear and Nonlinear Techniques*, 2e. Wiley.

3 Hewlett Packard (1967). HP test and application note 95-1, *mil*sarameter techniques. *Hewlett Packard Journal. February* **3**.

4 Low noise enhancement mode pseudomorphic HEMT in a surface mount plastic package. https://www.modelithics. com/models/Vendor/Avago/ATF-54143.pdf (accessed 27th August 2021).

Problems

Problem 10.1

Assume a cosinusoidal signal, $v_i(t) = \cos(\omega t)$, with 5 Hz frequency is applied to a linear amplifier which has output, as shown in Figure 10.56. (a) Obtain frequency domain representation of power spectra of the output signal of the amplifier. (b) Now assume it is applied to a nonlinear amplifier with an output response equal to $v_o(t) = \alpha_0 + \alpha_1 \beta \cos \omega t + \alpha_2 \beta^2 \cos^2 \omega t + \alpha_3 \beta^3 \cos^3 \omega t$ second and third harmonic distortion for nonlinear amplifier and THD when $\alpha_0 = 0$, $\alpha_1 = 1$, $\alpha_2 = 3$, $\alpha_3 = 1$.

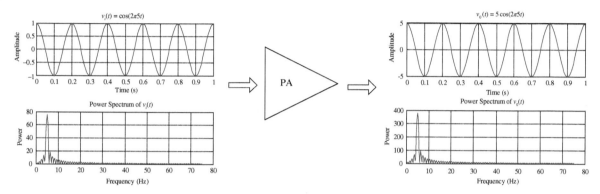

Figure 10.56   Problem 10.1.

Problem 10.2

The RF signal $v_i(t) = \beta_1 \cos \omega_1 t + \beta_2 \cos \omega_2 t$ is applied to an amplifier. The output response of the amplifier is obtained. Calculate IP_3, DR, $IMFDR_3$, and V_{in}, 1 dBc when $\alpha_0 = 0$, $\alpha_1 = 4$, $\alpha_2 = 2$, $\alpha_1 = 1$ and $\beta_1 = 1$, $\beta_2 = 3$.

Problem 10.3

Consider the biasing circuit in Figure 10.57. If, in the biasing circuit, the desired operating values of the voltage and current are given as $I_c = 10$ mA, $V_{CE} = 3$ V, and $V_{CC} = 5$ V, calculate the required values of R_1 and R_C. Assume the transistor has $\beta = 100$ and $V_{BE} = 0.8$ V. Confirm your results with ADS simulation.

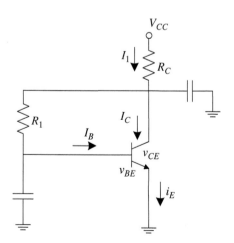

Figure 10.57   Problem 10.3.

Problem 10.4

Consider the biasing circuit in Figure 10.58. If, in the biasing circuit, the desired operating values of the voltage and current are given as $I_c = 10\,\text{mA}$, $V_{CE} = 3\,\text{V}$, and $V_{CC} = 5\,\text{V}$, (a) calculate the required values of R_1, R_2, R_3, and R_4. Assume the transistor has $\beta = 100$ and $V_{BE} = 0.8\,\text{V}$. (b) For stability purposes, RF $= 1\,\text{k}\Omega$ has been added between the base and the collector of the transistor in the circuit. Re-compute the values of all resistors in the biasing network with the same operating conditions. Confirm your results with ADS simulation.

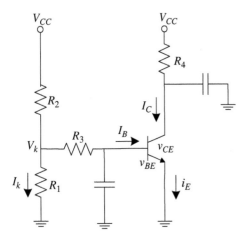

Figure 10.58 Problem 10.4.

Problem 10.5

Consider the biasing circuit in Figure 10.59. If, in the biasing circuit, the desired operating values of the voltage and current are given as $I_{C2} = 10\,\text{mA}$, $V_{CE2} = 3\,\text{V}$, and $V_{CC} = 5\,\text{V}$, calculate R_E, R_C, R_1, and R_2. Assume the transistor has $\beta 1 = 150$, $\beta 2 = 80$ and both transistors have $V_{BE} = 0.8\,\text{V}$. Confirm your results with ADS simulation.

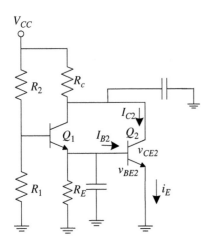

Figure 10.59 Problem 10.5.

Problem 10.6
Find the power outputs indicated in the RF power amplifiers system shown in Figure 10.60.

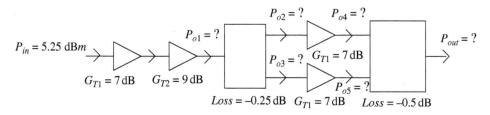

Figure 10.60 Problem 10.6.

Problem 10.7
Consider the amplifier system below in Figure 10.61. Calculate the power levels at each stage, including input and output.

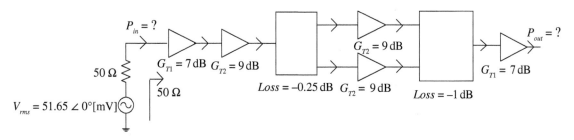

Figure 10.61 Problem 10.7.

Problem 10.8
The scattering parameters of a transistor 2N4223 at 200 MHz are given to be $S_{11} = 0.953 < -29.4°$, $S_{12} = 0.0936 < 70.4°$, $S_{21} = 0.345 < 127.6°$, and $S_{22} = 0.978 < -11.3°$. Determine the stability of the device and draw the stability circles if the device is potentially stable.

Problem 10.9
A transistor operating at $f = 2$ GHz is assumed to be unconditionally stable. If it is biased so that S_{11} is measured to be $S_{11} = 0.85 < 35°$, calculate the source gain circles, G_s, within a -1 to 2 dB range and obtain G_{smax}, radius, and center location values for the constant source gain circles.

Problem 10.10
A transistor's S parameters are given at three frequencies as
   At 50 MHz

$$s_{11} = 0.032 \angle 3.392°, s_{12} = 0.008 \angle 3.636°, s_{21} = 8.008 \angle -11.695°, s_{22} = 0.069 \angle 15.31°$$

At 500 MHz

$$s_{11} = 0.086\angle - 12.64°, s_{12} = 0.009\angle - 38.391°, s_{21} = 7.526\angle - 115.025°, s_{22} = 0.153\angle - 98.54°$$

At 1 GHz

$$s_{11} = 0.142\angle - 82.81°, s_{12} = 0.012\angle - 78.3°, s_{21} = 7.345\angle 137.07°, s_{22} = 0.297\angle - 152.38°$$

Investigate the stability of the transistor at these three frequencies and identify their stability accordingly. Plot the stability circles at 1 GHz and discuss your findings.

Design Challenge 10.1 Small Signal Amplifier Design

Design, simulate, and implement a small signal amplifier operating at 900 MHz. The design should include matching and bias networks to satisfy the specifications provided below. Use Agilent ADS as your simulator. Provide analytical formulation and results, simulation results, and measurement results. The recommended transistor is the BFP182W and a low loss substrate.

The amplifier has the following specifications

- Operating range = 900 MHz with min 15% bandwidth
- Gain > 10 dB
- Noise figure < 1.5 dB
- Return loss for source > 10 dB
- Return loss for load > 10 dB

11

Antennas

11.1 Introduction

An antenna can be defined as a device which can radiate and receive electromagnetic energy in an efficient and desired manner. Radiation from an antenna is possible when there is a time-varying current or an acceleration or deceleration of charges occurs. Several antenna structures are illustrated in Figure 11.1.

Radiation patterns can be mathematically and/or graphically represented. They exhibit power properties of an antenna versus spatial direction coordinates. There are three antenna regions that are defined based on the distance of the observation point and the maximum linear dimension of an antenna. These regions are defined as the reactive near field, radiative near field, and radiative far field. These regions are illustrated in Figure 11.2. We are mostly interested in fields that are sufficiently far from an antenna.

In the reactive near-field, the phases of E and H fields are almost quadrature, so impedance is highly reactive. This region can be defined by

$$R < 0.62\sqrt{\frac{D^3}{\lambda}} \tag{11.1}$$

In the radiative near field, fields are close to being in phase, but they don't yet have spherical wavefronts. Hence, the radiation fields change with distance. These regions are given within the following limits.

$$0.62\sqrt{\frac{D^3}{\lambda}} < R < \frac{2D^2}{\lambda} \tag{11.2}$$

In the radiative far field, E an H fields are in phase and a spherical wavefront is established. Hence, there is real power flow in this region. This region is defined by

$$R > \frac{2D^2}{\lambda} \tag{11.3}$$

A radiation pattern in a polar plot can be obtained taking the ratio of the far-field electric field vector to its maximum value, as given by (11.4). This is the normalized electric field pattern, and its corresponding polar plot is given in Figure 11.3a. The normalized electric field pattern in dB is given in Figure 11.3b. The normalized electric field pattern in dB can be obtained using (11.5) and is identical to the normalized power pattern given by Eq. (11.6).

$$E_n(\theta, \phi) = \frac{E(r, \theta, \phi)}{E_{\max}} \tag{11.4}$$

$$E_n(\theta, \phi)(dB) = 20 \log\left[E_n(\theta, \phi)\right] \tag{11.5}$$

RF/Microwave Engineering and Applications in Energy Systems, First Edition. Abdullah Eroglu.
© 2022 John Wiley & Sons Ltd. Published 2022 by John Wiley & Sons Ltd.
Companion website: www.wiley.com/go/eroglu/rfmicrowave

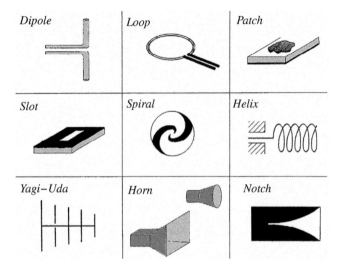

Figure 11.1 Several antenna types.

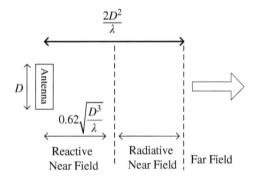

Figure 11.2 Illustration of field regions.

$$U_n(\theta, \phi) = \frac{U(r, \theta, \phi)}{U_{\max}} \tag{11.6}$$

Antennas can be classified based on their radiation patterns. Isotropic antennas radiate equally in all directions. They are ideal antennas and do not exist in real life. Omnidirectional antennas radiate equally in all directions in one plane only, such as dipoles, monopoles, loops, etc. Directional antennas have their major lobe or strong radiation in one direction.

11.2 Antenna Parameters

It is important to understand antenna parameters and their definitions to be able to analyze antennas. The fundamental antenna parameters that are used in antenna analysis are beam solid angle, radiated power, radiation power density, radiation intensity, directivity, beamwidth, efficiency, gain, radiation resistance, aperture area, polarization loss factor. These parameters are detailed in this section.

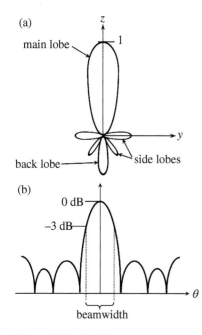

(a)

main lobe

side lobes

back lobe

(b)

0 dB

−3 dB

beamwidth

Figure 11.3 Radiation patterns:
(a) polar plot; (b) rectangular plot.

Beam Solid Angle

The radiation emitted by an antenna is constant in a cone of beam solid angle Ω and equal to the antenna's maximum radiation value. The beam solid angle, Ω, is defined by

$$\Omega = \int\int d\Omega = \int_0^{2\pi}\int_0^\pi \sin\theta\, d\theta\, d\phi = 4\pi\,(sr) \tag{11.7}$$

where *sr* is steradian.

Power, Power Density, and Radiation Intensity

Poynting vector can be written as

$$\mathbf{S} = \mathbf{E} \times \mathbf{H}\ \text{W/m}^2 \tag{11.8}$$

The total power then can be found from

$$P = \oiint \mathbf{S}\ \cdot\ d\mathbf{S}\ \text{W} \tag{11.9}$$

The time average Poynting vector is obtained as

$$\mathbf{S}_{avg} = \frac{1}{2}\,\text{Re}\,[\mathbf{E} \times \mathbf{H}^*]\ \ \text{W/m}^2 \tag{11.10}$$

Then, the average radiated power from (11.9) and (11.10) is

$$P_{\text{rad}} = \oiint \mathbf{S}_{avg} \cdot d\mathbf{S} = \oiint \left(\frac{1}{2}\,\text{Re}\,[\mathbf{E} \times \mathbf{H}^*]\right) \cdot d\mathbf{S}\ \text{W} \tag{11.11}$$

The radiation power density is found from

$$W(\theta,\phi) = \left|\mathbf{S}_{avg}\right|\ \ \text{W/m}^2 \tag{11.12}$$

The radiation intensity is expressed as

$$U(\theta,\phi) = r^2 W\ \ \text{W/Sr} \tag{11.13}$$

From (11.11) and (11.13), we can write the radiated power as

$$P_{\text{rad}} = \oiint U d\Omega\ \text{W} \tag{11.14}$$

Directivity

The directivity of an antenna is a parameter to identify the radiation of the antenna in a given direction. It can be found from

$$D(\theta,\phi) = \frac{U(\theta,\phi)}{U_{\text{avg}}} = \frac{4\pi U(\theta,\phi)}{P_{\text{rad}}} \tag{11.15}$$

The maximum directivity can be given by

$$D_{\text{max}} = \frac{U_{\text{max}}}{U_0} = \frac{4\pi U_{\text{max}}}{P_{\text{rad}}} = \frac{4\pi}{\Omega} \tag{11.16}$$

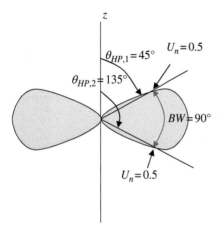

Figure 11.4   Illustration of HPBW.

Directivity in dB is

$$D(\text{dB}) = 10 \log_{10}(D) \tag{11.17}$$

Beamwidth

Beamwidth is an antenna parameter that can be used to identify the resolution capability of an antenna. It is used for the determination of the angular separation between two identical points on the opposite sides of the main lobe. The half-power (3 dB) beamwidth (HPBW) is the most common type of beamwidth. Figure 11.4 illustrates the HPBW on the radiation pattern.

The first-null beamwidth (FNBW) is another beamwidth that is used to measure the angular separation between the first nulls.

Antenna Efficiency, η

Power radiated by an antenna will be lost for several reasons, including impedance mismatch, ohmic loss, and dielectric loss. Antenna efficiency can be expressed as

$$\eta_{\text{ant}} = \eta_r \eta_c \eta_d \tag{11.18}$$

where η_c is used for ohmic losses, η_d, is used for dielectric losses. η_r is used for the reflection mismatch and expressed as

$$\eta_r = \left(1 - |\Gamma_{\text{in}}|^2\right) \tag{11.19}$$

Antenna radiation efficiency can be written as

$$\eta_{rad} = \eta_c \eta_d = \frac{P_{rad}}{P_{\text{in}}} \tag{11.20}$$

Antenna Gain

Antenna gain takes into account antenna loss, whereas directivity does not. Antenna gain is defined as

$$G(\theta, \phi) = \frac{4\pi U(\theta, \phi)}{P_{\text{in}}} = \eta_{ant} D(\theta, \phi) \tag{11.21}$$

When there is no loss or mismatch, directivity becomes equal to antenna gain. The maximum gain is

$$G_{\text{max}} = \eta_{ant} D_{\text{max}} \tag{11.22}$$

In dB, gain can be found as

$$G(\text{dB}) = 10 \log_{10}(G) \tag{11.23}$$

Radiation Resistance

Antenna radiation resistance can be found from radiated power with a known current flowing through the antenna as

$$P_{rad} = \frac{1}{2} \text{Re} \left\{ \oint_{\nabla s} \overline{E}(r) \times \overline{H}(r)^* \cdot ds \right\} = \frac{1}{2} \text{Re} \left\{ \int_0^{2\pi} d\phi \int_0^\pi r^2 \sin\theta \left(E_\theta H_\phi^* \right) d\theta \right\} = \frac{1}{2} I_o^2 R_r \tag{11.24}$$

Then, the radiation resistance from (11.24) is

$$R_{rad} = \frac{2P_{rad}}{I_0^2} \tag{11.25}$$

Antenna Effective Aperture Area

If we assume that the antenna is matched with the transmission line, the effective aperture area of the antenna is defined by

$$A_e = \frac{P_T}{S_{av}} = \frac{P_T}{W_i} \tag{11.26a}$$

where P_T is the power delivered to the matched load and W_i is the incident power density. The maximum effective aperture of any antenna, A_{em}, can be expressed by

$$A_{em} = \frac{\lambda^2}{4\pi} D_{\max} \tag{11.26b}$$

Equation (11.26b) is valid under the assumption that there are no losses, no mismatch, and no polarization loss.

Antenna Polarization Loss Factor

When the incident wave does not have the same polarization as the receiving antenna, there will be loss due to polarization mismatch. This polarization mismatch is called the polarization loss factor (PLF) and is expressed as

$$\text{PLF} = |\hat{e}_i \cdot \hat{e}_r|^2 = |\cos \chi|^2 \tag{11.27}$$

where $\hat{e}_i$ is the unit vector for the incident wave and $\hat{e}_i$ is the unit vector for the receiving antenna. The aperture area given in (11.25) considering all the effects, including polarization loss, can be expressed as

$$A_{em} = \left(1 - |\Gamma_{\text{in}}|^2\right) \eta_{cd} \left(\frac{\lambda^2}{4\pi}\right) D_{\max} |\hat{e}_i \cdot \hat{e}_r|^2 \tag{11.28}$$

Antenna Impedance and Equivalent Circuit

Consider the transmitter and receiver antenna configuration and their equivalent circuits given in Figure 11.5.

The antenna impedance is given as

$$Z_A = R_A + X_A \tag{11.29}$$

where

$$R_A = R_r + R_L \tag{11.30}$$

In (11.30), R_r is the radiation resistance and R_L is the loss (ohmic and dielectric) resistance for an antenna. The maximum power transfer from a generator to an antenna requires the following condition.

(a)   (b)

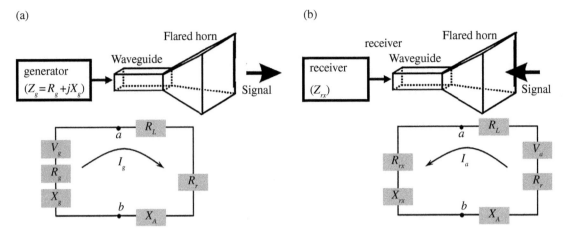

Figure 11.5 Equivalent circuits: (a) transmitter antenna; (b) receiver antenna.

$$R_A = R_r + R_L = R_g \quad \text{and} \quad X_A = -X_g \tag{11.31}$$

Under the maximum power transfer, half of the generator power is delivered to the antenna and the other half is consumed by the generator. The power radiated by an antenna is

$$P_r = \frac{|V_g|^2}{8} \frac{R_r}{(R_r + R_L)^2} \tag{11.32}$$

The power loss by the antenna is

$$P_L = \frac{|V_g|^2}{8} \frac{R_L}{(R_r + R_L)^2} \tag{11.33}$$

From (11.31)

$$P_r + P_L = \frac{|V_g|^2}{8} \frac{1}{(R_r + R_L)} = P_g \tag{11.34}$$

The total power is then

$$P_T = P_g + P_r + P_L = \frac{|V_g|^2}{4(R_r + R_L)} \tag{11.35}$$

The antenna radiation efficiency can then be simplified using (11.32) and (11.34) as

$$\eta_{rad} = \eta_c \eta_d = \frac{P_{rad}}{P_{in}} = \frac{R_r}{R_r + R_L} \tag{11.36}$$

The relationship between transmitter and receiver antennas shown in Figure 11.5 in terms of directivity and aperture areas can be expressed as

$$\frac{D_t}{A_{et}} = \frac{D_r}{A_{er}} \tag{11.37}$$

where D_t is the directivity of the transmitter antenna, D_r is the directivity of the receiver antenna, A_{et} is the aperture area of the transmitter antenna, and A_{er} is the aperture area of the receiver antenna.

Figure 11.6 Illustration for Friis relation.

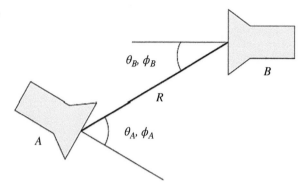

Friis Transmission Relation

The Friis transmission equation establishes the power relation between the receiver and transmitter antennas in the far field, $R > 2D^2/\lambda$. Consider the configuration illustrated in Figure 11.6 where antenna A transmits to antenna B and is located in the far-field region.

The Friis relation can be expressed for this configuration as

$$\frac{P_r}{P_t} = \eta_{cdt}\left(1 - |\Gamma_t|^2\right)\eta_{cdr}\left(1 - |\Gamma_r|^2\right)\frac{\lambda^2}{(4\pi R)^2}D_t D_r |\hat{e}_i \cdot \hat{e}_r|^2 \tag{11.38}$$

If the antennas are matched and there is no polarization loss, (11.38) simplifies to

$$\frac{P_r}{P_t} = \frac{G_t G_r \lambda^2}{(4\pi R)^2} = \frac{A_{e,t} A_{e,r}}{\lambda^2 R^2} \tag{11.39}$$

The Friis transmission relation can also be used to determine the radar range equation and radar cross section, σ, of a target. To calculate the radar cross section, consider the illustration given in Figure 11.7 for a bistatic radar configuration. Bistatic radars have separate transmitter and receiver antennas, whereas monostatic radars use the same antenna for both transmitting and receiving. The Friis relation to relating transmitter and receiver antenna power to the radar cross section shown in Figure 11.7 can be written as

$$\frac{P_r}{P_t} = \eta_{cdt}\left(1 - |\Gamma_t|^2\right)\eta_{cdr}\left(1 - |\Gamma_r|^2\right)\sigma\frac{\lambda^2}{(4\pi R_1 R_2)^2}\frac{D_t D_r}{4\pi}|\hat{e}_i \cdot \hat{e}_r|^2 \tag{11.40}$$

Example 11.1   Antenna Parameters

It is given that the antenna is radiating with the electric field vector given by

$$E = \frac{\eta_o I_o}{r} \sin\theta\, \mathbf{a}_\theta$$

Find W, P_{rad}, U_n, HPBW, beamwidth (BW), D_{max}.

Solution

H field is found from

$$\mathbf{H} = \frac{1}{\eta_o}\mathbf{a}_r \times \mathbf{E} = \frac{I_o}{r}\sin\theta\, \mathbf{a}_\phi$$

Transmitter Antenna

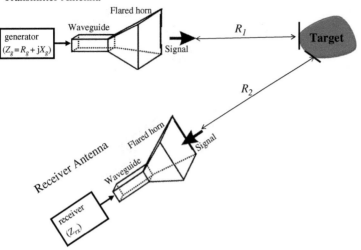

Figure 11.7 Illustration of Friis relation for radar cross section.

The average Poynting vector is found from (11.10)

$$\mathbf{S}_{avg} = \frac{1}{2} \text{Re} \left[\mathbf{E} \times \mathbf{H}^* \right] = \frac{1}{2} \text{Re} \left[\left(\frac{\eta_o I_s}{r} \sin\theta \, \mathbf{a}_\theta \right) \times \left(\frac{I_s}{r} \sin\theta \, \mathbf{a}_\phi \right)^* \right]$$

$$= \frac{1}{2} \text{Re} \left[\left(\frac{\eta_o I_o}{r} \sin\theta \, \mathbf{a}_\theta \right) \times \left(\frac{I_o}{r} \sin\theta \, \mathbf{a}_\phi \right)^* \right] = \frac{1}{2} \text{Re} \left[\left(\frac{\eta_o I_o}{r} \sin\theta \, \mathbf{a}_\theta \right) \times \left(\frac{I_o}{r} \sin\theta \, \mathbf{a}_\phi \right) \right]$$

So

$$\mathbf{S}_{avg} = \frac{1}{2} \text{Re} \left[\eta_o \frac{I_o^2}{r^2} \sin^2\theta \left(\mathbf{a}_\theta \times \mathbf{a}_\phi \right) \right] = \frac{1}{2} \eta_o \frac{I_o^2}{r^2} \sin^2\theta \, \mathbf{a}_r$$

The radiation power density from (11.12) is then equal to

$$W(\theta, \phi) = \left| \mathbf{S}_{avg} \right| = \frac{1}{2} \eta_o \frac{I_o^2}{r^2} \sin^2\theta$$

The radiated power is then found from (11.11) as

$$P_{rad} = \oint \mathbf{S}_{avg} \cdot d\mathbf{S} = \iint W r^2 \sin\theta \, d\theta \, d\phi$$

$$P_{rad} = \iint \left(\frac{1}{2} \eta_o \frac{I_o^2}{r^2} \sin^2\theta \right) r^2 \sin\theta \, d\theta \, d\phi$$

$$P_{rad} = \left(\frac{1}{2} \eta_o I_o^2 \right) \int_0^{2\pi} \int_0^{\pi} \sin^3\theta \, d\theta \, d\phi$$

$$P_{rad} = \left(\frac{1}{2} \eta_o \frac{I_o^2}{r^2} \right) \left(\int_0^{\pi} \sin^3\theta \, d\theta \right) \left(\int_0^{2\pi} d\phi \right)$$

$$P_{rad} = = \left(\frac{1}{2} \eta_o \frac{I_o^2}{r^2} \right) \left(\frac{4}{3} \right) (2\pi) = \frac{4}{3} \pi \eta_o I_o^2$$

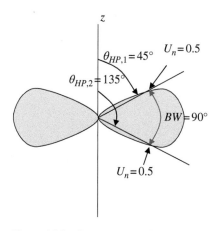

Figure 11.8 Power pattern for Example 11.1.

The normalized power patter is obtained from (11.6) and given by

$$U_n(\theta, \phi) = \frac{U(r, \theta, \phi)}{U_{\max}} = \sin^2 \theta$$

where $U = r^2 W = \frac{1}{2} \eta_o I_o^2 \sin^2 \theta$ and $U_{\max} = r^2 W = r^2 \left(\frac{1}{2} \eta_o \frac{I_o^2}{r^2} \sin^2 \theta \right)$

$$= \frac{1}{2} \eta_o I_o^2$$

The HPBW is found from

$$U_n(\theta, \phi) = \sin^2 \theta = 0.5 \rightarrow \sin \theta_{HP} = \frac{1}{\sqrt{2}}$$

Hence, $\theta_{HP,1} = 45°$ and $\theta_{HP,2} = 135°$. So, the beamwidth (BW) is

$$\text{Beamwidth(BW)} = 135° - 45° = 90°$$

The power pattern is illustrated in Figure 11.8.
$D_{\max}$ is found from (11.16) as

$$D_{\max} = \frac{U_{\max}}{U_0} = \frac{4\pi U_{\max}}{P_{\text{rad}}} = \frac{4\pi \left(\frac{1}{2} \eta_o I_o^2 \right)}{\left(\frac{4}{3} \pi \eta_o I_o^2 \right)} = \frac{3}{2} = 1.5$$

11.3 Wire Antennas

Wire antennas are commonly used because of their cost effectiveness and ease of manufacturing. In this section, some of the basic wire antenna structures are discussed and antenna parameters derived for them. The radiation calculation of antennas can be done using vector magnetic potentials once the current source is known. The vector magnetic potential can be calculated from

$$\nabla^2 A(r, t) - \frac{1}{c^2} \frac{\partial^2 A(r, t)}{\partial t^2} = -\mu_0 J(r, t) \tag{11.41}$$

The solution for A is obtained as

$$A(r, t) = \frac{\mu_0}{4\pi} \int_{\nabla v} \frac{J(r', t - R/c)}{R} dv' \tag{11.42}$$

where $J(r', t)$ is the current density.

11.3.1 Infinitesimal (Hertzian) Dipole $\left(l \leq \frac{\lambda}{50} \right)$

Consider a wire antenna of length $1 \leq \lambda/50$ centered at the origin along the z axis, as illustrated in Figure 11.9. Assume it has uniform current, I_o, along the wire. The stored charge at the ends of the wire resembles an electric dipole, and the short line of oscillating current is then referred to as a Hertzian dipole.

Find Vector Magnetic Potential

The vector magnetic potential is found from

$$\mathbf{A} = \frac{\mu_o}{4\pi} \int_{-\ell/2}^{\ell/2} \frac{\mathbf{J}(\mathbf{r}') \mathbf{a}_z e^{-j\beta R}}{R} dv' \tag{11.43}$$

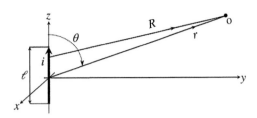

Figure 11.9 Hertzian dipole located at the origin.

Since

$$\mathbf{J} = \frac{I}{S}\mathbf{a_z} \quad \text{and} \quad dv' = Sdz' \tag{11.44}$$

Then, from (11.44)

$$\mathbf{J}dv' = Idz'\mathbf{a_z} \tag{11.45}$$

Substitution of (11.45) into (11.43) gives

$$\mathbf{A} = \frac{\mu_o}{4\pi} \int\limits_{-\ell/2}^{\ell/2} \frac{Idz'\mathbf{a_z}e^{-j\beta R}}{R} \tag{11.46}$$

Since the Hertzian dipole is very short, then

$$R \cong r \tag{11.47}$$

From (11.46) and (11.47), we have

$$\mathbf{A} = \frac{\mu_o I\ell}{4\pi} \frac{e^{-j\beta r}}{r}\mathbf{a_z} \tag{11.48}$$

We need to transform (11.48) into a spherical coordinate system using

$$\mathbf{a_z} = \cos\theta\mathbf{a_r} - \sin\theta\mathbf{a_\theta} \tag{11.49}$$

Then

$$\mathbf{A} = \frac{\mu_o I\ell}{4\pi} \frac{e^{-j\beta r}}{r}(\cos\theta\mathbf{a_r} - \sin\theta\mathbf{a_\theta}) \tag{11.50}$$

Find **E** and **H**

The magnetic field intensity is found from

$$\mathbf{H} = \frac{\mathbf{B}_{os}}{\mu_o} = \frac{1}{\mu_o}(\nabla \times \mathbf{A}) \tag{11.51}$$

which leads to

$$\mathbf{H} = \frac{I\ell}{4\pi} \frac{e^{-j\beta r}}{r}\left(j\beta + \frac{1}{r}\right)\sin\theta\,\mathbf{a_\phi} = \frac{I\ell\beta^2 e^{-j\beta r}}{4\pi}\left[\frac{j}{\beta r} + \frac{1}{(\beta r)^2}\right]\sin\theta\,\mathbf{a_\phi} \tag{11.52}$$

From Maxwell's equations

$$E(r) = \frac{1}{j\omega\varepsilon_0}\nabla \times H(r) = \frac{1}{j\omega\varepsilon_0}\left[\frac{1}{r\sin\theta}\frac{\partial(H_\phi \sin\theta)}{\partial\theta}\mathbf{a_r} - \frac{1}{r}\frac{\partial(rH_\phi)}{\partial r}\mathbf{a_\theta}\right] \tag{11.53}$$

Hence, the electric field intensity is calculated from

$$\mathbf{E}_r = \eta_o \frac{I\ell\beta}{2\pi} \frac{e^{-j\beta r}}{r} \cos\theta \left[\frac{1}{\beta r} - \frac{j}{(\beta r)^2} \right] \tag{11.54}$$

$$\mathbf{E}_\theta = \eta_o \frac{I\ell\beta}{4\pi} \frac{e^{-j\beta r}}{r} \sin\theta \left[j + \frac{1}{\beta r} - \frac{j}{(\beta r)^2} \right] \tag{11.55}$$

In the far-field region, $\beta >> r$, Eqs. (11.52), (11.54), and (11.55) reduce to

$$\mathbf{E} = j\eta_o \frac{I\ell\beta}{4\pi} \frac{e^{-j\beta r}}{r} \sin\theta \, \mathbf{a}_\theta \tag{11.56}$$

$$\mathbf{H} = j \frac{I\ell\beta}{4\pi} \frac{e^{-j\beta r}}{r} \sin\theta \, \mathbf{a}_\phi \tag{11.57}$$

Radiation Power Density

The time average Poynting vector is found from

$$\mathbf{S}_{avg} = \frac{1}{2} \operatorname{Re} \left[\mathbf{E} \times \mathbf{H}^* \right] = \left(\frac{\eta_o \beta^2 I^2 \ell^2}{32\pi^2 r^2} \right) \sin^2\theta \mathbf{a}_r \tag{11.58}$$

Radiation power density, W, is then equal to

$$W(\theta, \phi) = \left| \mathbf{S}_{avg} \right| = \left(\frac{\eta_o \beta^2 I^2 \ell^2}{32\pi^2 r^2} \right) \sin^2\theta \tag{11.59}$$

Radiation intensity is calculated using (11.13) as

$$U(\theta, \phi) = r^2 W = \left(\frac{\eta_o \beta^2 I^2 \ell^2}{32\pi^2} \right) \sin^2\theta \tag{11.60}$$

Maximum radiation intensity is then

$$U_{\max} = \left(\frac{\eta_o \beta^2 I^2 \ell^2}{32\pi^2} \right) \tag{11.61}$$

Radiated power is found from (11.11)

$$P_{\mathrm{rad}} = \oiint \mathbf{S}_{avg} \cdot d\mathbf{S} = \oiint \left(\frac{1}{2} \operatorname{Re} \left[\mathbf{E} \times \mathbf{H}^* \right] \right) \cdot d\mathbf{S} = \frac{1}{2}\eta_o \int_{\phi=0}^{2\pi} \int_{\theta=0}^{\pi} \left| H_\phi(r) \right|^2 r^2 \sin\theta d\theta d\phi$$

or

$$P_{\mathrm{rad}} = \frac{\eta_o \beta^2 (Il)^2}{16\pi} \int_0^\pi \sin^3\theta d\theta = -\frac{\eta_o \beta^2 (Il)^2}{16\pi} - \int_0^\pi (1 - \sin^2\theta) d(\cos\theta)$$

which can be simplified to

$$P_{rad} = \frac{\eta_o \beta^2 I^2 l^2}{12\pi} \tag{11.62}$$

Directivity is found from (11.14) as

$$D(\theta, \phi) = \frac{4\pi U(\theta, \phi)}{P_{rad}} = 4\pi \frac{\left(\frac{\eta_0 \beta^2 I^2 \ell^2}{32\pi^2}\right) \sin^2 \theta}{\frac{\eta_0 \beta^2 I^2 l^2}{12\pi}} = 1.5 \sin^2 \theta \tag{11.63}$$

The maximum directivity for Hertzian dipole is then

$$D_{max} = 1.5 \tag{11.64}$$

Radiation resistance is obtained as

$$P_{rad} = I^2 R_{rad} \rightarrow \frac{\eta_0 \beta^2 I^2 l^2}{12\pi} = I^2 R_{rad} \tag{11.65}$$

So

$$R_{rad} = 80\pi^2 \left(\frac{\ell}{\lambda}\right)^2 \tag{11.66}$$

Effective aperture is found from (11.26b)

$$A_{em} = \frac{\lambda^2}{4\pi} D_{max} = \frac{3\lambda^2}{8\pi} \tag{11.67}$$

The beam solid angle for Hertzian dipole is obtained using (11.16) as

$$\Omega = \frac{4\pi}{D_{max}} = \frac{4\pi}{3/2} = \frac{8\pi}{3} \tag{11.68}$$

11.3.2 Short Dipole $\left(\frac{\lambda}{50} \leq l \leq \frac{\lambda}{10}\right)$

We follow the same procedure and illustration on the wire antenna as illustrated in Figure 11.9, when its length is $\frac{\lambda}{50} \leq l \leq \frac{\lambda}{10}$ in the presence of the uniform triangular current distribution I, which is given by

$$I(z') = I\left(1 - \frac{2|z'|}{l}\right) a_z \tag{11.69}$$

The magnetic vector potential is found using (11.45) and (11.69) as

$$\mathbf{A} = \frac{\mu_0 I \ell}{8\pi} \frac{e^{-j\beta r}}{r} \mathbf{a_z} \tag{11.70}$$

Comparison of (11.48) and (11.70) shows that the vector magnetic potential of the short dipole for a uniform triangular current distribution is one-half the vector magnetic potential of a Hertzian dipole. Hence, the E and H radiated fields of the short dipole will also be one-half of the Hertzian dipole and can be expressed as

$$\mathbf{E} = j\eta_0 \frac{I\ell\beta}{8\pi} \frac{e^{-j\beta r}}{r} \sin\theta \, \mathbf{a}_\theta \tag{11.71}$$

$$\mathbf{H} = j\frac{I\ell\beta}{8\pi} \frac{e^{-j\beta r}}{r} \sin\theta \, \mathbf{a}_\phi \tag{11.72}$$

The power radiated for a short dipole will be one-fourth of a Hertzian dipole and is given by

$$P_{rad} = \frac{\eta_0 \beta^2 I^2 l^2}{48\pi} \tag{11.73}$$

Similarly, the radiation resistance will be one-fourth of a Hertzian dipole

$$R_{rad} = 20\pi^2 \left(\frac{\ell}{\lambda}\right)^2 \tag{11.74}$$

The directivity, maximum directivity, effective area, and beam solid angle of the short dipole are the same as the Hertzian dipole.

11.3.3 Half-wave Dipole $\left(l = \frac{\lambda}{2}\right)$

Assume the current distribution for the half-wave dipole is defined as

$$\vec{I}(x' = 0, y' = 0, z') = \hat{z} I_0 \cos(kz') - \lambda/4 \le z' \le \lambda/4 \tag{11.75}$$

This is a simplified version of the finite length dipole case

$$\vec{I}(x' = 0, y' = 0, z') = \begin{cases} \hat{z} I_0 \sin\left(k\left(\frac{1}{2} - z'\right)\right) & 0 \le z' \le 1/2 \\ \hat{z} I_0 \sin\left(k\left(\frac{1}{2} + z'\right)\right) & -1/2 \le z' \le 0 \end{cases} \tag{11.76}$$

In both cases, we are assuming the diameter is negligible ($6.35 \times 10^{-3} \ll 3$ m), or $a \ll \lambda$.
The first step is to determine the vector potential function A.
The following far-field approximations are made to derive a closed-form solution:

$$R \cong r - z' \cos\theta \text{ for phase terms}$$

$$R \cong r \text{ for amplitude terms}$$

$$R = \sqrt{(x - x')^2 + (y - y')^2 + (z - z')^2} = \sqrt{x^2 + y^2 + (z - z')^2} = \sqrt{(x^2 + y^2 + z^2) + (-2zz' + z'^2)}$$

$$= \sqrt{(r^2) + (-2rz' \cos\theta + z'^2)} \tag{11.77}$$

where

$$r^2 = x^2 + y^2 + z^2$$

$$z = r \cos\theta$$

$$R = r - z' \cos\theta + \frac{1}{r}\left(\frac{z'^2}{2} \sin^2\theta\right) + \frac{1}{r^2}\left(\frac{z'^3}{2} \cos\theta \sin^2\theta\right) + ...\text{Binomial expansion}$$

The third- and fourth-order terms will represent the maximum phase error. Hence, the vector magnetic potential can be expressed as

$$\vec{A} = \frac{\mu}{4\pi} \int\int\int_V \hat{j} \frac{e^{-jkR}}{R} dv' = \frac{\mu}{4\pi} \int_1 I \frac{e^{-jkR}}{R} dl' = \frac{\mu}{4\pi} \int_{z'} I \frac{e^{-jk(r - z' \cos\theta)}}{r} dz' \tag{11.78}$$

So

$$A = \frac{\mu e^{-jkr}}{4\pi r} \int_{z'} I e^{jkz' \cos(\theta)} dz' = \hat{z} \frac{\mu e^{-jkr}}{4\pi r} I_0 \int_{-\frac{\lambda}{4}}^{\frac{\lambda}{4}} \cos(kz') e^{jkz' \cos(\theta)} dz' \tag{11.79}$$

The integral in (11.79) is in the form of

$$\int e^{\alpha z'} \sin(\beta z' + \gamma) dz' = \frac{e^{\alpha z'}}{x^2 + \beta^2} (\alpha \sin(\beta z' + \gamma) - \beta \cos(\beta z' + \gamma)) \tag{11.80}$$

where $\alpha = jk\cos(\theta)$, $\beta = k$, $\gamma = \frac{\pi}{2}$

Performing the integration using (11.80) gives

$$A_z = \frac{\mu e^{-jkr}}{4\pi r} I_0 \left\{ \frac{e^{jkz' \cos(\theta)}}{j^2 k^2 \cos^2(\theta) + k^2} \left[jk\cos(\theta) \sin\left(kz' + \frac{\pi}{2}\right) - k\cos\left(kz' + \frac{\pi}{2}\right) \right]_{-\lambda/4}^{\lambda/4} \right\}$$

$$= \frac{\mu e^{-jkr}}{4\pi r} I_0 \left\{ \frac{e^{jk\frac{\lambda}{4}\cos(\theta)}}{k^2 \sin^2\theta} [jk\cos(\theta) \sin(\pi) - k\cos(\pi)] - \frac{e^{-jk\frac{\lambda}{4}\cos(\theta)}}{k^2 \sin^2\theta} [jk\cos(\theta) \sin(0) - k\cos(0)] \right\}$$

$$= \frac{\mu e^{-jkr}}{4\pi r} I_0 \left\{ \frac{ke^{jk\frac{\lambda}{2}\cos(\theta)}}{k^2 \sin^2\theta} + \frac{e^{-jk\frac{\lambda}{2}\cos(\theta)}}{k^2 \sin^2\theta} \right\} = \frac{\mu e^{-jkr}}{4\pi r} I_0 \left\{ \frac{2\cos\left(\frac{\pi}{2}\cos\theta\right)}{k\sin^2\theta} \right\} \tag{11.81}$$

where

$$k = \frac{2\pi}{\lambda} \rightarrow \frac{k\lambda}{4} = \frac{2\pi}{\lambda}\left(\frac{\lambda}{4}\right) = \frac{\pi}{2} \text{ and } A_z = \frac{\mu e^{-jkr}}{2\pi r} I_0 \left[\frac{\cos\left(\frac{\pi}{2}\cos(\theta)\right)}{k\sin(\theta)} \right]$$

Transforming from a Cartesian to a spherical coordinate system using

$$\begin{bmatrix} A_r \\ A_\theta \\ A_\phi \end{bmatrix} = \begin{bmatrix} \sin\theta\cos\phi & \sin\theta\sin\phi & \cos\theta \\ \cos\theta\cos\phi & \cos\theta\sin\phi & -\sin\theta \\ -\sin\phi & \cos\phi & 0 \end{bmatrix} \begin{bmatrix} 0 \\ 0 \\ A_z \end{bmatrix}$$

$$A_r = A_z \cos\theta$$
$$A_\theta = -A_z \sin\theta$$
$$A_\phi = 0$$

gives the vector magnetic potential for half-wave dipole as

$$A = A_z \cos\theta\hat{r} - A_z \sin\theta\hat{\theta} = \frac{\mu e^{-jkr}}{2\pi r} I_0 [\cos\theta\hat{r} - \sin\theta\hat{\theta}] \left[\frac{\cos\left(\frac{\pi}{2}\cos(\theta)\right)}{k\sin^2(\theta)} \right] \tag{11.82}$$

Note that this expression is only valid in the far-field region, owing to the far-field approximations made.

The radiated E and H fields can now be found. For the half-wave dipole case we are primarily interested in the far-field region.

$$\vec{H}_A = \frac{1}{\mu r} \nabla \times \vec{A} = \Phi \frac{1}{\mu r} \left[\frac{\partial}{\partial r}(rA_\theta) - \frac{\partial}{\partial\theta} A_r \right] \tag{11.83}$$

which can be expanded as

$$\frac{\partial}{\partial r}(rA_\theta) = \frac{\partial}{\partial r} \left\{ \frac{\mu e^{-jkr}}{2\pi} I_0 \left[\frac{\cos\left(\frac{\pi}{2}\cos\theta\right)}{k\sin\theta} \right] \right\} = -\frac{jk\mu e^{-jkr}}{2\pi} I_0 \left[\frac{\cos\left(\frac{\pi}{2}\cos\theta\right)}{k\sin\theta} \right]$$

$$\frac{\partial}{\partial \theta}(A_\theta) = \frac{\partial}{\partial \theta}\left\{ \frac{\mu e^{-jkr}}{2\pi r}I_0 \left[\frac{\cos\left(\frac{\pi}{2}\cos\theta\right)}{k\sin^2\theta}\right]\cos\theta \right\}$$

$$= \frac{\mu e^{-jkr}}{2\pi r}I_0 \left[\frac{\pi\cos\theta\sin^2\theta\sin\left(\frac{\pi}{2}\cos\theta\right) - (2\sin^2\theta + 4\cos^2\theta)\cos\left(\frac{\pi}{2}\cos\theta\right)}{2k\sin^3\theta}\right]\vec{H}_A$$

$$= \Phi\frac{e^{-jkr}}{2\pi r}I_0 \left[\frac{j\cos\left(\frac{\pi}{2}\cos\theta\right)}{\sin\theta} - \frac{1}{r}\frac{\pi\cos\theta\sin^2\theta\sin\left(\frac{\pi}{2}\cos\theta\right) - (2\sin^2\theta + 4\cos^2\theta)\cos\left(\frac{\pi}{2}\cos\theta\right)}{2k\sin^3\theta}\right]$$

$$(11.84)$$

For far field (kr ≫ 1) constant terms dominate and

$$H_\Phi \approx \frac{je^{-jkr}}{2\pi r}I_0 \left[\frac{\cos\left(\frac{\pi}{2}\cos\theta\right)}{\sin\theta}\right] \tag{11.85}$$

The E field by applying Maxwell's equation as

$$\vec{E}_A = \frac{1}{j\omega\varepsilon}\nabla\times\vec{H} = \frac{1}{j\omega\varepsilon}\left\{\hat{r}\frac{1}{r\sin\theta}\left[\frac{\partial}{\partial\theta}(H_\Phi\sin\theta)\right] + \hat{\theta}\frac{1}{r}\left[-\frac{\partial}{\partial r}(rH_\Phi)\right]\right\} \tag{11.86}$$

Expanding (11.86) gives

$$\frac{\partial}{\partial\theta}(H_\Phi\sin\theta) = \frac{\partial}{\partial\theta}\left\{\frac{e^{-jkr}}{2\pi r}I_0\left[j\cos\left(\frac{\pi}{2}\cos\theta\right) - \frac{1}{r}\frac{\pi\cos\theta\sin^2\theta\sin\left(\frac{\pi}{2}\cos\theta\right) - (2\sin^2\theta + 4\cos^2\theta)\cos\left(\frac{\pi}{2}\cos\theta\right)}{2k\sin^2\theta}\right]\right\}$$

$$= \frac{e^{-jkr}}{2\pi r}I_0\left[j\frac{\pi}{2}\sin\theta\sin\left(\frac{\pi}{2}\cos\theta\right) + \frac{1}{r}\frac{(4\pi\sin^4\theta + 4\pi\cos^2\theta\sin^2\theta)\sin\left(\frac{\pi}{2}\cos\theta\right) + (\pi^2\cos\theta\sin^4\theta - 16\cos\theta\sin^2\theta - 16\cos^3\theta)\cos\left(\frac{\pi}{2}\cos\theta\right)}{4k\sin^3\theta}\right]$$

and

$$-\frac{\partial}{\partial r}(rH_\Phi) = -\frac{\partial}{\partial r}\left\{\frac{e^{-jkr}}{2\pi}I_0\left[\frac{j\cos\left(\frac{\pi}{2}\cos\theta\right)}{\sin\theta} - \frac{1}{r}\frac{\pi\cos\theta\sin^2\theta\sin\left(\frac{\pi}{2}\cos\theta\right) - (2\sin^2\theta + 4\cos^2\theta)\cos\left(\frac{\pi}{2}\cos\theta\right)}{2k\sin^3\theta}\right]\right\}$$

$$= -\frac{e^{-jkr}}{2\pi}I_0\left[\frac{k\cos\left(\frac{\pi}{2}\cos\theta\right)}{\sin\theta} + \left(\frac{1}{r} + \frac{jk}{r}\right)\frac{\pi\cos\theta\sin^2\theta\sin\left(\frac{\pi}{2}\cos\theta\right) - (2\sin^2\theta + 4\cos^2\theta)\cos\left(\frac{\pi}{2}\cos\theta\right)}{2k\sin^3\theta}\right]$$

Hence, the E field can be written as

$$E_\theta = \frac{1}{j\omega\varepsilon}\left\{-\frac{e^{-jkr}}{2\pi r}I_0\left[\frac{k\cos\left(\frac{\pi}{2}\cos\theta\right)}{\sin\theta} + \left(\frac{1}{r} + \frac{jk}{r}\right)\frac{\pi\cos\theta\sin^2\theta\sin\left(\frac{\pi}{2}\cos\theta\right) - (2\sin^2\theta + 4\cos^2\theta)\cos\left(\frac{\pi}{2}\cos\theta\right)}{2k\sin^3\theta}\right]\right\}$$

$$(11.87)$$

For far field (kr ≫ 1) constant terms dominate and

$$E_\theta \approx \frac{je^{-jkr}}{\omega\varepsilon(2\pi r)}I_0\left[\frac{k\cos\left(\frac{\pi}{2}\cos\theta\right)}{\sin\theta}\right] \approx \frac{j\omega\mu e^{-jkr}}{2\pi r}I_0\left[\frac{\cos\left(\frac{\pi}{2}\cos\theta\right)}{k\sin\theta}\right] \tag{11.88}$$

In the above equations, k = β

The average Poynting vector, radiation density, and radiated power can be found as follows. The average Poynting vector is

$$S_{avg} = \frac{1}{2}(E \times H^*) = \frac{1}{2}\left(E_\theta \hat{\theta} \times H_\phi^* \hat{\phi}\right) = \frac{1}{2}\left(E_\theta \frac{E_\theta^*}{\eta} \hat{r}\right) = E_\theta \frac{|E_\theta|^2}{2\eta} \hat{r}$$

or

$$S_{avg} = \frac{\eta}{8\pi^2 r^2}|I_o|^2 \left(\frac{\cos^2\left(\frac{\pi}{2}\cos\theta\right)}{\sin^2\theta}\right) \hat{r} \tag{11.89}$$

The radiation density is obtained from

$$W = \left|\vec{S}_{avg}\right| = \frac{\eta}{8\pi^2 r^2}|I_o|^2 \left(\frac{\cos^2\left(\frac{\pi}{2}\cos\theta\right)}{\sin^2\theta}\right) \tag{11.90}$$

The radiation intensity is

$$U(\theta) = r^2 W = \frac{\eta}{8\pi^2}|I_o|^2 \left(\frac{\cos^2\left(\frac{\pi}{2}\cos\theta\right)}{\sin^2\theta}\right) \tag{11.91}$$

The radiated power is

$$P_{rad} = \int\int_{0\,0}^{2\pi\,\pi} U(\theta)\sin\theta d\theta d = \frac{2\pi\eta I_o^2}{8\pi^2}\int_0^\pi \left(\frac{\cos^2\left(\frac{\pi}{2}\cos\theta\right)}{\sin\theta}\right) d\theta = 29.98 I_o^2 \times 1.219 = 36.54 I_o^2\,W \tag{11.92}$$

The following parameters are of interest in the far-field region.
Maximum directivity and gain

$$D_0 = \frac{4\pi U_{max}}{P_{rad}} = \frac{4\pi\eta I_o^2}{8\pi^2(36.54)I_o^2} = 1.64 = 2.15dB \tag{11.93}$$

and gain is

$$G_0 = e_{cd}D_0(0.9964)(1.64) = 1.634 = 2.133\,dB \tag{11.94}$$

Radiation resistance

$$R_{rad} = \frac{2P_{rad}}{I_o^2} = \frac{2(36.54)I_o^2}{I_o^2} = 73.08\,\Omega \tag{11.95}$$

Antenna reactance

$$X_A = \frac{\eta_0}{4\pi}\left\{2S_i(kl) + \cos(kl)[2S_i(kl) - 2S_i(2kl)] - \sin(kl)[2C_i(kl) - 2C_i(2kl)] - C_i\left(\frac{2ka^2}{1}\right)\right\} = 42.5\,\Omega \tag{11.96}$$

Hence, the half-wave dipole antenna impedance is

$$\therefore Z_A = 73 + j42.5\,\Omega \tag{11.97}$$

Antenna loss resistance

$$R_L = \frac{1}{2}R_{hf} = \frac{1}{P}x\sqrt{\frac{\omega\mu_o}{2\sigma}} = \frac{1.5}{(6.35 \times 10^{-3})\pi}\sqrt{\frac{2\pi(100 \times 10^6)(4\pi \times 10^{-7})}{2(3.5 \times 10^7)}} = 252.5 \times 10^{-3}\,\Omega \tag{11.98}$$

Antenna radiation efficiency

$$\eta_{cd} = \frac{R_{rad}}{R_{rad} + R_l} = \frac{70}{70 + 2525.5 \times 10^{-3}} = 099.64 = 99.64\% \tag{11.99}$$

Beam solid angle

$$\Omega = \frac{4\pi}{D_0} = \frac{4\pi}{1.64} = 7.662 \text{ sr} \tag{11.100}$$

Maximum effective area

$$A_{em} = \frac{\lambda^2}{4\pi}D_0 = \frac{(3)^2}{4\pi}(1.64) = 1.175 \text{ m}^2$$

FNBW

$$\frac{\cos^2\left(\frac{\pi}{2}\cos\theta\right)}{\sin^2\theta} = 0 \text{ or } \cos\left(\frac{\pi}{2}\cos\theta\right) = 0$$

$$\frac{\pi}{2}\cos\theta = 1.571, \cos\theta = 1, \theta = 0°$$

So

$$\therefore \text{FNBW} = 180° \tag{11.101}$$

HPBW

$$\frac{\cos^2\left(\frac{\pi}{2}\cos\theta\right)}{\sin^2\theta} = \frac{1}{2} \text{ or } \theta_h = 0.8894 = 50.96° \text{ (solved numerically)}$$

So

$$HPBW = 2(90° - \theta_h) = 78.08° \tag{11.102}$$

Example 11.2 Wire Antenna Design

A conventional wire antenna design program development using MATLAB GUI is required. The program should follow the design steps and produce the required antenna parameters for the given parameters of

- Current distribution
  - Program should enable user to define the current distribution.
  - Constant uniform or any other form.
- f (MHz) Frequency
- Type of wire antenna
  - Infinitesimal, short, finite (half, dipole, monopole, etc.).
- h (m) height of the antenna from ground plane.
- Wire dimensions (diameter = a in mm). Assume copper is used.

Use your program to calculate the radiation efficiency of a half-wave dipole at 100 MHz if it is made of aluminum wire 6.35 mm (0.25 in.) in diameter. Assume the radiation resistance to be 70 Ω. Compare MATLAB results with HFSS simulation results.

Solution

This antenna design toolkit consists of the following user input button:

- Type of wire antenna: the user will select between {Infinitesimal, Short, Finite}.

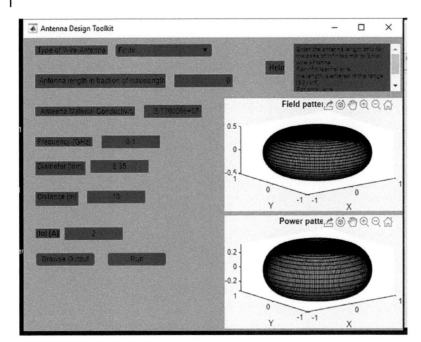

Figure 11.10 Input and output for the case of finite wire antenna.

- Antenna length in fraction of wavelength: this is special for the case of infinitesimal and small dipole where the range of lengths for the infinitesimal dipole is $[l < \frac{\lambda}{50}]$, and for the small antenna the range is in $[\frac{\lambda}{50} \le l \le \frac{\lambda}{10}]$.
- Antenna material conductivity: the user will enter the conductivity for the material the wire antenna is made of (i.e. for copper 5.69e7).
- Frequency: the frequency that the antenna operates along.
- Diameter: the diameter of the wire antenna (i.e. 6.35 mm).
- Distance: the distance from the antenna, which identifies the field region {reactive near field, radiating near field, far-field regions}.
- $|I_o|$: the current amplitude.
- Browse output: this button is used to make the user select the place to store the antenna far-field region parameters. (Rl, Rr, Xa e_{cd}, D_0, G_0, Γ, e_r, HPBW, FNBW).
- Run: when the user pushes this button, behind the scenes mathematical calculations and selections based on input data entered by the user will be performed.

The output of MATLAB GUI is shown in Figure 11.10.

The same antenna is simulated using HFSS, as shown in Figure 11.11.

The wire antenna geometry consists of two cylinders connected by a port that represent the source of excitation, as shown in Figure 11.12. After assigning the proper material to the wire, the boundaries and excitation to the design elements, one needs to set the setup frequency and frequency sweep, and finally far-field simulation needs to be set up from the radiation menu. The boundary cylinder is assigned radiation boundaries, while the two cylinders representing the half-wave dipole will be assigned PEC boundaries. One last step before the design is ready to be simulated is the validation check.

HFSS simulation results are illustrated in Figures 11.13–11.16.

Table 11.1 illustrates and compares MATLAB and HFSS results.

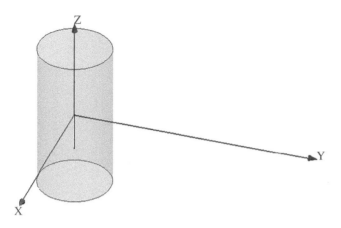

Figure 11.11 Half-wave dipole geometry.

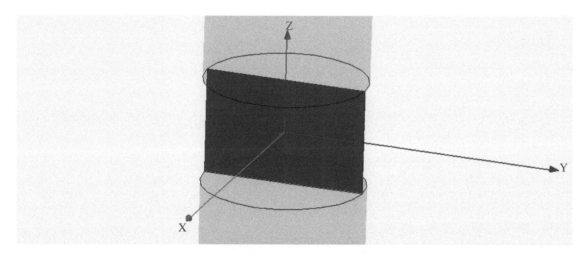

Figure 11.12 Excitation port of half-wave dipole.

11.4 Microstrip Antennas

Microstrip antennas are widely used in the microwave frequency region because of their simplicity and compatibility with printed-circuit technology, making them easy to manufacture as stand-alone elements [1]. Microstrip patch antennas are one of the most commonly used microstrip antenna designs. They generally consist of four parts (a patch layer, ground plane layer, substrate, and a feeding port), as shown in Figure 11.17.

The metallic patch is normally made of thin copper. The shapes of some patches are shown in Figure 11.18. The substrate layer thickness is 0.01–0.05 of free space wavelength (λ_0). It is also often used with high dielectric constant material to load the patch and reduce its size. The advantages of the microstrip antennas are their small size, low profile, and light weight, and that they are conformable to planar and nonplanar surfaces [2]. They require only a very little volume of the structure when mounting. They are simple and cheap to manufacture using modern printed- circuit technology. However, patch antennas have disadvantages. The main disadvantages of the microstrip antennas are: low efficiency, narrow bandwidth of less than 5%, and low RF power due to the small separation between the radiation patch and the ground plane (not suitable for high power applications) [2].

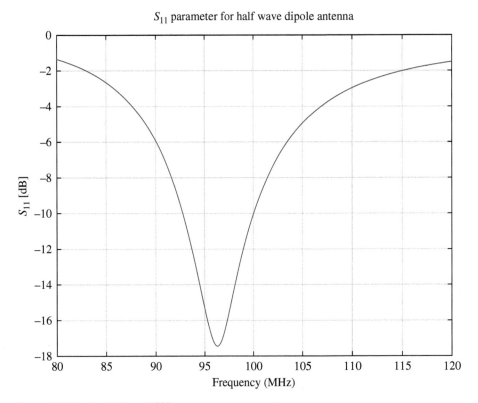

Figure 11.13 S_{11} dB from HFSS.

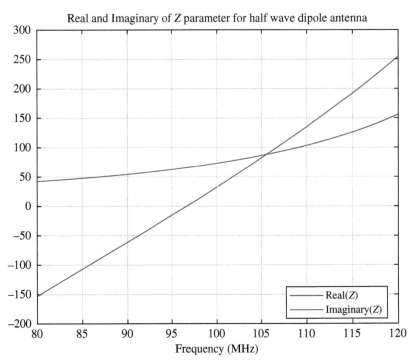

Figure 11.14 Real and imaginary parts of Z parameter from HFSS.

Normalized field and power pattern in 2-D

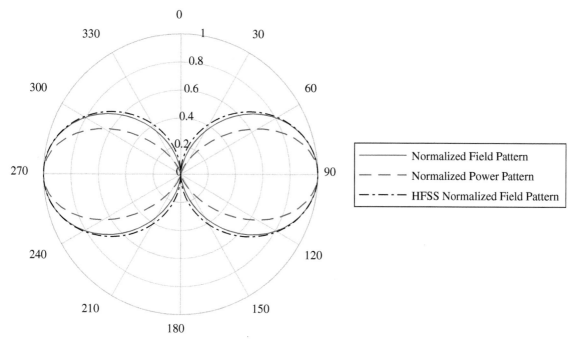

Figure 11.15 Two-dimensional radiation pattern result from HFSS.

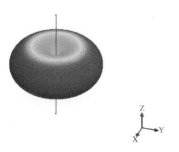

Figure 11.16 Three-dimensional polar plot of E along the half-wave dipole antenna.

11.4.1 Type of Patch Antennas

The radiating patch can be square, rectangular, thin strip (dipole), circular, elliptical, triangular, or any other configuration. Some of these configurations are illustrated in Figure 11.18. Square, rectangular, dipole (strip), and circular are the most common because of ease of analysis and fabrication, and their attractive radiation characteristics, especially low cross-polarization radiation [2]. The thickness of the substrate h has a big effect on the resonant frequency f_r and bandwidth BW of the antenna.

11.4.2 Feeding Methods

The commonly used feeding methods for microstrip antennas are microstrip line feed, coaxial probe, aperture coupling, and proximity coupling [3–6] which are shown in Figure 11.19. The selection of the feeding method is based on the efficiency of the power transfer between the radiating patch and the feed line, which depends on the impedance matching between them. These feeding methods can be classified into two categories: contact and noncontact. In the contact method, the RF source is fed directly to the radiating patch using a connecting element such as a microstrip line. In the noncontact method, electromagnetic field coupling is done with power transfer between the microstrip line and the radiating patch.

11.4.2.1 Microstrip Line Feed

The microstrip feed line shown in Figure 11.20 is also a conducting strip, and has a much smaller width than the patch. The microstrip line feed is easy to fabricate, simple to match by controlling the inset position, and rather simple to model. The width and the inset position of the strip line can be optimized to match the input impedance

Table 11.1 Antenna parameter comparison for aluminum wire half-wave dipole at f = 0.1 GHz.

Parameter	MATLAB	HFSS
Γ (dB)	−1.118286E + 01	−10.037
R_l (Ω)	1.216597e − 01	—
R_r (Ω)	7.305000e + 01	73.1434
X	4.250000e + 01	32.7794
D_0	1.642090e + 00	1.8938
G_0	1.639360e + 00	1.8902
e_{cd}	9.983373e − 01	0.9981
e_r	9.238422e − 01	0.9008

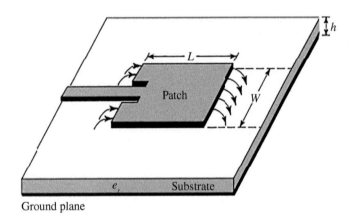

Figure 11.17 Geometry of microstrip antennas.

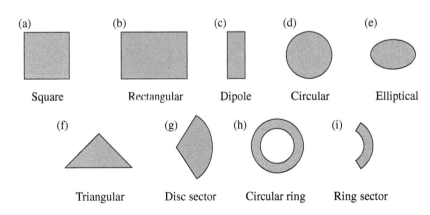

Figure 11.18 Microstrip patch antenna shapes [2].

(a)

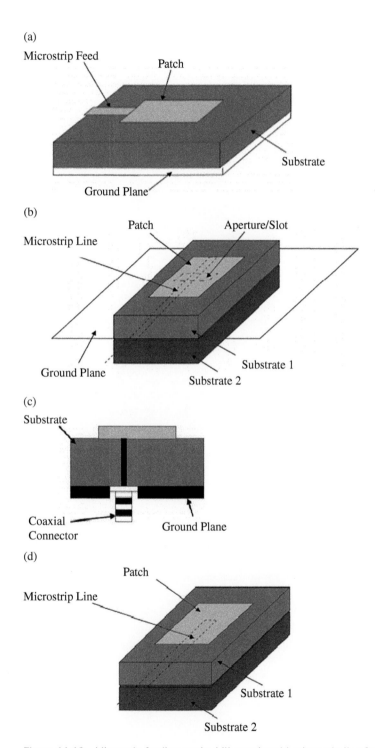

(b)

(c)

(d)

Figure 11.19 Microstrip feeding method illustration: (a) microstrip line feed; (b) aperture coupling; (c) coaxial probe; (d) proximity coupling.

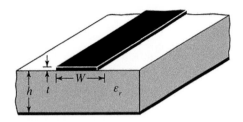

Figure 11.20 Microstrip feed line.

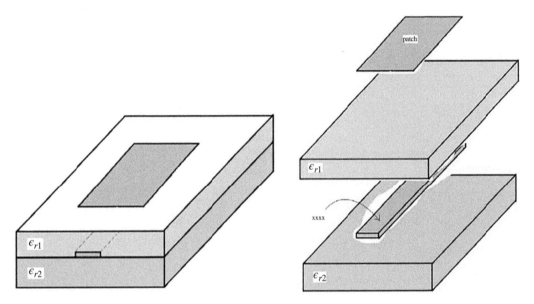

Figure 11.21 Proximity coupled feed.

without the need for any additional matching element. This makes fabrication, modeling, and impedance matching relatively straightforward. However, as the thickness of the dielectric substrate is increased, surface waves and spurious feed radiation increase, creating a limit to the bandwidth of the antenna [2].

11.4.2.2 Proximity Coupling

Proximity coupling is illustrated in Figure 11.21. The feed line is sandwiched between two dielectric substrates and the radiating patch is on top of the upper substrate. The advantages of this kind of feeding technique are the elimination of spurious feed radiation and high bandwidth due to an overall increase in the thickness of the microstrip patch antenna. Impedance matching is achieved by controlling the length of the feed line and the width-to-line ratio of the patch [2].

11.4.3 Microstrip Antenna Analysis – Transmission Line Method

The transmission line method is considered the simplest design method of those that are available. This model was developed for rectangular patch antennas but has been extended for generalized patch shapes. In this modeling method, the microstrip antenna is represented by two slots of width W and height h, separated by a transmission line with length L. The microstrip is a nonhomogeneous line of two dielectrics, typically the substrate and air. The width of the radiated patch is given by

$$W = \left(\frac{3*10^8}{2*f}\right)*sqrt\left(\frac{2}{\varepsilon_r + 1}\right)*1000 \tag{11.103}$$

In microstrip lines, owing to the dielectric–air interface, the effective dielectric constant ε_{reff} must be obtained in order to account for the fringing and the wave propagation in the line. The value of ε_{reff} is slightly smaller than ε_r since the fringing fields around the periphery of the patch are not confined in the dielectric substrate. The expression for ε_{reff} is given by

$$\varepsilon_{reff} = \left(\frac{\varepsilon_r + 1}{2}\right) + \left(\frac{\varepsilon_r - 1}{2}\right)*\left(1 + \left(\frac{12*h}{W}\right)\right)^{-0.5} \tag{11.104}$$

where ε_{reff} = effective dielectric constant, ε_r = dielectric constant of substrate, h = height of dielectric substrate, and W = width of the patch.

The fringing fields along the width can be modeled as radiating slots, and electrically the patch of the microstrip antenna looks greater than its physical dimensions. The dimension of the patch along its length have now been extended on each end by a distance of ΔL, which is given empirically by

$$\Delta L = \left(\frac{(\varepsilon_{reff} + 0.3)*\left(\left(\frac{W}{h}\right) + 0.264\right)}{(\varepsilon_{eff} - 0.258)*\left(\left(\frac{W}{h}\right) + 0.8\right)}\right)*(0.412*h) \tag{11.105}$$

The effective length of the patch L_{eff} now becomes

$$L_{eff} = L + 2\Delta L \tag{11.106}$$

For a given resonance frequency f_o, the effective length can be obtained by

$$L_{eff} = \left(\frac{3e8}{2*f*sqrt(\varepsilon_{reff})}\right)*1000 \tag{11.107}$$

Therefore, the actual physical length is found from

$$L = L_{eff} - 2\Delta L \tag{11.108}$$

The minimum required ground plane length and width are then equal to

$$L_g = 6h + L \tag{11.109}$$
$$W_g = 6h + W \tag{11.110}$$

11.4.4 Impedance Matching

The theory of maximum power transfer states that, for the transfer of maximum power from a source with fixed internal impedance to the load, the impedance of the load must be a complex conjugate of the source.

$$Z_S = Z_L^* \tag{11.111}$$

where Z_S = impedance of the source and Z_L^* = impedance of the load.

The configuration for the lumped element equivalent circuit for the microstrip patch antenna is illustrated in Figure 11.22.

The admittance is

$$Y = \frac{1}{Z_L} = G + jB \tag{11.112}$$

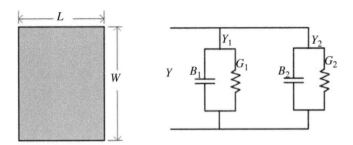

Figure 11.22 Rectangular patch and its transmission model equivalent.

where

$$Y_1 = Y_2 \tag{11.113}$$

$$G_1 = G_2 \tag{11.114}$$

$$B_1 = B_2 \tag{11.115}$$

Expression for G_1 is given by

$$G_1 = \frac{2P_{rad}}{|V_0|^2} \tag{11.116}$$

In (11.16), P_{rad} = radiation power and V_0 = voltage across the slot and they are defined by

$$P_{rad} = \frac{|V_0|^2}{2\pi\eta_0} \int_0^\pi \left[\frac{\sin\left(\frac{k_0 W}{2} \cos\theta\right)}{\cos\theta} \right]^2 \sin\theta^3 d\theta \tag{11.117}$$

Hence, (11.16) can be written as

$$G_1 = \frac{I_1}{120\pi^2} \tag{11.118}$$

where

$$I_1 = \int_0^\pi \left[\frac{\sin\left(\frac{k_0 W}{2} \cos\theta\right)}{\cos\theta} \right]^2 \sin\theta^3 d\theta \tag{11.119}$$

$$I_1 = -2 + \cos(x) + (x * sinint(x)) + \left(\frac{\sin(x)}{x}\right) \tag{11.120}$$

$$x = K_o * W \tag{11.121}$$

$$K_o = \frac{2\pi}{l_a} \tag{11.122}$$

$$l_a = \left(\frac{3e8}{f}\right) * 1000 \tag{11.123}$$

The parallel equivalent admittances Y_1 and Y_2 for the configuration shown in Figure 11.22 taking into account of mutual coupling effects is defined by

$$R_{in} = \frac{1}{2(G_1 \mp G_{12})} \tag{11.124}$$

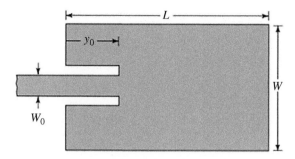

Figure 11.23 Inset feed point distance (y_0).

$$G_{12} = \frac{1}{|V_o|^2} \int\int_S E_1 \times H_2^* \, d \tag{11.125}$$

Where E_1 = electric field radiated by Y_1, H_2^* = magnetic field radiated by Y_2, and J_0 = Bessel function of the first kind of order zero. Equation (11.125) can be simplified as

$$G_{12} = \frac{1}{120\pi^2} \int_0^\pi \left[\frac{\sin\left(\frac{k_0 W}{2} \cos\theta\right)}{\cos\theta} \right]^2 *J_0(k_0 L \sin\theta) \sin\theta^3 \, d\theta \tag{11.126}$$

The position of the feed point of the patch where the impedance of the patch would be generally $50\,\Omega$ is found from

$$R_{in} = \frac{1}{2(G_1 \mp G_{12})} \cos\left(\frac{\pi}{L} y_o\right)^2 \tag{11.127}$$

Hence, the inset feed point distance shown in Figure 11.23 is

$$y_0 = \left(\frac{L}{\pi}\right) * \left(\mathrm{acos}\left(\mathrm{sqrt}\left(\frac{Z_0}{R_{in}}\right)\right)\right) \tag{11.128}$$

To match the patch antenna using a microstrip line feed, we use a transmission line whose characteristic impedance (Z_0) is equal to $50\,\Omega$ and calculate its transmission line width (W_0) from

$$W_0 = h * \left(\frac{2}{\pi}\right) * \left(B - 1 - m + \left(\left(\frac{\varepsilon_r - 1}{2*\varepsilon_r}\right) * \left(n + 0.39 * \left(\frac{0.61}{\varepsilon_r}\right)\right)\right)\right) \tag{11.129}$$

where

$$B = \frac{377\pi}{2*Z_0*\sqrt{\varepsilon_r}} \tag{11.130}$$

$$m = \log\left(2*B - 1\right) \tag{11.131}$$

$$n = \log\left(B - 1\right) \tag{11.132}$$

11.5 Engineering Application Examples

In this section, design examples including analytical design, and simulation and implementation of wire and microstrip antennas are given.

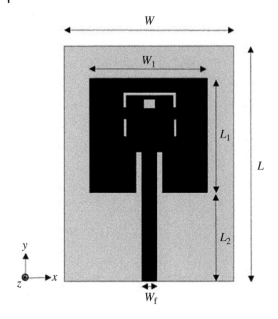

Figure 11.24 Layout of dual band slot loaded antenna layout.

Design Example 11.1 Dual Band Slotted Antenna Design

Design a dual band slotted antenna for energy harvesting operating around 900 MHz and 1.8 GHz for WLAN applications. The incorporation of a C-shaped slot, a rectangle slot, and a pair of I-shaped slots on patch antenna was supportive for excitation of dual distinctive resonance bands.

Solution

Analytical

Figure 11.24 shows the layout of a slot loaded antenna for dual band WLAN applications. The planar antenna is printed on Rogers RT/Duroid 5880 substrate with thickness $h = 1.575$ mm and dielectric constant $\varepsilon_r = 2.2$. The proposed dual band antenna is composed of a microstrip patch, which should operate at resonant frequency, $f_r = 900$ MHz and is loaded with multiple slots to generate an additional excitation path for higher band 1.8 GHz resonance. A positive 90° rotated C-shaped slot of 1 mm width is placed near the nonradiating top edge of the microstrip patch element. The total length of the C-shaped slot is designed to be half wavelength and corresponds to higher resonant frequency.

The width *W1* and length *L1* of microstrip patch are calculated from (11.132) to (11.136) as

$$W_1 = \frac{1}{2f_r\sqrt{(4\pi \times 10^{-7})(8.854 \times 10^{-12})}}\sqrt{\frac{2}{\varepsilon_r + 1}} \tag{11.133}$$

$$\varepsilon_{reff} = \frac{\varepsilon_r + 1}{2} + \frac{\varepsilon_r - 1}{2}\left[1 + 12\frac{h}{W_1}\right]^{-1/2} \tag{11.134}$$

$$\Delta L_1 = 0.412h\frac{(\varepsilon_{reff} + 0.3)\left(\dfrac{W_1}{h} + 0.264\right)}{(\varepsilon_{reff} - 0.258)\left(\dfrac{W_1}{h} + 0.8\right)} \tag{11.135}$$

$$L_1 = \frac{1}{2f_r\sqrt{\varepsilon_{reff}}\sqrt{(4\pi \times 10^{-7})(8.854 \times 10^{-12})}} - 2\Delta L_1 \tag{11.136}$$

Table 11.2 Slot antenna parameter calculations.

$W = 141.2116$ mm	$F_i = 39.2735$ mm
$L = 121.1556$ mm	$\varepsilon_r = 2.2$
$W_p = 131.7616$ mm	$H = 1.6$ mm
$L_p = 111.7616$	$f = 0.9$ GHz
$W_f = 4.8516$ mm	$Z_{input} = 50\,\Omega$
$G_{pf} = 0.77455$ mm	$W_{f/2} = 2.4$ mm

Characteristic impedance, input impedance, conductance, and mutual conductance for the inset feed are determined using the previously given formulations for the microstrip and tabulated in Table 11.2.

Simulation

The simulated return loss is demonstrated in Figure 11.25. The first resonance is seen around 900 MHz with −13.17 dB loss. At the second resonance, 1.8 GHz, the reflected power is merely 0.01% at return loss level of −12.1352 dB. The low loss characteristic achieved through the proper impedance matching in this design is crucial. As the antenna is able to capture more power, greater energy can be radiated away to the intended application. At lower resonances, the present thickness is consistently disseminated on the left and right area of the patch antenna along the emanating edge. At higher resonances, it can be seen that the present stream seriously in the center area of the patch antenna where slot located. Symmetrical current conveyance designs are observed for the proposed antenna structure at the two resonances.

The 3D radiation pattern is obtained and illustrated in Figure 11.26. The gain parameter is obtained by considering the total efficiency of the antenna. At the lower resonance 900 MHz band, the 5 dBi gain obtained in the peak direction of radiation. At the upper resonance of 1.8 GHz, a gain of 10 dBi indicates the proposed antenna possesses about twice as much power as the lower resonance isotropic antenna in the peak direction.

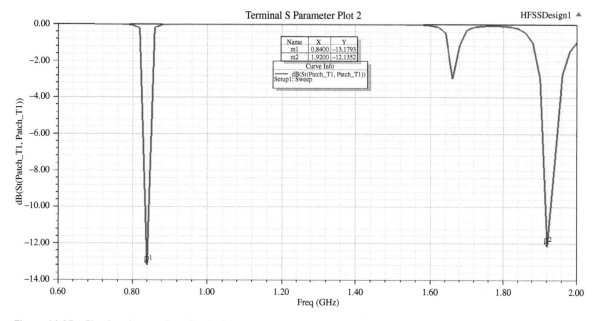

Figure 11.25 Simulated return loss for dual band antenna.

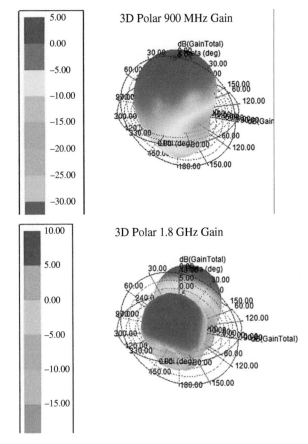

Figure 11.26 Gain (dBi) in 900 MHz and 1.8 GHz.

The 2D radiation patterns are shown in Figure 11.27 on the elevation and azimuthal planes.

The prototype has been built and measured, as illustrated in Figure 11.28.

The measured results showed that the return loss is −8.29 dB at 990.4 MHz and − 11.316 dB at 1.725 GHz.

Design Example 11.2   Wire Antennas Dipole Antenna Design

Design, simulate, and implement half-wave dipole antenna utilizing the VHF frequency band, which is 30–300 MHz. The center frequency of the antenna is to be 146 MHz.

Solution

Analytical

Frequency (target) = 146 MHz; antenna wire diameter = 3.175 mm

Wavelength:

$$\lambda = c/f; c = \text{speed of light in free space}$$

$$\lambda = \frac{3 \times 10^8 m/s}{146 \times 10^6 Hz} = 2.04479 \text{ m}$$

Antenna length:

$$l = \lambda/2 = \frac{2.04479\ m}{2} = 1.0274\ \text{m}$$

Directivity:

$$D_0 = \frac{4}{C_{in}(2\pi)} \approx \frac{4}{2.43765}$$

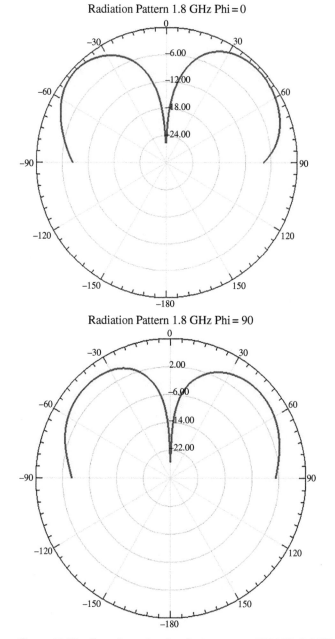

Radiation Pattern 1.8 GHz Phi = 0

Radiation Pattern 1.8 GHz Phi = 90

Figure 11.27 Two-dimensional radiation pattern 900 MHz/1.8 GHz (Phi = 0° and Phi = 90°).

Radiation Pattern 900 MHz Phi = 0

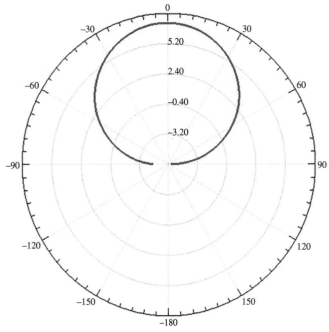

Radiation Pattern 900 Hz Phi = 90

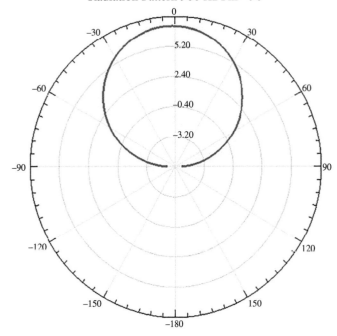

Figure 11.27 (Continued)

Figure 11.28   Prototype antenna built and measured.

$$D_0 \approx 1.64092 = 2.1511 \, \text{dB}$$

Antenna effective area:

$$A_{\text{em}} = \lambda \verb|^|2 * D_0 = \frac{2.04479^2 \times 1.64092}{4\pi}$$

$$A_{\text{em}} = 0.54578 \, \text{m}^2$$

Radiation resistance:

$$R_{\text{r}} = \frac{\eta}{4\pi} \times C_{\text{in}}(2\pi)$$

$$R_{\text{r}} \approx \frac{120\pi}{4\pi} \times 2.43765 \approx 73.12 \, \Omega$$

Input resistance:

$$R_{\text{in}} = R_{\text{r}} \approx 73.12 \, \Omega$$

Input impedance:

$$Z_{\text{in}} = 73 + \text{j}42.5$$

Loss resistance:

$$R_L = \frac{l}{4\pi b}\sqrt{\frac{\omega\mu_o}{2\sigma}}$$

$$= \frac{l}{4\pi \times 1.5875}\sqrt{\frac{2\pi \times 146 \times 10^6 \times 4\pi \times 10^{-7}}{2 \times 5.9 \times 10^6}}$$

$$R_L = 0.509033\,\Omega$$

Steel conductance $\sigma = 5.9{*}10^6$ S, free space permittivity $\mu_0 = 4\pi{*}10^{-7}$
Radiation efficiency:

$$\eta_{cd} = \frac{R_r}{R_r + R_L} = \frac{73.12}{73.12 + 0.509033}$$

$$\eta_{cd} = 0.993087$$

Reflection coefficient:

$$\Gamma = \frac{Z_{in} - Z_0}{Z_{in} + Z_0} = \frac{73.12 + j42.5 - 50}{73.12 + j42.5 + 50}$$

$$\Gamma = 0.371455 \,/\, (0.740185)\ \text{rad}$$

*By utilizing a coaxial connector, with an impedance of 50 Ω, Z_t has been defined.
Gain:

$$G_0 = \eta_{cd}D_o$$

$$G_0 = 0.993087{*}1.64092$$

$$G_0 = 1.62958 = 2.12075\ \text{dB}$$

HPBW: 78°.

MATLAB was utilized to plot the analytical results that were derived from the analytical design calculations shown in Section 11.3. A plot of the radiation power pattern of a half-wavelength dipole antenna is given in Figure 11.29.

Simulation

For the simulation's loss resistance of the antenna, iron was selected as the material to match the conductance of the physical antenna, which was constructed from steel. Steel was not listed as a selectable option, but we used iron as the steel rods are over 99% composed of iron. The 3D simulation layout of HFSS is given in Figure 11.30.

The antenna parameters obtained by HFSS are given in Figure 11.31. VSWR of simulated dipole antenna is given in Figure 11.32. S_{11} of the half dipole is obtained and shown in Figure 11.33. Two-dimensional radiation patterns are plotted and given in Figure 11.34.

Prototype
Parts List:

- 2 each ⅛″ steel rod, 3 ft long.
- Wooden board, ¼″ thick.
- 2 each ¼″ × 1″ (25 mm) machine screws.
- 2 each ¼″ wingnuts.
- 4 each ¼″ large flat washers.
- 2 each ¼″ split lock washers.
- 2 each ¼″ external tooth lock washers.
- 2 each ¼″ crimp-on ring connectors.
- 50 Ω coax cable with BNC connector.

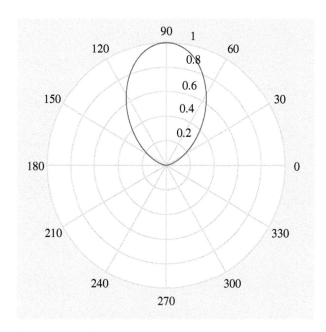

Figure 11.29 Radiation plot by MATLAB.

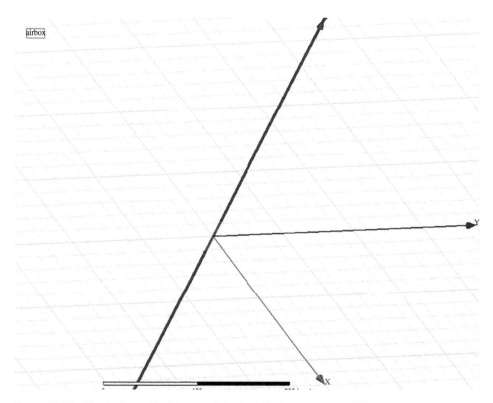

Figure 11.30 Three-dimensional layout of the half dipole antenna by HFSS.

Antenna Parameters:

	Quantity	Freq	Value	
	Max U	0.146GHz	103.55 mW/sr	
	Peak Directivity		1.7436	
	Peak Gain		1.3028	
	Peak Realized Gain		1.3012	
	Radiated Power		746.27 mW	
	Accepted Power		998.8 mW	
	Incident Power		1 W	
	Radiation Efficiency		0.74717	
	Front to Back Ratio		-N/A-	
	Decay Factor		0	

Figure 11.31 Antenna parameters from HFSS.

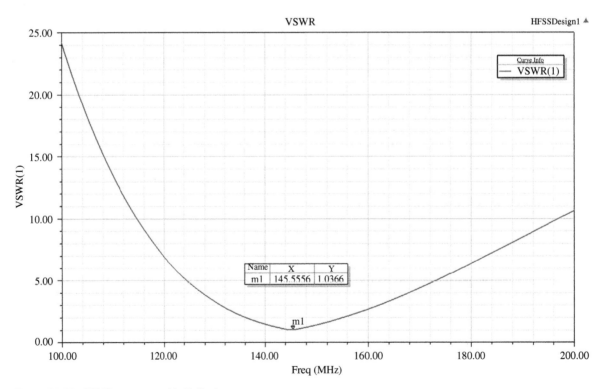

Figure 11.32 VSWR response of half dipole antenna.

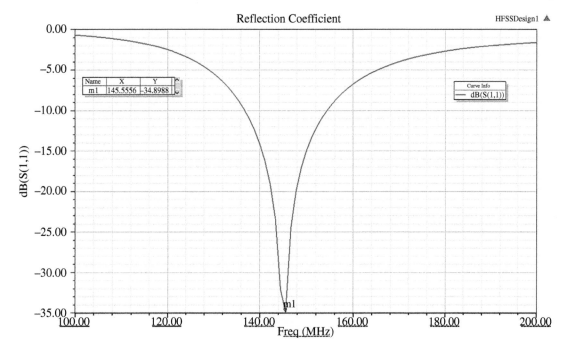

Figure 11.33 S_{11} of the half-wave dipole.

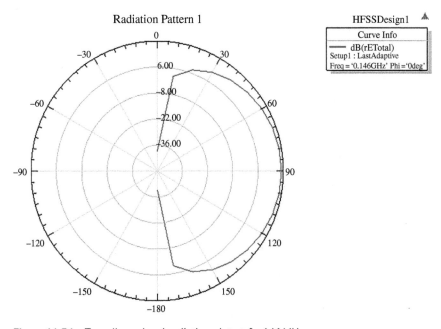

Figure 11.34 Two-dimensional radiation plot at f = 146 MHz.

Construction Procedure:

The antenna is made up of two 1/8″ carbon steel rod sections, each bent into a U-shape to accommodate the fastener materials. The rods making up the antenna are mounted onto the wooden board using the screws and washers. One end of the coaxial cable is taken apart to separate the center conductor from the braided shielding. Each part of the coaxial cable is soldered to one of the ring connectors and then secured to the screws using the wingnuts.

This antenna also has two smaller dipole elements that are approximately 17.5 cm long. This portion could be used as a 70 cm dipole which could tune into the 420–450 MHz portion of the UHF band, but this band was not used for this project. This design was chosen because the U shape of the dipole conductors allowed for easy adjustment of one or both dipole elements, if necessary. A simple loosening of a wing nut and sliding the dipole element could be done if adjustment is required. Figures 11.35 and 11.36 provide a visual image of the finished antenna construction.

The prototype illustrated in Figure 11.36 is measured using an Agilent 8753ES Vector Network Analyzer. S_{11} measurements were taken for our dipole antenna to compare with the simulation results. The magnitude of the reflection coefficient is shown in Figure 11.37, whereas its phase is illustrated in Figure 11.38. The radiation pattern is given in Figure 11.39.

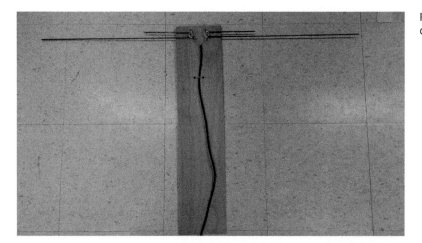

Figure 11.35 Dipole antenna construction.

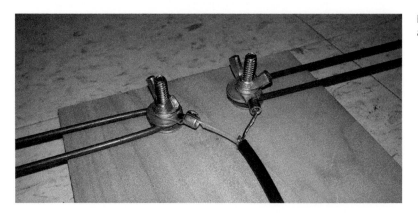

Figure 11.36 Close-up view of dipole antenna connections.

Figure 11.37 Measured S_{11} reflection coefficient magnitude.

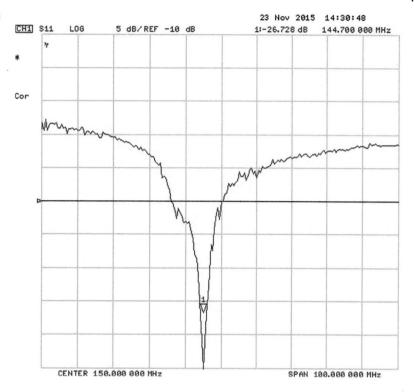

Figure 11.38 Measured phase of S_{11} reflection coefficient.

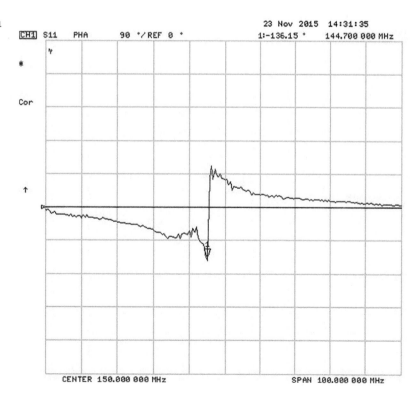

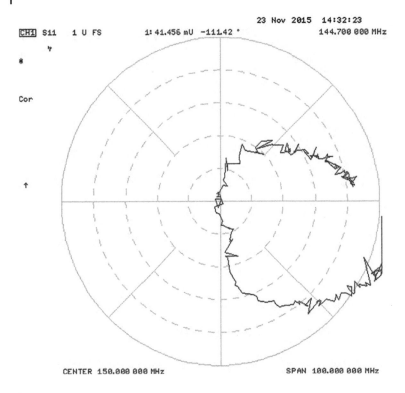

CH1 S11   1 U FS   1: 41.456 mU -111.42 °

23 Nov 2015 14:32:23
144.700 000 MHz

CENTER 150.000 000 MHz   SPAN 100.000 000 MHz

Figure 11.39 Measured radiation pattern.

References

1 Pozar, D.M. and Schaubert, D.H. (1995). *Microstrip Antennas: The Analysis and Design of Microstrip Antennas and Arrays*, 2e. Wiley-IEEE Press.
2 Balanis, C.A. (2005). *Antenna Theory Analysis and Design*, 4e. New Jersey: Wiley.
3 Gillette, M.R. and Wu, P.R. (1977). RF anechoic chamber design using ray tracing. In: *1977 International Symposium Digest IEEE/AP-S*, https://doi.org/10.1109/APS.1977.1147813, 246–252.
4 Kummer, W.H. and Gillespie, E.S. (1978). Antenna Measurements—1978. *Proceedings of the IEEE* **66** (4): 483–507. © (1978) IEEE.
5 Institute of Electrical and Electronics Engineers (1978). *ANSI/IEEE Standard Test Procedures for Antennas 149-1979*. New York: IEEE.
6 Huang, Y. and Boyle, K. (2008). *Antennas from Theory to Practice*. Wiley, ISBN 978-0-470-51028-5.

Problems

Problem 11.1
Calculate the radiated power for the electric field in the far field by

$$\overline{E} = \hat{\theta}\, \frac{\sin(\theta)\cos(\theta)}{2}\, e^{-jkr}$$

Problem 11.2
Calculate the directivity of an antenna if its normalized radiation intensity is given by

$$U = \begin{cases} 2\sin(\theta)\sin(\phi) & 0 < \theta < \pi \text{ and } 0 < \phi < \pi \\ 0 & anywhere\ else \end{cases}$$

Problem 11.3
An antenna with a directivity of 15 dB has an input resistance of 120 Ω. If its maximum radiation intensity is 80 W/sr, what is the rms current at its input? Assume antenna is lossless.

Problem 11.4
A vertical wire antenna operating at 2 MHz has a length of 10 m and wire radius of 1 cm. (a) Identify the type of wire antenna. (b) Calculate the radiation resistance. (c) Calculate the radiation efficiency. (d) If radiated power from the antenna at 10 km away is 1 kW, what is the maximum value of electric field intensity?

Problem 11.5
Calculate the radiation resistance of an antenna if its electric field of in the far field is given by

$$\overline{E} = \hat{\theta} I_o \frac{e^{-j\beta r}}{2\pi r} \sin\theta$$

Problem 11.6
Two element diploe antennas each with length $2\,l$ are excited with equal amplitude and phase and aligned collinearly along the z axis. They are spaced by d from their centers. Assume $2\,l < <\lambda$ and calculate a general expression for far field.

Problem 11.7
If a target reflects 10^{-12} W peak power to receiver radar when transmitted peak power is 75 kW for 10 GHz operational frequency and 2 km range, what is the radar cross section for that target? Assume there are no polarization and reflection losses and the gain of the transmitter and receiver is 20 dB.

Problem 11.8
A wire antenna operating at $f = 10^8$ MHz with a uniform current of 150 mA is 5 cm long. Calculate (a) E_r, (b) E, and (c) H_ϕ when $t = 1$ s, angle $= 90°$, and $r = 3$ m.

Problem 11.9
Find radiated power, directivity, and radiation resistance for a small loop antenna by assuming uniform current distribution of I_o.

Problem 11.10
A rectangular microstrip pact antenna with dimensions of $L = 1.72$ cm and $W = 2.22$ cm, substrate with height $h = 0.2$ cm, and dielectric constant of $\varepsilon_r = 4.4$ is operating at 4 GHz. The input impedance is desired to be 50 Ω. Assume the input impedance of the radiating patch at the feed point is 317.91 Ω. Calculate the inset position and width of the feed line.

Design Challenge 11.1

Design, simulate, and build a dual band slotted antenna for energy harvesting applications operating around 900 MHz and 1.8 GHz. The return loss at each frequency should be better than −20 dB with a 5 dB gain. Use a dielectric substrate such as Teflon with $\varepsilon_2 = 2.2$ and thickness 62 mil.

Design Challenge 11.2

Design, simulate, and build a wearable RFID antenna system using Pyralux AP flexible material with dielectric constant, $\varepsilon_r = 3.4$ and 5 mil thickness in UHF band. The desired return loss at $f = 433$ MHz is minimum −20 dB and no less than 2 dB gain.

12

RF Wireless Communication Basics for Emerging Technologies

12.1 Introduction

In this chapter, the current state of wireless communication technology is outlined, and a few commonly used wireless protocols are summarized. Basic technology trends for 6LoWPAN, ZigBee, Bluetooth, and Wi-Fi are considered. These trends include some emerging technologies consisting of smart grids, smart devices, modules vs. chip solutions, body area networks, Home automation and healthcare applications, etc. Radio frequency identification devices (RFIDs) and their applications are also discussed in this chapter.

12.2 Wireless Technology Basics

Wireless communication is essentially the communication of information or commands via a wireless connection. It can be broken into many different technologies or protocols, and these protocols vary in performance, complexity, cost, and features. Just a short list of current wireless protocols includes Wi-Fi, Bluetooth, ZigBee, 6LoWPAN, Ultra-Wide Band, Wireless USB, and Wireless IR. Wi-Fi, Bluetooth, ZigBee, and 6LoWPAN technologies are considered in depth here and basic information on sub-GHz proprietary protocols are also discussed.

Wireless technology and communication is becoming more attractive in everyday life. From wireless headsets and keyboards to wireless home automation systems, wireless technology is being integrated into all aspects of our life. Currently, there are many city-, company-, or campus-wide wireless networks in place that provide Internet access to residents, workers, and students.

The use of wireless technologies has exploded with the advent and increasing use of smart devices. Smart devices often use access to data plans or Wi-Fi to connect to the Internet and connect people via many mediums. Some of these mediums are: emails, cell phone calls, video conferences, text and picture messaging, Internet texting, and peer-to-peer communication for video and picture sharing. These smart devices are often Wi-Fi, Bluetooth, and cellular technology enabled, but are not ZigBee or 6LoWPAN enabled.

Wireless technology has also expanded with the new "smart home" concept. "Smart home" devices are used to automate the home environment, or provide a wireless interface for in-home appliances. Some examples of these devices include: power usage meters, Internet connected thermostats and cameras, light controls, universal remotes, and sensors. Most of these devices provide a wireless interface to a smart device or an Internet application to allow control or monitoring. IControl is currently working on a project called Open Home Converge that seeks to create a home network that can interface the user with Wi-Fi and Zigbee devices via the Internet in one comprehensive system.

RF/Microwave Engineering and Applications in Energy Systems, First Edition. Abdullah Eroglu.
© 2022 John Wiley & Sons Ltd. Published 2022 by John Wiley & Sons Ltd.
Companion website: www.wiley.com/go/eroglu/rfmicrowave

12.3   Standard Protocol vs Proprietary Protocol

There are two different types of protocols for wireless communication: standard protocols and proprietary protocols. Each type has its own advantages and disadvantages. The first type of solution is the proprietary solution. Within this solution are all the different protocols that comprise wireless technology as it is today. Some of these protocols include: ZigBee, Wi-Fi, Bluetooth, UWB, Wireless USB, and 6LoWPAN, among many others.

12.3.1   Standard Protocols

The advantages of standard protocols is that they are well defined, widely accepted, can easily communicate between other devices using the same protocol, and a large part of the software coding is already done. Some of the disadvantages are the required memory size to implement the protocol, the cost, and the backwards compatibility issue with some protocols.

Standard protocols are well known and well defined. A user can easily understand what the protocol is suitable for and select the best option. For instance, Wi-Fi is best suited for longer distance high bandwidth applications, while Bluetooth has a much shorter range and is used for lower bandwidth applications. ZigBee is a higher range, low data rate, and low power protocol that allows for the largest network structure.

Devices of the same protocol can easily communicate between other similar devices. For example, all ZigBee devices of a certain application profile have the basic structure to be recognized and communicate between other ZigBee devices of the same application profile. One thing to note is that not all protocols are backwards compatible with earlier versions of the same protocol. Backwards compatibility must be checked for each protocol.

Using standard protocols ensures that a large part of the software needed to communicate is already written, leaving less work for engineers. They would only be required to conform to the protocol rather than to create one from scratch. This will greatly reduce the amount of software coding required to get a working system. A small amount of coding may be required to package and unpackage the signals, but the networking code and other capabilities are already finished.

Some standard protocols also have a smaller version of themselves available for use. For instance, ZigBee has its Simple Media Access Coding (SMAC) that only uses about 4 kB of coding, compared to the 90–200 kB required for the full ZigBee stack. ZigBee's SMAC can be used as a wire replacement tool, but it is very limited in its capabilities when compared to the full ZigBee suite.

One disadvantage of these protocols is the large amount of memory needed to facilitate these protocols. For example, 6LoWPAN requires around 30–90 kB of memory, ZigBee requires around 90–200 kB of memory, while Bluetooth and Wi-Fi require even more memory than ZigBee. Wi-Fi will require the most memory, needing twice as much more memory than ZigBee. The extra memory needs really begin to increase the cost of single chip solutions for these protocols.

12.3.2   Proprietary Protocols

Proprietary solutions are customized protocols. The advantages are that these protocols are cheaper to implement, fully customizable, and often simpler and easier to work with. The disadvantages are that there isn't as much capability, it's very difficult to interface with devices not controlled or developed by engineers, and all or most of the protocol encoding must be done in house. Coding a full communication system with security, error correcting, and networking is not a trivial project.

Proprietary protocols can be custom designed for the desired modulation, the extent of encoding and decoding needed, extra security measures, and the amount of networking needed in the desired application. They are less complex, and this will result in smaller coding footprints and memory requirements. Proprietary solutions are also much cheaper to implement, owing to the large amount of work being done in house.

One of the downfalls of proprietary protocols is the amount of work and RF expertise required by the host company. For instance, engineers would be required to do most of the software for the proprietary protocol, and they would need to ensure that they follow all Federal Communications Commission (FCC) guidelines. This often discourages the use of proprietary protocols in most applications. They are still used in simpler systems, such as garage openers and key fobs among others. Another major downfall of using a proprietary protocol is its interface with other systems.

12.3.2.1 Physical Layer Only Approach

A physical layer only approach employs the most basic modulation technique, such as on–off keying (OOK). The communication will imitate the serial communication available and will add no extra capabilities. Physical layer only encoding will have the same problems as proprietary protocols. An agreement between companies will be needed to facilitate use of the protocol between devices and to ensure everyone complies with all FCC regulations.

The physical layer only approach is a bare bones approach that will not provide solutions in all cases. It will not be smart device capable, and it will be difficult to expand the coding capabilities to solve all use cases. It will not provide a readily accepted protocol solution, and may require extensive testing and FCC accreditation before implementation. Overall, the physical layer only approach is the cheapest and simplest solution, but it will not provide solutions in all cases.

12.4 Overview of Protocols

12.4.1 ZigBee

ZigBee is a specification for multiple communication standards based on the IEEE 802.15.4 standard [1]. ZigBee standards are developed and certified by a nonprofit association of member companies called the ZigBee Alliance. All commercial products that seek to use ZigBee must be certified. ZigBee was designed for low complexity and simpler implementation than Wi-Fi and Bluetooth.

ZigBee standards are designed for low cost, low power, wireless mesh network applications. Typical applications require short ranges, low data rates, and secure networking. A few examples of these applications include: ZigBee Home Automation, ZigBee Smart Energy, and ZigBee Health Care. Currently, a device using one application protocol cannot communicate with a device using a different protocol. ZigBee devices also cannot communicate to other IEEE 802.15.4 devices, smart devices, or the Internet.

There are two versions of the standards: ZigBee PRO and ZigBee. The most recent ZigBee versions (ZigBee 2007 and ZigBee Pro) are backwards compatible with ZigBee 2006. ZigBee PRO and ZigBee are very similar. The main differences are what features come as standard, and which are optimized. ZigBee PRO is optimized for large networks and supports additional network management. Basic ZigBee, on the other hand, doesn't come as standard with frequency agility and message fragmentation, but these functions are optional. Basic ZigBee is not optimized for large networks, but still works well with hundreds of nodes.

Applications using ZigBee communicate in three possible bands: 915 MHz (Americas), 868 (Europe), and 2.4 GHz (global). ZigBee communication links come with frequency agility and automatic channel selection [2]. Frequency agility means that the communication link can dynamically change channels within the target frequency to deal with interference as it comes and goes. Automatic channel selection means that the protocol automatically selects the best channel at startup.

ZigBee uses binary phase-shift keying (BPSK) in the 915 MHz and 868 MHz bands, and offset quadrature phase-shift keying (OQPSK) in the 2.4 GHz band [2]. Ranges for ZigBee products are usually around 10–100 m, but can be as high as 1600 m.

In the past, ZigBee has been known as the solution for the home automation and smart energy markets. This is no longer the situation as 6LoWPAN and Bluetooth Low Energy (BTLE) are being implemented inside the home. One major cause of this is because the ZigBee Alliance initially chose not to use an Internet protocol (IP) suite. ZigBee Alliance began creating the ZigBee IP stack. The ZigBee IP layer allows the ZigBee devices to communicate with the Internet.

12.4.2 LowPAN

6LoWPAN was created in 2005 by the Internet Engineering Task Forced (IETF). 6LoWPAN was designed to "enable the efficient use of IPv6 over low-power, low-rate wireless networks on simple embedded devices through an adaptation layer and the optimization of related protocols" [3]. IPv6 is a form of Internet addressing that was created to expand the number IP addresses for Internet devices. 6LoWPAN's main goal is to enable embedded devices to communicate with the Internet, and it is considered a chief competitor of ZigBee's.

Both 6LoWPAN and ZigBee are built atop the IEEE 802.15.4 protocol, which means their basic wireless communication is very similar. IEEE 802.15.4 sets up the modulation techniques, frequency bands, security, and most other settings for wireless communication. The strength of 6LoWPAN lies in its ability to be layered with other protocols to extend the system's capabilities. For instance, 6LoWPAN has been implemented using Bluetooth protocols to enable a direct connection from the 6LoWPAN network to smart devices. Details of how to access information from the Internet on an Android phone using 6LoWPAN is given in [4].

6LoWPAN was initially created due to the void caused when ZigBee chose not to include an IP stack in its protocol. Since then, 6LoWPAN has gained favor among engineers due to the IPv6 and IPv4 compatibility, and the royalty free implementation. NXP partnered with a lightbulb manufacturing company to create 6LoWPAN enabled lightbulbs. These lightbulbs each have an IPv6 address and can be individually controlled through an Internet enabled smart device. The lightbulbs are connected to a 6LoWPAN enabled Wi-Fi router that provides a gateway to the Internet, and then the Internet provides connections to any Internet connectable device. In this implementation, a direct connection from the smart device to the lightbulb is not realized. The range of the 6LoWPAN network will be extended with each lightbulb implemented, because 6LoWPAN enables multinode hopping to allow users to reach their destination.

12.4.3 Wi-Fi

Wi-Fi is a wireless standard based on the IEEE 802.11 standard, and it was designed for wire replacement of a high bandwidth cable such as Ethernet. Wi-Fi standards are managed by the Wi-Fi Alliance, which is a trade association created in 1999 comprising hundreds of members. The Wi-Fi Alliance's main purpose is to test and certify products for compliance with the 802.11 standards and compatibility. All products seeking to use the Wi-Fi protocol should be tested and certified by the Wi-Fi Alliance.

Wi-Fi devices can work at either 2.4 or 5 GHz. Communication is facilitated using a spread spectrum signal coupled with dynamically frequency selection, transmission power selection, and one of many different types of modulation schemes. These modulation schemes are outlined in Table 12.1 differentiating Wi-Fi, Bluetooth, and Zigbee.

Wi-Fi technology was designed for high bandwidth applications. These applications include video conferencing, Internet television, live camera feeds, and picture and video streaming. Wi-Fi has also been optimized for range. Most ranges are around 12–100 m, but if the most recent version, 802.11n, is used then the range could be doubled to 200 m. These two factors force Wi-Fi devices to have very high power consumption when compared to Bluetooth or ZigBee.

Data security has been one concern for Wi-Fi technology. Unencrypted Wi-Fi devices can be monitored using a "sniffer" program. The sniffer program is used to read communication packets that may contain personal information. Wired Equivalent Privacy (WEP), the most used security encryption can easily be broken to allow access

Table 12.1 Comparison of ZigBee, 6LoWPAN, Wi-Fi, and Bluetooth.

	ZigBee	6LoWPAN	Wi-Fi	Bluetooth
Maximum data rate	250 Kbits/s	250 Kbits/s	11–54 Mbits/s	1 Mbits/s
Nominal range	10–100 m	10–100 m	100 m	10 m
Maximum range	100 m	100 m	200–300 m	100 m
Networking topology	Ad hoc, p2p, star, or mesh	Mesh	Point-to-hub, or ad hoc	Point-to-point
Max networking size	65 000+	65 000+	2007	7
Operating frequency	868/915 MHz, 2.4 GHz	2.4 GHz	2.4 and 5 GHz	2.4 GHz
Power consumption	Low (~30 mW)	Low (~30 mW)	High (~750–2000 mW)	Low (~1–100 mW)
Complexity (device and application impact)	Low	Low	High	Medium
Security/ encryption	128-bit AES encryption plus Application Layer	128-bit AES encryption plus application layer	RC4 Cipher (WEP), AES	E0 stream cipher
Typical applications	Home and Industrial automation, sensor networks, light control etc.	Low power sensor networks, and Internet addressable devices	Broadband Internet access	Body area networks, PDAs, phones, headsets etc.
Channel bandwidth	0.3, 0.6, 2 MHz	0.3, 0.6, 2 MHz	22 MHz	1 MHz
Modulation techniques	BPSK (+ASK), OQPSK	BPSK (+ASK), OQPSK	BPSK, QPSK, CCK COFDM, M-QAM	GFSK, PSK
Interference avoidance	Dynamic frequency selection	Dynamic frequency selection	Dynamic frequency selection	Adaptive frequency hopping
Internet access	Yes (ZigBee IP only)	Yes (via bridge)	Yes	Yes (via bridge)
Chip royalties	Yes	No	No	No
Membership to certify products	Yes	No	Yes	Yes

unwanted users. Another type of encryption that causes problems is the Wi-Fi Protected Setup (WPS). WPS can be hacked in a matter of hours by a brute force attack, and many Wi-Fi devices have WPS automatically selected as the security encryption.

Interference can also be a large problem. Zigbee, Wi-Fi, and Bluetooth all use the same unlicensed ISM band, 2.4 GHz. Wi-Fi chooses a channel, but range and speed can be severely inhibited if that channel is being used by someone else. Wi-Fi does not have a specific way to handle interference other than by switching to a different channel, but this must be done manually.

Wi-Fi is attempting to spread more into the user electronics market. For instance, Wi-Fi is being implemented in refrigerators. These appliances have several applications, including cooking recipe applications, WeatherBug, and

Pandora that can be interfaced to via an 8 in. touch screen mounted on the front. Another example consumer product is the Nest thermostat. The Nest thermostat uses your home's Internet to allow for smart device control and claims to have learning software that can program itself. Overall, Wi-Fi seems to be gaining use among high end consumer electronics and big ticket items such as: thermostats, refrigerators, TVs, washers, dryers, and automobiles.

12.4.4 Bluetooth

Bluetooth is a proprietary wireless standard that defines a uniform structure for communication between wireless devices. Details of these protocols can be found by looking at the IEEE 802.15.1 standard [5]. It is designed as a low bandwidth wire placement solution. Bluetooth is managed by the Bluetooth Special Interest Group, which is a not-for-profit group comprising over 16 000 companies. These companies have banded together to create new initiatives and versions of Bluetooth. Bluetooth is used to create an ad hoc network between a maximum of eight devices. The basic cell of this network is called a piconet. Bluetooth communicates in the 2.4–2.485 GHz ISM band. Communication is facilitated using a spread spectrum, frequency hopping, full duplex signal.

The modulation technique is GFSK (Gaussian frequency shift keying) or PSK (phase shift keying). Interference is avoided using adaptive frequency hopping. Range and power usage vary for different classes of Bluetooth devices. Class 1 radios, used in industrial settings, are limited to 100 mW and have a range of 100 m. Class 2 radios, used in mobile applications, are limited to 2.5 mW of power and have a range of 10 m. Class 3 radios use 1 mW and have a range of only 1 m.

BTLE is gaining acceptance in body area networks. A body area network is a system of sensors located on the human body that communicates with a central hub. These sensors are used to monitor physiological changes in the body, and can be used to alert a doctor or medical staff if the need arises. They can also be used to sense changes in the body that are precursors to heart attacks and other such major medical problems. A few example sensors include heart rate devices and thermometers.

12.5 RFIDs

RFIDs have been used widely in industrial and medical applications because they bridge the real and virtual worlds and enable information transfer at a large scale in a cost-effective way. These devices use radio waves for noncontact reading and are effective in manufacturing and several other applications where bar code labels could not survive.

An RFID system includes mainly three components: a tag or transponder located on the object to be identified, an interrogator (reader) which may be a read or write/read device, and an antenna that emits radio signals to activate the tag and read/write data to it, as shown in Figure 12.1.

RFID tags are classified as passive, semiactive (or semipassive), and active. A passive tag operates on its own with no source and makes use of the incoming radio waves as an energy source. Active tags use the battery power for continuous operation. Semiactive tags use battery power for some functions but utilize the radio waves of the reader as an energy source for its own transmission just like passive tags. Passive tags are very popular because

Figure 12.1 Basic diagram of RFID system.

Table 12.2 RFID tags and communication distance.

RFID frequency band	Frequency band	Typical communication distance	Common application
125–134.2 and 140–148.5 kHz	Low frequency	Up to ~1/2 m	Animal tracking Access control Product authentication
13.553–13.567 MHz	HF	Up to ~1 m	Smart cards Shelve item tracking Airline baggage Maintenance
858–930 and 902–928 MHz, North America	UHF	Up to 10 m	Pallet tracking Carton tracking Electronic toll collection Parking lot access
2.45/5.8 MHz	Microwave	Up to 2 m	Electronic toll collection Airline baggage

they operate without the need of an external power source. One of the biggest challenges of passive RFID tags is their communication distance. The operating range of an RFID system is based on tag parameters such as tag antenna gain and radar cross section, distances between readers, operating frequency, transmission power from reader to the tag, and the gain of the reader antenna. As a result, tag antenna performance plays a very important role in the identification of the communication distance with operational frequency. Typical operational frequencies and their corresponding communication distances for RFID tags are given in Table 12.2. As seen from Table 12.2, the communication distance for ultrahigh frequency (UHF) range is much higher than other conventional RFID frequency bands [2].

Some type of unique identifier is usually stored in an RFID tag, and varying amounts of user defined data are available. This unique identifier could be the electronic product code (EPC), which is a 96-bit number used to universally identify any product [2]. The format of an EPC is shown in Figure 12.2. Other types of unique identifiers may be used as well. Some examples of user defined data may include: group number, serial number, date and time, shipping information, pallet number, firmware version, and factory information.

RFID devices come with a variety of features and designs available for the user. For example, a kill feature is sometimes added. This allows the RFID device to be disabled, or "killed," once the item is bought and transferred to the user. Some RFID devices come with user memory in addition to the EPC code data. Other RFID devices

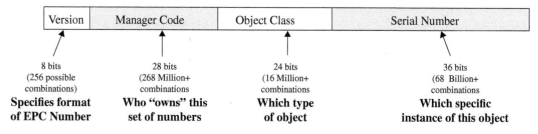

Figure 12.2 Electronic product code.

Table 12.3 Differences between active and passive RFID tags.

	Passive	Active
Memory	Normal ~1 Kb Max ~128 Kb	128 Kb
Range	6 in. –40 ft	>300 ft
Lifetime	10 yr	3–8 yr
Power source	RF signal 124 kHz	External 124 kHz
Frequency	13.56 MHz 4.23 MHz to 2.4 GHz	13.56 MHz 4.23 MHz to 2.4 GHz
Security	Password	Password, more features possible
Integrated sensors	No	Possible

have been designed to work extremely well under water, under high stress, or under a variety of environmental stresses.

12.5.1 Active RFID Tags

Active RFID tags require some type of power source. Often this power source is a battery, but the source may take power from solar or other sources. Active tags have the longest range, up to 300 ft (91 m) or more [4]. These types of tags are used to track highly expensive, important, or large products. Their cost is too high for consumer products.

Active tags usually use one of two modes: beacon mode or transponder mode [6]. In transponder mode, the RFID chip stays asleep until a wakeup signal is sent, and then it begins to transmit data. This is done to minimize power consumption, extend battery life, and is normally used in automatic pay at toll booths (EZ Pass). In beacon mode, the RFID chip will periodically send out a ping that will update the position of the RFID chip in a tracking system. This is done to keep the location of the RFID chip at all times, and is often used in hospitals to keep track of patients or equipment. An outline of the major differences between passive and active RFID technologies can be found in Table 12.3.

12.5.2 Passive RFID Tags

Passive tags require no power source to function. They work by harnessing power from the RF signal used to read it, and reflect that power back to the reader with the information requested. Owing to this, these tags are often much cheaper and are implemented far more often. Their size and makeup vary and can be as simple as a paper bar code, with the RFID antenna sandwiched between the paper and the adhesive. They can also be mounted to a substrate to form a tag or embedded into cards or key fobs.

Passive tags also have a much shorter range. Range varies greatly depending on the type of tag, the protocol used, the frequency of the tag, and the size and power of the reader. For example, the M24L64k from ST Microelectronics has a normal range of around 15 cm (6 in.), while the TM4NM from SimplyRFID has a range of 3.5 m (10.5 ft) based on the datasheet provided by the manufacturers.

12.5.3 RFID Frequencies

12.5.3.1 Low Frequency ~124 kHz and High Frequency ~13.56 MHz

Low frequency (LF) and high frequency (HF) RFID tags communicate using a process called inductive coupling. The antenna of the reader and the antenna of the RF device create an electromagnetic (EM) field, and the RF device uses the induced energy to power the tag. The tag then varies the load which modulates the EM field and the reader converts these modulations to binary numbers for communication [3]. The communication is done through the EM field created, and not necessarily through RF wave propagation. Inductive coupling is the basis of near-field communication (NFC), and certain types of RFID tags. Inductive coupling is only possible because the RFID or NFC device is within one radio wavelength of the reader. RFID tags using inductive coupling are limited in read range by the frequency they use to communicate. As the frequency of the tag increases, the wavelength decreases, and thus the RFID tags have shorter range.

Highly metallic environments or mounting the RFID tag to metal can cause a reduction in range for these frequencies. This is because eddy currents are created in the metal surfaces, which reduce the coupling factor.

12.5.3.2   Ultrahigh Frequency (UHF) Tags ~423 MHz–2.45 GHz

UHF devices work by a process known as propagation coupling, or backscatter. The reader emits RF waves to communicate with the RFID tag. The RFID tag then harvests this energy, and modulates it before sending it back. UHF has its own set of complications as well. Radio waves at 2.45 GHz are absorbed by water. For instance, if there is a lot of condensation in the enclosure, RF waves will be disrupted.

There might be reflections for UHF tags when they are in close proximity to metals. These will cause errors in the transmitted RF signal. These errors may not be correctable in the reader.

12.6   RF Technology for Implantable Medical Devices

Implantable medical devices (IMDs) replace absent biological structure, support deficient biological structure, or enhance biological function with partial/complete surgical or medical insertion. IMDs are intended to remain in the body following the procedure. In 1958, the pacemaker was the first IMD to be introduced to the public. Examples of other applications include artificial eyes, cochlear and retinal implants, brain and cardiac pacemakers, cardioverter defibrillators, temperature monitors, implantable drug pumps, neural recording microsystems, and deep brain stimulators [4, 6]. Efficient integrated antennas are essential for reliable IMD communication with the external-to-body environment, such as base stations, medical equipment, monitoring devices, etc. Figure 12.3 shows a diagram of an IMD system with a bidirectional, dual band wireless communication link to a base station controller and remote monitor.

IMDs continue to pique interest in medical and engineering research disciplines because millions of people depend on them to maintain and improve health and quality of life; they hold great potential for successful medical

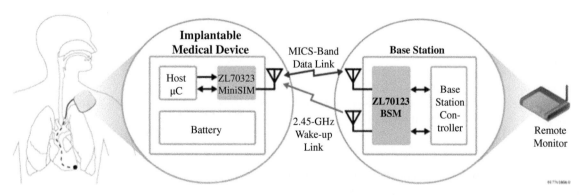

Figure 12.3   Example of IMD and communication system. *Source:* Microsemi.

diagnosis, treatment, and monitoring; they require development of innovative qualities; and they represent enormous market capital potential. Market analysts reported sales revenue of IMD at close to $74 billion in 2018, a 172%+ increase from approximately $43 billion in 2011.

12.6.1 Challenges with IMDs

Challenges with IMD technology include biocompatibility, efficacy, material requirements, power delivery and consumption, packaging dimension constraints, and wireless communication specifications. Integrated antennas with bidirectional communication capability are key components of the biomedical telemetry that enables IMDs to communicate with the external environment, i.e. base stations, medical equipment, etc. Implanted antennas must be safe and able to effectively work within specified and multiple frequency bands earmarked for medical communications. Radio frequency (RF) linked IMD research shows promise in addressing constraints encountered when using LF bands, i.e. substantial volume, low data rates, short distance range, and susceptibility to interference. Patch antenna designs readily lend themselves to this application as their flexibility in design, shape, and conformability meet biocompatibility, patient safety, and communication standards, and facilitate comparatively simple downsizing and assimilation into the shape of an IMD [6].

12.6.1.1 Biocompatibility

The human body may react to device implantation in numerous ways, which can result in unexpected risks for patients. Inflammatory response and foreign body reaction are signs of IMD incompatibility. Biocompatibility is determined on the basis of cytotoxicity, mutagenesis/carcinogenesis, and cell biofunction. To meet biocompatibility standards, IMDs must be nonpoisonous, noncarcinogenic, and able to support healthy cell function. Biocompatibility is a critical issue in IMD performance and longevity.

Various natural, synthetic, and semisynthetic materials are currently utilized in the fabrication of implantable devices. Biological reactions that are adverse for a material in one application may not be adverse for the same material in a different application. Similarly, a material found to be safe in one application may not be safe in another application [7].

The addition of a superstrate of high relative permittivity to an antenna design can be multipurpose. The superstrate also addresses the issues of biocompatibility, safety, and efficacy by separating the antenna's metal radiator from being short-circuited by conductive human tissue.

12.6.1.2 Frequency

Possible new solutions for current challenges in medical diagnosis, treatment, and monitoring are revealed as IMDs persistently become smaller. Owing to the high degree of device miniaturization, wireless communication antennas must be very small. Antenna size constraints push their communication range to the HF domain. Human body tissue attenuation, however, increases with frequency, making HF wireless link efficiency in IMDs challenging.

To facilitate improvements in the communication range of IMD systems, the portion of the RF spectrum between 402 and 405 MHz (with a maximum emission bandwidth of 300 kHz) has been allocated for MICS medical and meteorological applications. Use of RF energy is reserved internationally on several RF bands for ISM purposes other than telecommunications. The frequency 2.4–2.48 GHz was chosen for this use in the reference paper because of the commercial availability of a transceiver for the band.

12.6.1.3 Dimension Constraints

Most conventional microstrip antenna designs for low-frequency operation have substantial volume, but a variety of antenna miniaturization techniques exist. Those employed in this project include use of high permittivity dielectric substrate and insertion of a shorting pin between the radiator patch and ground planes. High permittivity dielectric substrate/superstrate materials reduce effective wavelength (λ_{eff}) to result in lower resonance frequencies, consequently aiding antenna miniaturization. Inserting a shorting pin between the ground and radiator patch planes usually reduces the required antenna size by half by increasing antenna electrical length and doubling the effective size for a given operating frequency [8]. Adding superstrate to antennas is a common technique to improve antenna bandwidth and gain efficiency. Loading the superstrate enhances impedance bandwidth, reduces resonant frequency, and decreases the resonant resistance [4].

12.7 Engineering Application Examples

In this section, two design examples for RFID and healthcare applications are given and discussed.

Design Example 1   Dual Band Microstrip Antenna for Biomedical Applications

Design and simulate differentially fed, dual band microstrip antenna that can operate around both 400 MHz and 2.4 GHz for UHF RFID and medical applications. These frequency bands are for the Medical Implant Communication Service (MICS) band for data transmission and Industrial, Scientific and Medical (ISM) band for the device wakeup signal, respectively.

Solution

The layout of the antenna which is a linear meandered antenna is shown in Figure 12.4.

The formulation of linear meandered antenna is detailed in [9]. The overall width, W, and length, L, as well as the patch width, W_1, and lengths of each turn, W_3, of the antenna are given by Eqs. (12.1)–(12.4).

$$L = 0.7\lambda_g \tag{12.1}$$

$$W = 0.42\lambda_g \tag{12.2}$$

$$W_1 = 0.05\lambda_g \tag{12.3}$$

$$W_3 = 0.16\lambda_g \tag{12.4}$$

where

$$\lambda_g = \frac{\lambda}{\sqrt{\varepsilon_{reff}}} \tag{12.5}$$

$$\varepsilon_{reff} = \frac{\varepsilon_r + 1}{2} + \frac{\varepsilon_r - 1}{2} \frac{1}{\sqrt{1 + 12\dfrac{h}{W}}} \tag{12.6}$$

h is the height of the substrate, ε_r is the dielectric constant of the substrate, and W is the overall width of the antenna. The final dimensions of the antenna are found from

$$W_r = W + \Delta L \tag{12.7}$$

$$L_r = L + \Delta L \tag{12.8}$$

$$W_{3r} = W_3 + \Delta L \tag{12.9}$$

where ΔL is the added length and given by

$$\Delta L = 0.5 \frac{c}{2f_r} \sqrt{\frac{2}{\varepsilon_r + 1}} \tag{12.10}$$

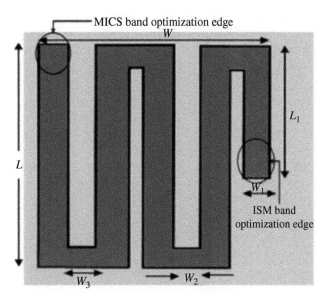

Figure 12.4   Layout of the meandered antenna.

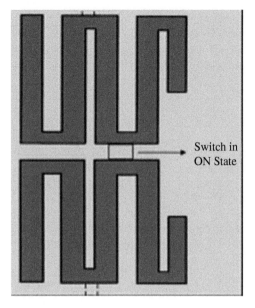

Figure 12.5 Dual linear meandered antenna for dual band operation.

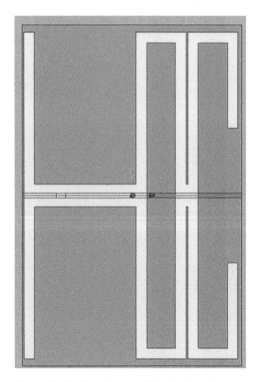

Figure 12.6 Final layout of the dual band antenna.

f_r is the resonant frequency. The formulations are used to design antennas at the first resonant frequency 400 MHz, with height and dielectric constant are 0.5 mm and $\epsilon_r = 11.9$, respectively. The dimensions are calculated as $W_r = 22.72$, $L_r = 16.5$, $W_{3r} = 10.84$ and $W_1 = 1.1$ cm. The 2.4 GHz band is obtained using a second linear meandered antenna and placing it next to the first one operating at $f = 400$ MHz and connected via a metallic switch, as shown in Figure 12.5.

L_1 and W_2 were obtained via optimization. L_1, specifically, controlled the 2.4 GHz band. The simulated antenna layout is shown in Figure 12.6.

The final design is scaled down by a 10 : 1 ratio due to the required low profile. The simulated simulation results are given in Figure 12.7.

Design Example 2 Implantable Dual Band Antenna Design

Design, simulate, and implement dual band implantable antenna with skin mimicking super substrate using meander antenna topology operating around 400 MHz and 2.4 GHz.

Solution
Analytical

Consider a meander antenna layout given in Figure 12.8.

The lumped element equivalent model using a T network is developed. This method considers each branch and corner of a meander line as a T network of inductance and capacitance [10], as shown in Figure 12.9.

The effective permittivity, ϵ_e, of the segments (excluding corners) is in terms of the substrate height, d, and width, w.

$$\epsilon_e = \frac{\epsilon_r + 1}{2} + \frac{\epsilon_r - 1}{2} \frac{1}{\sqrt{1 + 12\dfrac{d}{W}}}$$

The characteristic impedance of the transmission line model for each section is calculated as

$$Z_0 = \begin{cases} \dfrac{60}{\sqrt{\epsilon_e}} \ln\left(\dfrac{8d}{W} + \dfrac{W}{4d}\right) & \text{for } \dfrac{W}{d} \leq 1 \\[4ex] \dfrac{120\pi}{\sqrt{\epsilon_e}[W/d + 1.393 + 0.667\ln(W/d + 1.444)]} & \text{for } \dfrac{W}{d} \geq 1 \end{cases}$$

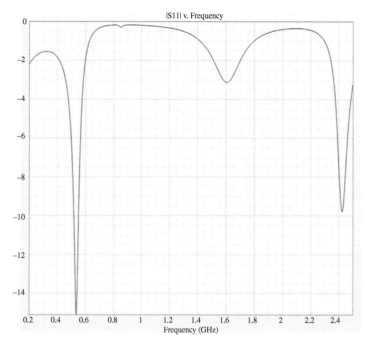

Figure 12.7 Return loss for dual band antenna.

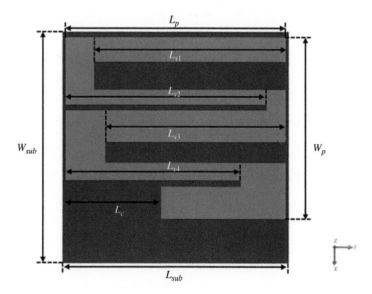

Figure 12.8 Meander antenna layout.

Then, the per unit length inductance and capacitance is calculated using

$$L = Z_0 \sqrt{\varepsilon_{eff}}/c \text{ and } C = \sqrt{\varepsilon_{eff}/cZ_0}$$

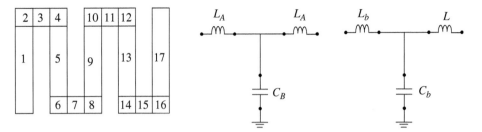

Figure 12.9 Modeling of meander line *T* network.

The lumped inductances and capacitances are calculated as

$$\text{Lumped inductance } (L_A) = \frac{Ll}{c}$$

$$\text{Lumped capacitance } (C_B) = Cl$$

$$\text{where } l = \text{length of the segment.}$$

Calculation of right-angle corners is

$$\frac{C_b}{W} \, (pF/m) = \begin{cases} \dfrac{(14\varepsilon_r + 12.5)W/h - (1.83\varepsilon_r - 2.25)}{\sqrt{W/h}} \ln\left(\dfrac{8d}{W} + \dfrac{W}{4d}\right) + \dfrac{0.02\varepsilon_r}{W/h} & \text{for } W/h < 1 \\[2ex] \dfrac{(9.5\varepsilon_r + 1.25)W}{h} + 5.2\varepsilon_r + 7.0 & \text{for } \dfrac{W}{d} \geq 1 \end{cases}$$

$$\frac{L_b}{h} \, (nH/m) = 100\left(4\sqrt{W/h} - 4.21\right)$$

where h = substrate height and w = sectionwidth.

There is a total of 17 branches for the meander line antenna prototype discussed in this report, which means there are also 17 *T* networks (Figure 12.10).

To calculate the *S* parameters for the total circuit, the *ABCD* matrices for each branch of the *T* network is calculated, as shown in Figure 12.11. The overall system <u>ABCD</u> matrix is the product of the individual *T* network branch matrices.

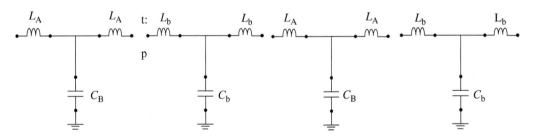

Figure 12.10 *T* network model of meander line antenna.

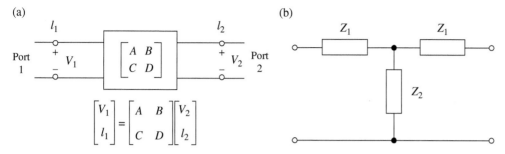

Figure 12.11 (a) *T* network *ABCD* matrix two-port model. (b) *T* network representation.

The *ABCD* matrix for a single *T* network is

$$V_1 = I_1(Z_1 + Z_2) + I_2Z_1$$
$$V_2 = I_1Z_2 + I_2(Z_1 + Z_2)$$

$$\begin{bmatrix} A & B \\ C & D \end{bmatrix} = \begin{bmatrix} 1 + \dfrac{Z_1}{Z_2} & 2Z_1 + \dfrac{Z_1^2}{Z_2} \\ \dfrac{1}{Z_2} & 1 + \dfrac{Z_1}{Z_2} \end{bmatrix}$$

The networks are then cascaded, as shown in Figure 12.12, and the overall network parameters are calculated. The *ABCD* parameters for the complete network are converted to *S* parameters.

$$S_{11} = \frac{A + B/Z_0 - CZ_0 - D}{A + B/Z_0 + CZ_0 + D}$$

$$S_{12} = \frac{2(AD - BC)}{A + B/Z_0 + CZ_0 + D}$$

$$S_{21} = \frac{2}{A + B/Z_0 + CZ_0 + D}$$

$$S_{22} = \frac{-A + B/Z_0 - CZ_0 + D}{A + B/Z_0 + CZ_0 + D}$$

Keysight Advanced Design System (ADS) electronic design automation (EDA) software is used to simulate the structure, as shown in Figure 12.13.

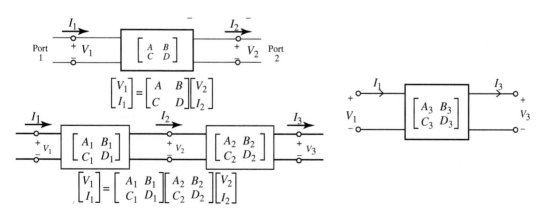

Figure 12.12 Final cascaded system.

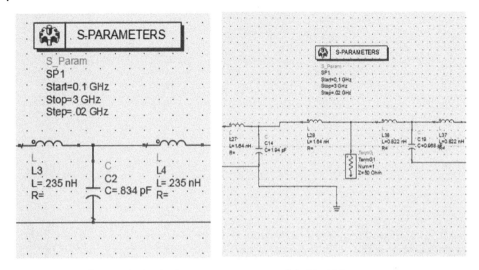

Figure 12.13 *T* network circuit of antenna prototype modeled in ADS.

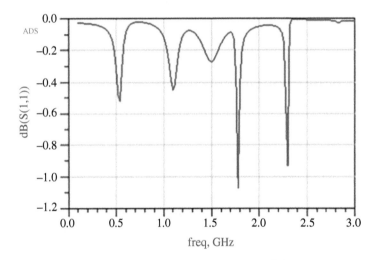

Figure 12.14 S_{11} parameter simulation result using ADS.

Figure 12.14 shows the S_{11} parameter of the antenna using the proposed lumped element equivalent model. The analytical model used in this paper is considered in free space, without the port and shorting pin. At this point, the 3D EM simulation of the antenna will be performed since the model seems to provide the desired resonant frequencies.

Three-dimensional EM Simulation

The antenna layout show in Figure 12.15 is simulated in HFSS. The antenna was simulated in free space and in a skin mimicking medium ($\varepsilon_r = 38.15$ and $\sigma = 2.27 \frac{S}{m}$) to mimic the electrical properties of human skin.

The S_{11} parameter values from HFSS for the antenna in free space are compared with the skin-mimicking medium and plotted against frequency in Figure 12.16.

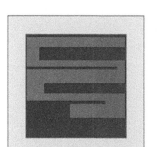

Skin-Mimicking Box
$\varepsilon_r = 38.15$

Figure 12.15 Top and side views of antenna prototype in skin-mimicking box.

Figure 12.16 S_{11} parameter values from HFSS for the antenna in free space and with the skin-mimicking medium.

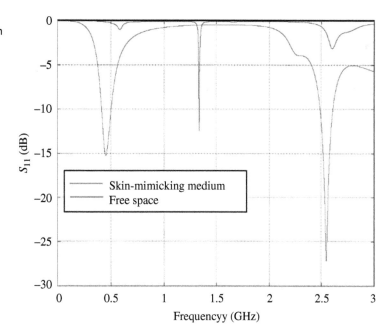

The 2D gain plots at $f = 0.4493$ GHz for the proposed antenna prototype with the skin-mimicking medium around the lower frequency are shown in Figure 12.17.

A polar plot comparison of the lower frequency gain in free space and with the skin-mimicking medium is shown in Figure 12.18.

The 2D gain plots for the proposed antenna prototype with the skin-mimicking medium around the higher frequency are shown in Figure 12.19.

A 3D polar plot at $f = 0.4493$ and $f = 2.5451$ GHz is shown in Figures 12.21 and 12.22, respectively.

A polar plot comparison of the higher frequency gain in free space and with the skin-mimicking medium is shown in Figure 12.20.

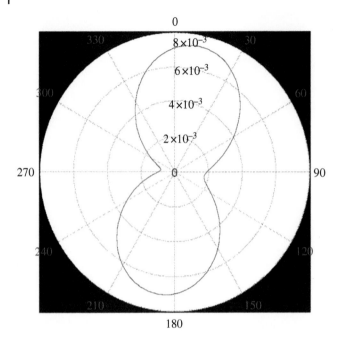

Figure 12.17 Two-dimensional polar plot of gain for implantable antenna in skin-mimicking medium at f = 0.4493 GHz.

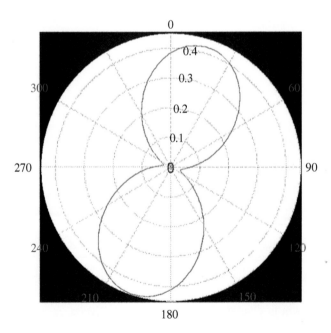

Figure 12.18 Two-dimensional polar plot comparison of gain for implantable antenna with and without skin-mimicking medium at lower frequency. Free space f = 0.5881 GHz, skin-mimicking at f = 0.4493 GHz.

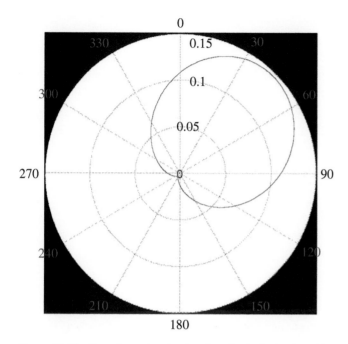

Figure 12.19   Two-dimensional polar plot of gain for implantable antenna in skin-mimicking medium at f = 2.5451 GHz.

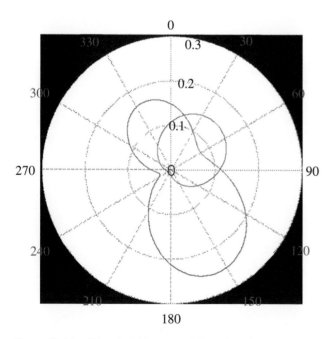

Figure 12.20   Skin-mimicking at f = 2.5441 GHz, free space at f = 2.5998 GHz.

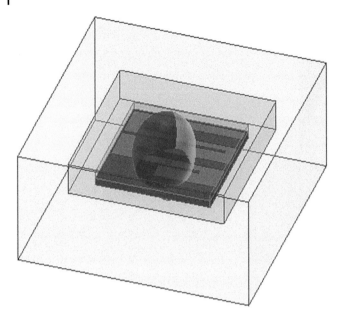

Figure 12.21 Three-dimensional polar plot of gain at f = 0.4493 GHz in skin-mimicking medium.

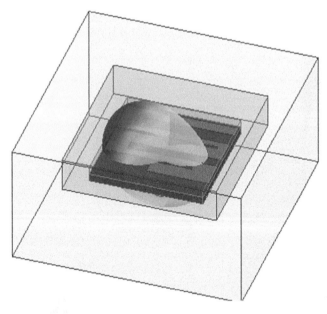

Figure 12.22 Three-dimensional polar plot of gain at f = 2.5451 GHz in skin-mimicking medium.

Prototype

The Gerber files were required to prototype antenna structure so superstrate was generated with Autodesk EAGLE EDA software according to the HFSS model. The Gerber files were imported into Bantam Tools desktop milling machine software.

The antenna prototype structure and superstrate are milled from Rogers RO3210 laminate. The material has a dielectric constant (ε_r) of 10.2 and dissipation factor $(\tan \delta)$ of 0.0027. The shorting pin is composed of a copper conductor. The port is an SMA female RF coaxial adapter/connector. The shorting pin and port were attached to the antenna prototype structure at a soldering station with as little solder and flux used as possible. The superstrate was attached to cover the antenna structure top layer with J-B Weld MinuteWeld instant setting epoxy.

The procedure for prototyping can be outlined as follows.

1) Generate Gerber files for antenna prototype structure and superstrate in Autodesk EAGLE EDA software according to HFSS model.
2) Reduce dimensions of Rogers RO3210 raw material to fit Othermill Desktop PCB Milling Machine.
3) Import Gerber files into Bantam Tools desktop milling machine software.
4) Mill antenna prototype structure and superstrate on desktop PCB milling machine.
5) Install shorting pin and port on antenna prototype structure with soldering equipment and materials, using as little soldering material as possible.
6) Cover antenna structure top layer securely with superstrate using epoxy.

The fabricated antenna prototype structures for top layer and superstrate are shown in Figure 12.23a,b.

Measurements of S_{11} parameters were performed using a Keysight E5071C ENA series network analyzer. Figure 12.24 is a plot comparing the S_{11} parameter values of the antenna prototype top layer (without superstrate installed) without and with superstrate when a shorting pin is installed, and measured in free space. The simulated and measured results are close with superstrate and shorting pin.

(a) (b)

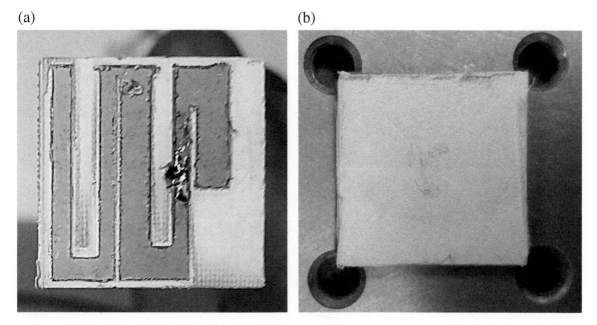

Figure 12.23 (a) Antenna patch and substrate-top layer. (b) Superstrate.

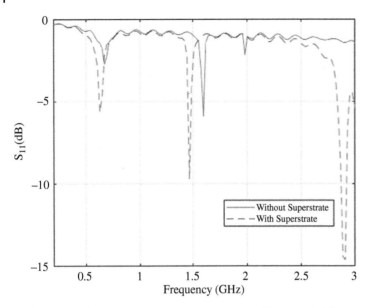

Figure 12.24 Comparison of S_{11} parameter values without and with superstrate installed.

References

1 IEEE 802.15.4-2011 (2011). *IEEE Standard for Local and metropolitan area networks: Part 15.4: Low-Rate Wireless Personal Area Networks (LR-WPANs)*. IEEE.

2 Omni-ID® Ultra (2020). An introduction to RFID. https://www.omni-id.com/pdfs/Intro-to-Radio-Frequency-Identification-Systems-and-RFID-Tags.pdf (accessed 1 November 2021).

3 A. Eroglu, *Wave Propagation and Radiation in Gyrotropic and Anisotropic Media*, 1, Springer, ISBN: 978-1-4419-6023-8 2010.

4 T. Karacolak, A. Z. Hood, and E. Topsakal, "Design of a dual-band implantable antenna and development of skin mimicking gels for continuous glucose monitoring," *IEEE Transactions on Microwave Theory and Techniques*. Vol. **56**, No. 4. 2008. 1001–1008.

5 IEEE 802.15.1-2005 (2005). *IEEE Standard for Information technology: Local and metropolitan area networks: Specific requirements: Part 15.1a: Wireless Medium Access Control (MAC) and Physical Layer (PHY) specifications for Wireless Personal Area Networks (WPAN)*. IEEE.

6 Chauhan, A., Chauhan, G.K., and Kaur, G. (2015). Implantable antennas in biomedical applications. *2015 International Conference on Computational Intelligence and Communication Networks (CICN)* (December 2015).

7 Anderson, J.M. and Langone, J.J. (1999). Issues and perspectives on the biocompatibility and Immunotoxicity evaluation of implanted controlled release systems. *Journal of Controlled Release* **57** (2): 107–113.

8 Kiourti, A. and Nikita, K.S. (2021). A review of implantable patch antennas for biomedical telemetry: challenges and solutions. *IEEE Antennas and Propagation Magazine* **54** (3): 210–228.

9 Bhattacharjee, S., Maity, S., Metya, S.K., and Tilak Bhunia, C. (2016). Performance enhancement of implantable medical antenna using differential feed technique. *Engineering Science and Technology, an International Journal* **19** https://doi.org/10.1016/j.jestch.2015.09.001.

10 Das, A., Dhar, S., and Gupta, B. (2011). Lumped circuit model analysis of meander line antennas. *11th Mediterranean Microwave Symposium (MMS)* (8–11 September 2011) https://doi.org/10.1109/MMS.2011.6068520

13

Energy Harvesting and HVAC Systems with RF Signals

13.1 Introduction

Energy harvesting is a process by which energy found in external sources (solar, thermal, wind, electromagnetics) is converted into usable electrical energy. This energy can be conditioned for either immediate use or processed and stored for future use. Energy harvesting provides an alternative source of power for applications intended in the locations where there is no access to grid power, and it is inefficient to place wind turbines or solar panels. Even if the harvested energy is low and inadequate for powering a device, it can still help to extend the life of a battery and so guarantees the continuous operation of a single or network of devices. There exist various sources to extract or scavenge energy from, including solar and wind powers, electromagnetics, ocean waves, piezoelectricity, thermo-electricity, and physical motions, as shown in Figure 13.1 [1].

Among the available energy harvesting sources, energy harvesting from ambient electromagnetic waves has been a good choice for researchers [2]. It relies on the energy of electromagnetic or RF signals transmitted by various communication systems, e.g. TV, radio, and cellular systems, radar, etc. ABI Research and iSupply estimate the number of mobile phone subscriptions has recently surpassed five billion, and the ITU estimates there are over one billion subscriptions for mobile broadband [3]. Mobile phones serve as RF transmitters or sources from which to harvest RF energy. Therefore, this will potentially allow users to provide power on demand for a variety of close-range sensing applications.

Heating, ventilation, and air conditioning (HVAC) systems are present in almost every residential and commercial places. A typical HVAC system without control electronics is illustrated in Figure 13.2. Application of wireless communication within a HVAC system is getting a lot of attention. A home HVAC wireless sensor network (WSN) could provide information to the user on status of the system, energy use of the system, potential maintenance problems within the system, and a variety of other information. Research shows that implementing a WSN to monitor and control HVAC systems can result in overall energy savings [4–6].

13.2 RF Energy Harvesting

Radio frequency energy harvesting (RFEH) harvests the electromagnetic energy in the ambient. Many RF circuit designers target a single frequency band, as this results in a more feasible antenna design as well as reducing the complexity of impedance matching between the antenna and the rectification circuitry. There are several examples

RF/Microwave Engineering and Applications in Energy Systems, First Edition. Abdullah Eroglu.
© 2022 John Wiley & Sons Ltd. Published 2022 by John Wiley & Sons Ltd.
Companion website: www.wiley.com/go/eroglu/rfmicrowave

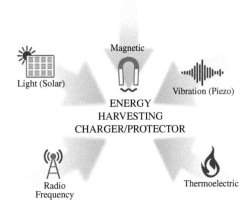

Figure 13.1 Sources of energy harvesting [1].

of RFEH in the literature. RFEH designed in [7] harvests energy from the ambient surroundings at the GSM-900 band, where GSM stands for global system for mobile communications and the 900 represents the 900 MHz frequency. The design provides an alternative source of energy for energizing low power devices, and it consists of three circuits: a single wideband 377 Ω E-shaped patch antenna, a Pi matching network and a seven-stage voltage doubler circuit. These three circuits were integrated and fabricated on a double-sided FR4 printed circuit board (PCB). The patch antenna and Pi matching network have been optimized through the electromagnetic simulation software Agilent Advanced Design System (ADS) 2009 environment. The proposed RFEH had produced a 2.9 Vdc at an approximate distance of 50 m from GSM cell tower, and it was enough to power the STLM20 temperature sensor.

Researchers reported energy harvester design using typical ambient RF power levels found within urban and semi-urban environments in [8]. A city-wide RF spectral survey was undertaken from outside all the 270 London Underground stations at street level. Using the results from the survey, researchers in [8] designed four narrow-band RFEHs (comprising antenna, impedance-matching network, rectifier, maximum power point tracking [MPPT] interface, and storage element) with each one designed just to cover a single frequency band from the largest RF contributors (digital television, GSM900, GSM1800, and 3G) within the ultrahigh frequency spectrum (0.3–3 GHz).

The system overview of RFEH system is given in Figure 13.3.

Electromagnetic or RF signals are transmitted by various communication systems, e.g. TV, radio and cellular systems, radar, etc. Those energy signals are available in ambient. An antenna will capture signal and feed it through an impedance (Z) matching network, which would guarantee a maximum transfer of energy between the antenna and the rectifier. The output from an antenna is a very low alternating current (AC) signal, which would then need to be converted to a DC signal with rectifier.

The RF–DC conversion efficiency (η) is given by:

$$\eta = \frac{P_{DC}}{P_{RF}} \tag{13.1}$$

where P_{DC} is the DC output power of the rectifier and P_{RF} is the RF input power to the rectifier. The rectifier's output voltage is low. Therefore, a booster needs to be used to increase the voltage level, and then store the energy in a supercapacitor for later use or feed directly to a sensor load or any other type of load depending on the application.

13.3   RF Energy Harvesting System Design for Dual Band Operation

As described in Section 13.2, there are several components in RF Energy Harvester Systems which are illustrated in Figure 13.3. In this section, these components will be detailed and design of matching network, rectifier, and booster for a dual band RF energy harvester operating at 900 MHz and 2.4 GHz will be discussed.

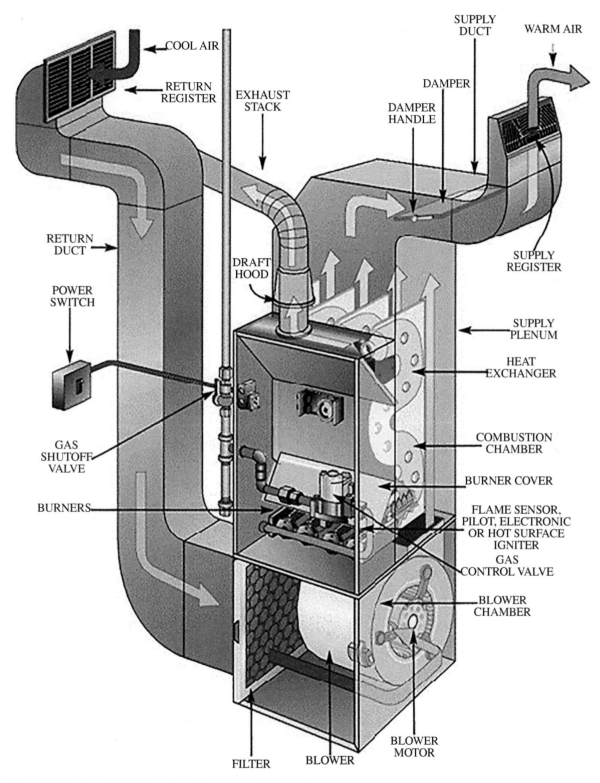

Figure 13.2 Standard HVAC system without control electronics.

Energy Harvester

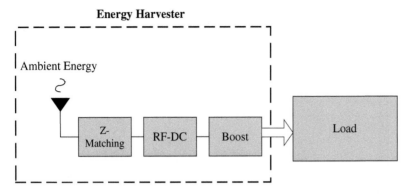

Figure 13.3 System overview of radiofrequency energy harvesting system.

SOURCE   LOAD

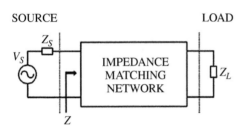

Figure 13.4 Impedance matching network implementation for antenna

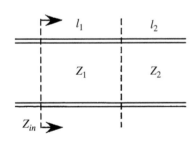

Figure 13.5 Two-section dual band transformer.

13.3.1 Matching Network for Energy Harvester

An impedance matching network ensures maximum energy transfer from the antenna to the load, as shown in Figure 13.4.

The design of the matching network will be done based on a two-section dual band transformer. Impedance transformers include lumped networks, tapered sections, shorting plungers, and double-stub matching. A quarter-wave impedance transformer, often referred to as a $\lambda/4$ transformer, is one of the most common types of transmission lines to match impedance at a single frequency. The RFEH system design proposed in this section operates at 900 MHz–2.4 GHz and must be matched to the antenna port at these two frequencies. A dual band two-section 1/3-wavelength transformer that works at the fundamental frequency (f_1) and its second harmonic ($2f_1$) is proposed in [9]. However, the transformer's design in [9] does not work for any two arbitrary frequencies. Monzon [10] reports a method to transform the transformer in [9] to work for any two arbitrary frequencies. A two-section dual band transformer is shown in Figure 13.5

The input impedance of the two-section TL line shown in Figure 13.5 is given by

$$Z_{in} = Z_1 \frac{Z_l' + jZ_1 \tan(\beta l_1)}{Z_1 + j Z_l' \tan(\beta l_1)} \tag{13.2}$$

where

$$Z_l' = Z_2 \frac{R_l + jZ_2 \tan(\beta l_2)}{Z_2 + jR_l \tan(\beta l_2)} \tag{13.3}$$

From (13.2) and (13.3), we obtain

$$Z_l' = Z_1 \frac{Z_0 - jZ_1 \tan(\beta l_1)}{Z_1 - jZ_0 \tan(\beta l_1)} \tag{13.4}$$

(13.3) and (13.4) lead to

$$\left(Z_1{}^2 R_l - Z_2{}^2 Z_0\right)\tan\left(\beta l_1\right)\tan\left(\beta l_2\right) = Z_1 Z_2 (R_l - Z_0) \tag{13.5}$$

$$\left(Z_0 Z_2 R_l - Z^2{}_1 Z_2\right)\tan\left(\beta l_1\right) = Z_1 \left(Z_2{}^2 - Z_0 R_l\right)\tan\left(\beta l_2\right) \tag{13.6}$$

From (13.5) to (13.6).

$$\tan\left(\beta l_1\right)\tan\left(\beta l_2\right) = \alpha = \frac{Z_1 Z_2 (R_l - Z_0)}{Z_1{}^2 R_l - Z_2{}^2 Z_0} \tag{13.7}$$

$$\frac{\tan\left(\beta l_1\right)}{\tan\left(\beta l_2\right)} = \gamma = \frac{Z_1 \left(Z_2{}^2 - Z_0 R_l\right)}{Z_2 \left(Z_0 R_l - Z_1{}^2\right)} \tag{13.8}$$

α and γ parameters are defined from (13.7) to (13.8) as

$$\left(\tan\left(\beta l_1\right)\right)^2 = \alpha\gamma \tag{13.9}$$

$$\left(\tan\left(\beta l_2\right)\right)^2 = \frac{\alpha}{\gamma} \tag{13.10}$$

Equations (3.9) and (3.10), when applied to frequencies f_1 and f_2, result in the following four equations.

$$\left(\tan\left(\beta_1 l_1\right)\right)^2 = \alpha\gamma \tag{13.11}$$

$$\left(\tan\left(\beta_2 l_1\right)\right)^2 = \alpha\gamma \tag{13.12}$$

$$\left(\tan\left(\beta_1 l_2\right)\right)^2 = \frac{\alpha}{\gamma} \tag{13.13}$$

$$\left(\tan\left(\beta_2 l_2\right)\right)^2 = \frac{\alpha}{\gamma} \tag{13.14}$$

From (13.11) to (13.12), we can obtain

$$\tan\left(\beta_1 l_1\right) = \pm\tan\left(\beta_2 l_1\right) \tag{13.15}$$

which leads to

$$\beta_1 l_1 \pm \beta_2 l_1 = n\pi \tag{13.16}$$

Similarly, from (3.13) to (3.14) we obtain

$$\beta_2 l_2 \pm \beta_1 l_2 = m\pi \tag{13.17}$$

m in (3.17) is an arbitrary integer number. Since we are interested in a small transformer, we can choose the + sign in (3.16) and (3.17), with $m = n = 1$ (we assume that $f_2 \geq f_1$). This leads to

$$l_1 = l_2 \tag{13.18}$$

$$l_2 = \frac{\pi}{\beta_1 + \beta_2} \tag{13.19}$$

With the transmission line lengths known, we can determine α *and* γ from (13.7) to (13.8) at either f_1 or f_2. For instance, for f_1 using (3.8) and (3.18), we obtain

$$\frac{\tan\beta_1 l_1}{\tan\beta_1 l_2} = \gamma = 1 \tag{13.20}$$

From (13.11), we obtain:

$$(\tan (\beta_1 l_1))^2 = \alpha \tag{13.21}$$

Substitute $\gamma = 1$ in Eq. (13.8), we obtain

$$Z_1\left(Z_2{}^2 - Z_0 R_l\right) = Z_2\left(Z_0 R_l - Z_1{}^2\right) \tag{13.22}$$

which leads to

$$Z_l' = Z_2 \frac{R_l + jZ_2 \tan (\beta l_2)}{Z_2 + jR_l \tan (\beta l_2)} \tag{13.23}$$

Equation in (13.23) is valid under $(Z_1 + Z_2 \neq 0)$. By substituting Eq. (13.23) into Eq. (13.7), we obtain

$$\frac{Z_0 R_l (R_l - Z_0)}{Z_1{}^2 R_l - Z_2{}^2 Z_0} = \alpha \tag{13.24}$$

Which can be rewritten as

$$Z_1{}^2 R_l - Z_2{}^2 Z_0 = \frac{Z_0 R_l (R_l - Z_0)}{\alpha} \tag{13.25}$$

Solving (13.25) for Z_1 (impedance of the first section of the TL) gives

$$Z_1 = \sqrt{\frac{Z_0}{2\alpha}(R_l - Z_0) + \sqrt{\frac{Z_0}{2\alpha}(R_l - Z_0)^2 + Z_0{}^3 R_l}} \tag{13.26}$$

From (13.23), we can obtain the impedance of the second section of the transmission line as

$$Z_2 = \frac{Z_0 R_l}{Z_1} \tag{13.27}$$

Total length of the transmission line is found from (13.16) to (13.17) by adding them as

$$\beta_1 (l_1 + l_2) + \beta_2 (l_1 + l_2) = 2\pi \tag{13.28}$$

which can be simplified to

$$l_1 + l_2 = \frac{\lambda_1 \beta_1}{\beta_1 + \beta_2} \tag{13.29}$$

13.3.2 RF–DC Conversion for Energy Harvester

The first element in the RFEH system is an antenna, which captures electromagnetic waves in the ambient. The output signal from an antenna (the RF signal) needs to be converted to a DC voltage. The conversion from RF to DC is done by a rectifier circuit.

13.3.3 Clamper and Peak Detector Circuits

A clamper circuit is an electronic circuit that can change the DC level of a signal to the desired level without modifying the shape of the applied signal. Generally, the clamper circuit shifts the whole signal up or down to set either the positive or the negative peak of the applied signal to the desired level. The DC component is simply added or subtracted from the applied or input signal. A positive clamper circuit, shown in Figure 13.6, adds the positive DC component to the input signal to shift it to the positive side. Therefore, the negative peak will coincide with the zero

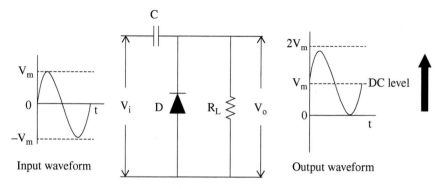

Figure 13.6 Positive clamper circuit.

level. Similarly, a negative clamper circuit adds the negative DC component to the input signal to shift it to the negative side, and so the positive peak will coincide with the zero level. The construction of the clipper circuit is almost similar to the clamper circuit. The only difference is that the clamper circuit has an extra added element (a capacitor), which is used to provide a DC offset from the stored charge.

The clamping voltage in Figure 13.6 is equal to

$$V_{clamp} = 2V_p - V_{th} \tag{13.30}$$

To generate a steady voltage from the rectifier, the peak detector circuit shown in Figure 13.7 is interfaced with the clamper circuit. Generally, the clamper circuit clamps the waveform either in a positive or negative direction using a diode with a capacitively coupled signal. The capacitor stores the peak voltage and the diode will prevent the capacitor from discharging. In the desired peak detector circuit, a diode with a low threshold voltage is needed. A voltage doubler rectifier circuit rectifies the full-wave peak-to-peak voltage. The output of the peak detector can be expressed by

$$V_{PD} = V_{clamp} - V_{th} \tag{13.31}$$

To improve the conversion efficiency of the RFEH system, a voltage doubler rectifier circuit is beneficial since multiple stages can be cascaded to increase the output voltage.

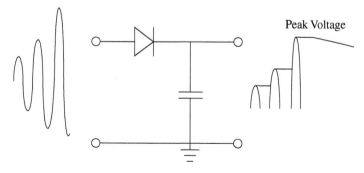

Figure 13.7 Peak detector circuit.

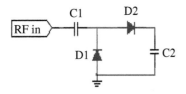

Figure 13.8 Single stage cascaded rectifier.

13.3.4 Cascaded Rectifier

Clamper and peak detector circuits can be cascaded to form a single-stage rectifier, which can boost the DC output voltage, so each individual stage can be said to act as a passive voltage amplifier. Multiple stages can then be added to reach the required level of output voltage. Figure 13.8 shows a single-stage cascaded rectifier.

In Figure 13.8, C1 and D1 form a clamper circuit. C2 and D2 form a peak detector circuit. Once the RF wave is captured, C1 starts building up charge. The charge stored in capacitor C1 clamps output across D1. The clamped output voltage is calculated from Eq. (13.30). Then, the output of each individual stage is clamped to a higher value by a subsequent stage. In an ideal condition, without losses or voltage drop, output would be doubled at each stage.

13.3.5 Villard Voltage Multiplier

A single-stage Villard voltage multiplier circuit consists of two diodes and two capacitors, as shown in Figure 13.9. Ideally, the output voltage of this circuit is double its input voltage.

The operation of a Villard voltage doubler can be explained by assuming that all diodes are ideal and capacitor values are high. So, the voltage across capacitors won't change significantly in a short amount of time during charging and discharging. During the first negative half cycle, the voltage at the anode of D1 is larger than its cathode voltage and current flows through D1 and charges up C1. By the end of the first cycle of the input voltage, exactly at the negative peak, $V_{C1} = V_m$ where $V_{in} = V_m$. When the second cycle starts, the anode voltage of D1 starts decreasing while the cathode voltage increases. So, D1 will be reversed biased and zero current flows through it. Hence, D2 is forward biased and C2 will start charging up from what we can call a dual source (V_{in} and C1). At the peak voltage of the second cycle, $V_{C2} = 2V_m$, and the same process continues for the next cycles. The output voltage can be increased to the required level by cascading multiple stages into the Villard voltage multiplier circuit. In the ideal situation, by adding N stages, the voltage across the output capacitor C2 will be $V_{out} = 2V_m * N$. However, this is not possible, owing to nonavoidable loss in the circuit and nonideal diode characteristics. By considering the effect of diodes threshold (V_{th}), Karthaus and Fischer [11] show that the output voltage of an N-stage Villard voltage rectifier can be expressed by

$$V_{out} = N(V_m - V_{th}) \tag{13.32}$$

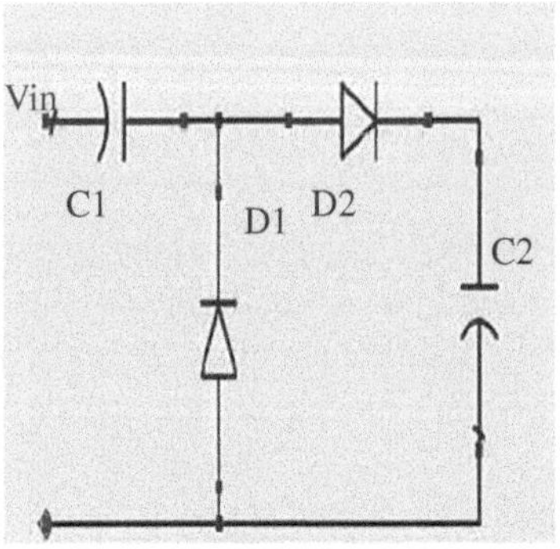

13.3.6 RF–DC Rectifier Stages

As discussed in Section 0, as the number of stages of a rectifier increases, the output voltage will increase. However, as the number of stages increases, the number of diodes and other components also increases, resulting in an increase of threshold (V_{th}) of the rectifier circuit. This will require a higher voltage at the rectifier's input for the circuit to operate. Another drawback of adding multiple stages is that they diminish the effect of the quality factor.

In real life, RF energy in the ambient is low, and so the input voltage at the rectifier will also be low. Owing to the diode threshold voltage, there will be a voltage drop across diodes and other circuit components, so residual resistance increases, which makes it a low power circuit. Therefore, after a certain number of stages, the overall output voltage of the rectifier will deteriorate. An RF–DC rectifier (one to eight stages) is simulated by ADS, with the transient analysis simulation controller, to examine the output voltage and the

Figure 13.9 Single stage Villard voltage doubler.

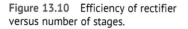
Figure 13.10 Efficiency of rectifier versus number of stages.

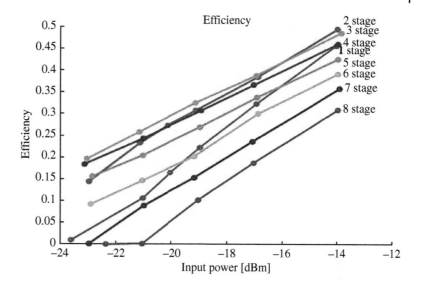

efficiency versus number stages. Since the power available in the ambient is low (−30 to −20 dBm) in real life, so the input power can be swept within that range. The efficiency of the rectifier with varying stages is shown in Figure 13.10. From Figure 13.10, it is evident that the three-stage system has a higher output voltage and efficiency compared to others.

13.4 Diode Threshold V_{th} Cancellation

In conventional RF–DC rectifiers, power conversion efficiency ($\eta = \frac{P_{DC}}{P_{RF}}$) is limited by reverse leakage in the diodes that are used for the RF to DC conversion. The diodes also have a fixed turn-on voltage that creates a dead zone, where the received power is wasted as the generated voltage will be too low to power up [12]. Several techniques can be used to reduce the turn-on voltage for diodes, including Schottky diodes (titanium silicon), low V_{th} transistors, and dual-poly floating gate transistors [12]. This section outlines some of the commonly used techniques to reduce the threshold voltage, known as V_{th} cancelation.

13.4.1 Internal V_{th} Cancellation

In internal V_{th} cancelation (IVC), the DC output voltage from the rectifier is used to bias the gates so that the effective threshold voltage is reduced. The IVC circuit normally operates under large input power conditions. It also relatively consumes extra power, which degrades efficiency. Figure 13.11 illustrates the IVC circuit [12].

The diode-connected MOS transistor M2 generates the bias voltage V_b, which decreases the threshold voltage of the diode-connected PMOS transistor M1. Thus, the difference in gate bias between M1 and M2 determines the diode voltage, V_D. By increasing V_b, the effective threshold of M1is decreased. If the effective threshold becomes too small, the reverse leakage of the diode rapidly becomes excessively large. Based on a theoretical model for ITC MOS diodes [10], the diode voltage V_D is given by

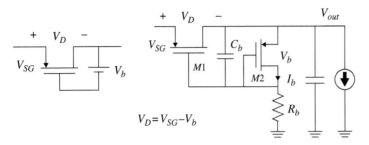

Figure 13.11 Diode connected PMOS transistor with internal threshold cancellation.

$$V_D \approx \sqrt{\frac{2\pi}{\beta_1} I_{out}} + \eta U_T \ln \frac{I_S}{I_b}, \text{where } I_S > 10\, I_b \tag{13.33}$$

In (13.33), $\beta = \mu_P C_{ox} \frac{W}{L}$ is the gain factor, $I_S = 2n\beta_2 U_T^2$ is the specific current, and U_T is the thermal voltage at the room temperature, which is about 26mV. The expression for the reverse leakage I_L is given by

$$I_L = I_b \frac{W_1}{W_2} \approx \frac{\hat{V}_{in} - V_{SG1}}{R_b} \frac{W_1}{W_2} \tag{13.34}$$

In Eq. (13.34), I_b is the bias current of M2. W_1 and W_2 indicate the widths of M1 and M2, respectively. According to Eqs. (13.33) and (13.24), if the bias current increased, the diode voltage will decrease, and the reverse leakage current will increase. Apparently, this would cause excessive losses in the ITC diode, and lead to low power conversion efficiency. To avoid that, transistor M2 should be biased so that $V_b < V_{th}$ to partially cancel the threshold voltage of M1. Typically, this would limit the voltage drop of MOS diodes to 200–300 mV with a power conversion efficiency of above 60% [13].

13.4.2 External V_{th} Cancellation

In the external V_{th} cancelation (EVC), a switched-capacitor circuit generates a gate bias voltage. Since the MOSFET does the switching, an external power supply and clock circuitry are required. Figure 13.12 illustrates the EVC circuit [14].

In Figure 13.12

$$V_{G1} = V_m + V_b \tag{13.35}$$

$$V_{out} = 2(V_{GS,max} - V_T) = 2(V_m + V_b - V_{th}) \tag{13.36}$$

In Eq. (13.36), when $V_b = V_{th}$; $V_{out} = 2\,V_m$, it would eliminate the efficiency loss caused by the threshold voltage of the MOSFET. An external voltage source and switched capacitor array can be used to obtain the gate bias voltage V_b for all the transistors [14].

13.4.3 Self-V_{th} Cancellation

The self-V_{th} cancelation (SVC) has an equivalent topology to the diode-connected complementary metal oxide semiconductor (CMOS) rectifier circuit except for gates of the N-channel metal oxide semiconductor (NMOS) and P-

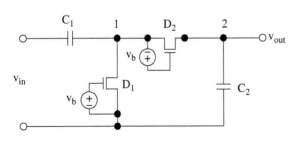

Figure 13.12 Voltage multiplier with external threshold voltage cancellation

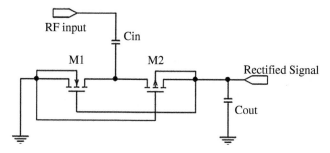

Figure 13.13 Self V_{th} cancellation with CMOS rectifier circuit.

channel metal oxide semiconductor (PMOS) transistors are connected to the output and ground terminal. This configuration boosts the gate source voltage of NMOS and PMOS transistors, and with the decrease of DC output voltage, threshold voltages are also decreased. Therefore, SVC provides better efficiency than EVC and IVC topology [15]. Figure 13.13 illustrates the SVC CMOS rectifier circuit.

EVC and IVC systems use switched capacitor mechanisms to generate the gate bias voltage, while the SVC system is much simpler and does not require an external power, hence it provides better efficiency. In addition, the SVC system can achieve the best V_{th} cancellation efficiency at low DC output voltage and RF input power levels.

13.5 HVAC Systems

Remote control and monitoring of HVAC systems carries great importance for manufacturers and homeowners because of several factors, including energy efficiency, cost, maintenance, reliability, and on time service. The main component under consideration is the electrical machine used to force air throughout the system. Parameters of the electrical motor that must be sensed are the torque, speed, and temperature of the motor. Sensors are used to measure these data which are then sent wirelessly from the motor to a separate wireless receiver, as shown in Figure 13.14. HVAC systems are currently not using wireless communication system. Certain aftermarket thermostats are capable of wireless communication, but the format and communication protocol used is not yet standardized. Specific research into a home sensor network (HSN) or integrated smart house may be useful for guidance in protocol selection [17–19]. Most research is focused on implementing home networks using only ZigBee and IEEE 802.15.4 [20], while some is focused on implementing 6LoWPAN home networks [21, 22].

A wireless thermostat provides capabilities to monitor and control the HVAC system, but it does not provide motor information or motor operation information to the user. An HVAC motor capable of wireless communication could cover deficiencies in the current system, and reduce the control wiring required. A few sample use cases include: wireless programming or reprogramming of the motor, tracking, and reporting maintenance needs to the user, monitoring energy consumption for smart

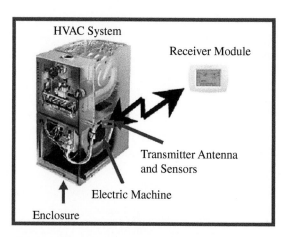

Figure 13.14 Wireless sensors and antennas implementation for HVAC systems [16].

energy, and implementing a WSN to monitor a complete HVAC system operation and to minimize energy consumption.

Implementing a wireless motor and WSN in an HVAC system has a variety of unique challenges. First, the propagation of signal in a highly metallic environment is difficult. The HVAC enclosure is wholly metallic and works like a faraday cage by blocking most RF waves in and out of the system. Signals that do penetrate into the enclosure often have multipath reflections causing destructive interference. Although some research has been done to model HVAC ventilation ducts for high speed Internet capabilities [23–25], this has yet to be expanded for use in providing easier wireless communication for HSNs or HVAC systems.

13.6 Engineering Application Examples

In this section, design, simulation, and implementation of complete energy harvesting system and energy harvesting antenna are given and discussed.

Design Example 13.1   Dual Band Energy Harvesting System
Design, simulate, and implement a dual band energy harvesting system operating at 900 MHz and 2.4 GHz. Assume the ambient energy is between −20 and − 30 dBm. Include the design, simulation, and implementation of all the components illustrated in Figure 13.3.

Solution
Design and Simulation
In the design of the energy harvester, the reference input was taken to be −20 dBm. As discussed in Section 13.3.6, three stages of rectifier configuration is more efficient for the desired energy harvester circuit. Figure 13.15 represents a three-stages dual band rectifier for the proposed RFEH. The excitation port has an impedance of 50 Ω, and operating at frequencies of interest (900 MHz–2.4 GHz). A simulation has been carried out individually at each frequency.

Figure 13.16a shows the simulated output of a three-stage rectifier with a −20 dBm input signal. It can be seen that as the number of stages increases the output voltage also increases. Each curve on the graph shows the output

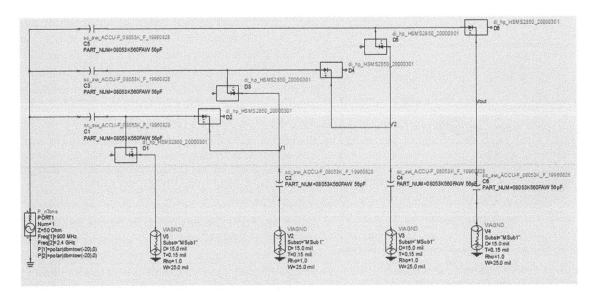

Figure 13.15   RF–DC rectifier based on Dickson's topology.

(a) 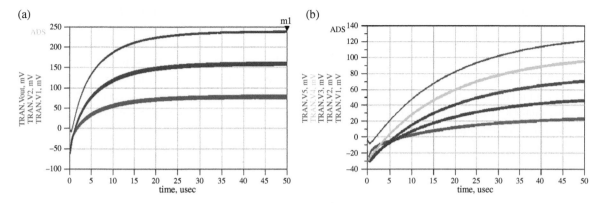 (b)

Figure 13.16 (a) Three stages rectifier. (b) Five stages rectifier output in transient simulation for RF_{in} = −20 dBm.

Figure 13.17 Transient analysis of three stages rectifier output voltage.

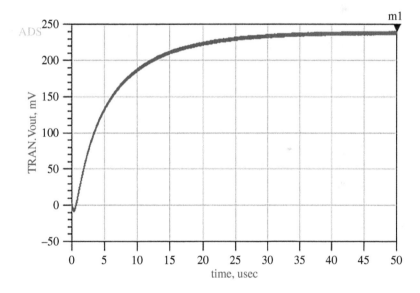

voltage form each stage. Figure 13.16b shows the output voltage for a five-stage rectifier. As expected, the output voltage is lower than the three-stage rectifier circuits owing to nonlinear characteristics and the impact of the additive threshold voltage.

Figure 13.17 shows the transient response of the third stage of the rectifier circuit output voltage as a standalone output voltage.

DC–DC Booster Circuit Design

Two DC–DC booster circuits have been designed to interface with the rectifier in order to boost its output voltage. Two different types of booster circuits should be designed and simulated.

Self-resonant transformer-based DC–DC boost converter

Self-resonant transformer-based DC–DC boost converter is a step-up DC–DC converter that is driven via junction-gate field-effect transistor (JFET), self-starting circuit which does not require an external oscillator to drive the field-effect transistor (FET) transistor. The simulation for this design has been carried out by ADS software. A transformer LPR6235 from Coilcraft, two J177 JFETs, and a set of different capacitor sizes has been used to simulate the boost converter. Figure 13.18 shows the booster circuit simulated by ADS software.

Figure 13.19 illustrates the simulation result of the DC–DC boost converter. The DC output voltage from the booster circuit is 1.88 V based on the simulation result. It is worth mentioning that the transformer used in the simulation is an ideal transformer with inductances added for the primary and secondary side. Also, a mutual inductance factor $K = 0.95$ is counted into the simulation based on the specification sheet of the transformer from Coilcraft LPR6235 transformer. A crucial factor in the design is the type of the JFET used. In the design, J177 is selected and it has a very low turn-on voltage (about 50 mV) and very low drain-gate and source-gate capacitances ($C_{gd} = C_{gs} = 5.5\ PF$).

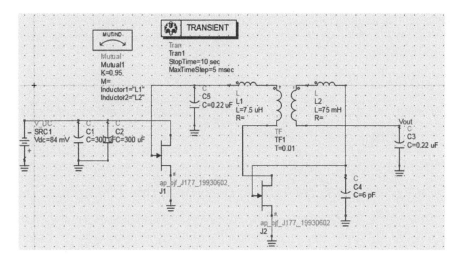

Figure 13.18 Self-resonant transformer-based DC–DC boost converter circuit simulated by ADS.

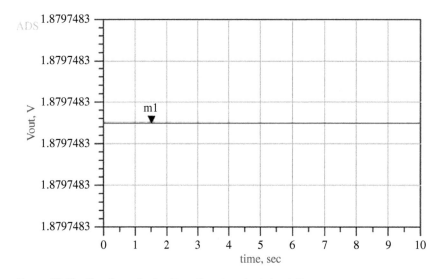

Figure 13.19 Booster output voltage from transient simulation.

Switch mode (MOSFET-based) DC–DC boost converter

This design utilizes an N channel MOSFET which is driven via an external oscillator (1 Vpp, $f = 3$ MHz, $D = 25\%$), where D represents the duty cycle of pulse signal. A dual gate MOSFET NTE222 from NTE electronics, 100 µH inductor RFC1010B-104KE from Coilcraft, diode, and 10 µF capacitor have been used to carry out this design for a booster converter. Figure 13.20 shows the circuit simulated by ADS.

Prototype

The charge pump rectifier is fabricated into an RF-35 board with a 3.5 dielectric constant. Other components include Schottky diodes HSMS2850 from Broadcom Inc. and 56 pF capacitors from Wurth Electronics. Figure 13.21 shows the top and bottom view of the rectifier circuit.

The rectifier was fed separately at 900 MHz and −20 dBm input power from a vector signal generator (Agilent E4438C), and the output was measured using a digital multimeter (HP 34401A). Figure 13.22 shows the actual DC output voltage from the rectifier (87 mV).

A self-resonant transformer-based DC–DC boost converter was fed separately with 84 mV from a DC power supply to check its output voltage before interfacing it with the rectifier circuit. Figure 13.23 shows the output voltage of the self-resonant boost converter (2.638 V).

Figure 13.24 shows the output voltage of the self-resonant transformer-based DC–DC boost converter when interfaced with the rectifier (0.8 V).

After testing subsections in the energy harvester, we integrated all sections, as shown in Figure 13.25 for a complete testing.

The RF signal is generated by a signal generator (Agilent N9310A, 9 KHz–3.0 GHz) at the frequencies of interest (900 MHz–2.4 GHz), and then transmitted by an Altelix AL0727G12-NF antenna which is placed at 34 in. apart from the receiving antenna (Altelix AL0727G12-NF) to provide −20 dBm RF signal in the ambient. The output of Altelix AL0727G12-NF antenna is connected to the input of the rectifier. Figure 13.26 shows antennas set up for the measurement of the energy harvester system.

The output of the receiving antenna (input to the rectifier) is measured at frequencies of interest with a signal analyzer (Agilent N9000A, 9 kHz–7.5 GHz) and an oscilloscope (Keysight MSOX3054A) before it's connected with the rectifier. Figures 13.27a,b show measurements for the input signal at the rectifier at 900 MHz.

Figures 13.28a,b show the same measurements for the input signal to the rectifier at 2.4 GHz.

The output of the rectifier is connected to the switch mode (MOSFET based) DC–DC boost converter. Figure 13.29 shows the final output of the boost converter (7.42 V), which was measured by a digital multimeter (Agilent 34401A).

Figure 13.20 Switch mode DC–DC boost converter using N channel MOSFET.

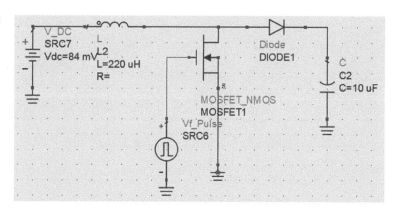

(a)

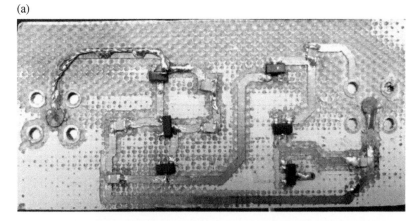

Figure 13.21   (a) Top and (b) bottom view of the rectifier.

(b)

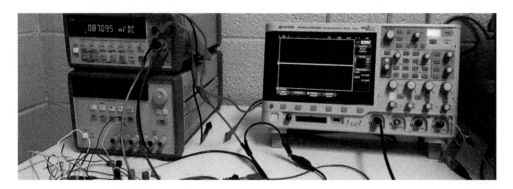

Figure 13.22   RF–DC rectifier output voltage.

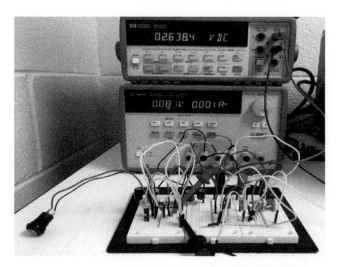

Figure 13.23 Self-resonant transformer-based DC–DC boost converter output voltage.

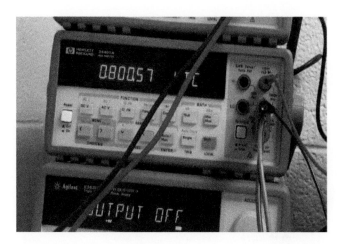

Figure 13.24 Output voltage of the self-resonant transformer based booster when interfaced with the rectifier.

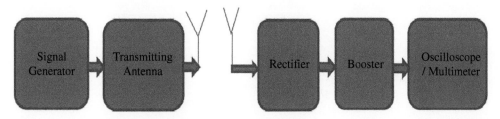

Figure 13.25 Block diagram of the RFEH output measurement.

Figure 13.26 Energy harvester test setup.

(a)

(b)

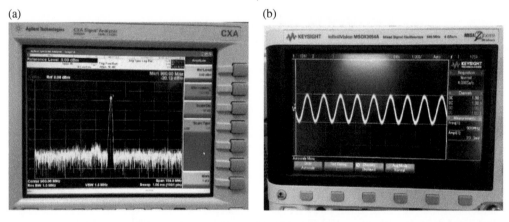

Figure 13.27 Signals at the input of the rectifier at 900 MHz: (a) RF power in dBm; (b) measured voltage.

(a)

(b)

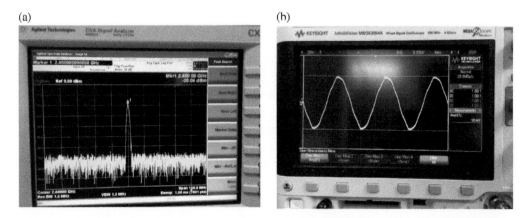

Figure 13.28 Signals at the input of the rectifier at 2.4 GHz: (a) RF power in dBm; (b) measured voltage.

Figure 13.29 Final measured output voltage of the RFEH by DMM.

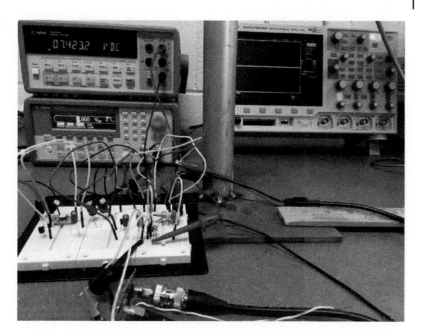

Design Example 13.2   Dual Band Antenna Design for Energy Harvesters

Design, simulate, and implement a dual band meandered printed dipole antenna for RF energy harvesting applications operating at around 900 MHz to 1.8 GHz.

Solution

Analytical

Meandering is a technique used to compress the length of the dipole antenna to fit designs into smaller areas; this technique becomes specifically critical at lower frequencies, which normally require larger antennas.

There are several key dimensions of the meanders that will affect the resonant frequency of the antenna:

- Number of meanders.
- Line height of the meander (h).
- Ratio of the meander width to the conducting line's diameter (w/b).
- Total conducting line length (s).

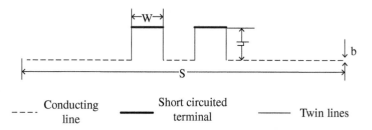

Through simulation and measurement, it is that there's an inverse relationship with the parameters above and the resonant frequency. The proposed design following the guidelines in [12] to have resonances at 900 MHz to 2.4 GHz is shown in Figure 13.30. Figure 13.30a shows the dimensions and Figure 13.30b illustrates the layout simulated.

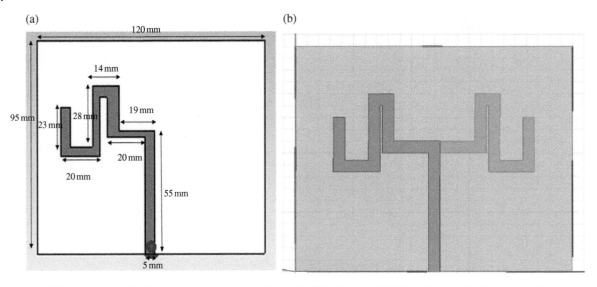

Figure 13.30 Proposed dual band meander antenna design for 900 MHz and 2.4 GHz: (a) physical dimensions; (b) simulated layout.

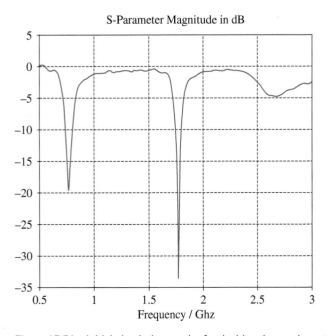

Figure 13.31 Initial simulation results for dual band meander antenna.

The HFSS simulation results in Figure 13.31 show that there was a shift in the desired resonant frequencies. The resonant frequency and return loss obtained were $f_1 = 795$ MHz (-20 dB) and $f_2 = 1795$ MHz (-34.55 dB). Hence, further improvement is needed. For this, the lumped element equivalent model has been studied. Each meander section can be modeled as a series inductor and a shunt capacitor T network, as shown in Figure 13.32.

Figure 13.32 (a) Meander sections. (b) *T* network model for a meander section.

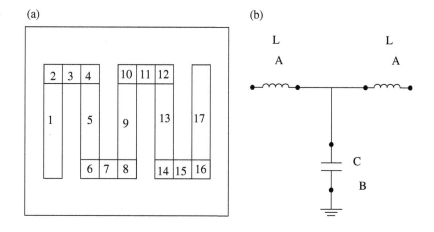

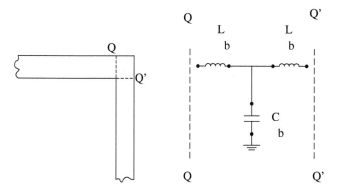

Figure 13.33 *T* network model of the square bend join.

In a similar fashion, the square bend junctions of each meander section can also be modeled as a *T* network, as shown in Figure 13.33.

The *T* network equivalent model can be determined with the following formulation.

For Meander Segments

$$\varepsilon_{eff} = \frac{\varepsilon_r + 1}{2} + \frac{\varepsilon_r - 1}{2} \times \frac{1}{\sqrt{1 + 12\dfrac{h}{w}}}$$

$$L = Z_0 \sqrt{\varepsilon_{eff}}/c$$

$$C = \sqrt{\varepsilon_{eff}}/c\, Z_0$$

Lumped inducatance $L_A = Ll/2$

Lumped capacitance $C_B = Cl$

For Square Bend Joints

$$\frac{C_b}{W}\,(pF/m) = \begin{cases} \dfrac{(14\varepsilon_r + 12.5)W/h - (1.83\varepsilon_r - 2.25)}{\sqrt{W/h}} + \dfrac{0.02\varepsilon_r}{W/h}\,[w/h < 1] \\ (9.5\varepsilon_r + 1.25)W/h + 5.2\varepsilon_r + 7.0\,[w/h \geq 1] \end{cases}$$

$$\frac{L_b}{h}\,(nH/m) = 100\left(4\sqrt{W/h} - 4.21\right)$$

The overall lumped model equivalent circuit is given in Figure 13.34 and the calculated inductance and capacitance values are given in Table 13.1.

The physical dimensions shown in Figure 13.35 are calculated and illustrated in Table 13.1. The resulting T network model was simulated on ADS to verify if the model would agree with the frequency of resonance of interest. Resonance was observed to be close to the desired frequencies at 1 GHz and 1.89 GHz. ADS simulation results are given in Figures 13.36 and 13.37.

MATLAB was used to plot the directivity of the microstrip antenna. The expression for directivity has the following general form.

$$D = D_2 F(\theta, \varphi)$$

D_2 is given by the expression

$$D_2 = D_0 D_{AF}$$

where D_{AF} is assumed to be 2. D_0 is given by the expression

$$D_0 = (k_0 W)^2 \frac{1}{I_1}$$

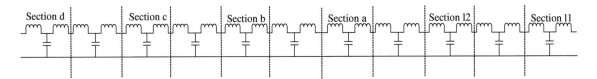

Figure 13.34 Complete T network model of the antenna design.

Table 13.1 Calculated dimensions for the meander antenna.

Section	d	c	b	a	l2	l1	Square bend
Section length (mm)	23	20	28	11	20	20	5
Inductance, L (nH)	2.774	1.03	2.695	0.159	1.427	1.11	0.512 99
Capacitance, C (pF)	2.148	0.798	2.087	0.123	1.105	0.859	0.978 25

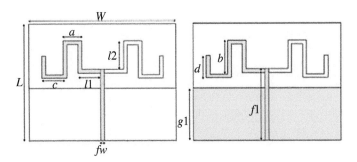

Figure 13.35 Physical dimension calculation for the layout (a) Top view (b) Bottom view.

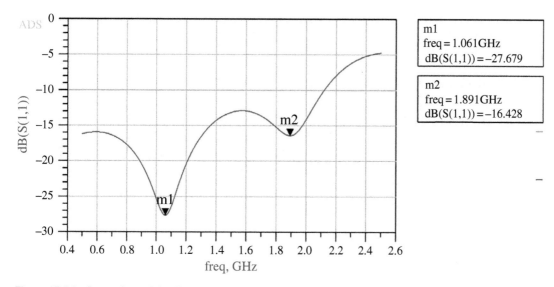

Figure 13.36 Return loss of the *T* network model.

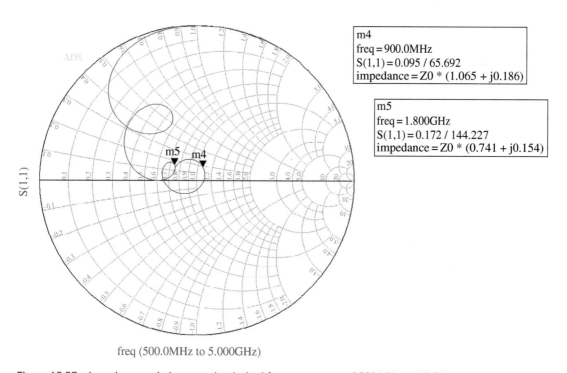

Figure 13.37 Impedance variation over the desired frequency range of 900 MHz to 1.8 GHz.

where I_1 is

$$I_1 = -2 + \cos{(k_0 W)} + (k_0 W) S_i(k_0 W) + \frac{\sin{(k_0 W)}}{k_0 W}$$

The 2D radiation pattern is given in Figure 13.38.

Simulation

The final layout that is simulated in HFSS is given in Figure 13.39. The simulation results are illustrated in Figure 13.40.

The comparison of the return loss at the resonance frequency from both the lumped element model and the HFSS simulation of the antenna are:

Lumped Element Model – Simulation Results

- F1 = 1.061 GHz (−27.68 dB)
- F2 = 1.891 GHz (−16.43 dB)

HFSS – Simulation Results

- F1 = 861.1 MHz (−24.96 dB)
- F2 = 1.85 GHz (−19.39 dB)

Hence, it is proven that the lumped element equivalent model is a good start to design the dual band energy harvester antenna. After obtaining the resonance frequencies close to the desired frequencies, HFSS optimetrics are used to finetune the resonance frequency to obtain resonance closer to of 900 MHz and 1.8 GHz. The results are given in Figure 13.41.

The gain and directivity were also obtained from HFSS simulations after optimization and are shown in Figure 13.42a,b.

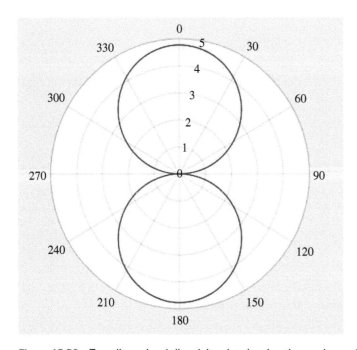

Figure 13.38 Two-dimensional directivity plot showing the maximum directivity = 4.88.

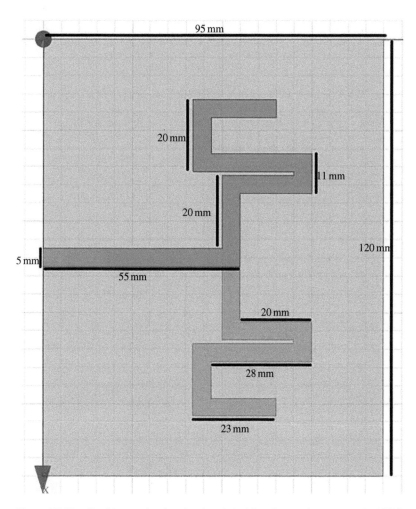

Figure 13.39 Final layout for the simulated dual band meander antenna in HFSS.

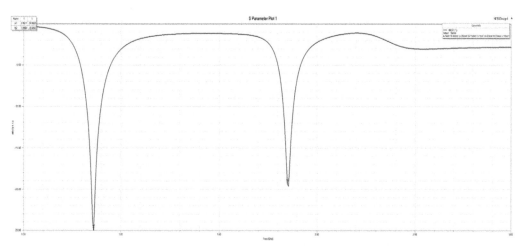

Figure 13.40 Simulation results of the final dual band energy harvester antenna design.

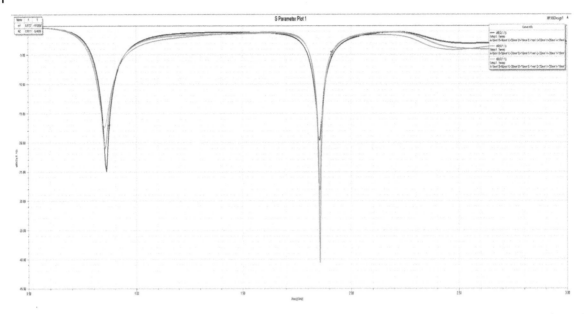

Figure 13.41 Simulation results after all dimensions are parametrized and changed slightly with HFSS optimetrics for tuning the design.

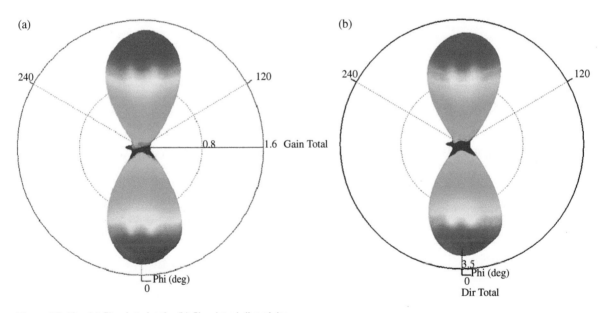

Figure 13.42 (a) Simulated gain. (b) Simulated directivity.

Prototype

The final antenna design was created on KiCAD EDA for the PCB layout based on the dimensions of final design and is shown in Figure 13.43.

After the design was created on KiCAD, the design was exported in Gerber files to be fabricated. The setup was:

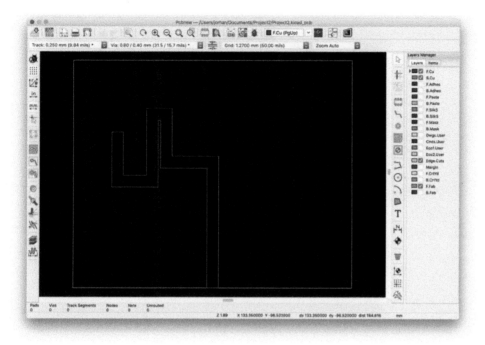

Figure 13.43 KiCAD EDA layout of the antenna design.

- Fabrication setup.
- Bantam Tools desktop PCB milling machine.
- 6 × 6 PCB with FR4 substrate and double-sided copper layers.
- The size was cut down to 4 × 5 with a Dremel tool to match the design's overall dimensions.
- A female SMA connector was soldered at the input of the antenna to be able to apply an RF signal and measure the return loss of the antenna.

The top and bottom layer of the PCB design for antenna are shown in Figure 13.44.

(a)   (b)

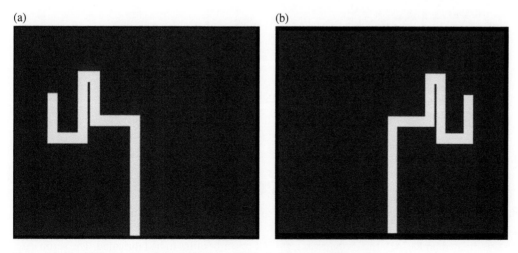

Figure 13.44 (a) Front side of the PCB layout. (b) Back side of the PCB design.

(a)   (b)

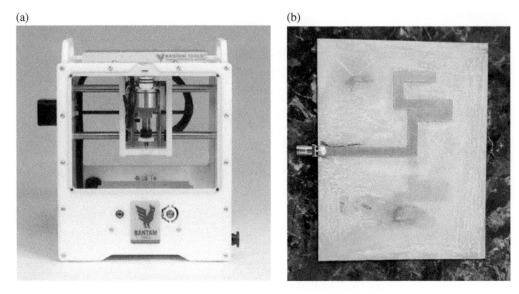

Figure 13.45   (a) Bantam PCB milling tool used to fabricate the antenna. (b) Antenna prototype.

The milling tool is shown in Figure 13.45a and the final prototype that was built is illustrated in Figure 13.45b.

A vector network analyzer, Agilent Technologies N5242A, was used to measure the return loss of the antenna while sweeping the input frequency from 10 MHz to 3 GHz. The measurement setup and results are shown in Figures 13.46 and 13.47, respectively.

The resonance frequencies measured on the final design were close to the desired frequencies, especially for 900 MHz. For the higher end of the dual band operation (1.8 GHz), the fabricated antenna was slightly off target. The

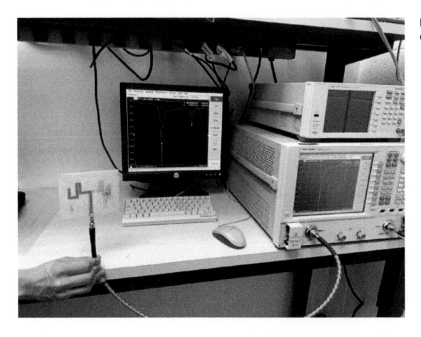

Figure 13.46   Measurement setup for dual band energy harvester antenna.

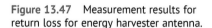

Figure 13.47 Measurement results for return loss for energy harvester antenna.

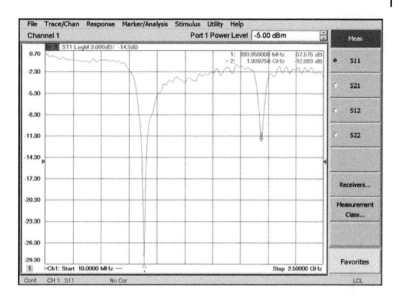

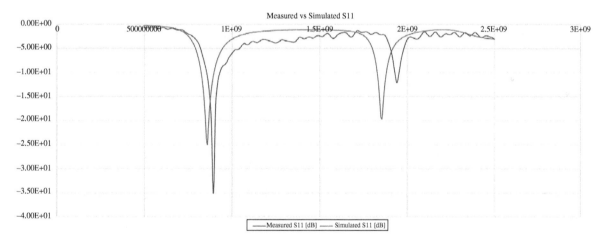

Figure 13.48 Comparison of measurement and simulated results.

measured return loss at $f_1 = 893.95$ MHz $(-37.58$ dB$)$ and $f_2 = 1.94$ GHz $(-12.08$ dB$)$. The comparison of the simulated and measured results are illustrated in Figure 13.48.

Design Example 13.3   Antenna Design for HVAC Systems

Investigate the implementation of a wireless antenna operating at 2.4 GHz for electronically commutated motors (ECMs) for HVAC systems. Consider a typical HVAC enclosure given in Figure 13.49 representing the cover of ECM where the desired antenna will be located.

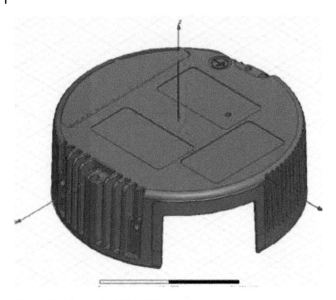

Figure 13.49 Typical HVAC ECM enclosure.

Solution

The best location the antenna can be placed for the ECM enclosure is illustrated in Figure 13.50. The two possible antenna configurations that can be used for communication are shown in Figure 13.51. The simulation and measurement results for an inverted F antenna are illustrated in Figure 13.52.

The simulation and measurement results for an F antenna are given in Figure 13.53.

Three-dimensional radiation of the simulated antenna is given in Figure 13.54b when the cover also has a dielectric encapsulation for the interface connector, as shown in Figure 13.54a.

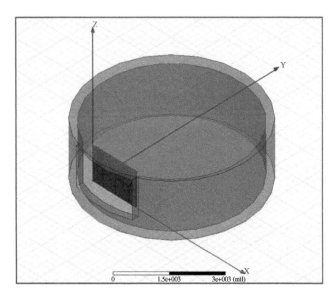

Figure 13.50 Antenna location for ECM.

(a)

(b)

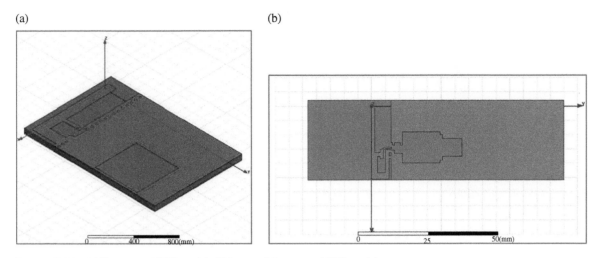

Figure 13.51 (a) F antenna HFSS model. (b) inverted F antenna HFSS model.

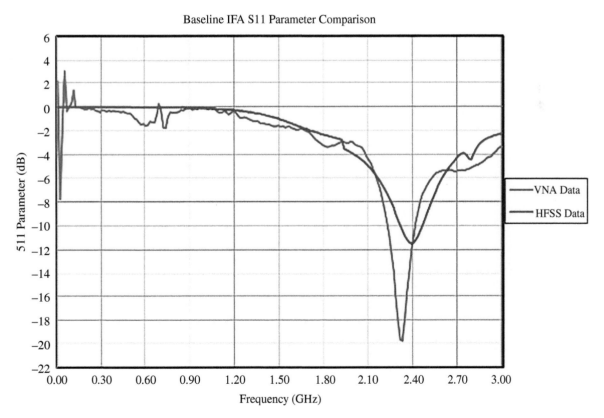

Figure 13.52 Comparison of measurement and simulation results for inverted F antenna.

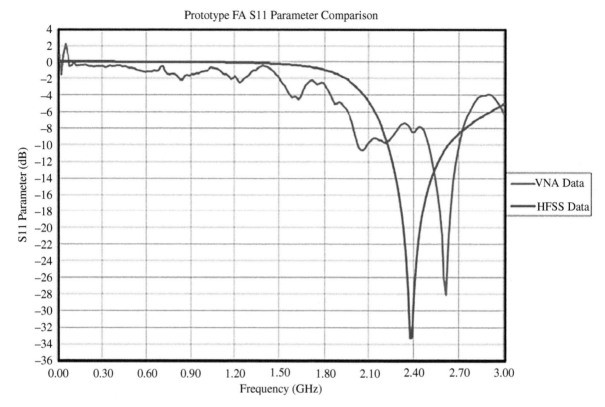

Figure 13.53 Comparison of measurement and simulation results for F antenna.

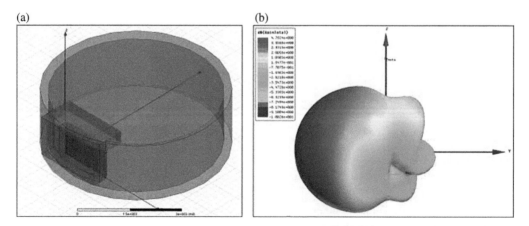

Figure 13.54 (a) ECM cover with dielectric encapsulation for interface connector. (b) Three-dimensional simulated radiation pattern.

References

1 Maxim Integrated Products (2012). Application Note 5259: Energy Harvesting Systems Power the Powerless. https://www.maximintegrated.com/en/design/technical-documents/app-notes/5/5259.html (accessed 22 September 2021).

2 Mekid, S., Qureshi, A., and Baroudi, U. (2017). Energy harvesting from ambient radio frequency: is it worth it? *Arabian Journal for Science and Engineering* **42** (7): 2673–2683.

3 Ostaffe, H. *RF-based wireless charging and energy harvesting enables new applications and improves product design.* Energy Harvesting Technologies. https://www.mouser.com/pdfdocs/rf-basedwireless-charging.pdf (accessed 31 August 2021).

4 Tachwali, Y., Refai, H., and Fagan, J. (2007). Minimizing HVAC energy consumption using a wireless sensor network. In: *IECON Proceedings (Industrial Electronics Conference)*, 439–444. IEEE https://doi.org/10.1109/IECON.2007.4460329.

5 Ahmadi, S., Shames, I., Scotton, F. et al. (2012). Towards more efficient building energy management systems. In: *Seventh International Conference on Creativity Support Systems (KICSS)* (8–10 November), 118–126. IEEE.

6 Sultan, S., Khan, T., and Khatoon, S. (2010). Implementation of HVAC System Through Wireless Sesnor Network. In: *2010 Second International Conference on Communication Software and Networks*, 52–56. IEEE https://doi.org/10.1109/ICCSN.2010.64.

7 Md Din, N., Chakrabarty, C.K., Bin Ismail, A. et al. (2012). Design of RF energy harvesting system for energizing low power devices. *Progress in Electromagnetics Research* **132**: 49–69.

8 Piñuela, M., Mitcheson, P.D., and Lucyszyn, S. (2013). Ambient RF energy harvesting in urban and semi-urban environments. *IEEE Transactions on Microwave Theory and Techniques* **61** (7): 2715–2726. https://doi.org/10.1109/TMTT.2013.2262687.

9 Chow, Y.L. and Wan, K.L. (2002). A transformer of one-third wavelength in two sections – for a frequency and its first harmonic. *IEEE Microwave and Wireless Components Letters* **12** (1): 22–23. https://doi.org/10.1109/7260.975723.

10 Monzon, C. (2003). A small dual-frequency transformer in two sections. *IEEE Transactions on Microwave Theory and Techniques* **51** (4): 1157–1161. https://doi.org/10.1109/TMTT.2003.809675.

11 Karthaus, U. and Fischer, M. (2003). Fully integrated passive UHF RFID transponder IC with 16.7-μW minimum RF input power. *IEEE Journal of Solid-State Circuits* **38** (10): 1602–1608.

12 Rabén, H., Borg, J., and Johansson, J. (2012). An active MOS diode with Vth-cancellation for RFID rectifiers. In: *2012 IEEE International Conference on RFID (RFID)* (3–5 April 2012), 54–57.

13 Kotani, K. and Ito, T. (2009). High efficiency CMOS rectifier circuits for UHF RFIDs using Vth cancellation techniques. In: *2009 IEEE 8th International Conference on ASIC* (20–23 October 2009), 549–552.

14 Yuan, F. and Soltani, N. (2008). Design techniques for power harvesting of passive wireless microsensors. In: *2008 51st Midwest Symposium on Circuits and Systems* (10–13 August 2008), 289–293.

15 Chaurasia, S. and Singhal, P.K. (2017). Design and analysis of dual band meandered printed dipole antenna for RF energy harvesting system at GSM bands. *Oaijse* **2** (4): 20–24.

16 Straub, A., Eroglu, A., Pomalaza-Ráez, C., and Becerra, R. (2013). Optimized UHF antenna design, simulation, implementation methods of HVAC systems. *Proceedings of the International Conference on Electromagnetic in Advanced Applications* (ICEAA'13, IEEE APWC'13, EMS'13) (9–13 September), Torino, Italy.

17 Nakajima, T., Satoh, I., and Aizu, H. (2002). A virtual overlay network for integrating home appliances. In: *Applications and the Internet, 2002. (SAINT 2002). Proceedings. 2002 Symposium on*, 246–253.

18 Suh, C. and Ko, Y.-B. (2008). Design and implementation of intelligent home control systems based on active sensor networks. *Consumer Electronics, IEEE Transactions on* **54** (3): 1177–1184.

19 Papadopoulos, N., Meliones, A., Economou, D. et al. (2009). A connected home platform and development framework for smart home control applications. In: *Industrial Informatics, 2009. INDIN 2009. 7th IEEE International Conference on*, 402–409. IEEE.

20 Han, D.-M. and Lim, J.-H. (2010). Design and implementation of smart home energy management systems based on Zigbee. *Consumer Electronics, IEEE Transactions on* **56** (3): 1417–1425.

21 Dorge, B.M. and Scheffler, T. (2011). Using IPv6 and 6LoWPAN for home automation networks. In: *Consumer Electronics - Berlin (ICCE-Berlin), 2011 IEEE International Conference on*, 44–47. IEEE.

22 Enjian, B. and Xiaokui, Z. (2012). Performance evaluation of 6LoWPAN gateway used in actual network environment. In: *Control Engineering and Communication Technology (ICCECT), 2012 International Conference on*, 1036–1039. IEEE Computer Society.

23 Tonguz, O.K., Stancil, D.D., Xhafa, A.E. et al. (2002). An empirical path loss model for HVAC duct systems. In: *Global Telecommunications Conference, 2002. GLOBECOM '02*, vol. **2**, 1850–1854. IEEE.

24 Xhafa, A.E., Tonguz, O.K., Cepni, A.G. et al. (2005). On the capacity limits of HVAC duct channel for high-speed Internet access. *Communications, IEEE Transactions on* **53** (2): 335–342.

25 Nikitin, P.V., Stancil, D.D., Tonguz, O.K. et al. (2002). RF propagation in an HVAC duct system: impulse response characteristics of the channel. In: *Antennas and Propagation Society International Symposium, 2002*, vol. **2**, 726–729. IEEE.

Index

RF/Microwave Engineering and Applications in Energy Systems, First Edition. Abdullah Eroglu.
© 2022 John Wiley & Sons Ltd. Published 2022 by John Wiley & Sons Ltd.
Companion website: www.wiley.com/go/eroglu/rfmicrowave